Calculations in Chemistry: An Introduction

CALCULATIONS IN CHEMISTRY

An Introduction

Donald J. Dahm
Rowan University

Eric A. Nelson
Falls Church, VA

W. W. NORTON & COMPANY
New York · London

W. W. Norton & Company has been independent since its founding in 1923, when William Warder Norton and Mary D. Herter Norton first published lectures delivered at the People's Institute, the adult education division of New York City's Cooper Union. The firm soon expanded its program beyond the Institute, publishing books by celebrated academics from America and abroad. By midcentury, the two major pillars of Norton's publishing program—trade books and college texts—were firmly established. In the 1950s, the Norton family transferred control of the company to its employees, and today—with a staff of four hundred and a comparable number of trade, college, and professional titles published each year—W. W. Norton & Company stands as the largest and oldest publishing house owned wholly by its employees.

Editor: Erik Fahlgren
Editorial assistant: Renee Cotton
Production manager: Eric Pier-Hocking
Marketing manager, chemistry: Stacy Loyal
Electronic media editor: Rob Bellinger
Associate editor, emedia: Matthew A. Freeman
Editorial assistant, emedia: Paula Iborra
Design director: Chris Welch
Composition: cMPreparé
Manufacturing: Transcontinental

ISBN: 978-0-393-91286-9 (pbk.)

W. W. Norton & Company, Inc., 500 Fifth Avenue, New York, NY 10110
www.wwnorton.com
W. W. Norton & Company Ltd., Castle House, 75/76 Wells Street, London W1T 3QT

1 2 3 4 5 6 7 8 9 0

Brief Contents

Contents

Preface

The Research behind This Book

Several years ago, at a meeting on chemistry education, we (the authors) were participants in a discussion among instructors in which there was broad agreement that each year, students seemed to be less prepared for the *math* of first-year chemistry. Following the meeting, we decided to seek out data that might evaluate these anecdotal observations.

Standardized test scores for high school students were available for one of our states, and the findings revealed that since 1995 scores in math computation—the math of calculations in the sciences—had decreased year after year.[1]

Looking beyond our own states, we found a paper by Dr. Tom Loveless of the Brookings Institution documenting steep declines in grades 9–12 scores in math computation on both state and national (NAEP) testing.[2] Loveless attributed this decline in part to a requirement in 48 of 50 states that students be taught to use calculators to do arithmetic on state tests—starting in *third* grade.

We were also fortunate to discover a series written for educators by University of Virginia cognitive scientist Daniel Willingham that summarized recent research on how the mind works and how it learns.[3] Willingham explains that humans possess an enormous long-term memory—needed in a brain that evolved to manage speech. However, for cognition in topics that are not speech, we must rely on our working memory: the space where the brain solves problems.

Working memory has a strong capacity to manipulate information that we *know* (that we can recall automatically from long-term memory) but a very limited ability to work with information that has not been *memorized* (moved into long-term memory by repeated practice).

Cognitive research has also established that although experts can solve problems in their discipline by reasoning, until graduate-level study, students are algorithmic problem solvers: advancing toward reasoning as they gradually commit to long-term memory the core facts, algorithms, and concepts of a discipline. This learning is most efficient when it combines *overlearning* (repeated memorization to

---

[1] Eric Nelson, "Student Test Scores in Math Computation and the Implications for Chemistry Instruction" (lecture, Cognition Symposium of the ACS Biennial Conference on Chemistry Education, August 2, 2010, available at: www.ChemReview.Net/BCCEPost.pdf).

[2] Tom Loveless, *The 2002 Brown Center Report on American Education* (Brookings Institution Press, 2002), 6–12.

[3] See, for example, Daniel T. Willingham, "Ask the Cognitive Scientist: Practice Makes Perfect—but Only If You Practice Beyond the Point of Perfection," *American Educator* 28, no. 1 (Spring 2004): 44–46, available at www.aft.org/newspubs/periodicals/ae/spring2004/willingham.cfm (accessed 4/12/12).

perfection) with activities that build mental models: the *conceptual* frameworks that organize knowledge.[4]

While calculators are powerful tools, cognitive science has established that memorized "automaticity" in math and science fundamentals is essential during problem solving to reduce "cognitive load" in working memory.[5]

This scientific finding that "both concepts *and* memorization are required" is not an outcome we might have preferred, but it is consistent with the experience that we have all shared in learning chemistry.

What Makes This Book Different

Based on this research, we began an experiment. Our goal was to design homework that, by applying the findings of cognitive science, would effectively review the fundamentals and the *math* needed in first-year chemistry. In addition, by shifting this review to study time, we hoped to free more time in lecture for guided inquiry, demonstrations, and discussion of the concepts of chemistry.

Over a four-year period, the lessons that would become *Calculations in Chemistry: An Introduction* were used with success in classes at Rowan University and then on other campuses.

Calculations in Chemistry is designed to serve as a primary text in preparatory chemistry or as a supplemental text in GOB/allied health chemistry.

The lessons can be assigned either to prepare students for lecture or to review and reinforce skills after lecture. To encourage the timely completion of this homework, editable quizzes covering the content of *Calculations in Chemistry* are available to instructors online in the Norton Resource Library.

In *Calculations in Chemistry*, students first review and practice the math that is prerequisite for each topic. Their reinforced math skills are then applied to problems in chemistry.

The **Try It** feature of *Calculations in Chemistry* puts into practice *completion problems*, a strategy that cognitive science has found to be especially effective for developing student problem-solving skills.[6] As students work on an introductory exercise, each Try It provides checks and hints along the way.

Another feature of *Calculations in Chemistry* is an emphasis on "mental math." To build students' sense of numeracy, many problems are included with simple arithmetic—and students are encouraged to solve without a calculator.

Finally, these lessons include "*how*-to-study" strategies that cognitive science has found to be especially effective. In each chapter, flashcards are suggested when needed to assist in mastering content.

---

[4]John Sweller, Jeroen J. G. van Merrienboer, and Fred G.W.C. Paas, "Cognitive Architecture and Instructional Design," *Educational Psychology Review* 10, no. 3 (1998): 251–296; Willingham, "Ask the Cognitive Scientist."

[5]Walter Kintsch, "Learning and Constructivism," in *Constructivist Instruction: Success or Failure?*, ed. Sigmund Tobias and Thomas M. Duffy (New York: Routledge, 2009), 223–241; J. D. Fletcher, "From Behaviorism to Constructivism: A Philosophical Journey from Drill and Practice to Situated Learning," in *Constructivist Instruction*, 242–263.

[6]Sweller et al., "Cognitive Architecture."

Early Results

In beta tests of these lessons, instructors have reported significant gains in student achievement when using *Calculations in Chemistry* as assigned homework. We would attribute a substantial share of this success to the time that *Calculations in Chemistry* creates in lecture for guided inquiry.

We hope that instructors will experiment with these lessons and share with us their recommendations, insights, and results.

Acknowledgments

We would like to thank our reviewers, who helped greatly in increasing the clarity and accuracy of this text.

Reviewers

Doyle Barrow, Jr., Georgia State University

Nathan Barrows, Grand Valley State University

Ivana Bozidarevic, San Francisco City College

Claire Cohen, University of Toledo

Bernadette Corbett, Metropolitan Community College

Jennifer Coym, University of South Alabama

Steven Davis, University of Mississippi

Nell Freeman, St. Johns River State College

Aaron Fried, Los Angeles Community College

Daniel Groh, Grand Valley State University

Tamara Hanna, Texas Tech University

Donna Ianotti, Brevard Community College

Roy Kennedy, Massachusetts Bay Community College

Carol Martinez, Central New Mexico Community College

David Nachman, Mesa Community College

Randa Roland, University of California Santa Cruz

Mary Setzer, University of Alabama in Huntsville

Matthew Smith, Walters State Community College

Steve Trail, Elgin Community College

Marie Villarba, Seattle Central Community College

Donald J. Dahm and Eric A. Nelson

Note to Students

The goal of these lessons is to help you solve *calculations* in first-year chemistry. This is only one part of a course in chemistry, but it can be the most challenging.

We suggest purchasing a *spiral notebook* as a place to write your work when solving problems in these lessons. You will also need

- Two packs of 100 3 × 5-inch index cards (two or more colors are preferred) plus a small assortment of rubber bands, and
- A pack of long sticky notes (4 × 6-inch notes are recommended) for use as cover sheets

It is important that you use the *same* calculator to solve homework problems that you will be allowed to use during tests, in order to learn and practice the rules for that calculator before tests.

Many courses will not allow the use of a graphing calculator or other calculators with extensive memory during tests. If no type of calculator is specified for your course, any inexpensive calculator with a $1/x$ *or* x^{-1}, y^x *or* $\wedge$, $\log$ *or* 10^x, *and* $\ln$ functions will be sufficient for most calculations in first-year chemistry.

How to Use These Lessons

1. ***Read*** **the lesson and** ***work*** **the** ▬▬▬▶ **Try It questions.** Use this method:

   - As you start a new page, *if* you see a *stop* sign STOP on the page, *cover* the text *below* the *stop sign*. As a cover sheet, use either a sticky note or a folded sheet of paper.

   - In the space provided in the text or in your problem notebook, write your answer to the question (**Q.**) that is above the stop sign. Then move your cover sheet down to the next stop sign and check your answer. If you need a hint, read a *part* of the answer, then re-cover the answer and try the problem again.

2. ***First*** **learn the rules,** ***then*** **do the Practice.** The goal of learning is to move rules and concepts into *memory*. To begin, when working Try It questions, you may look back at the rules, but make an effort to commit the rules to memory before starting the Practice sets.

   Answers to the Practice problems are at the end of each chapter. If you need a hint, read a part of the answer and try again.

3. **How many Practice problems should you do?** It depends on your background. These lessons are intended to

   • Refresh your memory on topics you once knew, and

   • Fill in the gaps for topics that are less familiar

   If a topic is familiar, read the lesson for reminders and review, then do a *few* problems in each Practice set. Be sure to do the last problem (usually the most challenging).

   If a topic is unfamiliar, do more problems.

4. **Work Practice problems at least three days a week.** Chemistry is cumulative. What you learn in initial topics you will need in memory later. To retain what you learn, *space* your study of a topic out over several days.

   Cognitive research has found that your memory tends to retain only what you practice over *several* days. If you wait until a quiz deadline to study, what you learn may remain in memory for a day or two, but on later tests and exams, it will tend to be forgotten.

   Begin lessons on new topics early, preferably before the topic is covered in a lecture.

5. **Memorize what must be memorized.** Use flashcards and other memory aids.

The key to success in chemistry is to *study* rules and concepts and *practice* solving problems at a *steady* pace.

About the Authors

Donald J. Dahm is a professor at Rowan University. In 2004, Dr. Dahm received the Outstanding Achievement in Near Infrared Spectroscopy Award from the Eastern Analytical Symposium. He was selected in 2011 by Rowan students as the professor they most wanted to give "The Last Lecture," an annual event based on Randy Pausch's book. Don's most recent book, coauthored with his son Kevin, is *Interpreting Diffuse Reflectance and Transmittance: A Theoretical Introduction to Absorption Spectroscopy of Scattering Materials* (2007).

Eric A. Nelson's career spanned work as a chemistry instructor and as an elected officer and legislative representative for his faculty organization. His professional interests include research on reading comprehension in the sciences and measures of student mathematics achievement in preparation for study in chemistry. Prior to his retirement, he served on committees at the state and regional level working to raise academic standards and encourage the adoption of research-based curricula in STEM fields.

1

Scientific Notation

Calculators and Exponential Notation

To multiply 492×7.36, the calculator is a useful tool. However, when using exponential notation, you will make fewer mistakes if you do as much exponential math as you can without a calculator. These lessons will review the rules for doing exponential math "in your head."

The majority of problems in Chapter 1 will *not* require a calculator. Problems that require a calculator will be clearly identified. You are encouraged to try complex problems with the calculator *after* you have tried them without one. This should help you to decide when, and when not, to use a calculator.

Notation Terminology

In science, we often deal with very large and very small numbers.
Examples:

- A drop of water contains about **1,500,000,000,000,000,000,000** molecules.
- An atom of neon has an empirical radius of about **0.000 000 007 0** centimeters.

When values are expressed as "regular numbers," such as 123, 0.024, or the numbers above, they are said to be in **fixed decimal** or **fixed notation**.

Very large and very small numbers are more clearly expressed in **exponential notation**: writing a *number* times **10** to *an integer* power. For the measurements above, we can write:

- A drop of water contains about 1.5×10^{21} molecules.
- An atom of neon gas has an empirical radius of about 7.0×10^{-9} centimeters.

Values represented in exponential notation can be described as having three parts. Example: In -6.5×10^{-4},

- The $-$ in front is the *sign*,
- The **6.5** is termed the *significand*, or *decimal*, or *digit*, or *mantissa*, or *coefficient*,
- The 10^{-4} is the *exponential* term: the *base* is **10** and the *exponent* (or *power*) is **−4**.

Because decimal, mantissa, and coefficient have other meanings, in these lessons we will refer to the parts of exponential notation as the **sign**, **significand**, and **exponential** term.

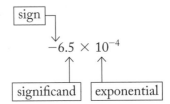

You should also learn (and use) any alternate terminology preferred in your course.

Lesson 1.1   Moving the Decimal

PRETEST   Do *not* use a calculator. If you get a perfect score on this Pretest, skip to Lesson 1.2. Otherwise, complete Lesson 1.1. Answers are provided at the end of this *chapter*.

1. In the space provided, write these answers in *scientific* notation:

    a. $9{,}400 \times 10^3 =$

    b. $0.042 \times 10^6 =$

    c. $-0.0067 \times 10^{-2} =$

    d. $-77 =$

2. Write these answers in fixed decimal notation:

    a. $14/10{,}000 =$ b. $0.194 \times 1000 =$

    c. $47^0 =$

Powers of 10

Below are the numbers that correspond to powers of 10. Note the relationship between the *exponents*, the number of zeros, and the position of the *decimal point* in the fixed decimal numbers as you go down the sequence.

$$10^6 = 1{,}000{,}000$$

$$10^3 = 1000 = 10 \times 10 \times 10$$

$$10^2 = 100$$

$$10^1 = 10$$

$$10^0 = 1 \qquad (Anything \text{ to the } zero \text{ power equals } one.)$$

$$10^{-1} = 0.1$$

$$10^{-2} = 0.01 = 1/10^2 = 1/100$$

$$10^{-3} = 0.001$$

When converting from powers of 10 to fixed decimal numbers, use these steps:

1. To convert a *positive* power of 10 to a fixed decimal *number*, write **1**, then move the decimal to the *right* by the number of places equal to the number in the exponent.

$$\text{Example: } 10^2 = \mathbf{100}$$

2. To convert a *negative* power of 10 to a fixed decimal number, move the decimal to the *left* by the number of places equal to the number after the negative sign in the exponent.

TRY IT

(See "How to Use These Lessons," point 1, p. xv.)

Q. Convert to a fixed decimal number: $10^{-2} =$

 Answer:

$$10^{-2} = \mathbf{0.01}$$

PRACTICE   A

Convert the following to fixed decimal notation. Write your answers in the spaces below and then check them at the end of this chapter.

1. $10^4 =$ 2. $10^{-4} =$ 3. $10^7 =$

4. $10^{-5} =$ 5. $10^0 =$

Multiplying and Dividing by 10, 100, and 1000

To multiply or divide by *numbers* that are positive whole-number powers of 10, such as 100 or 10,000, use these rules:

1. When *multiplying* a number by a 10, 100, 1000, etc., move the decimal to the *right* by the number of *zeros* in the 10, 100, or 1000.
      Examples: $72 \times 100 = 7{,}200$; $-0.0624 \times 1000 = -62.4$

2. When *dividing* a number by 10, 100, or 1000, etc., move the decimal to the *left* by the number of *zeros* in the 10, 100, or 1000.

 TRY IT

Q. Convert values to fixed decimal numbers:

  a. $34.6/1000 =$ b. $0.47/100 =$

Answers:

  a. $34.6/1000 = 0.0346$ b. $0.47/100 = 0.0047$

3. When writing a number that has a value between -1 and 1 (a number that "begins with a decimal point"), always place a *zero* in *front* of the decimal point.
      Example: Do *not* write .42 or $-.74$; *do* write **0.42** or $-\mathbf{0.74}$
      During your written calculations, the zero in front helps you see your decimals.

PRACTICE   B

Write your answers, then check them at the end of this chapter.

1. When dividing by 1000, move the decimal to the _____ by _____ places.

2. Write these answers as fixed decimal numbers:

   a. $0.42 \times 1000 =$ b. $63/100 =$ c. $-74.6/10,000 =$

Converting Exponential Notation to Fixed Decimal Notation

When working with numbers in chemistry, we often need to convert between fixed decimal notation and exponential notation, or move the decimal in the significand of exponential notation.

A key rule in all of these conversions is: The sign in *front* never changes. The sign identifies whether the *value* is positive or negative. When moving the decimal, the sign of the *exponential* term may change, but positive values remain positive and negative values remain negative.

To convert from exponential notation (such as -4×10^3) to fixed decimal notation ($-4,000$), use these rules:

1. The sign in front never changes.

2. In the exponential notation, if the significand is multiplied by a *positive* power of 10, move the decimal point in the significand to the *right* by the same number of places as the value of the exponent:
   Examples: $2 \times 10^2 = \mathbf{200}$; $-0.0033 \times 10^3 = \mathbf{-3.3}$

3. If the significand is multiplied by a *negative* power of 10, move the decimal point in the significand to the *left* by the same number of places as the number *after* the minus sign of the exponent.
   Examples: $2 \times 10^{-2} = \mathbf{0.02}$; $-7,653.8 \times 10^{-3} = \mathbf{-7.6538}$

PRACTICE   C

Convert these values to fixed decimal notation:

1. $3 \times 10^3 =$ 2. $5.5 \times 10^{-4} =$

3. $0.77 \times 10^6 =$ 4. $-95 \times 10^{-4} =$

Changing Exponential to Scientific Notation

In chemistry, when we work with very large or very small numbers, it is preferable to write the numbers in **scientific notation**: a special form of exponential notation that makes numeric values easier to compare. There are many equivalent ways to write a value in exponential notation, but there is only one correct way to write a value in scientific notation.

The general rule is:

> To convert a value from exponential notation to *scientific* notation, move the decimal so that the significand is *1 or greater*, but *less* than *10*, then adjust the power of 10 to keep the same numeric value.

Another way to say this:

> To convert from exponential to *scientific* notation,
>
> - Move the decimal point in the significand to *after* the first digit that is not a zero, then
> - Adjust the exponent to keep the same numeric value.

To apply these rules, the specific steps are:

1. Do not change the + or − *sign* in *front* of the significand.

2. If moving the decimal Y times to make the significand *larger*, make the power of 10 *smaller* by a *count* of Y.

   Example: Convert exponential to scientific notation:

   $$0.045 \times 10^5 = 4.5 \times 10^3$$

   The decimal must be after the 4. Move the decimal two places to the right. This makes the significand 100 times larger. To keep the same numeric value, lower the exponent by 2, making the 10^x value 100 times smaller.

TRY IT

Apply the rules above to the following problem.

Q. Convert this expression to scientific notation:

$$-0.0057 \times 10^{-2} =$$

 Answer:

The value *must* be written as -5.7×10^{-5}.

To convert to scientific notation, the decimal must be moved to after the first number that is not a zero: the 5. That move of three places makes the significant 1000 times larger. To keep the same value, make the exponential term 1000 times smaller.

The logic of the math is:

$$-0.0057 \times 10^{-2} = -[(5.7 \times 10^{-3}) \times 10^{-2}] = \mathbf{-5.7 \times 10^{-5}}$$

---

3. When moving the decimal Y times to make the significand *smaller*, make the power of 10 *larger* by a count of Y.

TRY IT

Q. Convert the following to scientific notation:

$$-8{,}544 \times 10^{-7} =$$

Answer:

$$-8{,}544 \times 10^{-7} = -8.544 \times \mathbf{10^{-4}}$$

In scientific notation, the decimal must be after the 8, so you must move the decimal three places to the left. This makes the significand 1000 times smaller. To keep the same numeric value, increase the exponent by 3, making the 10^x value 1000 times larger.

Remember, 10^{-4} is 1000 times *larger* than 10^{-7}.

Every time you move a decimal, it helps to *recite*, for the terms *after* the sign in front:

If one gets *smaller*, the other gets *larger*. If one gets *larger*, the other gets *smaller*.

PRACTICE   D

Convert these values to scientific notation:

1. $5{,}420 \times 10^3 =$

2. $0.0067 \times 10^{-4} =$

3. $0.020 \times 10^3 =$

4. $-870 \times 10^{-4} =$

5. $0.00492 \times 10^{-12} =$

6. $-602 \times 10^{21} =$

Converting Numbers to Scientific Notation

To convert regular (fixed decimal) numbers to *exponential* or *scientific* notation, use the following rules.

- Any number to the zero power equals one. $2^0 = \mathbf{1}$; $42^0 = \mathbf{1}$; exponential notation most often uses $\mathbf{10^0 = 1}$
- Because any number can be multiplied by 1 without changing its value, any number can be multiplied by 10^0 without changing its value.
    Example:

$$42 = 42 \times 1 = 42 \times \mathbf{10^0} \text{ in } exponential \text{ notation}$$

$$= \mathbf{4.2} \times 10^1 \text{ in } scientific \text{ notation}$$

To convert fixed notation to *scientific* notation, the steps are:

1. Add "$\times \mathbf{10^0}$" after the number.

2. Apply the rules that convert exponential to scientific notation:
    - Do not change the sign in *front*.
    - Write the decimal in the significand after the first digit that is not a zero.
    - Adjust the power of 10 to compensate for moving the decimal.

TRY IT

Q. Convert these numbers to scientific notation:

  a. 943 =   b. −0.00036 =

STOP   **Answers:**

  a. $943 = 943 \times 10^0 = \mathbf{9.43 \times 10^2}$

  b. $-0.00036 = -0.00036 \times 10^0 = \mathbf{-3.6 \times 10^{-4}}$

When converting to scientific notation, a positive fixed decimal number that is

- *Larger* than *one* has a *positive* exponent (zero and above) in scientific notation;
- *Between* zero and one (such as 0.25) has a *negative* exponent in scientific notation; and
- The number of *places* that the decimal in a number moves is the *number* after the sign in its exponent.

For negative values, these same rules apply to numbers *after* the negative sign in front. The sign in front is independent of the numbers after it.

Note how these rules apply to the two answers above.

Note also that in both exponential and scientific notation, whether the sign in front is positive or negative has no relation to the sign of the *exponential* term. The sign in front shows whether the value is positive or negative. The exponential term indicates only the position of the decimal point.

PRACTICE   E

1. Which lettered parts in problem 2 below must have exponentials that are negative when written in scientific notation?

2. Change these values to scientific notation.

    a. 6,280 =   b. 0.0093 =

    c. 0.741 =   d. −1,280,000 =

3. Complete the problems in the *Pretest* at the beginning of this lesson.

In your problem notebook,

- Write a list of rules in this lesson that were unfamiliar or you found helpful, and
- Condense your wording, number the points, and write and recite your rules until you can write them from memory.

Then, complete the problems below.

PRACTICE   F

Check (✓) and do every *other* letter. If you miss one, do another letter for that set. Save a few parts for review before your quiz or test on this material.

1. Write these answers in fixed decimal notation:

    a. $924/10{,}000 =$ b. $24.3 \times 1000 =$ c. $-0.024/10 =$

2. Convert these to scientific notation:

    a. $0.55 \times 10^5 =$ b. $0.0092 \times 100 =$

    c. $940 \times 10^{-6} =$ d. $0.00032 \times 10^1 =$

3. Write these numbers in scientific notation:

    a. $7{,}700 =$ b. $160{,}000{,}000 =$

    c. $0.023 =$ d. $0.00067 =$

Lesson 1.2   Calculations Using Exponential Notation

PRETEST   If you can answer these two questions correctly, you may skip to Lesson 1.3. Otherwise, complete Lesson 1.2. Answers are at the end of this chapter. Do *not* use a calculator. Convert final answers to scientific notation.

1. $(2.0 \times 10^{-4})(6.0 \times 10^{23}) =$

2. $\dfrac{10^{23}}{(100)(3.0 \times 10^{-8})} =$

Mental Arithmetic

In chemistry, most of the processes that we study are based on simple *whole-number* ratios. Research on how we think has confirmed that if we can do simple "math-fact" calculations based on mental arithmetic rather than a calculator, it frees cognitive resources that are needed for reasoning during problem solving.

It is also essential to develop ways to check your calculated answers to be sure that they are reasonable. In the case of a complex calculation, a calculator answer must be checked either by solving using different calculator keys or by estimating the answer.

Mental arithmetic can be especially helpful to simplify calculations and estimate answers in calculations using exponential notation. In this lesson, we will review the rules for solving exponential calculations "in your head."

Multiplying and Dividing Powers of 10

The following rules should be recited until they can be recalled from memory.

1. When you *multiply* exponentials, you *add* the exponents.
   Examples: $10^3 \times 10^2 = 10^5$; $10^{-5} \times 10^{-2} = 10^{-7}$; $10^{-3} \times 10^5 = 10^2$

2. When you *divide* exponentials, you *subtract* the exponentials.
   Examples: $10^3/10^2 = 10^1$; $10^{-5}/10^2 = 10^{-7}$; $10^{-5}/10^{-2} = 10^{-3}$
   When subtracting, remember: Minus a minus is a plus.
   Example: $10^{6-(-3)} = 10^{6+3} = 10^9$

3. When you take the reciprocal of an exponential, change the exponential's sign.
   This rule is often remembered as:
   When you take an exponential term from the bottom to the top, change its sign.
   Example:
   $$\frac{1}{10^3} = 10^{-3}; \quad 1/10^{-5} = 10^5$$

   Why does this work? Rule 2:
   $$\frac{1}{10^3} = \frac{10^0}{10^3} = 10^{0-3} = 10^{-3}$$

4. For any X: $1/(1/X) = X$ The reasoning? $(X^{-1})^{-1} = X^{+1} = X$
   When you take an exponential term to a power, you multiply the exponentials. We will practice "power to a power" rules in later lessons, but for now, know rule 4.

Tip: When fractions include several terms, it may help to simplify the top and bottom *separately*, then divide.
   Example:
   $$\frac{10^{-3}}{10^5 \times 10^{-2}} = \frac{10^{-3}}{10^3} = 10^{-6}$$

 TRY IT

Q. Without using a calculator, write the simplified top, then the simplified bottom, then divide:
$$\frac{10^{-3} \times 10^{-4}}{10^5 \times 10^{-8}} = \underline{\hspace{2cm}} =$$

STOP

Answer:

$$\frac{10^{-3} \times 10^{-4}}{10^5 \times 10^{-8}} = \frac{10^{-7}}{10^{-3}} = 10^{-7-(-3)} = 10^{-7+3} = 10^{-4}$$

The Role of Practice

Do as many **Practice** problems as necessary until you feel "quiz ready."

- If the material in a lesson is easy review, do the *last* problem on each series of similar problems.
- If the lesson is less easy, put a check (✓) by every other problem, then work that half of the problem set. If you miss one, do additional problems in the set.
- Save a *few* problems for your next study session—and quiz/test review.

During **examples** and **Try Its** you *may* look back at the rules, but practice writing and recalling new rules from memory *before* starting the Practice set.

If you use Practice to learn the rules, it will be difficult to find time for all of the problems you will need to do. If you use Practice to *apply* rules that are in memory, you will need to solve fewer problems to be "quiz ready."

PRACTICE   A

Write answers as 10 to a power. Work on this page if possible. Do *not* use a calculator. Check your answers at the end of the chapter.

1. $10^6 \times 10^2 =$

2. $10^{-5} \times 10^{-6} =$

3. $\dfrac{10^{-5}}{10^{-4}} =$

4. $\dfrac{10^{-3}}{10^5} =$

5. $\dfrac{1}{1/10^4} =$

6. $1/10^{23} =$

7. $\dfrac{10^3 \times 10^{-5}}{10^{-2} \times 10^{-4}} =$

8. $\dfrac{10^5 \times 10^{23}}{10^{-1} \times 10^{-6}} =$

9. $\dfrac{100 \times 10^{-2}}{1000 \times 10^6} =$

10. $\dfrac{10^{-3} \times 10^{23}}{10 \times 1000} =$

Multiplying and Dividing in Exponential Notation

This is the rule we use most often:

> When multiplying and dividing using exponential notation, handle the *significands* and *exponents separately*.

Another way to say this rule is:

> Do number math using number rules, and exponential math using exponential rules. Then combine the two parts.

➡ TRY IT

Apply the rule to the following three problems.

Q1. Do not use a calculator: $(2 \times 10^3)(4 \times 10^{23}) =$

STOP   **Answer:**

For numbers, use number rules: 2 times 4 is **8**

For exponentials, use exponential rules: $10^3 \times 10^{23} = 10^{3+23} = 10^{26}$

Then combine the two parts: $(2 \times 10^3)(4 \times 10^{23}) = \mathbf{8 \times 10^{26}}$

Q2. Do the *significand* math on a calculator, but try the exponential math in your head for: $(2.4 \times 10^{-3})(3.5 \times 10^{23}) =$

STOP   **Answer:**

Handle significands and exponents separately.

Use a calculator for the numbers: $2.4 \times 3.5 = \mathbf{8.4}$

Do the exponentials in your head: $10^{-3} \times 10^{23} = 10^{20}$

Then combine:

$$(2.4 \times 10^{-3})(3.5 \times 10^{23}) = (2.4 \times 3.5) \times (10^{-3} \times 10^{23}) = \mathbf{8.4 \times 10^{20}}$$

We will review how much to round answers in Chapter 3. Until then, round numbers and significands in your answers to *two* digits unless otherwise noted.

Q3. Do significand math on a calculator, but exponential math with*out* a calculator.

$$\frac{6.5 \times 10^{23}}{4.1 \times 10^{-8}} =$$

STOP   **Answer:**

$$\frac{6.5 \times 10^{23}}{4.1 \times 10^{-8}} = \frac{6.5}{4.1} \times \frac{10^{23}}{10^{-8}} = 1.585 \times [10^{23-(-8)}] = \mathbf{1.6 \times 10^{31}}$$

Here's one more rule that will help with problem solving:

> When dividing, if an exponential term does *not* have a significand, add a "**1×**" in front of the exponential so that the number–number division is clear.

➡ TRY IT

Q. Apply the rule to answer the following. Try it with*out* a calculator.

$$\frac{10^{-14}}{2.0 \times 10^{-8}} =$$

STOP   **Answer:**

$$\frac{10^{-14}}{2.0 \times 10^{-8}} = \frac{\mathbf{1} \times 10^{-14}}{2.0 \times 10^{-8}} = 0.50 \times 10^{-6} = \mathbf{5.0 \times 10^{-7}}$$

PRACTICE   B

Study the two rules above, then apply them from memory to these problems. To allow room for careful work, solve these in your notebook. Do the odd-numbered problems first, then the evens if you need more practice. Try these *first* without a calculator, then check your mental arithmetic with a calculator if needed. Write final answers in scientific notation, rounding significands to two digits.

1. $(2.0 \times 10^1)(6.0 \times 10^{23}) =$

2. $(5.0 \times 10^{-3})(1.5 \times 10^{15}) =$

3. $\dfrac{3.0 \times 10^{-21}}{-2.0 \times 10^3} =$

4. $\dfrac{6.0 \times 10^{-23}}{2.0 \times 10^{-4}} =$

5. $\dfrac{10^{-14}}{-5.0 \times 10^{-3}} =$

6. $\dfrac{10^{14}}{4.0 \times 10^{-4}} =$

7. Complete the two problems in the *Pretest* at the beginning of this lesson.

SUMMARY

In your problem notebook, write a list of rules in Lesson 1.2 that were unfamiliar, needed reinforcement, or you found helpful. Then condense your list. Add this new list to your numbered points from Lesson 1.1. Write and recite the combined list until you can write all of the points from memory. Then, do the problems below.

PRACTICE   C

Start by doing every *other* letter. If you get those right, go to the next number. If not, do a few more of that number. Save one part of each question for your next study session.

1. Try these with*out* a calculator. Convert your final answers to scientific notation.

    a. $3 \times (6.0 \times 10^{23}) =$

    b. $1/2 \times (6.0 \times 10^{23}) =$

    c. $0.70 \times (6.0 \times 10^{23}) =$

    d. $10^3 \times (6.0 \times 10^{23}) =$

    e. $10^{-2} \times (6.0 \times 10^{23}) =$

    f. $(-0.5 \times 10^{-2})(6.0 \times 10^{23}) =$

    g. $\dfrac{1}{10^{12}} =$

    h. $1/(1/10^{-9}) =$

    i. $\dfrac{3.0 \times 10^{24}}{6.0 \times 10^{23}} =$

    j. $\dfrac{2.0 \times 10^{18}}{6.0 \times 10^{23}} =$

    k. $\dfrac{1.0 \times 10^{-14}}{4.0 \times 10^{-5}} =$

    l. $\dfrac{10^{10}}{2.0 \times 10^{-5}} =$

(continued)

2. Use a calculator for the numbers, but not for the exponents.

     a. $\dfrac{2.46 \times 10^{19}}{6.0 \times 10^{23}} =$ b. $\dfrac{10^{-14}}{0.0072} =$

3. Try these with*out* a calculator. Write answers as a power of 10.

     a. $\dfrac{10^{7} \times 10^{-2}}{10 \times 10^{-5}} =$ b. $\dfrac{10^{-23} \times 10^{-5}}{10^{-5} \times 100} =$

Lesson 1.3   Estimating Exponential Calculations

PRETEST   If you can solve both of the following problems correctly, skip this lesson. Convert your final answers to scientific notation. Check your answers at the end of this chapter.

   1. Solve this problem with*out* a calculator:

$$\frac{(10^{-9})(10^{15})}{(4 \times 10^{-4})(2 \times 10^{-2})} =$$

   2. For this problem, use a calculator as needed:

$$\frac{(3.15 \times 10^{3})(4.0 \times 10^{-24})}{(2.6 \times 10^{-2})(5.5 \times 10^{-5})} =$$

Choosing a Calculator

If you have not already done so, please read "Note to Students" at the front of this book.

Complex Calculations

The prior lessons covered the fundamental rules for exponential notation. For longer calculations, the rules are the same. The challenges are keeping track of the numbers and using the calculator correctly. The steps below will help you to simplify complex calculations, minimize data-entry mistakes, and quickly *check* your answers.

▬▬▬▶ **TRY IT**

Let's try the following calculation two ways.

$$\frac{(7.4 \times 10^{-2})(6.02 \times 10^{23})}{(2.6 \times 10^{3})(5.5 \times 10^{-5})} =$$

Method 1: Do Numbers and Exponents Separately Work the calculation on the previous page using the following steps:

1. **Do the numbers on the calculator.** Ignoring the exponentials, use the calculator to multiply all of the *significands* on top. Write the result. Then, multiply all the significands on the bottom and write the result. Divide, write your answer rounded to two digits, and then check below.

Answer:

$$\frac{7.4 \times 6.02}{2.6 \times 5.5} = \frac{44.55}{14.3} = \mathbf{3.1}$$

2. **Then handle the exponents.** Starting from the original problem, look only at the powers of 10. Try to solve the exponential math "in your head" with*out* the calculator. Write the answer for the top, then the bottom, then divide.

Answer:

$$\frac{10^{-2} \times 10^{23}}{10^{3} \times 10^{-5}} = \frac{10^{21}}{10^{-2}} = 10^{21-(-2)} = \mathbf{10^{23}}$$

3. **Now combine** the significand and exponential and write the final answer.

Answer:

$$\mathbf{3.1 \times 10^{23}}$$

Note that by grouping the numbers and exponents separately, you did not need to enter the exponents into your calculator. To multiply and divide powers of 10, you simply add and subtract whole numbers.

TRY IT

Let's try the calculation a second way.

Method 2: All on the Calculator Starting from the original problem, enter *all* of the numbers and exponents into your calculator. (Your calculator manual, which is usually available online, can help.) Write your final answer in scientific notation. Round the significand to two digits.

On most calculators, you will need to use an $\boxed{\text{E } or \text{ EE } or \text{ EXP}}$ key, rather than the *multiplication* key, to enter a "10 to a power" term.

Answer:

Your calculator answer, rounded, should be the same as with Method 1:

$$\mathbf{3.1 \times 10^{23}}$$

Note how your calculator *displays* the *exponential* term in answers. The exponent may be set apart at the right, sometimes with an **E** in front.

Which way was easier: "Numbers, then exponents," or "all on the calculator"? How you do the arithmetic is up to you, but "numbers, then exponents" is often quicker and easier.

Checking Calculator Results

Whenever a complex calculation is done on a calculator, to check for errors in calculator use, you *must* do the calculation a *second* time using different steps.

One way to check calculator answers is to calculate again using a different key sequence. However, "mental arithmetic estimation" is often a faster way to check a calculator answer. To learn the estimation method, let's use the calculation that we used in the first section of this lesson:

$$\frac{(7.4 \times 10^{-2})(6.02 \times 10^{23})}{(2.6 \times 10^{3})(5.5 \times 10^{-5})} =$$

Apply the following steps to the numbers above.

 TRY IT

1. *Estimate* **the numbers first.** Ignoring the exponentials, *round* and then multiply all of the top significands, and write the result. *Round* and multiply the bottom significands. Write the result. Then write a *rounded estimate* of the answer when you divide those two numbers, and then check below.

STOP **Answer:**

Your rounding might be

$$\frac{7 \times 6}{3 \times 6} = \frac{7}{3} \approx 2 \qquad \text{(The } \approx \text{ sign means } approximately\ equals.\text{)}$$

If your mental arithmetic is good, you can estimate the number math on the paper without a calculator. The estimate needs to be fast, but does *not* need to be exact. If needed, evaluate the *rounded* top and bottom numbers on the calculator, but try to practice the arithmetic "in your head."

2. **Evaluate the exponents.** The exponents can be solved by adding and subtracting whole numbers. Try the exponential math without the calculator.

STOP **Answer:**

$$\frac{10^{-2} \times 10^{23}}{10^{3} \times 10^{-5}} = \frac{10^{21}}{10^{-2}} = 10^{21-(-2)} = \mathbf{10^{23}}$$

3. **Combine** the estimated number and exponential answers. Compare this estimate to the answer found when you did this calculation in the section above using a calculator. Are they close?

STOP **Answer:**

The estimate is **2 × 10²³**. The answer with the calculator was **3.1 × 10²³**. Allowing for rounding, the two results are close.

If your fast, rounded, done-in-your-head answer is *close* to the calculator answer, it is likely that your calculator answer is correct. If the two answers are far apart, check your work.

Estimating Number Division

If you know your multiplication tables, and if you memorize these simple **decimal equivalents** to help in estimating division, you should be able to do many numeric estimates without a calculator.

1/2 = 0.50	1/3 = 0.33	1/4 = 0.25	1/5 = 0.20
2/3 = 0.67	3/4 = 0.75	1/8 = 0.125	

The method used to get your *final* answer should be slow and careful. Your *checking* method should use different steps or calculator keys, or, if time is a factor, should use rounded numbers and quick mental arithmetic.

On timed tests, you may want to do the exact calculation first, and then go back at the end, if time is available, and use rounded numbers as a check. When doing a calculation the second time, try not to look back at the first answer until after you write the estimate. If you look back, by the power of suggestion, you will often arrive at the first answer whether it is correct or not.

For complex operations on a calculator, work each calculation a second time using rounded numbers and/or different steps or keys.

PRACTICE

You will need to memorize the above fraction-to-decimal-equivalent conversions. If you need practice, try this:

- On a sheet of paper, draw 5 columns and 7 rows. List the fractions down the middle column.

		1/2		
		1/3		
		1/4		
		1/5		
		2/3		
		3/4		
		1/8		

- Write the decimal equivalents of the fractions at the far right.
- Fold over those answers and repeat at the far left. Fold over those and repeat.

After learning those conversions, do problems 1–4 without a calculator. Convert final answers to scientific notation. Round the significand in the answer to two digits.

(continued)

1. $\dfrac{4 \times 10^3}{(2.00)\,(3.0 \times 10^7)} =$

2. $\dfrac{1}{(4.0 \times 10^9)\,(2.0 \times 10^3)} =$

3. $\dfrac{(3 \times 10^{-3})\,(8.0 \times 10^{-5})}{(6.0 \times 10^{11})\,(2.0 \times 10^{-3})} =$

4. $\dfrac{(3 \times 10^{-3})\,(3.0 \times 10^{-2})}{(9.0 \times 10^{-6})\,(2.0 \times 10^1)} =$

Complete problems 5–8 below in your notebook as follows.

* First write an *estimate* based on rounded numbers, then exponentials. *Try* to do this estimate with*out* using a calculator.
* Then, calculate a more precise answer. You may
  * Plug the entire calculation into the calculator, or
  * Use the "numbers on calculator, exponents on paper" method, or
  * Experiment with both approaches to see which is best for you.

Convert both the estimate and the final answer to *scientific notation*. Round the significand in the answer to two digits. Use the calculator that you will be allowed to use on quizzes and tests.

To start, complete the even-numbered problems. If you get those right, go to the next lesson. If you need more practice, do the odds.

5. $\dfrac{(3.62 \times 10^4)\,(6.3 \times 10^{-10})}{(4.2 \times 10^{-4})\,(9.8 \times 10^{-5})} =$

6. $\dfrac{10^{-2}}{(750)\,(2.8 \times 10^{-15})} =$

7. $\dfrac{(1.6 \times 10^{-3})\,(4.49 \times 10^{-5})}{(2.1 \times 10^3)\,(8.2 \times 10^6)} =$

8. $\dfrac{1}{(4.9 \times 10^{-2})\,(7.2 \times 10^{-5})} =$

9. For additional practice, do the two *Pretest* problems at the beginning of this lesson.

Lesson 1.4   Special Project—The Atoms (Part 1)

At the center of chemistry are **atoms**: the building blocks of matter. There are 91 different kinds of atoms found in the Earth's crust. When a substance is in its most stable form at room temperature and pressure and contains only one kind of atom, the substance is said to be an **element** and the atoms are in their **elemental** form.

The periodic table helps in predicting the properties of the elements and atoms. In first-year chemistry, about 40 of the atoms are frequently encountered. Quick, automatic conversion between the names and symbols of those atoms "in your head" will speed problem solving. In addition, if you know where to *look* for an atom in the table, it will speed finding additional data about the atom that a more detailed periodic table will provide.

To begin to learn those atoms, your assignment is:

- For the 12 atoms below, memorize the name, symbol, and position of the atom in the table.
- For each atom, given either its symbol or name, be able to write the other.
- Be able to fill in an empty chart like the one below with these *names* and *symbols* in their correct locations.

To complete this assignment, draw several charts like the one below in your problem notebook. Practice until you can fill in all of the first 12 names and symbols from memory.

PERIODIC TABLE

1A	2A		3A	4A	5A	6A	7A	8A
1 **H** Hydrogen								2 **He** Helium
3 **Li** Lithium	4 **Be** Beryllium		5 **B** Boron	6 **C** Carbon	7 **N** Nitrogen	8 **O** Oxygen	9 **F** Fluorine	10 **Ne** Neon
11 **Na** Sodium	12 **Mg** Magnesium							

SUMMARY

1. When writing a number between −1 and 1, place a *zero* in *front* of the decimal point. Do *not* write .42 or −.74; *do* write **0.42** or **−0.74**

2. *Exponential* notation represents numeric values in three parts:
   - A *sign* in front showing whether the value is positive or negative;
   - A *number* (the significand);
   - Times a *base* taken to a *power* (the exponential term).

3. In *scientific* notation, the significand must be a number that is 1 or greater, but less than 10, and the exponential term must be 10 to an integer power. This places the decimal point in the significand after the first number that is not a zero.

4. When moving a decimal in exponential notation, the sign in front never changes.

5. To keep the same numeric value when moving the decimal of a number in base 10 exponential notation, if you
   - Move the decimal Y *times* to make the significand *larger*, make the exponent *smaller* by a *count* of Y;
   - Move the decimal Y times to make the significand smaller, make the exponent larger by a count of Y.

(continued)

When moving the decimal, for the significand and exponential term *after* the sign in front, recite and repeat to remember:

If one gets *smaller*, the other gets *larger*. If one gets *larger*, the other gets *smaller*.

6. In *multiplication and division* using scientific or exponential notation, handle numbers and exponential terms separately. Recite and repeat to remember:
   - Do numbers by number rules and exponents by exponential rules.
   - When you multiply exponentials, you add the exponents.
   - When you divide exponentials, you subtract the exponents.
   - When you take an exponential *term* to a power, you multiply the exponents.
   - To take the reciprocal of an exponential, change the sign of the exponent.
   - For *any* X: $1/(1/X) = X$

7. In calculations using exponential notation, try the significands on the calculator but the exponents on paper.

8. For complex operations on a calculator, do each calculation a *second* time using rounded numbers or a different key sequence on the calculator.

ANSWERS

(To make answer pages easy to locate, use a sticky note.)

Lesson 1.1

Pretest 1a. 9.4×10^6 1b. 4.2×10^4 1c. -6.7×10^{-5} 1d. -7.7×10^1

2a. 0.0014 2b. 194 2c. 1

Practice A 1. 10,000 2. 0.0001 3. 10,000,000 4. 0.00001 5. 1

Practice B 1. When dividing by 1000, move the decimal to the left by three places.

2a. 420 2b. 0.63 (Must have a zero in front.) 2c. −0.00746

Practice C 1. 3,000 2. 0.00055 3. 770,000 4. −0.0095

Practice D 1. 5.42×10^6 2. 6.7×10^{-7} 3. 2.0×10^1

4. -8.7×10^{-2} 5. 4.92×10^{-15} 6. -6.02×10^{23}

Practice E 1. b and c 2a. 6.28×10^3 2b. 9.3×10^{-3} 2c. 7.41×10^{-1} 2d. -1.28×10^6

Practice F 1a. 0.0924 1b. 24,300 1c. −0.0024

2a. 5.5×10^4 2b. 9.2×10^{-1} 2c. 9.4×10^{-4} 2d. 3.2×10^{-3}

3a. 7.7×10^3 3b. 1.6×10^8 3c. 2.3×10^{-2} 3d. 6.7×10^{-4}

Lesson 1.2

Pretest   1. 1.2×10^{20} 2. 3.3×10^{28}

Practice A   1. 10^8 2. 10^{-11} 3. 10^{-1} 4. 10^{-8} 5. 10^4 6. 10^{-23} 7. 10^4 8. 10^{35}

9. $\dfrac{100 \times 10^{-2}}{1000 \times 10^6} = \dfrac{10^2 \times 10^{-2}}{10^3 \times 10^6} = \dfrac{10^0}{10^9} = \mathbf{10^{-9}}$ 10. $\dfrac{10^{-3} \times 10^{23}}{10 \times 1000} = \dfrac{10^{20}}{10^4} = \mathbf{10^{16}}$

(For problems 9 and 10, you may use different steps, but you must arrive at the same answer.)

Practice B   1. 1.2×10^{25} 2. 7.5×10^{12} 3. -1.5×10^{-24} 4. 3.0×10^{-19} 5. -2.0×10^{-12} 6. 2.5×10^{17}

7. See Pretest answers.

Practice C   1a. $3 \times (6.0 \times 10^{23}) = 18 \times 10^{23} = \mathbf{1.8 \times 10^{24}}$ 1b. $1/2 \times (6.0 \times 10^{23}) = \mathbf{3.0 \times 10^{23}}$

1c. $0.70 \times (6.0 \times 10^{23}) = \mathbf{4.2 \times 10^{23}}$ 1d. $10^3 \times (6.0 \times 10^{23}) = \mathbf{6.0 \times 10^{26}}$

1e. $10^{-2} \times (6.0 \times 10^{23}) = \mathbf{6.0 \times 10^{21}}$ 1f. $(-0.5 \times 10^{-2})(6.0 \times 10^{23}) = \mathbf{-3.0 \times 10^{21}}$

1g. $\dfrac{1}{10^{12}} = \mathbf{1.0 \times 10^{-12}}$ 1h. $1/(1/10^{-9}) = \mathbf{1.0 \times 10^{-9}}$

1i. $\dfrac{3.0 \times 10^{24}}{6.0 \times 10^{23}} = \dfrac{3.0}{6.0} \times \dfrac{10^{24}}{10^{23}} = 0.50 \times 10^1 = \mathbf{5.0 \times 10^0 \, (= 5.0)}$

1j. $\dfrac{2.0 \times 10^{18}}{6.0 \times 10^{23}} = 0.33 \times 10^{-5} = \mathbf{3.3 \times 10^{-6}}$ 1k. $\dfrac{1.0 \times 10^{-14}}{4.0 \times 10^{-5}} = 0.25 \times 10^{-9} = \mathbf{2.5 \times 10^{-10}}$

1l. $\dfrac{10^{10}}{2.0 \times 10^{-5}} = \dfrac{1}{2.0} \times \dfrac{10^{10}}{10^{-5}} = 0.50 \times 10^{15} = \mathbf{5.0 \times 10^{14}}$ 2a. $\dfrac{2.46 \times 10^{19}}{6.0 \times 10^{23}} = 0.41 \times 10^{-4} = \mathbf{4.1 \times 10^{-5}}$

2b. $\dfrac{10^{-14}}{0.0072} = \dfrac{1.0 \times 10^{-14}}{7.2 \times 10^{-3}} = \dfrac{1.0}{7.2} \times \dfrac{10^{-14}}{10^{-3}} = 0.14 \times 10^{-11} = \mathbf{1.4 \times 10^{-12}}$

3a. $\dfrac{10^7 \times 10^{-2}}{10^1 \times 10^{-5}} = \dfrac{10^5}{10^{-4}} = \mathbf{10^9}$ 3b. $\dfrac{10^{-23} \times 10^{-5}}{10^{-5} \times 10^2} = \mathbf{10^{-25}}$

Lesson 1.3

Pretest   1. 1.25×10^{11} or 1.3×10^{11} 2. 8.8×10^{-15}

Practice   You may do the arithmetic using different steps than shown below, but you must get the same answer.

1. $\dfrac{4 \times 10^3}{(2.00)(3.0 \times 10^7)} = \dfrac{4}{6} \times 10^{3-7} = \dfrac{2}{3} \times 10^{-4} = \mathbf{0.67 \times 10^{-4}} = \mathbf{6.7 \times 10^{-5}}$

2. $\dfrac{1}{(4.0 \times 10^9)(2.0 \times 10^3)} = \dfrac{1}{8 \times 10^{12}} = \dfrac{1}{8} \times 10^{-12} = \mathbf{0.125 \times 10^{-12}} = \mathbf{1.3 \times 10^{-13}}$

3. $\dfrac{(3 \times 10^{-3})(8.0 \times 10^{-5})}{(6.0 \times 10^{11})(2.0 \times 10^{-3})} = \dfrac{8}{4} \times \dfrac{10^{-3-5}}{10^{11-3}} = 2 \times \dfrac{10^{-8}}{10^8} = 2 \times 10^{-8-8} = \mathbf{2.0 \times 10^{-16}}$

4. $\dfrac{(3 \times 10^{-3})(3.0 \times 10^{-2})}{(9.0 \times 10^{-6})(2.0 \times 10^1)} = \dfrac{9}{18} \times \dfrac{10^{-3-2}}{10^{-6+1}} = 0.50 \times \dfrac{10^{-5}}{10^{-5}} = 0.50 = \mathbf{5.0 \times 10^{-1}}$

5. First the *estimate*. The rounding for the *numbers* might be $\dfrac{4 \times 6}{4 \times 10} = \mathbf{0.6}$

For the *exponents*: $\dfrac{10^4 \times 10^{-10}}{10^{-4} \times 10^{-5}} = \dfrac{10^{-6}}{10^{-9}} = 10^9 \times 10^{-6} = \mathbf{10^3}$

Estimate in scientific notation: $0.6 \times 10^3 = \mathbf{6 \times 10^2}$

For the *precise* answer, doing numbers and exponents separately,

$$\frac{(3.62 \times 10^4)\,(6.3 \times 10^{-10})}{(4.2 \times 10^{-4})\,(9.8 \times 10^{-5})} = \frac{3.62 \times 6.3}{4.2 \times 9.8} = \mathbf{0.55}$$

The exponents are done as in the estimate above. Final in scientific notation: $0.55 \times 10^3 = \mathbf{5.5 \times 10^2}$

This is close to the estimate; a check that the more precise answer is correct.

6. Estimate: $\dfrac{1}{7 \times 3} = \dfrac{1}{20} = \mathbf{0.05}$; $\dfrac{10^{-2}}{(10^2)\,(10^{-15})} = 10^{-2-(-13)} = \mathbf{10^{11}}$

   Estimate in scientific notation: $0.05 \times 10^{11} = \mathbf{5 \times 10^9}$

   Numbers on calculator: $\dfrac{1}{7.5 \times 2.8} = \mathbf{0.048}$. Exponents: same as in estimate.

   Final in scientific notation: $0.048 \times 10^{11} = \mathbf{4.8 \times 10^9}$. Close to the estimate.

7. Estimate: $\dfrac{1.6 \times 4.49}{2.1 \times 8.2} = \dfrac{2 \times 4}{2 \times 8} = \mathbf{0.5}$; $\dfrac{10^{-3} \times 10^{-5}}{10^3 \times 10^6} = \dfrac{10^{-8}}{10^9} = \mathbf{10^{-17}}$

   Estimate in scientific notation: $0.5 \times 10^{-17} = \mathbf{5 \times 10^{-18}}$

   Numbers on calculator: $\dfrac{1.6 \times 4.49}{2.1 \times 8.2} = \mathbf{0.42}$. The exponents are done as in the estimate above.

   Final in scientific notation: $0.42 \times 10^{-17} = \mathbf{4.2 \times 10^{-18}}$. This is close to the estimate. Check!

8. Estimate: $\dfrac{1}{5 \times 7} = \dfrac{1}{35} = \mathbf{0.03}$; $\dfrac{1}{(10^{-2})\,(10^{-5})} = 1/(10^{-7}) = \mathbf{10^7}$

   Estimate in scientific notation: $0.033 \times 10^7 = \mathbf{3 \times 10^5}$

   Numbers on calculator: $\dfrac{1}{4.9 \times 7.2} = \mathbf{0.028}$. Exponents: see estimate.

   Final in scientific notation: $0.028 \times 10^7 = \mathbf{2.8 \times 10^5}$. Close to the estimate.

9. See Pretest answers.

2

The Metric System

Lesson 2.1   Metric Fundamentals

Have you previously mastered the metric system? If you get a perfect score on the following Pretest, you may skip to Lesson 2.2. If not, complete Lesson 2.1.

> **PRETEST**   Write answers to these, then check your answers at the end of the chapter.
>
> 1. What is the mass, in kilograms, of 150 cm^3 of liquid water?
>
> 2. How many cm^3 are in a liter?
>
> 3. How many dm^3 are in a liter?
>
> 4. 2.5 pascals is how many millipascals?
>
> 5. 3,500 centigrams is how many kilograms?

The Importance of Units

The fastest and most effective way to solve problems in chemistry is to focus on the **units** used to measure quantities. In science, measurements and calculations are done using the **metric system**.

All measurement systems begin by defining **base units** that measure the **fundamental quantities**, including **distance**, **mass**, and **time**.

Distance Units The metric base unit for distance is the **meter**, abbreviated **m**. It is about 39.3 inches, slightly longer than one yard. A meter stick is usually numbered in **centimeters**.

Just as a dollar can be divided into 100 *cent*s, and a *cent*ury is 100 years, a meter is divided into 100 *centi*meters. The centimeter, abbreviated **cm**, is 1/100th of a meter. As equalities, we can write

1 **centi**meter ≡ 1/100th of a meter ≡ 10^{-2} meters *and* 1 meter ≡ **100** *centi*meters

The symbol ≡ means "is defined as equal to" and/or "is exactly equal to."

A centimeter can be divided into 10 **milli**meters (**mm**). Each millimeter is 1/1000th of a meter. In relation to the meter,

1 **milli**meter ≡ 1/1000th of a meter ≡ 10^{-3} meters *and* 1 meter ≡ **1000** *milli*meters

A meter stick can also be divided into 10 **deci**meters (**dm**). As equalities,

1 **deci**meter ≡ 1/10th of a meter ≡ 10^{-1} meters *and* 1 meter ≡ **10** *deci*meters

One decimeter is also equal to 10 centimeters.

Long distances are usually measured in **kilo**meters (**km**): **1 kilometer ≡ 1000 meters**.

What do you need to remember from the above? You will need to be able to write from memory the following two rules.

1. **The "meter-stick" equalities are**

   1 *meter* $\equiv$ 10 **deci**meters $\equiv$ 100 **centi**meters $\equiv$ 1000 **milli**meters

   1000 meters $\equiv$ 1 **kilo**meter

2. **The "1-prefix" definitions are**

   1 **milli**meter $\equiv 10^{-3}$ meters ($\equiv$ 1/1000th meter $\equiv$ 0.001 meters)

   1 **centi**meter $\equiv 10^{-2}$ meters ($\equiv$ 1/100th meter $\equiv$ 0.01 meters)

   1 **deci**meter $\equiv 10^{-1}$ meters ($\equiv$ 1/10th meter $\equiv$ 0.1 meters)

   1 **kilo**meter $\equiv 10^{3}$ meters ($\equiv$ 1000 meters)

To help in remembering the meter-stick equalities, visualize a meter stick. Recall what the numbers and marks on a meter stick mean. Use that image to help you to write the equalities above.

To help in remembering the kilometer definition, visualize 1000 meter sticks in a row. That's a distance of one *kilo*meter. 1 kilometer $\equiv$ 1000 meter sticks.

Rule 1 uses the "1 meter =" format and rule 2 uses the "1-prefix" format. Because both formats are used in calculations, we will need to be familiar with both ways of defining the prefix relationships. However, once rule 1 is in memory, rule 2 should be easier to write and check because it is mathematically equivalent.

Rules 1 and 2 are especially important because of rule 3.

3. You may substitute *any unit* for *meter* in the equalities above.

Rule 3 means that the prefix relationships that are true for meters are true for *any* units of measure. The three rules above allow us to write a wide range of equalities that we can use to solve science calculations, such as

$$1 \; liter \equiv 1000 \; milli liter s$$

$$1 \; centi gram \equiv 10^{-2} \; grams$$

$$1 \; kilo calorie \equiv 10^{3} \; calories$$

To use *kilo-*, *deci-*, *centi-*, or *milli-* with *any* units, you simply need to be able to recall and write the metric equalities in rules 1 and 2.

What do you need to be able to recall *from memory* so far in this lesson? Rules 1, 2, and 3.

PRACTICE A

Write rules 1 and 2 until you can do so from memory. Learn rule 3. Then complete these problems without looking back at the rules.

1. From memory, add exponential terms to these blanks.

   a. 1 millimeter = _____ meters   b. 1 deciliter = _____ liter

2. From memory, add full metric prefixes to these blanks.

   a. 1000 grams = 1 _____ gram   b. 10^{-2} liters = 1 _____ liter

Volume Units **Volume** is the amount of three-dimensional space that a material or shape occupies. Volume is termed a **derived quantity**, rather than a fundamental quantity, because it is derived from distance. Any volume unit can be converted to a distance unit cubed.

A cube that is 1 centimeter *wide* by 1 cm *high* by 1 cm *long* has a volume of one **cubic** centimeter (1 **cm**^{3}). In biology and medicine, a cm^3 is often abbreviated as a "**cc**," but cm^3 is the abbreviation used in chemistry.

In chemistry, cubic centimeters are usually referred to as **milliliters**, abbreviated as **mL**. One milliliter is defined as exactly one cubic centimeter. Based on this definition, because

- 1000 milli*meters* ≡ 1 *meter* and 1000 milli*anythings* ≡ 1 *anything*
- 1000 milli*liters* is defined as 1 **liter (1 L)**

The mL is a convenient measure for smaller volumes, while the liter (about 1.1 quarts) is preferred when measuring larger volumes.

One liter is the same as **one cubic decimeter (1 dm^3)**. Note how these units are related.

- The volume of a cube that is

$$10 \text{ cm} \times 10 \text{ cm} \times 10 \text{ cm} = 1000 \text{ cm}^3 = 1000 \text{ mL}$$

- Because 10 cm ≡ 1 dm, the volume of this same cube can be calculated as

$$1 \text{ dm} \times 1 \text{ dm} \times 1 \text{ dm} \equiv 1 \text{ cubic } decimeter \equiv 1 \text{ dm}^3$$

Based on the above, by definition, all of the following terms are *equal.*

$$1000 \text{ cm}^3 \equiv 1000 \text{ mL} \equiv 1 \text{ L} \equiv 1 \text{ dm}^3$$

What do you need to remember about volume? For now, just two more sets of equalities:

> 4. 1 milliliter (mL) ≡ 1 cm^3
>
> 5. 1 liter ≡ 1000 mL ≡ 1000 cm^3 ≡ 1 dm^3

Mass Units **Mass** measures the amount of matter in an object. If you have studied physics, you know that mass and weight are not the same. In chemistry, however, unless stated otherwise, we assume that mass is measured at the constant gravity of the Earth's surface. In that case, mass and weight are directly proportional and can be measured with the same instruments.

The metric base-unit for mass is the gram. One **gram (g)** was originally defined as the mass of *one cubic centimeter* of *liquid water* at 4 degrees Celsius, the temperature at which water has its highest density. The modern definition for one gram is a bit more complex, but it is still very close to the historic definition. We will often use that historic definition of a gram in calculations involving liquid water if high precision is not required.

For most calculations involving *liquid* water at or near room temperature, the following *approximation* may be used.

> 6. 1 **cm**^{3} H_2O(*liquid*) ≡ 1 **mL** $H_2O(\ell)$ ≈ 1.00 **gram** $H_2O(\ell)$
>   (≈ means *approximately.*)

The substance H_2O is solid when it is ice, liquid when it is water, and gaseous when it is steam. The notation (*liquid*), abbreviated as (ℓ), written after the chemical formula, means that this rule is true only if the water is in its liquid state.

Temperature Metric temperature scales are defined by the properties of water. Temperature in the metric system can be measured by **degrees Celsius** (°C).

    0°C = the freezing point of water

    100°C = the boiling point of water at a pressure of one atmosphere

    Room temperature is generally between 20°C (which is 68°F) and 25°C (which is 77°F).

Time Units The base unit for time in the metric system is the **second**.

Unit and Prefix Abbreviations

The following list of abbreviations for metric units should also be committed to memory. The order is not important, but given the unit, you need to be able to write the abbreviation, and given the abbreviation, you need to be able to write the unit.

    Unlike other abbreviations, abbreviations for metric units do *not* have periods at the end.

Units

    m = meter   **g** = gram   **s** = second

    L = liter = $\mathbf{dm^3}$ = cubic decimeter

    $\mathbf{cm^3}$ = cubic centimeter = "cc" = **mL**

Metric *prefix* abbreviations may be written in front of any metric unit abbreviation.

The most frequently used prefixes are:

        k- = kilo-; **d-** = deci-; **c-** = centi-; **m-** = milli-

---

PRACTICE **B**

Write rules 1 through 6 until you can do so from memory. Learn the unit and prefix abbreviations as well. Then complete the following problems without looking back at the above.

1. Fill in the prefix abbreviations: 1 m = 10 _____ m = 100 _____ m = 1000 _____ m

2. From memory, add metric prefix abbreviations to these blanks.

    a. 10^3 g = 1 _____ g   b. 10^{-3} s = 1 _____ s

3. From memory, add fixed decimal numbers to these blanks.

    a. $1000 \, cm^3$ = _____ mL   b. $100 \, cm^3 \, H_2O(\ell)$ ≈ _____ grams $H_2O(\ell)$

4. Add fixed decimal numbers: 1 liter = _____ mL = _____ cm^3 = _____ dm^3

SI Units

The modern metric system (*Le Système International d'Unités*) is referred to as the **SI system** and is based on what are termed the **SI units**. SI units are a subset of metric units that chooses one preferred metric unit as the standard for measuring each physical quantity.

The SI standard unit for distance is the meter, for mass is the kilogram, and for time is the second. Historically, the SI system is derived from what in physics was termed the **mks system** because it measured in units of meters, kilograms, and seconds.

In physics, and in many chemistry calculations that are based on relationships derived from physics, the use of SI standard units is essential to simplify calculations. However, for dealing with laboratory-scale quantities, measurements and calculations in chemistry frequently use units that are not SI but are metric.

Example: In chemistry we generally measure *volume* in liters or milliliters instead of cubic meters and *mass* in grams instead of kilograms.

In Chapters 4 and 5, you will learn to convert between the non-SI units often used in chemistry and the SI units that are required for some types of chemistry calculations.

Learning the Metric Fundamentals

A strategy that can help in problem solving is to start each homework assignment, quiz, or test by writing *recently* memorized rules at the top of your paper. By writing the rules at the beginning, you avoid having to remember them under time pressure later in the test.

We will use *equalities* to solve most problems. The rules above define the equalities that we will use most often.

A Note on Memorization   A goal of these lessons is to minimize what you must memorize. However, it is not possible to eliminate memorization from science courses. When there are facts that you must memorize in order to solve problems, these lessons will tell you. This is one of those times. Memorize the metric basics in the table on the next page. You will need to write them automatically, from memory, as part of most assignments in chemistry.

Memorization Tips   When you memorize, it helps to use as many *senses* as you can.

- *Say* the rules out loud, over and over, as you would learn lines for a play.
- *Write* the equations several times, in the same way and order each time.
- *Organize* the rules into patterns, rhymes, or mnemonics.
- *Number* the rules so you know which rule you forgot, and when to stop.
- *Picture* real objects:
  - Sketch a meter stick, then write the first two metric rules and compare them to your sketch.
  - For volume, mentally picture a 1 cm × 1 cm × 1 cm = **1 cm³** cube. Call it *one milliliter*. Fill it with water to have a *mass* of 1.00 *grams*.

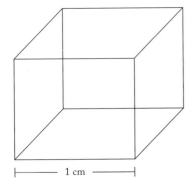

After repetition, you will recall new rules *automatically*. That's the goal.

Metric Basics

1. 1 *meter* $\equiv$ 10 decimeters
             $\equiv$ 100 centimeters
             $\equiv$ 1000 millimeters
   1000 meters $\equiv$ 1 kilometer

2. **1 milli**meter $\equiv$ **1 mm** $\equiv$ 10^{-3} meter
   1 centimeter $\equiv$ **1 cm** $\equiv$ 10^{-2} meter
   1 decimeter $\equiv$ **1 dm** $\equiv$ 10^{-1} meter
   1 kilometer $\equiv$ **1 km** $\equiv$ 10^{3} meter

3. Any unit can be substituted for *meter* above.

4. $1 \text{ mL} \equiv 1 \text{ cm}^3$

5. $1 \text{ liter} \equiv 1000 \text{ mL} \equiv 1000 \text{ cm}^3 \equiv 1 \text{ dm}^3$

6. $1 \text{ cm}^3 \text{ H}_2\text{O}(liquid) \equiv 1 \text{ mL H}_2\text{O}(\ell)$
              $\approx 1.00 \text{ gram H}_2\text{O}(\ell)$

7. meter = m; gram = g; second = s

PRACTICE C

Study the seven rules in the metric basics table above, then write the table on paper from memory. Repeat until you can write all parts of the table from memory. Then cement your knowledge by doing the following problems.

1. In your mind, picture a kilometer and a millimeter. Which is larger?

2. Which is larger, a kilojoule or a millijoule?

3. Name four units that can be used to measure volume in the metric system.

4. How many centimeters are on a meter stick?

5. How many liters are in a kiloliter?

6. What is the mass of 15 milliliters of liquid water?

7. One liter of liquid water has what mass in grams?

8. Fill in the portion of the periodic table below for the first 12 atoms.

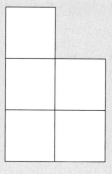 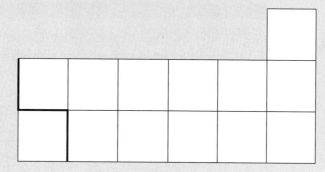

Lesson 2.2   Metric Prefixes

Additional Prefixes

Prefix	Abbreviation	Means
tera-	T-	$\times 10^{12}$
giga-	G-	$\times 10^{9}$
mega-	M-	$\times 10^{6}$
kilo-	k-	$\times 10^{3}$
deci-	d-	$\times 10^{-1}$
centi-	c-	$\times 10^{-2}$
milli-	m-	$\times 10^{-3}$
micro-	μ- (mu) *or* u-	$\times 10^{-6}$
nano-	n-	$\times 10^{-9}$
pico-	p-	$\times 10^{-12}$

For measurements of very large or very small quantities, prefixes larger than *kilo-* and smaller than *milli-* may be used. Ten prefixes encountered frequently are listed in the table at left.

Note that

- Outside the range between −3 and 3, metric prefixes are abbreviations of powers of 10 that are divisible by 3.
- When the full prefix name is written, the first letter is not normally capitalized.
- For prefixes above **k-** (kilo-), the abbreviation for a prefix *must* be *capitalized*.
- For the prefixes **k-** and below, the abbreviation *must* be lowercase.

Using Prefixes

A metric prefix is interchangeable with the exponential term it represents. For example, during measurements and calculations,

- An exponential term can be *substituted* for a prefix based on what the prefix means, as defined in the table above.

  Examples:

  7.0 *milli*liters = 7.0×10^{-3} liters

  5.6 kg = 5.6×10^{3} g

  43 nanometers = 43 nm = 43×10^{-9} meters

- A metric *prefix* can be substituted for its equivalent exponential term.

  Examples:

  3.5×10^{-12} meters = 3.5 **pico**meters = 3.5 pm

  7.2×10^{6} watts = 7.2 **mega**watts

In calculations, we will often need to convert between a prefix and its equivalent exponential term. One way to do this is to apply the prefix definitions.

▶ **TRY IT**

(See "How to Use These Lessons," point 1, p. xv.)

Q1. From memory, fill in these blanks with full prefixes (not the prefix abbreviations).

    a. 10^{3} grams = 1 _____ gram

    b. 2×10^{-3} meters = 2 _____ meters

Q2. From memory, fill in these blanks with prefix *abbreviations*.

    a. 2.6×10^{-1} L = 2.6 ____ L   b. 6×10^{-2} g = 6 ____ g

Q3. Fill in these blanks with exponential terms (consult the table on page 00 if needed).

    a. 1 gigajoule = $1 \times$ _____ joules   b. 9 μm = $9 \times$ _____ m

Answers:

1a. 10^3 grams = 1 **kilo**gram   1b. 2×10^{-3} meters = 2 **milli**meters

2a. 2.6×10^{-1} L = 2.6 **dL**   2b. 6×10^{-2} g = 6 **cg**

3a. 1 **giga**joule = 1×10^9 joules   3b. 9 μm = 9×10^{-6} m

From the prefix definitions, even if you are not yet familiar with the quantity that a unit is measuring, you can convert between a *prefix* and its equivalent exponential.

Science versus Computer-Science Prefixes Computer science, which calculates based on powers of 2, uses slightly different definitions for prefixes, such as *kilo-* = 2^{10} = 1,024 instead of 1000. However, in chemistry and all other sciences, the prefix to power-of-10 relationships in the metric prefix table are *exact* definitions.

Learning the Additional Prefixes

To solve calculations, you will need to recall the rows in the table of 10 metric prefixes quickly and automatically. To begin, practice writing the table from memory. To help, look for patterns and use memory devices. Note

tera- = T- = 10^{twelve}

nano- (which connotes *small*) = n- = 10^{-nine}

Focusing on those two can help to "anchor" the prefixes near them in the table. Then make a self-quiz: On a sheet of paper, draw a table 3 columns across and 11 rows down. In the top row, write

Prefix   Abbreviation   Means

Then fill in the table. Repeat writing the table until you can do so from memory, without looking back. Once you can do so, try to do the problems below without looking back at your table.

PRACTICE   **A**

Use a sticky note to mark the answer page at the end of this chapter.

1. From memory, add exponential terms to these blanks.

    a. 7 microseconds = 7 × _____ seconds   b. 9 kg = 9 × _____ g

    c. 8 cm = 8 × _____ m   d. 1 ng = 1 × _____ g

2. From memory, add full metric prefixes to these blanks.

    a. 6×10^{-2} amps = 6 _____ amps   b. 45×10^9 watts = 45 _____ watts

(continued)

3. From memory, add prefix abbreviations to these blanks.

   a. 10^{12} g = 1 _____ g

   b. 10^{-12} s = 1 _____ s

   c. 6×10^{-9} m = 6 _____ m

   d. 5×10^{-1} L = 5 _____ L

   e. 4×10^{-6} L = 4 _____ L

   f. 16×10^{6} Hz = 16 _____ Hz

4. When writing prefix abbreviations *by hand*, write so that you can distinguish between (add a prefix abbreviation): 5×10^{-3} g = 5 _____ g and 5×10^{6} g = 5 _____ g

5. For which prefix abbreviations is the first letter always capitalized?

6. Convert "0.30 gigameters/second" to a value without a prefix in scientific notation.

Converting between Prefix Formats

To solve calculations, we will often use conversion factors that are constructed from metric prefix definitions. For those definitions, we have learned two types of equalities.

- Our "meter stick" equalities are based on what *one unit* is equal to:

1 meter ≡ 10 **deci**meters ≡ 100 **centi**meters ≡ 1000 **milli**meters

- Our prefix definitions are based on what *one prefix* equals, such as *nano = 10^{-9}*

It is essential to be able to correctly write *both* forms of the metric definitions, because work in science often uses both.

Example: When converting between milliliters and liters, you may see either

- 1 mL = 10^{-3} L, based on what *1 milli-* means, *or*
- 1000 mL = 1 L, which is an easy-to-visualize definition of one liter

Those two equalities are equivalent. The second is simply the first with the numbers on both sides multiplied by 1000.

However, note that **1** mL = 10^{-3} L, but **1** L = 10^{3} mL. The numbers in the equalities change depending on whether the 1 is in front of the prefix or the unit. Which format should we use? How do we avoid errors?

In these lessons, we will generally use the 1-*prefix* equalities to solve problems. After learning the definitions for the 10 prefixes in the table, such as 1 milli- = **10^{-3}**, recalling those definitions makes conversions easier to check.

Once those prefix, abbreviation, and meaning correlations are in memory, we will then need to "watch where the 1 is."

If you need to write or check prefix equalities in the "one unit =" format, you can derive them from the 1-*prefix* definitions in the prefix definition table, if needed.

Example: 1 gram = ? **micrograms**

- Because, from the prefix table, 1 **micro**-anything = **10^{-6}** anythings, then 1 **microgram** = **10^{-6}** grams.

- To get a 1 in front of *gram*, we multiply both sides by 10^6, so **10^6 micrograms** = 1 gram.

The steps above can be summarized as the *reciprocal* rule for prefixes:

> If **1** *prefix-* = **10^a**, **1** *unit* = **10^{-a}** *prefix* units.

Another way to state the reciprocal rule for prefixes:

> To change a prefix definition between the "1 *prefix* =" format and the "1 *unit* =" format, change the sign of the exponent.

If you need to check your logic, write a familiar example: Because **1 milli**liter = **10^{-3}** liter, then 1 liter = **10^3** milliliters = 1000 mL.

TRY IT

Q. Using those rules, provide the missing exponential terms below.

1. 1 nanogram = 1 × _____ grams, so

   1 gram = 1 × _____ nanograms.

2. 1 dL = 1 × _____ liters, so 1 L = 1 × _____ dL.

STOP

Answers:

1. 1 nanogram = 1 × **10^{-9} grams**, so 1 gram = 1 × **10^9** nanograms.

2. 1 dL = 1 × **10^{-1}** liters, so 1 L = 1 × **10^1** dL = 10 dL.

To summarize:

- When using metric prefix definitions, be careful to note whether the **1** is in front of the prefix or the unit.
- A way to avoid confusing the signs of the exponential terms is to memorize the table of prefix definitions. Then, if you need an equality with the "*1 unit* = **10^x** prefix-unit" format, reverse the sign of the prefix definition.

PRACTICE   B

Write the table of the 10 metric prefixes until you can do so from memory; then try answering these without consulting the table.

1. Fill in the blanks with exponential terms.

   a. 1 terasecond = 1 × _____ seconds, so 1 second = 1 × _____ teraseconds.

   b. 1 ng = 1 × _____ grams, so 1 g = 1 × _____ ng.

(continued)

2. Apply the reciprocal rule to add exponential terms to these *one unit* equalities.

    a. 1 gram = _____ centigrams   b. 1 meter = _____ picometers

    c. 1 s = _____ ms   d. 1 s = _____ Ms

3. Add exponential terms to these blanks. Watch where the 1 is!

    a. 1 micromole = _____ moles   b. 1 g = 1 × _____ Gg

    c. 1 kilogram = 1 × _____ grams   d. _____ kg = 1 g

    e. _____ ns = 1 s   f. 1 pL = _____ L

Lesson 2.3   Cognitive Science—and Flashcards

In this lesson, you will learn a *system* that will help you automatically recall the vocabulary needed to read science with comprehension and the facts needed to solve calculations.

Cognitive science studies how the mind works and how it learns. The model that science uses to describe learning includes the following fundamentals.

The purpose of learning is to solve problems. You solve problems in your brain, using information from your environment and your memory.

The human brain contains different types of memory, including

- **Working** memory: the part of your brain where you solve problems.
- **Short-term** memory: information that you remember for only a few days.
- **Long-term** memory: information that you can recall for many years.

Working memory is limited, but human *long*-term memory has enormous capacity. The goal of learning is to move new information from short-term into long-term memory so that it can be recalled by working memory for years after initial study. If new information is not moved into long-term memory, useful long-term learning has not taken place.

Children learn speech naturally, but most other learning requires repeated *thought* about the meaning of new information, plus *practice* at recalling new facts and using new skills that is *timed* in ways that encourage the brain to move new learning from short- to long-term memory.

The following principles of cognitive science will be helpful to keep in mind during your study of chemistry and other disciplines.

1. **Learning is cumulative.** Experts in a field learn new information quickly because they already have in long-term memory a storehouse of knowledge about the context surrounding new information. That storehouse must be developed over time, with practice.

2. **Learning is incremental** (done in small pieces). Especially for an unfamiliar subject, there is a limit to how much new information you can move into long-term memory in a short amount of time. Knowledge is extended and refined gradually. In learning, steady wins the race.

3. **Your brain can do parallel processing.** Though adding information to long-term memory is a gradual process, studies indicate that your brain can work on separately remembering what something looks like, where you saw it, what it sounds like, how you say it, how you write it, and what it means, all at the same time. The cues associated with each separate type of memory can help to trigger the recall of information needed to solve a problem, so it helps to use multiple memory strategies. When learning new information, listen, see, say, write, and try to connect it to other information that will help you to remember what it means.

4. **The working memory in your brain is limited.** Working memory is where you think. Try multiplying 556 by 23 in your head. Now try it with a pencil, a paper, and your head. Because of limitations in working memory, manipulating multiple pieces of new information in your head is difficult. Learning stepwise procedures (standard algorithms) is one way to reduce "cognitive load" during problem solving.

5. **"Automaticity in the fundamentals"** is another way to overcome limitations in working memory. Cognitive science has found that information that can be recalled automatically from long-term memory does not create the cognitive load that other information does. That's why being able to remember key facts and processes is so important when solving problems: When you can recall core knowledge automatically because of repeated practice, more working memory is available for higher-level thought.

6. **Concepts are crucial.** Your brain works to construct a "conceptual framework" to categorize knowledge being learned so that you can recall facts and procedures when you need them. The brain tends to store information in long-term memory only if the new information is in agreement with your mental model of how things work. In addition, if you have a more complete and accurate understanding of the big picture, your brain is better able to judge which information should be selected to solve a problem.

    Learning concepts does not replace the need to move key facts and procedures into your long-term memory. However, concepts speed the initial learning, recall, and appropriate application of knowledge in long-term memory.

7. **"You can always look it up" is a poor strategy for problem solving.** Your working memory is quite limited in how much information it can manipulate that is not in your long-term memory. The more information you must stop to look up, the less likely you will be able to follow your train of thought to the end of a complex problem.

Moving Knowledge into Memory

How can you promote the retention of needed fundamentals? It takes practice, but some forms of practice are more effective than others. The following strategies will improve your success in solving problems.

1. **Overlearning.** If you practice recalling new information until you can recall it perfectly only one time, you will tend to recall the new information for only a few days. To be able to recall new facts and skills for more than a few days, *repeated* practice to perfection (which scientists call overlearning) is necessary.

2. **The spacing effect.** To *retain* what you learn, 20 minutes of study spaced over 3 days is more effective than 1 hour of study for 1 day. Studies of "massed versus distributed practice" show that if the initial learning of facts and vocabulary is practiced over 3–4 days, then revisited weekly for 2–3 weeks, then monthly for 3–4 months, it can often be recalled for decades thereafter.

3. **Effort.** Experts in a field usually attribute their success to "hard work over an extended period of time" rather than to talent.

4. **Core skills.** The facts and processes you should practice most often are those needed most often in the discipline.

5. **Get a good night's sleep.** There is considerable evidence that while you sleep, your brain reviews the experience of your day to decide what to store in long-term memory. Sufficient sleep promotes retention of what you learn.

For additional science that relates to learning, see Daniel Willingham, *Cognition: The Thinking Animal* (Upper Saddle River, NJ: Pearson/Prentice Hall, 2007), and John T. Bruer, *Schools for Thought: A Science of Learning in the Classroom* (Cambridge, MA: MIT Press, 1994).

PRACTICE **A**

Answer these questions in your notebook.

1. What is "overlearning?"

2. What is the "spacing effect?"

Flashcards

Which is more important in learning: knowing the facts or knowing the concepts? Cognitive studies have found that the answer is both. However, to "think as an expert," you need a storehouse of factual information in memory that you can apply to new and unique problems.

In these lessons, we will use the following flashcard system to master fundamentals that need to be recalled automatically in order to solve problems efficiently. Using this system, you will make two types of flashcards:

- "One-way cards" for questions that make sense in *one* direction
- "Two-way" cards for facts that need to be recalled in both directions

If you have access to about 30 3 × 5-in. index cards, you can get started now. Plan to buy about 100–200 additional index cards, lined or unlined, tomorrow. (A variety of colors is helpful but not essential.) With your first 30 cards, complete these steps:

1. On 12–15 cards (of the same color, if possible), cut a triangle off the top-right corner, making cards like this:

These cards will be used for questions that go in *one* direction. Keeping the notch at the *top right* will identify the *front* side.

2. Using the following table, cover the answers in the right column with a folded sheet or index card. For each question in the left column, verbally answer, then slide the cover sheet down to check your answer. Put a check beside questions that you answer accurately and without hesitation. When done, write the questions and answers with*out* checks onto the notched cards.

Front Side of Cards (with Notch at Top Right)	Back Side—Answers
To convert to scientific notation, move the decimal to _____	After the first number that is not a zero
If you make the significand larger, _____	Make the exponent smaller
42^0	Any number to the zero power = 1
Simplify $1/(1/X)$	X
To divide exponentials, _____	Subtract the exponents
To bring an exponent from the bottom of a fraction to the top	Change its sign
1 cc ≡ 1 _____ ≡ 1 _____	1 cc ≡ 1 cm^3 ≡ 1 mL
0.0018 in scientific notation =	1.8×10^{-3}
1 L ≡ _____ mL ≡ _____ dm^3	1 L ≡ 1000 mL ≡ 1 dm^3
To multiply exponentials that have the same base	Add the exponents
Simplify $1/10^x$	10^{-x}
74 in scientific notation =	7.4×10^1
The historic definition of 1 gram is	The mass of 1 cm^3 of liquid water at 4°C
8 × 7	56
42/6	7

If there are any multiplication or division facts up to 12s that you cannot answer *instantly*, add them to your list of one-sided cards. If you need a calculator to do number math, parts of chemistry such as balancing an equation will be frustrating. With flashcard practice, you will quickly learn to remember what you need to know.

3. To make two-way cards, use the index cards as they are, with*out* a notch cut.

    For the following cards, first cover the *right* column, then put a check on the left if you can answer the left column question quickly and correctly. Then cover the *left* column and check the right side if you can answer the right side *automatically*.

    When done, if a row does not have *two* checks, make the flashcard.

Two-Way Cards (with*out* a Notch)

10^3 g or 1000 g = 1 ___ g	1 kg = ___ g
Boiling temperature of water	100 degrees Celsius—if 1 atm pressure
1 nanometer = 1 × ___ meters	1 ___ meter = 1 × 10^{-9} meters
Freezing temperature of water	0 degrees Celsius
4.7×10^{-3} = ___ (number)	$0.0047 = 4.7 \times 10^?$

1 GHz = $10^?$ Hz	10^9 Hz = 1 ___ Hz	2/3 = 0.?	0.666 . . . = ?/?
1 pL = $10^?$ L	10^{-12} L = 1 ___ L	1/80 = 0.?	0.0125 = 1/?
3/4 = 0.?	0.75 = ?/?	1 dm^3 = 1 ___	1 L = 1 ___
1/8 = 0.?	0.125 = 1/?	1/4 = 0.?	0.25 = 1/?

More Two-Way Cards (with*out* a Notch) for the Metric Prefix Definitions

kilo = × $10^?$	× 10^3 = ? prefix	d = × $10^?$	× 10^{-1} = ? abbr.	micro = ? abbr.	μ = ? pref.
nano = × $10^?$	× 10^{-9} = ? pref.	m = × $10^?$	× 10^{-3} = ? abbr.	mega = ? abbr.	M = ? pref.
giga = × $10^?$	× 10^9 = ? prefix	T = × $10^?$	× 10^{12} = ? abbr.	kilo = ? abbrev.	k = ? pref.
milli = × $10^?$	× 10^{-3} = ? pref.	k = × $10^?$	× 10^3 = ? abbr.	pico = ? abbr.	p = ? pref.
deci = × $10^?$	× 10^{-1} = ? pref.	n = × $10^?$	× 10^{-9} = ? abbr.	deci = ? abbr.	d = ? pref.
tera = × $10^?$	× 10^{12} = ? pref.	μ = × $10^?$	× 10^{-6} = ? abbr.	centi = ? abbr.	c = ? pref.
pico = × $10^?$	× 10^{-12} = ? pref.	G = × $10^?$	× 10^9 = ? abbr.	tera = ? abbr.	T = ? pref.
mega = × $10^?$	× 10^6 = ? prefix	M = × $10^?$	× 10^6 = ? abbr.	milli = ? abbr.	m = ? pref.
micro = × $10^?$	× 10^{-6} = ? pref.	p = × $10^?$	× 10^{-12} = ? abbr.	nano = ? abbr.	n = ? pref.
centi = × $10^?$	× 10^{-2} = ? pref.	c = × $10^?$	× 10^{-2} = ? abbr.	giga = ? abbr.	G = ? pref.

Which cards you need will depend on your prior knowledge, but when in doubt, make the card. On fundamentals, you need quick, confident, accurate recall—every time.

4. **Practice** with one *type* of card at a time.
   • **For front-sided cards,** if you get a card right quickly, place it in the *got-it* stack. If you miss a card, say the content. Close your eyes. Say it again.

And again. If needed, write it several times. Return that card to the bottom of the *do* stack. Practice until every card is in the *got-it* stack.

- **For two-sided cards**, do the same steps as above in one direction, then the other.

5. Master the cards at least once, then apply them to the Practice on the topic of the new cards. Treat the Practice as a practice test.

6. **For three days in a row**, repeat those steps. Repeat them again before working assigned problems, before your next quiz, and before your next test that includes this material.

7. Make cards for new topics early: before the lectures on a topic, if possible. Mastering fundamentals first will help in understanding lecture.

8. Rubber band and carry these new cards. Practice during "down times."

9. After a few chapters or topics, change card colors.

This system requires an initial investment of time, but in the long run it will save time and improve achievement. The above flashcards are examples. Add cards of your design and choosing as needed.

Flashcards, Charts, or Lists? What is the best strategy for learning new information? Use *multiple* strategies: numbered lists, mnemonics, phrases that rhyme, flashcards, reciting, and writing what must be remembered. Practice repeatedly, spaced over time.

For complex information, automatic recall may be less important than being able to methodically write out a *chart* for information that falls into *patterns*.

For the metric system, learning flashcards *and* the prefix chart *and* picturing the meter stick relationships all help to fix these fundamentals in long-term memory.

PRACTICE B

Run your set of flashcards until all cards are in the *got-it* stack. Then try these problems. Make additional cards, if needed. Run the cards again in a day or two.

1. Fill in the following blanks with an exponential term.

Format: 1 Prefix	1 Base Unit
1 micrometer = _____ meters	1 meter = _____ micrometers
1 gigawatt = _____ watts	1 watt = _____ gigawatts
1 nanoliter = _____ liters	_____ nanoliters = 1 liter

2. Add exponential terms to these blanks. Watch where the 1 is!

a. 1 picocurie = _____ curies   b. 1 megawatt = _____ watts

c. 1 cg = _____ g   d. 1 mole = _____ millimoles

e. 1 m = _____ nm   f. 1 kPa = _____ Pa

(continued)

3. Do these with*out* a calculator.

    a. $10^{-6}/10^{-8}$ =

    b. 1/5 = ___.___ ___

    c. 1/50 = ___.___ ___ ___

4. Write each atom's symbol.

    a. Helium =

    b. Hydrogen =

    c. Sodium =

5. For the following symbols, write the atom name.

    a. N =

    b. Ne =

    c. B =

Lesson 2.4   Calculations with Units

Try doing this lesson without a calculator, except as noted.

Adding and Subtracting with Units

Many calculations in mathematics consist of numbers without units. In science, however, calculations are nearly always based on measurements of physical quantities. A measurement consists of a numeric value *and* its unit.

When doing calculations in science, it is essential to write the *unit* after numbers. Why?

- Units give physical meaning to a quantity.
- Units are the best indicators of what steps are needed to solve problems.
- Units provide a check that you have done a calculation correctly.

When solving calculations, the math must take into account *both* the numbers and their units. To do so, apply the following three rules.

> **Rule 1.** The *units* must be the *same* in quantities being *added and subtracted*, and those same units must be added to the answer.

TRY IT

Q. Rule 1 is logical. Apply it and fill in the blanks for these two examples.

1. 5 apples + 2 apples =

2. 5 apples + 2 oranges =

 Answers:

Example 1 is 7 apples. Example 2 cannot be added. It makes sense that you *can* add two numbers that refer to apples, but you *can't* add apples and oranges.

By rule 1, you can add numbers that have the same units, but you can*not* add numbers directly that do *not* have the same units.

> **▷ TRY IT**
>
> **Q.** Apply rule 1 to this problem:
>
> $$14.0 \text{ grams}$$
> $$- \ 7.5 \text{ grams}$$
>
> **Answer:**
>
> $$14.0 \text{ grams}$$
> $$- \ 7.5 \text{ grams}$$
> $$\mathbf{6.5 \text{ grams}}$$
>
> *If* the units are all the same, you can add or subtract numbers, but you must add the common unit to the answer.

Multiplying and Dividing with Units

The rule for *multiplying* and *dividing* with units is different, but logical.

> **Rule 2.** When multiplying and dividing *units*, the units multiply and divide.

> **▷ TRY IT**
>
> **Q.** Complete this example of unit math: cm $\times$ cm =
>
> **Answer:**
>
> $$\text{cm} \times \text{cm} = \mathbf{cm^2}$$

Units obey the laws of algebra.

> **▷ TRY IT**
>
> **Q.** Complete this example of unit math:
>
> $$\frac{\text{cm}^5}{\text{cm}^2} =$$
>
> **Answer:**
>
> Solve as:
>
> $$\frac{\text{cm} \cdot \text{cm} \cdot \text{cm} \cdot \text{cm} \cdot \text{cm}}{\text{cm} \cdot \text{cm}} = \mathbf{cm^3}$$
>
> or by using the rules for exponential terms

$$\frac{cm^5}{cm^2} = cm^{5-2} = \mathbf{cm^3}$$

Both methods arrive at the same answer (as they must).

> **Rule 3.** When multiplying and dividing, *group* the numbers, exponentials, and units separately, solve the three parts separately, then recombine the terms.

Apply rule 3 to this problem.

TRY IT

Q. If a postage stamp has dimensions of 2.0 cm $\times$ 4.0 cm, the surface area of one side of the stamp = _____.

Answer:

$$\text{Area of a rectangle} = l \times w = 2.0\ cm \times 4.0\ cm$$
$$= (2.0 \times 4.0) \times (cm \times cm) = \mathbf{8.0\ cm^2}$$
$$= 8.0\ \textit{square}\ centimeters$$

By rule 2, the units must obey the rules of multiplication and division. By rule 3, the unit math is done *separately* from the number math.

Units follow the familiar laws of multiplication, division, and powers, including "like units cancel."

TRY IT

Q. Apply rule 3 to the following.

a. $\dfrac{8.0\ L^6}{2.0\ L^2} =$

b. $\dfrac{9.0\ m^6}{3.0\ m^6} =$

Answer:

a. $\dfrac{8.0\ L^6}{2.0\ L^2} = \dfrac{8.0}{2.0} \cdot \dfrac{L^6}{L^2} = 4.0\ \mathbf{L^4}$

b. $\dfrac{9.0\ \cancel{m^6}}{3.0\ \cancel{m^6}} = \mathbf{3.0}$ (with no unit)

In science, the *unit math* must be done as part of calculations. A *calculated* unit *must* be included as part of a calculated answer (except in rare cases, such as *part b* above, when all of the units cancel).

━━━▶ TRY IT

Q. To the following problem, apply separately the math rules for numbers, exponential terms, and units.

$$\frac{12 \times 10^{-3}\,\text{m}^4}{3.0 \times 10^2\,\text{m}^2} =$$

STOP **Answer:**

$$\frac{12 \times 10^{-3}\,\text{m}^4}{3.0 \times 10^2\,\text{m}^2} = \frac{12}{3.0} \cdot \frac{10^{-3}}{10^2} \cdot \frac{\text{m}^4}{\text{m}^2} = \mathbf{4.0 \times 10^{-5}\,\text{m}^2}$$

In science calculations, you often need to use a calculator to do the number math, but both the *exponential* and *unit* math nearly always can (and should) be done with*out* a calculator.

If more than one unit is being multiplied or divided, the math for each unit is done separately.

For this problem, use a calculator for the numbers, but do the exponential math and unit cancellation using pencil and paper, not the calculator.

━━━▶ TRY IT

Q. Simplify:

$$4.8\ \frac{\text{g}\cdot\text{m}}{\text{s}^2}\cdot 3.0\ \text{m}\cdot\frac{6.0\ \text{s}}{9.0 \times 10^{-4}\,\text{m}^2} =$$

STOP **Answer:**

Do the math for numbers, exponentials, and then *each* unit separately.

$$\frac{86.4}{9.0}\cdot\frac{1}{10^{-4}}\cdot\frac{\text{g}\cdot\cancel{\text{m}}\cdot\cancel{\text{m}}}{\text{s}\cdot\cancel{\text{s}}}\cdot\frac{\cancel{\text{s}}}{\cancel{\text{m}^2}} = 9.6 \times 10^4\ \frac{\text{g}}{\text{s}}$$

This answer unit can also be written as $\text{g}\cdot\text{s}^{-1}$, but you will find it helpful to use the x/y unit format until we work with mathematical equations later in the course.

PRACTICE **A**

Do *not* use a calculator except as noted. Start by doing the even-numbered problems. If you need more practice, do more. Save a few problems for your pre-quiz and/or pre-test review. After completing each problem, check your answer. If you miss a problem, review the rules to find out why before continuing.

1. 16 cm − 2 cm =

2. 12 cm · 2 cm =

3. $(\text{m}^4)(\text{m}) =$

4. $\text{m}^4/\text{m} =$

5. $\dfrac{10^5}{10^{-2}} =$

6. $\dfrac{\text{s}^{-5}}{\text{s}^2} =$

7. 3.0 meters · 9.0 meters =

8. 3.0 g/9.0 g =

(continued)

9. $\dfrac{24\,L^5}{3.0\,L^{-4}} =$

10. $\dfrac{18 \times 10^{-3}\,g \cdot m^5}{3.0 \times 10^1\,m^2} =$

11. $12 \times 10^{-2}\,\dfrac{L \cdot g}{s} \cdot 2.0\,m \cdot \dfrac{2.0\,s^3}{6.0 \times 10^{-5}\,L^2} =$

12. A rectangular box has dimensions of 2.0 cm × 4.0 cm × 6.0 cm. Calculate its volume.

13. Using a calculator, find the volume of a sphere that is 4.0 cm in diameter ($V_{sphere} = 4/3\pi r^3$).

14. With*out* a calculator, multiply:

$$2.0\,\dfrac{g \cdot m}{s^2} \cdot \dfrac{3.0\,m}{4.0 \times 10^{-2}} \cdot 6.0 \times 10^2\,s =$$

Additional Metric Practice

Run your metric flashcards one more time until you can quickly and correctly answer each card in both directions. Then, without looking back at the tables or rules, try every other problem below. Do more if you need more practice.

PRACTICE   B

1. From memory, add exponential terms to these blanks.

    a. 4 centigrams = 4 × _____ grams

    b. 3 pm = 3 × _____ m

    c. 5 dL = 5 × _____ L

    d. 1 THz = 1 × _____ Hz

2. From memory, add full metric *prefixes* to these blanks.

    a. 7×10^{-6} liters = 7 _____ liters

    b. 5×10^3 grams = 5 _____ grams

3. From memory, add prefix *abbreviations* to these blanks.

    a. 10^6 g = 1 ____ g

    b. 10^{-3} g = 1 ____ g

    c. 6×10^{-9} L = 6 ____ L

    d. 2×10^{-1} m = 2 ____ m

4. From memory, add exponential terms to these blanks. Watch where the 1 is!

    a. 1 micrometer = _____ meters

    b. 1 watt = _____ gigawatts

    c. 1 MHz = _____ Hz

    d. 1 mole = _____ nanomoles

    e. 1 m = _____ km

    f. 1 ms = _____ s

SUMMARY

1. 1 *meter*   ≡ 10 decimeters
             ≡ 100 centimeters
             ≡ 1000 millimeters
    1000 meters ≡ 1 kilometer

2. **1 milli**meter ≡ **1 mm** = 10^{-3} meter
    1 centimeter ≡ **1 cm** = 10^{-2} meter
    1 decimeter ≡ **1 dm** = 10^{-1} meter
    1 kilometer ≡ **1 km** = 10^{3} meter

3. Any unit can be substituted for *meter* above.

4. $1\ cm^3 \equiv 1\ mL \equiv 1\ cc$

5. $1\ liter \equiv 1000\ mL \equiv 1\ dm^3$

6. $1\ cm^3\ H_2O(\ell) \equiv 1\ mL\ H_2O(\ell) \approx 1.00\ g$ $H_2O(\ell)$

7. meter = m; gram = g; second = s

8. If prefix- = 10^a, 1 *unit* = 10^{-a} *prefix*-units

9. To change a prefix definition from the "1 prefix- = " format to the "1 unit = " format, change the exponent sign.

10. **Rules for units in calculations**
    a. When adding or subtracting, the units must be the *same* in the numbers being added and subtracted, and those same units must be added to the answer.
    b. When multiplying and dividing units, the units multiply and divide.
    c. When multiplying and dividing, *group* the numbers, exponentials, and units separately. Solve the separate parts, then recombine the terms.

Prefix	Abbreviation	Means
tera-	T	$\times 10^{12}$
giga-	G	$\times 10^{9}$
mega-	M	$\times 10^{6}$
kilo-	k	$\times 10^{3}$
deci-	d	$\times 10^{-1}$
centi-	c	$\times 10^{-2}$
milli-	m	$\times 10^{-3}$
micro-	μ (mu) or u	$\times 10^{-6}$
nano-	n	$\times 10^{-9}$
pico-	p	$\times 10^{-12}$

ANSWERS

Lesson 2.1

Pretest   1. 0.15 kg   2. $1000\ cm^3$ 3. $1\ dm^3$ 4. 2,500 millipascals   5. 0.035 kg

Practice A   1a. 1 millimeter = $\mathbf{10^{-3}}$ meters   1b. 1 deciliter = $\mathbf{10^{-1}}$ liter

2a. 1000 grams = 1**kilo**gram   2b. 10^{-2} liters = 1 **centi**liter

Practice B   1. 1 m = 10 **dm** = 100 **cm** = 1000 **mm**   2a. $10^{3}\ g$ = 1 **kg**   2b. $10^{-3}\ s$ = 1 **ms**

3a. $1000\ cm^3$ = **1000** mL   3b. $100\ cm^3\ H_2O(\ell) \approx$ **100** grams $H_2O(\ell)$

4. 1 liter ≡ **1000** mL ≡ **1000** cm^3 ≡ **1** dm^3

Practice C   1. A kilometer   2. A kilojoule

3. Possible answers include cubic centimeters, milliliters, liters, cubic decimeters, cubic meters, and any metric distance unit cubed.

4. 100 cm   5. 1000 liters   6. 15 grams   7. 1000 grams   8. See the periodic table.

Lesson 2.2

Practice A   1a. 7 microseconds = $7 \times \mathbf{10^{-6}}$ seconds   1b. 9 kg = $9 \times \mathbf{10^{3}}$ g

1c. 8 cm = $8 \times \mathbf{10^{-2}}$ m   1d. 1 ng = $1 \times \mathbf{10^{-9}}$ g

2a. 6×10^{-2} amps = 6 **centi**amps   2b. 45×10^{9} watts = 45 **giga**watts

3a. $10^{12}\ g$ = 1 **Tg**   3b. $10^{-12}\ s$ = 1 **ps**   3c. 6×10^{-9} m = 6 **nm**

3d. $5 \times 10^{-1}\ L$ = 5 **dL**   3e. $4 \times 10^{-6}\ L$ = 4 **μL**   3f. 16×10^{6} Hz = 16 **MHz**

4. 5 **mg** and 5 **Mg**   5. **M-, G-**, and **T-**   6. $\mathbf{3.0 \times 10^{8}}$ meters/second

Practice B 1a. 1 terasecond $= 1 \times 10^{12}$ seconds , so 1 second $= 1 \times 10^{-12}$ teraseconds.

1b. 1 ng $= 1 \times 10^{-9}$ grams, so 1 g $= 1 \times 10^{9}$ ng.

2a. 1 gram $= 10^{2}$ centigrams   (For "1 *unit* =" take reciprocal [reverse sign] of prefix definition.)

2b. 1 meter $= 10^{12}$ picometers   2c. 1 s $= 10^{3}$ ms   2d. 1 s $= 1 \times 10^{-6}$ Ms

3a. 1 micromole $= 1 \times 10^{-6}$ moles   3b. 1 g $= 1 \times 10^{-9}$ Gg

3c. 1 kilogram $= 1 \times 10^{3}$ grams   3d. 10^{-3} kg $= 1$g

3e. 10^{9} ns $= 1$s   3f. 1 pL $= 10^{-12}$ L

Lesson 2.3

Practice A   1. Repeated practice to perfection.

2. Study over several days results in better retention than "cramming."

Practice B   1.

1 micrometer $= 10^{-6}$ meters	1 meter $= 10^{6}$ micrometers
1 gigawatt $= 10^{9}$ watts	1 watt $= 10^{-9}$ gigawatts
1 nanoliter $= 10^{-9}$ liters	10^{9} nanoliters $= 1$ liter

2a. 1 picocurie $= 10^{-12}$ curies   2b. 1 megawatt $= 10^{6}$ watts   2c. 1 cg $= 10^{-2}$ g

2d. 1 mole $= 10^{3}$ millimoles   2e. 1 m $= 10^{9}$ nm   2f. 1 kPa $= 10^{3}$ Pa

3a. $10^{-6}/10^{-8} = 10^{-6+8} = 10^{2}$ 3b. 1/5 $= 0.20$ 3c. 1/50 $= 0.020$

4a. Helium $=$ **He**   4b. Hydrogen $=$ **H**   4c. Sodium $=$ **Na**

5a. N $=$ **Nitrogen**   5b. Ne $=$ **Neon**   5c. B $=$ **Boron**

Lesson 2.4

Practice A Both the *number* and the *unit* must be written and correct.

1. **14 cm**   2. **24 cm^2**   3. $m^{(4+1)} =$ **m^5**   4. $m^{(4-1)} =$ **m^3**   5. **10^7**

6. **s^{-7}**   7. **27 meters^2**   8. **0.33** (*no* unit)   9. **8.0 L^9**   10. **6.0 $\times$ 10^{-4} g · m^3**

11. **8.0 $\times$ 10^3** $\dfrac{\textbf{g·m·s}^2}{\textbf{L}}$

12. $V_{\text{rectangular solid}} =$ length *times* width *times* height $=$ **48 cm^3**

13. Diameter $= 4.0$ cm, *radius* $= 2.0$ cm

$V_{\text{sphere}} = 4/3\pi r^3 = 4/3\pi(2.0 \text{ cm})^3 = 4/3\pi(8.0 \text{ cm}^3) = (32/3)\pi \text{cm}^3 = 33.51 \text{ cm}^3 =$ **34 cm^3**

(If you use $\pi = 3.14$, your answer would be 33.49 cm$^3 =$ **33 cm^3**. That's OK, too.)

14. $\dfrac{2.0 \text{ g·m}}{\text{s}^2} \cdot \dfrac{3.0 \text{ m}}{4.0 \times 10^{-2}} \cdot 6.0 \times 10^2 \text{ s} = \dfrac{(2.0)(3.0)(6.0)}{4.0} \cdot 10^4 \cdot \dfrac{\text{g·m·m·s}}{\text{s}^2}$

$= 9.0 \times 10^4 \dfrac{\textbf{g·m}^2}{\textbf{s}}$

Practice B 1a. 4 centigrams $= 4 \times 10^{-2}$ grams   1b. 3 pm $= 3 \times 10^{-12}$ m

1c. 5 dL $= 5 \times 10^{-1}$ L   1d. 1 THz $= 1 \times 10^{12}$ Hz

2a. 7×10^{-6} liters $=$ **micro**liters   2b. 5×10^3 grams $= 5$ **kilo**grams

3a. 10^6 g $= 1$ **Mg**   3b. 10^{-3} g $= 1$ **mg**

3c. 6×10^{-9} L $= 6$ **nL**   3d. 2×10^{-1} m $= 2$ **dm**

4a. 1 micrometer $= 10^{-6}$ meters   4b. 1 watt $= 10^{-9}$ gigawatts

4c. 1 MHz $= 10^6$ Hz   4d. 1 mole $= 10^9$ nanomoles

4e. 1 m $= 10^{-3}$ km   4f. 1 ms $= 10^{-3}$ s

3

Significant Figures

Introduction

Nearly all measurements have *uncertainty*. In science, we need to express

- How much uncertainty exists in measurements
- The uncertainty in calculations based on measurements

One method for expressing uncertainty is **significant figures**, abbreviated in these lessons as *s.f.* or *S.F.*

Other methods measure uncertainty more accurately, but significant figures provide an approximation of uncertainty that, compared to other methods, is easy to use in calculations. In first-year chemistry, significant figures in nearly all cases will be the method of choice to indicate the uncertainty in measurements and calculations.

Lesson 3.1   Rules for Significant Figures

Use these rules when recording measurements and rounding calculations.

1. **Recording a measurement using significant figures:** Write all the digits you are sure of, plus the *first* digit that you must *estimate* in the measurement: the first **doubtful digit** (the first **uncertain digit**). Then *stop*.

> When writing a measurement using significant figures, the *last* digit is the first *doubtful* digit. Round measurements to the highest place with doubt.

   Example: If a balance reads mass to the thousandths place, but under the conditions of the experiment the uncertainty in the measurement is ±0.02 grams, we can write

12.432 g ±0.02 g using **plus-minus notation** to record uncertainty

   However, in a calculation, if we need to multiply or divide by that measured value, the math to include the ± can be time-consuming. So, for use in calculations, we convert the measurement to *significant figures* notation by writing

12.43 g

When using significant figures to indicate uncertainty, the last place written in a measurement is the first place with doubt. The ± showed that the highest *place* with doubt is the hundredths place. To convert to significant figures, we *round* the recorded digits back to that place, then remove the ±.

We convert the measurement to significant figures notation because in calculations, the math using numbers like 12.43 follows familiar rules.

2. **Adding and subtracting using significant figures:**

   a. First, add or subtract as you normally would.

   b. Next, search the numbers for the doubtful digit in the *highest place*. The answer's *doubtful* digit must be in that *place*. *Round* the answer to that *place*.

Example:

$$23.\underline{1} \quad \leftarrow$$
$$+ \ 16.01$$
$$+ \ \underline{1.008}$$
$$40.\underline{1}18 = \mathbf{40.1}$$

This answer must be rounded to **40.1** because the tenths place is the highest *place* with doubt among the numbers added.

Recall that the tenths place is *higher* than the hundredths place, which is higher than the thousandths place.

c. *The logic*: If you add a number with doubt in the tenths place to a number with doubt in the hundredths place, the answer has doubt in the tenths place.

A doubtful digit is significant, but numbers after it are not.

In a measurement, if the number in a given place is doubtful, it means *any* numbers could be after that place. We allow one *doubtful* digit in answers, but omit numbers that have no reliability.

d. Another way to state this rule: When adding or subtracting, round your answer back to the last *full column* on the right. This will be the first column of numbers, moving right to left ($\leftarrow$), with no *blanks* above.

The blank space *after* a doubtful digit indicates that we have no idea what that number is, so we cannot get a significant number in the answer for that column.

e. When adding or subtracting using a calculator, use this rule: In the numbers being added and subtracted, identify and underline the doubtful digit in the number that has its doubt in the highest *place*. Round your calculator answer to that *place*.

 TRY IT

(See "How to Use These Lessons," point 1, p. xv.)

Q. Using a calculator, apply the rule to

$$43 + 1.00 - 2.008 =$$

STOP

Answer:

$43 + 1.00 - 2.008 = 41.992$ on the calculator $= \mathbf{42}$ in significant figures

Among these numbers being added and subtracted, the *highest* doubt is in the ones place. In science calculations, you must round your final answer to that place.

SUMMARY

When **adding or subtracting**, round your final answer back to the

- Highest *place* with doubt

When adding or subtracting in columns, this is also the

- *Leftmost place* with doubt, which is also the
- Last full column on the right, which is also the
- Last column to the right with*out* a blank space

PRACTICE   A

First, memorize the boxed rules above. Then do the problems. When finished, check your answers at the end of the chapter.

1. Convert these from plus-minus notation to significant figures (*s.f.*) notation.

   a. 65.316 mL $\pm$ 0.05 mL

   b. 5.2 cm $\pm$ 0.1 cm

   c. 1.8642 km $\pm$ 0.2 km

   d. 16.8°C $\pm$ 1°C

2. Add these numbers without using a calculator. Round your final answer to the proper number of significant figures.

$$\begin{array}{r} 23.1 \\ 23.1 \\ + \ 16.01 \end{array}$$

3. Use a calculator. Round your final answer to the proper number of significant figures.

   a. 2.016 + 32.18 + 64.5 =

   b. 16.00 − 4.034 − 1.008 =

3. **Counting significant figures:** When multiplying and dividing, we need to *count* the number of significant figures in a measurement. To count the number of *s.f.*, count the sure digits *plus* the doubtful digit. The doubtful digit is significant.

   This rule means that for numbers that do not include zeros, the count of *s.f.* is simply the number of digits shown.

   Examples: 123 meters has **three** *s.f.*; 14.27 grams has **four** *s.f.*

   In exponential notation, to find the number of *s.f.*, look only at the significand. The exponential term does not affect the number of significant figures.

   Example: 2.99×10^8 meters/second has **three** *s.f.*

   We call it the significan*d* because it contains the significan*t* figures.

4. **Multiplying and dividing using significant figures:** This is the rule we will use most often.

   a. First multiply or divide as you normally would.

   b. Then count the number of *s.f.* in each of the numbers you are multiplying or dividing.

   c. Your answer can have no more *s.f.* than the measurement with the least *s.f.* that you multiply or divide by. Round the answer back to that number of *s.f.*

   Example:

$$3.1865 \text{ cm} \times \textbf{8.8} \text{ cm} = 28.041 = \textbf{28} \text{ cm}^2$$

$$\quad (5 \text{ s.f.}) \qquad (2 \text{ s.f.}) \qquad\qquad\qquad (2 \text{ s.f.})$$

(Must *round* to **two** *s.f.*)

SUMMARY

When multiplying and/or dividing a 10-*s.f.* number and a 9-*s.f.* number and a 2-*s.f.* number, you must round your answer to two *s.f.*

5. **When moving the decimal,** do not change the number of *s.f.* in a significand.

TRY IT

Q. Convert 424.7×10^{-11} to scientific notation.

STOP **Answer:**

$$4.247 \times 10^{-9}$$

Both the original and the scientific notation have four *s.f.*

In scientific notation, all digits shown in the signifi*cand* are signifi*cant*.

6. **Using significant figures in calculations with steps or parts:** The rules for *s.f.* should be applied at the *end* of a calculation.

In problems that have several separate parts (1a, 1b, etc.), and earlier answers are used for later parts, many instructors prefer that you carry one extra *s.f.* until the end of a calculation, then round to proper *s.f.* at the final step. This method minimizes changes in the final doubtful digit owing to rounding in the steps.

PRACTICE B

First memorize the preceding rules, *then* do the problems.

1. Multiply and divide using a calculator. Write the first six digits of the calculator result, then write the final answer with units and the proper number of *s.f.*

   a. 3.42 cm *times* 2.3 cm^2 =

   b. 74.3 L^2 divided by 12.4 L =

2. Convert to scientific notation.

   a. 0.0060×10^{-15} =

   b. $1{,}027 \times 10^{-1}$ =

3. Solve this two-part question using the rules for significant figures.

   a. 9.76573×1.3 = part a answer =

   b. (Part a answer)/2.5 =

Lesson 3.2   Special Cases

When using significant figures to express uncertainty, there are special rules for rounding a 5, zeros, and exact numbers.

1. **Rounding:** If the number *beyond* the place to which you are rounding is

   a. *Less* than 5: Drop it; round *down*.
      Example: 1.342 rounded to *tenths* = 1.3

   b. *Greater* than 5: Round *up*.
      Example: 1.748 = 1.75 (Rounded to <u>underlined</u> place.)

   c. A 5 followed by *any non-zero* digits: Round *up*.
      Example: 1.02502 = 1.03

2. **Look only *one* place past** the place to which you are rounding.
   Example: Rounding 7.649 to tenths = 7.6
   When rounding the 6, look *only* at the 4. A value of 4 rounds down.
   Why not use the 9 to round the 4 up to 5, which then rounds the 6 up? That would give the 9 credibility it does not deserve. If tenths is doubtful, hundredths is used as a basis for rounding to tenths, but numbers past the hundredths place have no reliability and should not affect the answer.

3. **Rounding a lone 5:** A lone 5 is a 5 without following digits *or* a 5 followed by zeros. To round off a lone 5, some instructors prefer the simple "round 5 up" rule. Others prefer a slightly more precise "engineer's rule," which rounds 5 up half the time, and down half the time.
   Rounding off a lone 5 or a 0.1500 is not a case that occurs often in calculations, but when it does, in these lessons, we will use the "always round 5 up" rule.
   You should use the rounding rule preferred by the instructor in *your* course.

PRACTICE **A**

Round to the <u>underlined</u> *place*. Check answers at the end of this chapter.

1. 0.00212	2. 0.0994	3. 20.0561
4. 23.25	5. 0.1950	6. 2.648 × 10⁻³

---

4. **Zeros:** *Non*-zero digits are always significant, but *zeros* may or may not be significant. When are zeros significant and when are they not significant? There are four cases.

   a. *Leading* zeros (zeros in *front* of *all* other digits) are *never* significant.
      Example: 0.0006 has one *s.f.* (Zeros in front never count.)

   b. Zeros embedded *between* other digits *are* always significant.
      Example: 300.07 has five *s.f.* (Zeros sandwiched by *s.f.* count.)

   c. Zeros *after all other* digits *as well as after* the decimal point *are* significant.
      Example: 565.0 has four *s.f.* You would not need to include that zero if it were not significant.

   d. Zeros *after* all other digits but *before* the decimal, unless specially marked, are assumed to be *not* significant.

Example: 300 is assumed to have one *s.f.*, meaning "give or take a few hundred."

When a number is written as 300, or 250, it is not clear whether the zeros are significant. Many science textbooks address this problem by using this rule:

- "500 meters" means one *s.f.*, but
- "500. meters" with an *unneeded decimal point* added after a zero means three *s.f.*

These lessons will use that convention as well. However, the best way to avoid ambiguity in the number of significant figures is to write numbers in scientific notation.

$$4 \times 10^2 \text{ has one } s.f.; \textbf{4.00} \times 10^2 \text{ has three } s.f.$$

In *exponential* notation, only the significand contains the significant figures.

In *scientific* notation, all of the digits in the significan*d* are significan*t*.

Why are zeros complicated? Zero has multiple uses in our numbering system.

In cases 4a and 4d above, the zeros are simply "indicating the place for the decimal." In that role, they are *not* significant as measurements. In the other two cases, the zeros represent numeric values. When the zero represents "a number between a 9 and a 1 in a measurement," it is significant.

5. **Exact numbers:** Measurements with *no* uncertainty have an *infinite* number of *s.f.* Exact numbers do not add uncertainty to calculations.

- If you multiply a three *s.f.* number by an *exact* number, round your answer to three *s.f.*

This rule means that *exact* numbers are *ignored* when deciding the *s.f.* in a calculated answer. In chemistry, we use this rule in situations including the following:

a. Numbers in *definitions* are exact.
Example: The relationship "1 kilometer = 1000 meters," is an exact definition of *kilo-* and not a measurement with uncertainty. Both the 1 and the 1000 are exact numbers. Multiplying or dividing by those *exact* numbers will not limit the number of *s.f.* in your answer.

b. The number **1** in nearly all cases is *exact*.
Example: The conversion "**1** kilometer = 0.62 miles" is a legitimate approximation, but it is not a *definition* (≡) and is not *exactly* correct. The 1 is therefore assumed to be exact, but the 0.62 has uncertainty and has two *s.f.*

c. Whole numbers (such as 2 or 6), *if* they are a count of exact quantities (such as 2 people or 6 molecules), are also exact numbers with infinite *s.f.*

d. *Coefficients* and *subscripts* in chemical formulas and equations are exact.
Example:

$$2\,H_2 + 1\,O_2 \rightarrow 2\,H_2O \qquad \text{(All of those } numbers \text{ are exact.)}$$

You will be reminded about these exact-number cases as we encounter them. For now, simply remember that exact numbers have infinite *s.f.*, and do not limit the *s.f.* in an answer.

PRACTICE **B**

Write the number of *s.f.* in each of these values.

1. 0.0075

2. 600.3

3. 178.40

4. 4640.

5. 800

6. 2.06×10^{-9}

7. 0.060×10^3

8. 0.02090×10^5

9. 3 (exact)

Lesson 3.3   Significant Figures Practice

If you have not already done so, make a set of flashcards that cover the rules in this chapter. Include points in the summary at the end of the chapter. Some flashcards are suggested after the summary. Practice your flashcards and/or list of rules until they are firmly in your long-term memory, then try the problems below.

PRACTICE

Try every *other* problem on day 1. Try the rest on day 2 of your practice.

1. Write the number of *s.f.* in each of these values.

    a. 107.42

    b. 10.04

    c. 13.40

    d. 0.00640

    e. 0.043×10^{-4}

    f. 1,590.0

    g. 320×10^9

    h. 14 (exact)

    i. 2,500

    j. 4,200.

2. Round to the place indicated.

    a. 5.15 cm (tenths place)

    b. 31.84 meters (three *s.f.*)

    c. 0.819 mL (hundredths place)

    d. 0.06349 cm^2 (two *s.f.*)

    e. 0.04070 g (two *s.f.*)

    f. 6.255 cm (tenths place)

3. Calculate with or without a calculator, then write *final* answers with proper *s.f.*

    a.
    $$\begin{array}{r} 1.008 \\ 1.008 \\ + 32.00 \end{array}$$

    b.
    $$\begin{array}{r} 17.65 \\ - 9.7 \end{array}$$

    c. $39.1 + 124.0 + 14.05 =$

4. Use a calculator. Write final answers using proper *s.f.* and proper units.

    a. 13.8612 cm $\times$ 2.02 cm =

    b. 4.4 meters $\times$ 8.312 meters^2 =

    c. 2.03 cm^2/1.2 cm =

    d. 0.5223 cm^3/0.040 cm =

5. Use a calculator. Answer in scientific notation with proper *s.f.*

   a. $(2.25 \times 10^{-2})(6.0 \times 10^{23}) =$

   b. $(6.022 \times 10^{23}) / (1.50 \times 10^{-2}) =$

6. Convert these to *s.f.* notation.

   a. $2.0646 \text{ m} \pm 0.050 \text{ m}$ b. $5.04 \text{ nm} \pm 0.1 \text{ nm}$

   c. $12.675 \text{ g} \pm 0.20 \text{ g}$ d. $24.81°\text{C} \pm 1.0°\text{C}$

7. Answer in scientific notation with proper units and *s.f.*

$$5.60 \times 10^{-2} \frac{\text{L}^2 \cdot \text{g}}{\text{s}} \cdot 0.090 \text{ s}^{-3} \cdot \frac{4.00 \text{ s}^4}{6.02 \times 10^{-5} \text{ L}^3} \cdot (\text{an exact 2}) =$$

Lesson 3.4 — The Atoms (Part 2)

To continue to learn the atoms encountered most often, complete this assignment.

- For the **20** atoms below, memorize the name, symbol, and position in this table. For each atom, given its symbol or name, be able to write the other.
- Be able to fill in an empty table of this shape with these atom names and symbols in their proper places. (The "atomic numbers" shown above each symbol are optional.)

PERIODIC TABLE

1A	2A		3A	4A	5A	6A	7A	8A
1 **H** Hydrogen								2 **He** Helium
3 **Li** Lithium	4 **Be** Beryllium		5 **B** Boron	6 **C** Carbon	7 **N** Nitrogen	8 **O** Oxygen	9 **F** Fluorine	10 **Ne** Neon
11 **Na** Sodium	12 **Mg** Magnesium		13 **Al** Aluminum	14 **Si** Silicon	15 **P** Phosphorus	16 **S** Sulfur	17 **Cl** Chlorine	18 **Ar** Argon
19 **K** Potassium	20 **Ca** Calcium							

Lesson 3.5 — Review Quiz for Chapters 1–3

Assume that you will not be able to use a calculator, the book, notes, or tables. Be sure that your chosen correct answer has the proper number of significant figures.

To answer multiple-choice questions, it is suggested that you

- Solve as if the question is *not* multiple choice, then
- Circle your answer among the choices provided

Set a 20-minute limit, then check your answers after the Summary that follows.

1. From Lesson 1.3:

$$\frac{10^{23}}{(1.25 \times 10^{10})(4.0 \times 10^{-6})} =$$

   a. 2.0×10^{18} b. 5.0×10^{18} c. 0.20×10^{19}

   d. 2.0×10^{20} e. 5.0×10^{-19}

2. Lesson 2.1: 1.5 kg of liquid water has what volume in mL?

   a. 1.5×10^{4} mL   b. 1.5×10^{3} mL   c. 1.5×10^{-3} mL

   d. 1.5×10^{-4} mL   e. 1.5×10^{2} mL

3. Lesson 2.2: 6.6 GHz =

   a. 6.6×10^{-6} Hz   b. 6.6×10^{-9} Hz   c. 6.6×10^{3} Hz

   d. 6.6×10^{6} Hz   e. 6.6×10^{9} Hz

4. Lesson 2.2: Fill in the blank: 1 g = _____ mg

   a. 10^{-6} b. 10^{-9} c. 10^{3} d. 10^{6} e. 10^{9}

5. Lesson 2.4:

$$5.00 \times 10^{-2} \frac{L^3 \cdot m}{s} \cdot 2.00 \, m \cdot \frac{2.0 \, s^3}{8.00 \times 10^{-5} L^2} \cdot (\text{an exact } 2) =$$

   a. $1.00 \times 10^{-4} \, m^2 \cdot s^2 \cdot L$ b. $5.00 \times 10^{3} \, m^2 \cdot s^2 \cdot L$

   c. $5.0 \times 10^{3} \, m^2 \cdot s^2 \cdot L$ d. $1.0 \times 10^{-3} \, m \cdot s^2 \cdot L$

   e. $5.0 \times 10^{-3} \, m^2 \cdot s^2 \cdot L$

6. Lesson 3.2: State your answer in proper significant figures, but do not convert to scientific notation.

$$\begin{array}{r} 1.008 \\ 238.00 \\ + \, 16.00 \\ \hline \end{array}$$

   a. 255.00   b. 255.0   c. 255.008   d. 255.1   e. 255.01

7. Lessons 1.2 and 3.2: Solve:

$$\frac{(\text{Exact } 2)(3.000 \times 10^{23})}{(12.0 \times 10^{-5})} =$$

   a. 5.0×10^{18} b. 5×10^{17} c. 5.0×10^{-18}

   d. 5.00×10^{27} e. 5.00×10^{28}

8. Lesson 3.4: For the following symbols, write the atom name.

   a. K =   b. S =

   c. Na =

SUMMARY

1. When expressing a measurement in significant figures, include the *first doubtful* digit, then stop. Round measurements to the doubtful digit's place.

2. When counting significant figures, include the doubtful digit.

3. When adding and subtracting *s.f.*,

   a. Find the measurement that has *doubt* in the *highest place*.

   b. *Round* your answer to that place.

4. When multiplying and dividing,

   a. Find the number in the calculation that has the least number of *s.f.*

   b. Round your answer to that number of *s.f.*

5. In exponential notation, the *s.f.* are in the significand.

6. When moving a decimal, keep the same number of *s.f.* in the significand.

7. When solving a problem with parts, carry an extra *s.f.* until the final step.

8. To round off a lone 5, use the rule preferred by your instructor. In these lessons, we will round up.

9. For zeros,

   a. Zeros in front of all other numbers are never significant.

   b. Sandwiched zeros are always significant.

   c. Zeros after the other numbers and after the decimal are significant.

   d. Zeros after all numbers but before the decimal place are not significant, but if an unneeded decimal point is shown after a zero, that zero is significant.

10. Exact numbers have infinite *s.f.*

For reinforcement, select and make the flashcards you need using the method in Lesson 2.3. Add additional cards or omit cards as needed.

Front Side (with Notch at Top Right)	Back Side—Answers
Writing measurements in *s.f.*, stop where?	At the first doubtful digit
Counting the number of *s.f.* which digits count?	All the sure, plus the doubtful digit
Adding and subtracting, round to where?	The *column* with doubt in highest place = last full column
Multiplying and dividing, round how?	Least number of *s.f.* in calculation = number *s.f.* allowed
In counting *s.f.*, zeros in front	Never count
Sandwiched zeros	Count
Zeros after numbers and after decimal	Count
Zeros after numbers but before decimal	Probably don't count
Zeros followed by un-needed decimal	Count
Exact numbers have	Infinite *s.f.*

ANSWERS

Lesson 3.1

Practice A   Your answers must match these exactly.

1a. 65.316 mL $\pm$0.05 mL   **65.32 mL**   The highest place with doubt is hundredths.

When converting to *s.f.*, write all the sure digits. At the first place with doubt, round and stop.

1b. 5.2 cm $\pm$0.1 cm   **5.2 cm**   Highest doubt is in tenths place. Round to tenths.

1c. 1.8642 km $\pm$0.2 km   **1.9 km**   The highest doubt is in tenths place. Round to back tenths.

1d. 16.8°C $\pm$1°C   **17°C**   Doubt in the ones place. Round back to the highest place with doubt.

2.
$$\begin{array}{r} 23.1 \\ 23.1 \\ +\ 16.01 \\ \hline 62.21 \end{array}$$
    Must round to **62.2**

3a. 2.016 + 32.18 + 64.**5** = 98.696   Must round to **98.7**

3b. 16.0**0** − 4.034 − 1.008 = 10.9**58**   Must round to **10.96**

Practice B   1a. **7.9 cm^3 (two *s.f.*)**   1b. **5.99 L (three *s.f.*)**   2a. **6.0 $\times$ 10^{-18}**   2b. **1.027 $\times$ 10^2**

3a. **12.7**   If this answer were not used in part b, the proper answer would be 13 (two *s.f.*), but because we need the answer in part b, it is preferred to carry an extra *s.f.*

3b. 12.7/2.5 = **5.1**

Lesson 3.2

Practice A   1. 0.0021**2** rounds to **0.0021**   2. 0.09**9**4 rounds to **0.10**   3. 20.05**6**1 rounds to **20.06**

4. 23.2**5** rounds to **23.3** by the "round lone 5 up" rule.   5. 0.19**5**0 rounds to **0.20**

6. 2.6**4**8 $\times$ 10^{-3} rounds to **2.6 $\times$ 10^{-3}**. Look *one* place past the place to which you are rounding.

Practice B   1. 0.0075 has **two** *s.f.*   Zeros in front never count.

2. 600.3 has **four** *s.f.*   Sandwiched zeros count.

3. 178.40 has **five** *s.f.*   Zeros after the decimal and after all the numbers count.

4. 4640. has **four** *s.f.*   Zeros after the numbers but before a written decimal count.

5. 800 has **one** *s.f.*   Zeros after all numbers but before the decimal place usually don't count.

6. 2.06 $\times$ 10^{-9} has **three** *s.f.*   The significand in front contains and determines the *s.f.*

7. 0.060 $\times$ 10^3 has **two** *s.f.*   The significand contains the *s.f.* Leading zeros never count.

8. 0.02090 $\times$ 10^5 has **four** *s.f.*   The significand contains the *s.f.* Leading zeros never count. The rest here do.

9. 3 (exact)   **Infinite *s.f.***   Exact numbers have no uncertainty and infinite *s.f.*

Lesson 3.3

Practice   1a. 107.42   **five** *s.f.*   Sandwiched zeros count.

1b. 10.04   **four**   Sandwiched zeros count.

1c. 13.40   **four**   Zeros after numbers and after the decimal count.

1d. 0.00640   **three**   Zeros in front never count, but zeros both after numbers and after the decimal count.

1e. 0.043×10^{-4} **two**   Zeros in front never count. The significand contains and determines the *s.f.*

1f. 1,590.0   **five**   The last 0 counts because it comes after numbers and after the decimal. This sandwiches the first 0.

1g. 320×10^{9} **two**   Zeros after numbers but before the decimal *usually* don't count.

1h. 14 (exact)   **infinite**   Exact numbers have infinite *s.f.*

1i. 2,500   **two**   Zeros at the end before the decimal usually don't count.

1j. 4,200.   **four**   The *decimal* at the end means the 0 before it counts, and first 0 is sandwiched.

2a. 5.15 cm (tenths place)   **5.2 cm**   Round up by lone 5 rule.

2b. 31.84 meters (three *s.f.*)   **31.8 meters**   Third digit is last digit: rounding off a 4, round down.

2c. 0.819 mL (hundredths place)   **0.82 mL**   9 rounds up.

2d. 0.06349 cm^2 (two *s.f.*)   **0.063 cm^2**   Leading zeros never count. Round to two *s.f.* based only on the third.

2e. 0.04070 g (two *s.f.*)   **0.041 g**   Zeros in front never count.

2f. 6.255 cm (tenths place)   **6.3 cm**   Rounding a 5 *followed* by other digits, always round up.

3a.
$$
\begin{aligned}
1.008 \\
1.008 \\
+ 32.00 \\
\hline
34.01\underline{6} = \mathbf{34.02}
\end{aligned}
$$

3b.
$$
\begin{aligned}
17.65 \\
- 9.7 \\
\hline
7.9\underline{5} = \mathbf{8.0}
\end{aligned}
$$
Round 5 up.

3c. $39.\underline{1} + 124.\underline{0} + 14.05 = 177.\underline{1}5 = \mathbf{177.2}$ Highest doubt in tenths place.

4a. $13.8612 \text{ cm} \times 2.02 \text{ cm} = 27.9996 = \mathbf{28.0 \text{ cm}^2}$ (**three *s.f.***)   For help with unit math, see Lesson 2.4.

4b. $4.4 \text{ meters} \times 8.312 \text{ meters}^2 = 3\underline{6}.5728 = \mathbf{37 \text{ meter}^3}$ (**two *s.f.***)   5 plus following digits, always round up.

4c. $2.03 \text{ cm}^2/1.2 \text{ cm} = 1.69166 = \mathbf{1.7 \text{ cm}}$ (**two *s.f.***)

4d. $0.5223 \text{ cm}^3/0.040 \text{ cm} = 13.0575 \text{ cm}^2 = \mathbf{13 \text{ cm}^2}$ (**two *s.f.***)

5a. $(2.25 \times 10^{-2})(6.0 \times 10^{23}) = 1\underline{3}.5 \times 10^{21} = \mathbf{1.4 \times 10^{22}}$ in scientific notation (**two *s.f.***)

5b. $(6.022 \times 10^{23})/(1.50 \times 10^{-2}) = \mathbf{4.01 \times 10^{25}}$ (**three *s.f.***)

6a. $2.0646 \text{ m} \pm 0.050 \text{ m}$ **2.06 m**   The highest doubt is in the hundredths place. Round to that place.

6b. $5.04 \text{ nm} \pm 0.1 \text{ nm}$ **5.0 nm**   The highest doubt is in the tenths place. Round to that place.

6c. $12.675 \text{ g} \pm 0.20 \text{ g}$ **12.7 g**   The highest doubt is in the tenths place. Round to that place.

6d. $24.81°\text{C} \pm 1.0°\text{C}$ **25°C**   The highest doubt is in the ones place. Round to that place.

7. $= \dfrac{5.60 \cdot 0.090 \cdot 4.00 \cdot (\text{an exact } 2)}{6.02} \cdot \dfrac{10^{-2}}{10^{-5}} \cdot \dfrac{L^2 \cdot g \cdot s^{-3} \cdot s^4}{s \cdot L^3}$

$= 0.67 \times 10^3 = \mathbf{6.7 \times 10^2} \dfrac{\mathbf{g}}{\mathbf{L}}$

The 0.090 limits the answer to two *s.f.* Exact numbers do not affect *s.f.* For unit cancellation, see Lesson 2.4. Group and handle numbers, exponentials, and units *separately*.

Lesson 3.5

Only *partial* solutions are provided.

1. **a. 2.0×10^{18}** $1/5 \times 10^{23-10+6} = 0.20 \times 10^{19} = 2.0 \times 10^{18}$

2. **b. 1.5×10^3 mL**   1.5 kg = 1,500 g; 1.00 g H_2O liquid $\approx$ 1 mL H_2O liquid

3. **e. 6.6×10^9 Hz**

4. **c. 10^3**

5. **c. 5.0×10^3 m$^2 \cdot$ s$^2 \cdot$ L** (two *s.f.*)

6. **e. 255.01**   Adding and subtracting, round to highest place with doubt.

7. **d. 5.00×10^{27}** $0.500 \times 10^{23-(-5)} = 0.500 \times 10^{28} = 5.00 \times 10^{27}$ (three *s.f.*)

8a. K = **Potassium**

8b. S = **Sulfur**

8c. Na = **Sodium**

4

Conversion Factors

Lesson 4.1   Conversion Factor Basics

Conversion factors can be used to change from one unit of measure to another, or to find equivalent measurements of substances or processes. A conversion factor is a *ratio* (a *fraction*) made from two measured quantities that are *equal* or *equivalent* in a problem. A conversion factor is a fraction that equals *one*.

Conversion factors have a value of unity (1) because if a fraction has a **numerator** (top) and **denominator** (bottom) that are equal or equivalent, its value is *one*.

Example:

$$\frac{7}{7} = 1$$

Or, because 1 milliliter $= 10^{-3}$ liters,

$$\frac{10^{-3}\,L}{1\;mL} = \mathbf{1} \quad \text{and} \quad \frac{1\;mL}{10^{-3}\,L} = \mathbf{1}$$

These last two fractions are typical conversion factors. Any fraction that equals *one* rightside up will also equal *one* upside down. Any conversion factor can be inverted (flipped over) for use if necessary, and it will still equal one.

When converting between liters and milliliters, *all* of these are legal conversion factors:

$$\frac{1\;mL}{10^{-3}\,L} \quad \frac{1000\;mL}{1\;L} \quad \frac{10^{3}\;mL}{1\;L} \quad \frac{3{,}000\;mL}{3\;L}$$

All are equal to *one*.

All of those fractions are mathematically equivalent because they all represent the same *ratio*. Upside down, each fraction is also a legitimate conversion factor because the top and bottom are equal, so its value is one.

In solving calculations, the conversions that are *preferred* (if available) are those that are made from fundamental definitions, such as "milli- $= 10^{-3}$." However, all of the first three forms above are familiar ways to define *milli*units, and any of the first three forms may be encountered during calculations shown in textbooks.

If a series of terms are equal, any *two* of the terms can be used as a conversion factor.

Example: Because 1 meter = 10 decimeters = 100 centimeters = 1000 millimeters, each of the following (and others) is a legitimate conversion factor:

$$\frac{1000\;mm}{1\;m} \quad \frac{1\;mm}{10^{-3}\;m} \quad \frac{10^{2}\;cm}{1\;m} \quad \frac{100\;cm}{10\;dm} \quad \frac{10\;cm}{1\;dm}$$

Each conversion factor represents a way to write the *ratio* between two terms in the series of terms that are equal.

➤ TRY IT

(See "How to Use These Lessons," point 1, p. xv.)

Let's work an example of conversion-factor math. Show your work on this page, then check your answer below.

Q. Multiply the following two terms.

$$7.5\;km \cdot \frac{10^{3}\;m}{1\;km} =$$

Answer:

$$7.5 \, \cancel{km} \cdot \frac{10^3 \, m}{1 \, \cancel{km}} = \frac{(7.5 \cdot 10^3)}{1} \, m = \mathbf{7.5 \times 10^3 \, m}$$

When these terms are multiplied, the "like units" on the top and bottom cancel, leaving meters as the unit on top.

Because the conversion factor multiplies the *given* quantity by *one*, the answer equals the *given* amount. This answer means that 7,500 meters is the same as 7.5 kilometers.

Multiplying a given quantity by a conversion factor changes the *units* that measure the quantity, but does not change its *amount*. The result is the original amount, measured in different units. This process answers a question posed in many science problems: From the units we are given, how can we obtain the units we want?

Our method of solving calculations will focus on finding equal or equivalent quantities. We will then use those equalities to construct conversion factors that solve problems.

SUMMARY

- Conversion factors are made from two measured quantities that are either defined as equal *or* are equivalent or equal in the problem.
- Conversion factors have a value of *one* because the top and bottom terms are equal or equivalent.
- Any *equality* can be made into a conversion (a *fraction* or *ratio*) equal to *one*.
- When the units are set up to cancel correctly, the *given* numbers and units, multiplied by conversions, will result in the WANTED numbers and units.
- *Units* tell you where to write the numbers to solve a calculation correctly.

To check the metric conversion factors that are encountered often, use these rules.

When writing the most common metric conversions, be sure that
- 1 *milli*unit or 1 **m**(unit abbreviation) is above or below 10^{-3} units
- 1 *centi*- or 1 **c**- is above or below 10^{-2}
- 1 *kilo*- or 1 **k**- is above or below 10^3

PRACTICE

Try every other lettered problem. Check your answers frequently. If you miss one in a section, try a few more.

For time abbreviations, use min (minute), hr (hour), and yr (year).

1. Multiply these terms. Cancel units that cancel, then group the numbers and do the math. Write the answer in scientific notation, and include the unit.

   a. $225 \, cg \cdot \frac{10^{-2} \, g}{1 \, cg} \cdot \frac{1 \, kg}{10^3 \, g} =$

   b. $1.5 \, hr \cdot \frac{60 \, min}{1 \, hr} \cdot \frac{60 \, s}{1 \, min} =$

(continued)

2. To be legal, the top and bottom of conversion factors must be equal. Under each conversion, label these as *legal* or *illegal*.

   a. $\dfrac{1000\ \text{mL}}{1\ \text{L}}$

   b. $\dfrac{1000\ \text{L}}{1\ \text{mL}}$

   c. $\dfrac{1.00\ \text{g H}_2\text{O}}{1\ \text{mL H}_2\text{O}}$

   d. $\dfrac{10^{-2}\ \text{volt}}{1\ \text{centivolt}}$

   e. $\dfrac{1\ \text{mL}}{1\ \text{cm}^3}$

   f. $\dfrac{10^3\ \text{cm}^3}{1\ \text{L}}$

   g. $\dfrac{10^3\ \text{kW}}{1\ \text{W}}$

   h. $\dfrac{1\ \text{kilocalorie}}{10^3\ \text{calories}}$

3. Place a 1 in front of the unit with a prefix, then complete the conversion factor.

   a. $\dfrac{\text{g}}{\text{kg}}$

   b. $\dfrac{\text{mole}}{\text{nanomole}}$

   c. $\dfrac{\text{picocurie}}{\text{curie}}$

4. Add numbers to make legal conversion factors, with at least one of the numbers in each conversion factor being a 1.

   a. $\dfrac{\text{centijoules}}{\text{joules}}$

   b. $\dfrac{\text{L}}{\text{cm}^3}$

   c. $\dfrac{\text{cm}^3}{\text{mL}}$

5. Finish the following calculations.

   a. $\dfrac{27\text{A} \cdot 2\text{T} \cdot 4\text{W}}{8\text{A} \quad 3\text{T}} =$

   b. $2.5\ \text{m} \cdot \dfrac{1\ \text{cm}}{10^{-2}\ \text{m}} =$

   c. $\dfrac{95\ \text{km}}{\text{hr}} \cdot \dfrac{0.621\ \text{mi.}}{1\ \text{km}} =$

   d. $\dfrac{27\ \text{m}}{\text{s}} \cdot \dfrac{60\ \text{s}}{1\ \text{min}} \cdot \dfrac{1\ \text{km}}{10^3\ \text{m}} =$

Lesson 4.2   Single-Step Conversions

In the previous lesson, conversion factors were supplied. In this lesson, you will learn to make your own conversion factors to solve problems. Let's learn the method with a simple example.

 **TRY IT**

Q. How many years is 925 days?

In your notebook, write an answer to each step below.

Steps for Solving with Conversion Factors

1. Begin by writing a question mark (?) and then the *unit* you are *looking for* in the problem: the *answer unit* that is WANTED in the problem.

2. Next write an equals (=) sign. It means, "OK, that part of the problem is done. From here on, leave the answer unit alone." You don't cancel the answer unit, and you don't multiply by it.

3. After the = sign, write the number and unit you are *given* (the known quantity).

Answer:

At this point, in your notebook should be

$$\textbf{? years } = \textbf{ 925 days}$$

4. Next, write a · and a line _____ for a conversion factor to multiply by.

5. A key step: Write the *unit* of the *given* quantity in the denominator (on the bottom) of the conversion factor. Leave room for a number in front.
   Do *not* put the given *number* in the conversion factor—just the given *unit*.

$$? \text{ years } = 925 \text{ days} \cdot \frac{\quad\quad\quad}{\textbf{days}}$$

This step puts the given unit where it must be to cancel. It also tells you one part that the next conversion must include.

6. Next, write the answer unit on the top of the conversion factor.

$$\textbf{? years } = 925 \text{ days} \cdot \frac{\textbf{year}}{\text{days}}$$

7. Add *numbers* that make the numerator and denominator of the conversion factor *equal*. In a legal conversion factor, the top and bottom quantities must be equal or equivalent.

8. Cancel the units that you set up to cancel.

9. *If* the unit on the right side after cancellation is the answer unit, stop adding conversions. Write an = sign. Multiply the *given* quantity by the conversion factor. Write the number and the un-canceled unit. Done!

Finish the above steps, then check your answer below.

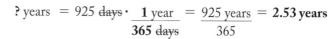

$$? \text{ years } = 925 \; \cancel{\text{days}} \cdot \frac{1 \text{ year}}{365 \; \cancel{\text{days}}} = \frac{925 \text{ years}}{365} = \textbf{2.53 years}$$

S.F.: **1** is *exact*, 925 has three *s.f.*, 365 has three *s.f.* (1 year = 365.24 days is more precise). Round the answer to three *s.f.*

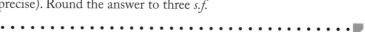

You may need to look back at the above steps, but you should not need to memorize them. By doing the following problems, you will quickly learn what you need to know.

PRACTICE

After each numbered problem, check your answers. Look back at the steps if needed. For the problems in this practice section, write conversions in which one of the numbers (in the numerator or the denominator) is a 1. Round answers to the correct number of significant figures.

If these are easy, do the last letter on each numbered problem. If you miss one, do another part of that problem.

(continued)

1. Add numbers to make these conversion factors legal, cancel the units that cancel, multiply the *given* by the conversion, and write your answer.

    a. $? \text{ days} = 96 \text{ hr} \cdot \dfrac{\text{days}}{24 \text{ hr}} =$

    b. $? \text{ mL} = 3.50 \text{ L} \cdot \dfrac{1 \text{ mL}}{\text{L}} =$

2. To start these, put the *unit* of the *given* quantity where it will cancel. Then finish the conversion factor, do the math, and write your answer with its unit.

    a. $? \text{ s} = 0.25 \text{ min} \cdot \dfrac{ \text{s}}{1} =$

    b. $? \text{ kg} = 250 \text{ g} \cdot \dfrac{ \text{kg}}{10^3} =$

    c. $? \text{ days} = 2.73 \text{ yr} \cdot \dfrac{365}{} =$

    d. $? \text{ yr} = 200. \text{ days} \cdot \dfrac{1}{} =$

3. You should not need to memorize the written rules for arranging conversion factors; however, it is helpful to use this "single-unit starting template."

    > When solving for single units, begin with
    >
    > $? \text{ unit }\textsc{wanted} = \text{number and unit } given \cdot \dfrac{}{\textbf{unit } \textit{\textbf{given}}}$

    The template emphasizes that your first conversion factor puts the given *unit* (but *not* the given *number*) where it will cancel. Apply the template rule to these.

    a. $? \text{ months} = 5.0 \text{ yr} \cdot \underline{} =$

    b. $? \text{ L} = 350 \text{ mL} \cdot =$

    c. $? \text{ min} = 5.5 \text{ hr}$

4. Use the starting template to find how many hours equal 390 minutes.

    ?

5. $? \text{ mg} = 0.85 \text{ kg} \cdot \dfrac{\text{g}}{\text{kg}} \cdot \dfrac{}{\text{g}} =$

Lesson 4.3   Multi-Step Conversions

In problem 5 at the end of the previous lesson, we did not know a direct conversion from kilograms to milligrams. However, we knew a conversion from kilograms to grams and another from grams to milligrams.

In most problems, you will not know a single conversion from the *given* to the WANTED unit, but there will be known conversions that you can *chain together* to solve the calculation.

 TRY IT

Q. Complete this two-step conversion, as done in problem 5 above. Answer in scientific notation.

? milliseconds = 0.25 minutes

STOP

Answer:

$$? \text{ ms} = 0.25 \text{ min} \cdot \frac{60 \text{ s}}{1 \text{ min}} \cdot \frac{1 \text{ ms}}{10^{-3} \text{ s}} = 15 \times 10^3 \text{ ms} = \mathbf{1.5 \times 10^4 \text{ ms}}$$

The 0.25 has two *s.f.*, and both conversions are exact definitions that do not affect the significant figures in the answer, so the answer is written with two *s.f.*

The rules are

Solving for Single Units Using Multiple Conversions

- If a unit remaining on the right side after you cancel units is *not* the answer unit, write it in the next conversion factor where it will cancel.
- Finish the next conversion by writing a known conversion, one that either *includes* the answer unit or gets you *closer* to the answer unit.
- In writing a conversion, first set up a *unit* to cancel, then finish by adding the second unit and *numbers* in front of both units that make the conversion legal.

PRACTICE

These problems are in pairs. If part a is easy, go to part a of the next question. If you need help with part a, do part b for more practice. Answers must *always* include an answer *unit* and proper *s.f.*

1a. ? Gg = 760 mg · _____ g · _____ =

1b. ? cg = 4.2 kg · _____ · _____ =
                                              g

(continued)

2a. $? \text{ yr} = 2.63 \times 10^4 \text{ hr} \cdot$ _____ $\cdot$ _____ =

2b. $? \text{ s} = 1.00 \text{ day} \cdot$ _____ $\text{hr} \cdot$ _____ $\cdot$ _____ =

3a. $? \text{ } \mu\text{g H}_2\text{O}(\ell) = 1.5 \text{ cm}^3 \text{ H}_2\text{O}(\ell) \cdot$ _____ $\text{g H}_2\text{O}(\ell) \cdot$ _____ =

3b. $? \text{ kg H}_2\text{O} = 5.5 \text{ L H}_2\text{O}(\ell)$ _____ $\cdot$ _____ $\cdot$ _____ =

Lesson 4.4 — English/Metric Conversions

Using Familiar Conversions

Most of the conversion factors we have used so far have had an exact number **1** on either the top or the bottom; however, a one is not required in a legal conversion.

Both "1 kilometer = 1000 meters" and "3 kilometers = 3,000 meters" are true equalities and both equalities could be used to make legal conversion factors. In most cases, however, conversions with a 1 are preferred.

Why? We want conversions to be *familiar*, so that we can write them automatically and quickly check that they are correct. Definitions are usually based on *one* component, such as "1 kilometer = 10^3 meters." Definitions are the most familiar equalities and are therefore preferred in conversions.

However, some conversions may be familiar even if they do not include a **1**. For example, many cans of soft drinks are labeled "12.0 fluid ounces (355 mL)." This supplies an equality for English-to-metric volume units: 12.0 fluid ounces = 355 milliliters. That is a legal conversion and, because its numbers and units are seen often, it is easy to remember and check.

Bridge Conversions

Science calculations often involve a key *bridge* conversion between one unit system, quantity, or substance, and another. Conversions between the metric and English systems provide practice in the bridge-conversion methods that we will use to solve chemical reaction calculations.

A *bridge* conversion between metric and English-system *distance* units is

2.54 centimeters ≡ 1 inch

In countries that use English units, this is now the exact definition of an inch. Using this equality, we can convert between metric and English measurements of distance.

Any metric-English distance equality can be used to convert between distance measurements in the two systems. Another metric-English distance conversion that is frequently used (but is not exact) is 0.621 mile = 1 kilometer. (When determining the significant figures for conversions based on equalities that are not exact definitions, assume that an integer **1** is *exact*, but the other number is precise only to the number of *s.f.* shown.)

In problems that require bridge conversions, our strategy will be to "head for the bridge," to begin by converting to one of the two units in the bridge conversion.

When a problem needs a bridge conversion between systems, use these steps.

1. First, convert the *given* unit to the unit in the *bridge* conversion that is in the *same system* as the *given* unit.

2. Next, multiply by the bridge conversion. In the bridge conversion, place the *given* system on the bottom and the WANTED system on top.

3. Multiply by other conversions in the WANTED system to get the answer *unit* wanted.

Add these English distance-unit definitions to your list of memorized conversions:

$$12 \text{ inches} \equiv 1 \text{ foot} \qquad 3 \text{ feet} \equiv 1 \text{ yard} \qquad 5{,}280 \text{ feet} \equiv 1 \text{ mile}$$

(Abbreviations: inch = in., foot = ft., yard = yd., and mile = mi.)
Also commit to memory this metric-to-English bridge conversion for distance.

$$\textbf{2.54 centimeters} \equiv \textbf{1 inch}$$

▷ TRY IT

Q. Apply the steps and conversions above to solve this problem:

? feet = 1.00 meter

Answer:

Because the WANTED unit is English and the *given* unit is metric, an English/metric bridge is needed.

Step 1: Head for the bridge. Because the *given* unit (meters) is metric system, convert to the *metric unit* used in the bridge conversion: *centimeters*.

$$? \text{ ft.} = 1.00 \text{ m} \cdot \frac{1 \text{ cm}}{10^{-2} \text{ m}} \cdot \frac{}{\text{cm}}$$

Note the start of the next conversion. Because centimeters is not the wanted answer unit, centimeters *must* be put in the next conversion where it will cancel. If you *start* the "next unit to cancel" conversion automatically after finishing the prior conversion, it helps to arrange and choose the next conversion.
Adjust and complete the steps if needed.

Step 2: Complete the bridge that converts to the *system* of the answer: English units.

$$? \text{ ft.} = 1.00 \text{ m} \cdot \frac{1 \text{ cm}}{10^{-2} \text{ m}} \cdot \frac{1 \text{ in.}}{2.54 \text{ cm}} \cdot \frac{}{\text{in.}}$$

Step 3: Get rid of the unit you've got. Get to the unit you WANT.

$$? \text{ ft.} = 1.00 \text{ m} \cdot \frac{1 \text{ cm}}{10^{-2} \text{ m}} \cdot \frac{1 \text{ in.}}{2.54 \text{ cm}} \cdot \frac{1 \text{ ft.}}{12 \text{ in.}} = 3.28 \text{ ft.}$$

The answer tells us that 1.00 meter (the *given* quantity) is equal to 3.28 feet.

Some science problems take 10 or more conversions to solve. However, if you know that a bridge conversion is needed, "heading for the bridge" breaks the problem into pieces that will simplify your navigation to the answer.

PRACTICE

Use the inch-to-centimeter bridge conversion above. Start by doing every other problem. Do more if you need more practice.

1. ? cm = 12.0 in. · _____ =

2. ? in. = 1.00 m · _____ · _____

3. For the problem: ? in. = 760. mm

   a. To what unit do you aim to convert the *given* in the initial conversions? Why?

   b. Solve: ? in. = 760. mm

4. ? mm = 0.500 yd.

5. For ? km = 1.00 mi., to convert using 1 in. = 2.54 cm,

   a. To what unit do you aim to convert the *given* in the *initial* conversions? Why?

   b. Solve: ? km = 1.00 mi.

6. As a bridge for metric mass and English weight units, use 1 kg = 2.2 lb. (lb. is the abbreviation for "pounds").

   ? g = 7.7 lb.

7. Use the "soda can" volume conversion (12.0 fl. oz. = 355 mL).

   ? fl. oz. = 2.00 L

8. For the following symbols, write the atom name.

   a. C =   b. Cl =   c. Ca =

Lesson 4.5   Ratio-Unit Conversions

Long-Distance Cancellation

The order in which numbers are multiplied does not affect the result. For example, $1 \times 2 \times 3$ has the same answer as $3 \times 2 \times 1$.

The same is true when multiplying symbols or units. While some sequences may be easier to set up or understand, from a mathematical perspective the order of multiplication does not affect the answer.

The following problem is an example of how units can cancel in separated as well as adjacent conversions.

TRY IT

Q. Multiply. Cancel units that cancel. Write the answer number and its unit.

$$\frac{12\ m}{s} \cdot \frac{60\ s}{1\ min} \cdot \frac{60\ min}{1\ hr} \cdot \frac{1\ km}{1000\ m} \cdot \frac{0.62\ mi.}{1\ km} =$$

Answer:

$$\frac{12\ \cancel{m}}{s} \cdot \frac{60\ \cancel{s}}{1\ \cancel{min}} \cdot \frac{60\ \cancel{min}}{1\ hr} \cdot \frac{1\ \cancel{km}}{1000\ \cancel{m}} \cdot \frac{0.62\ mi.}{1\ \cancel{km}} = \frac{27\ mi.}{hr}$$

This answer means that a speed of 12 meters/second is the *same* as 27 miles/hour.

Ratio Units in the Answer

In these lessons, we will use the term *single unit* to describe a unit that has one base unit in the numerator but no denominator (which means the denominator is 1). Single units measure *amounts*. Meters, grams, minutes, milliliters, and cubic centimeters are all single units. These base units may have prefixes or powers, but they must be the same as or equivalent to a unit that measures a fundamental quantity (such as distance, mass, or time).

We will use the term *ratio unit* to describe a fraction that has *one* kind of base unit in the numerator and *one* different base unit in the denominator. If a problem asks you to find

$$\text{meters per second} \quad or \quad \text{meters/second} \quad or \quad \frac{\text{meters}}{\text{second}} \quad or \quad m \cdot s^{-1}$$

all of those terms are identical, and the problem is asking for a ratio unit. During *conversion* calculations, ratio units should be written in the *fraction* form (with a top and bottom) rather than the form using units with negative exponents.

In Chapter 11, we will address in detail the different characteristics of single units and ratio units. For now, the distinctions above will allow us to solve problems.

Converting the Denominator

In solving for single units, we have used a starting template that includes canceling a *given* single unit.

When solving for single units, begin with:

$$? \text{ unit WANTED} = \# \text{ and unit } \textit{given} \cdot \underline{\hspace{4cm}}$$
$$\textit{unit given}$$

When solving for ratio units, we may need to cancel a denominator (bottom) unit to start a problem. To do so, we will loosen our *starting* rule to say the following.

Solving with Conversion Factors

If a unit to the right of the equals (=) sign, in or after the *given*, on the top or the bottom

- *Matches* a unit in the answer unit, in both what it is and where it is, (circle) that unit on the right side and do not convert it further
- Is *not* what you WANT, put it where it will cancel, and convert until it matches what you WANT

    After canceling units, if the unit or units to the right of the equal sign *match* the WANTED answer unit, no more conversions are needed. Write an = sign, do the math, and write the answer.

➤ TRY IT

Q. Use the rule above to solve this problem.

$$? \frac{cm}{min} = 0.50 \frac{cm}{s} \cdot \underline{\hspace{2cm}} =$$

Answer:

$$? \frac{cm}{min} = 0.50 \frac{\text{(cm)}}{s} \cdot \frac{60 \text{ s}}{1 \text{ (min)}} = \frac{30. \text{ cm}}{min}$$

Start by comparing the WANTED units to the *given* units.

Because you WANT cm on top, and are *given* cm on top, circle (cm) to say, "The top is done. Leave the top alone."

On the bottom, you have seconds, but you WANT minutes. Put seconds where it will cancel. Convert to minutes on the bottom.

When the units on the right of the equals sign match the units you WANT on the left in the answer unit, stop conversions and do the math.

PRACTICE

Do problem 2. Then do problem 1 if you need more practice.

1. $? \dfrac{g}{dL} = 355 \dfrac{g}{L} \cdot \underline{\hspace{2cm}} =$

2. $? \dfrac{m}{s} = \dfrac{4.2 \times 10^{5} \text{ m}}{hr} \cdot \underline{\hspace{2cm}} \cdot \underline{\hspace{2cm}} =$

Converting Both Top and Bottom Units

Many problems require converting both numerator and denominator units. In the following problem, an order to convert both units is specified.

▸ **TRY IT**

Q. Write what must be placed in the blanks to make legal conversions, cancel units, do the math, and write the answer.

$$? \frac{m}{s} = 740 \frac{cm}{min} \cdot \underline{\hspace{2cm}} \cdot \frac{min}{cm} =$$

STOP **Answer:**

$$? \frac{m}{s} = 740 \frac{cm}{min} \cdot \frac{10^{-2} \text{ m}}{1 \text{ cm}} \cdot \frac{1 \text{ min}}{60 \text{ s}} = 0.12 \frac{m}{s}$$

In the *given* on the right, cm is *not* the unit WANTED on top, so put it where it will cancel, and convert to the unit you want on top.

Next, because minutes are on the bottom on the right, but seconds are WANTED, put minutes where it will cancel. Convert to the seconds WANTED.

When chaining conversions, which unit you convert first—the top or bottom unit—makes no difference. The order in which you multiply factors does not change the answer.

▸ **TRY IT**

On the following problem, no order for the conversions is specified.

Q. Add legal conversions in any order, solve, then check your answer below. Before doing the math, double-check each conversion to make sure it is legal.

$$? \frac{cg}{L} = 0.550 \times 10^{-2} \frac{g}{mL}$$

STOP **Answer:**

Your conversions may be in a different order.

$$? \frac{cg}{L} = 0.550 \times 10^{-2} \frac{g}{mL} \cdot \frac{1 \text{ cg}}{10^{-2} \text{ g}} \cdot \frac{1 \text{ mL}}{10^{-3} \text{ L}} = 5.50 \times 10^{2} \frac{cg}{L}$$

PRACTICE B

Do every other part, and more if you need more practice.

1. On these, an order of conversion is specified. Write what must be placed in the blanks to make legal conversions, then solve.

    a. $? \frac{mi.}{hr} = 80.7 \frac{ft.}{s} \cdot \underline{\hspace{1cm}} mi. \cdot \underline{\hspace{1cm}} \cdot \underline{\hspace{1cm}} =$

    b. $? \frac{m}{s} = 250. \frac{ft.}{min} \cdot \underline{\hspace{1cm}} min \cdot \underline{\hspace{1cm}} \cdot \underline{\hspace{1cm}} \cdot \underline{\hspace{1cm}} =$
                                                                     1 in.

(continued)

2. Add conversions in any order and solve.

a. $\dfrac{?\ \text{km}}{\text{hr}} = \dfrac{1.17 \times 10^4\ \text{mm}}{\text{s}}$

b. $\dfrac{?\ \text{ng}}{\text{mL}} = \dfrac{47 \times 10^2\ \text{mg}}{\text{dm}^3}$

c. $\dfrac{?\ \text{ft.}}{\text{s}} = \dfrac{95\ \text{m}}{\text{min}}$

Lesson 4.6   The Atoms (Part 3)

To continue to learn the most often encountered atoms, your assignment is

- For the first 20 atoms, **plus** the first and last two *columns* in the periodic table, memorize the name, symbol, and position of the atom. For each atom, when given either its symbol or name, be able to write the other.
- For the first and last two columns, focus on being able to write the symbols in the column in order from the top down.
- Be able to fill in a blank table with the names and symbols of these atoms in their correct positions.

PERIODIC TABLE

1A	2A		3A	4A	5A	6A	7A	8A
1 **H** Hydrogen								2 **He** Helium
3 **Li** Lithium	4 **Be** Beryllium		5 **B** Boron	6 **C** Carbon	7 **N** Nitrogen	8 **O** Oxygen	9 **F** Fluorine	10 **Ne** Neon
11 **Na** Sodium	12 **Mg** Magnesium		13 **Al** Aluminum	14 **Si** Silicon	15 **P** Phosphorus	16 **S** Sulfur	17 **Cl** Chlorine	18 **Ar** Argon
19 **K** Potassium	20 **Ca** Calcium						**Br** Bromine	36 **Kr** Krypton
Rb Rubidium	**Sr** Strontium						**I** Iodine	54 **Xe** Xenon
Cs Cesium	**Ba** Barium						**At** Astatine	86 **Rn** Radon
Fr Francium	**Ra** Radium							

SUMMARY

1. Conversion factors are fractions or ratios made from two entities that are equal or equivalent. Conversion factors have a value of one.

2. An *equality* can be written as a *conversion* or *fraction* or *ratio* that is equal to *one*.

3. Units determine the placement of numbers that result in the correct answer.

4. When solving a problem, first write the *unit* WANTED, then an = sign.

5. Solving for single units, start conversion factors with

$$? \text{ unit WANTED} = \# \text{ and } unit\ given \cdot \frac{\rule{2cm}{0.4pt}}{unit\ given}$$

6. Finish each conversion factor with the answer unit or with a unit that takes you closer to the answer unit.

7. In making conversions, set up units to cancel, but add numbers that make legal conversions.

8. Chain your conversions so that the units cancel to get rid of the unit you've got and get to the unit you WANT.

9. When the unit on the right is the unit of the answer on the left, stop conversion factors. Complete the number math. Write the answer with its unit.

If you plan on a career in a science-related field, add these to your flashcard collection.

Front Side (with Notch at Top Right)	Back Side—Answers
1 in. = ? cm	2.54 cm
1 kg = ? lb.	2.20 lb.
12 fl. oz. = ? mL	355 mL

ANSWERS

Lesson 4.1

Practice   1a. $225 \text{ cg} \cdot \dfrac{10^{-2}\ g}{1\ cg} \cdot \dfrac{1\ kg}{10^3\ g} = \dfrac{225 \times 10^{-2} \times 1}{1 \times 10^3} \textbf{ kg} = \textbf{2.25} \times \textbf{10}^{-3} \textbf{ kg}$

The answer means that 2.25×10^{-3} kg is *equal* to 225 cg.

1b. $1.5 \text{ hr} \cdot \dfrac{60\ min}{1\ hr} \cdot \dfrac{60\ s}{1\ min} = \dfrac{1.5 \times 60 \times 60}{1} \text{ s} = \textbf{5,400 s} \ or \ \textbf{5.4} \times \textbf{10}^3 \textbf{ s}$

Recall that **s** is the abbreviation for seconds. This answer means that 1.5 hours is equal to 5,400 seconds.

2a. $\dfrac{1000 \text{ mL}}{1 \text{ L}}$	2b. $\dfrac{1000 \text{ L}}{1 \text{ mL}}$	2c. $\dfrac{1.00 \text{ g } H_2O}{1 \text{ mL } H_2O}$	2d. $\dfrac{10^{-2} \text{ volt}}{1 \text{ centivolt}}$
Legal	**Illegal**	**Legal if *liquid* water**	**Legal**

2e. $\dfrac{1 \text{ mL}}{1 \text{ cm}^3}$ 2f. $\dfrac{10^3 \text{ cm}^3}{1 \text{ L}}$ 2g. $\dfrac{10^3 \text{ kW}}{1 \text{ W}}$ 2h. $\dfrac{1 \text{ kilocalorie}}{10^3 \text{ calories}}$

    Legal   **Legal**   **Illegal**   **Legal**

3a. $\dfrac{\mathbf{10^3}\ \mathbf{g}}{\mathbf{1}\ \mathbf{kg}}$ 3b. $\dfrac{\mathbf{10^{-9}}\ \text{mole}}{\mathbf{1}\ \text{nanomole}}$ 3c. $\dfrac{\mathbf{1}\ \text{picocurie}}{\mathbf{10^{-12}}\ \text{curie}}$

4a. $\dfrac{\text{1 centijoule}}{10^{-2}\ \text{joules}}$ *or* $\dfrac{\mathbf{100}\ \text{centijoules}}{\mathbf{1}\ \text{joule}}$ 4b. $\dfrac{\mathbf{1\ L}}{\mathbf{1000}\ \mathbf{cm^3}}$ *or* $\dfrac{\mathbf{10^{-3}\ L}}{\mathbf{1}\ \mathbf{cm^3}}$ 4c. $\dfrac{\mathbf{1\ cm^3}}{\mathbf{1\ mL}}$

     Either fixed decimal numbers (such as 100) or equivalent exponentials (10^2) may be used in conversions.

5a. $27A \cdot \dfrac{2T}{8A} \cdot \dfrac{4W}{3T} = 27A \cdot \dfrac{2\cancel{T}}{8\cancel{A}} \cdot \dfrac{4W}{3\cancel{T}} = \dfrac{27 \cdot 2 \cdot 4}{8 \cdot 3} \cdot W = \mathbf{9W}$

5b. $2.5\ \cancel{\text{m}} \cdot \dfrac{1 \text{ cm}}{10^{-2}\ \cancel{\text{m}}} = 2.5 \times 10^2 \text{ cm} = \mathbf{250\ cm}$

5c. $\dfrac{95\ \cancel{\text{km}} \cdot 0.621 \text{ mi.}}{\text{hr} \quad 1\ \cancel{\text{km}}} = \dfrac{95 \cdot 0.621}{1} \dfrac{\text{mi.}}{\text{hr}} = \mathbf{59\ \dfrac{mi.}{hr}}$

5d. $\dfrac{27\ \cancel{\text{m}}}{\text{s}} \cdot \dfrac{60 \text{ s}}{1 \text{ min}} \cdot \dfrac{1 \text{ km}}{10^3\ \cancel{\text{m}}} = \dfrac{27 \cdot 60}{10^3} \dfrac{\text{km}}{\text{min}} = \mathbf{1.6\ \dfrac{km}{min}}$

Lesson 4.2

Practice   Some but not all unit cancellations are shown. For your answer to be correct, it must include its unit.

     Your conversions may be in different formats, such as 1 meter = 100 centimeters *or* 1 centimeter = 10^{-2} meters, as long as the conversion's top and bottom are equal and your answer is the same as below.

1a. $? \text{ days} = 96\ \cancel{\text{hr}} \cdot \dfrac{\mathbf{1}\ \mathbf{day}}{24\ \cancel{\text{hr}}} = \dfrac{96}{24} \text{ days} = \mathbf{4.0\ days}$

1b. $? \text{ mL} = 3.50\ \cancel{\text{L}} \cdot \dfrac{\mathbf{1}\ \mathbf{mL}}{\mathbf{10^{-3}}\ \cancel{\mathbf{L}}} = 3.50 \cdot 10^3 \text{ mL} = \mathbf{3.50 \times 10^3\ mL}$

     (*S.F.*: 3.50 has three *s.f.*, prefix definitions are exact with infinite *s.f.*, answer is rounded to three *s.f.*)

2a. $? \text{ s} = 0.25\ \cancel{\text{min}} \cdot \dfrac{\mathbf{60}\ \mathbf{s}}{1\ \cancel{\mathbf{min}}} = (0.25 \cdot 60) \text{ s} = \mathbf{15\ s}$

     (*S.F.*: 0.25 has two *s.f.*, 1 min = 60 s is a definition with infinite *s.f.*, answer is rounded to two *s.f.*)

2b. $? \text{ kg} = 250\ \cancel{\text{g}} \cdot \dfrac{\mathbf{1}\ \mathbf{kg}}{10^3\ \cancel{\mathbf{g}}} = \dfrac{250}{10^3} \text{ kg} = \mathbf{0.25\ kg}$

2c. $? \text{ days} = 2.73\ \cancel{\text{yr}} \cdot \dfrac{365\ \mathbf{days}}{1\ \cancel{\mathbf{yr}}} = (2.73 \cdot 365) \text{ days} = \mathbf{996\ days}$

2d. $? \text{ yr} = 200.\ \cancel{\text{days}} \cdot \dfrac{\mathbf{1}\ \mathbf{yr}}{365\ \cancel{\mathbf{days}}} = \dfrac{200}{365} \text{ yr} = \mathbf{0.548\ yr}$

3a. $? \text{ months} = 5.0\ \cancel{\text{yr}} \cdot \dfrac{\mathbf{12}\ \mathbf{months}}{1\ \cancel{\mathbf{yr}}} = \mathbf{60.\ months}$

     (*S.F.*: 5.0 has two *s.f.*, 12 months = 1 yr is a definition with infinite *s.f.*, round to two *s.f.*, the 60. decimal means two *s.f.*)

3b. $? L = 350 \text{ mL} \cdot \dfrac{\mathbf{10^{-3}\,L}}{\mathbf{1\,mL}} = 350 \times 10^{-3}\,L = \mathbf{0.35\,L}$

(*S.F.*: 350 has two *s.f.*, prefix definitions are *exact* with infinite *s.f.*, round to two *s.f.*)

3c. $? \min = 5.5 \text{ hr} \cdot \dfrac{\mathbf{60\,min}}{\mathbf{1\,hr}} = \mathbf{330\,min}$

4. $\mathbf{?\,hr} = 390 \text{ min} \cdot \dfrac{\mathbf{1\,hr}}{60\,\mathbf{min}} = \mathbf{6.5\,hr}$

5. $?\,mg = 0.85 \text{ kg} \cdot \dfrac{\mathbf{10^3\,g}}{1\,\text{kg}} \cdot \dfrac{\mathbf{1\,mg}}{10^{-3}\,\text{g}} = 0.85 \times 10^6 \text{ mg} = \mathbf{8.5 \times 10^5\,mg}$

Lesson 4.3

Practice For visibility, not all cancellations are shown, but cancellations should be marked on your paper.
Your conversions may be different (for example, you may use **1000 mL = 1 L** *or* **1 mL = 10⁻³ L**), but you must arrive at the same answer as below.

1a. $?\,Gg = 760 \text{ mg} \cdot \dfrac{10^{-3}\,g}{1\,\text{mg}} \cdot \dfrac{1\,Gg}{10^9\,g} = 760 \times 10^{-12}\,Gg = \mathbf{7.6 \times 10^{-10}\,Gg}$

1b. $?\,cg = 4.2 \text{ kg} \cdot \dfrac{10^3\,g}{1\,\text{kg}} \cdot \dfrac{1\,cg}{10^{-2}\,g} = \mathbf{4.2 \times 10^5\,cg}$

2a. $?\,yr = 2.63 \times 10^4 \text{ hr} \cdot \dfrac{1\,\text{day}}{24\,\mathbf{hr}} \cdot \dfrac{1\,\mathbf{yr}}{365\,\text{days}} = \dfrac{2.63 \times 10^4}{24 \cdot 365}\,yr = \mathbf{3.00\,yr}$

2b. $?\,s = 1.00 \text{ day} \cdot \dfrac{24\,\mathbf{hr}}{1\,\text{day}} \cdot \dfrac{60\,\mathbf{min}}{1\,\mathbf{hr}} \cdot \dfrac{60\,s}{1\,\mathbf{min}} = \mathbf{8.64 \times 10^4\,s}$

3a. $?\,\mu g\ H_2O(\ell) = 1.5 \text{ cm}^3\ H_2O(\ell) \cdot \dfrac{1.00\,g\ H_2O(\ell)}{1\,\text{cm}^3\ H_2O(\ell)} \cdot \dfrac{1\,\mu g}{10^{-6}\,g} = \mathbf{1.5 \times 10^6\,\mu g\ H_2O(\ell)}$

3b. $?\,kg\ H_2O(\ell) = 5.5 \text{ L}\ H_2O(\ell) \cdot \dfrac{1\,\text{mL}}{10^{-3}\,L} \cdot \dfrac{1.00\,g\ H_2O(\ell)}{1\,\text{mL}\ H_2O(\ell)} \cdot \dfrac{1\,kg}{10^3\,g} = \dfrac{5.5}{10^0} = \mathbf{5.5\,kg\ H_2O(\ell)}$

Lesson 4.4

Practice In these answers, some but not all of the unit cancellations are shown. The definition 1 centimeter = 10 millimeter may be used for *mm* to *cm* conversions. Doing so will change the number of conversions but not the answer.
To be correct, answers must be written that include correct units.

1. $?\,cm = 12.0 \text{ in.} \cdot \dfrac{2.54\,cm}{1\,\text{in.}} = 12.0 \cdot 2.54 \text{ cm} = \mathbf{30.5\,cm}$ (Check how many cm are on a 12-inch ruler.)

2. $?\,\text{in.} = 1.00 \text{ m} \cdot \dfrac{1\,\text{cm}}{10^{-2}\,\text{m}} \cdot \dfrac{\mathbf{1\,in.}}{\mathbf{2.54\,cm}} = \dfrac{1}{2.54} \times 10^2 \text{ in.} = 0.394 \times 10^2 \text{ in.} = \mathbf{39.4\,in.}$

3a. Aim to convert the *given* unit (mm) to the one unit in the *bridge* conversion that is in the same system (English or metric) as the *given*. The bridge unit **cm** is in the same measurement *system* as mm.

3b. $?\,\text{in.} = 760. \text{ mm} \cdot \dfrac{10^{-3}\,\text{m}}{1\,\text{mm}} \cdot \dfrac{1\,\text{cm}}{10^{-2}\,\text{m}} \cdot \dfrac{\mathbf{1\,in.}}{\mathbf{2.54\,cm}} = \dfrac{760 \times 10^{-1}}{2.54} \text{ in.} = \mathbf{29.9\,in.}$

(*S.F.*: 760., with the *decimal* after the 0, means three *s.f.* Metric definitions and one have infinite *s.f.* The answer must be rounded to three *s.f.*; see Chapter 3.)

4. $? \text{ mm} = 0.500 \text{ yd.} \cdot \dfrac{3 \text{ ft.}}{1 \text{ yd.}} \cdot \dfrac{12 \text{ in.}}{1 \text{ ft.}} \cdot \dfrac{\mathbf{2.54 \, cm}}{\mathbf{1 \, in.}} \cdot \dfrac{10^{-3} \, \cancel{m}}{1 \text{ cm}} \cdot \dfrac{1 \text{ mm}}{10^{-3} \, \cancel{m}} = \mathbf{457 \, mm}$

5a. Aim to convert the *given* unit (miles) to the bridge unit in the same system (English or metric) as the *given*. **Inches** is in the same system as miles.

5b. $? \text{ km} = 1.00 \, mi. \cdot \dfrac{5.280 \, \cancel{\text{ft.}}}{1 \, mi.} \cdot \dfrac{12 \text{ in.}}{1 \, \cancel{\text{ft.}}} \cdot \dfrac{\mathbf{2.54 \, cm}}{\mathbf{1 \, in.}} \cdot \dfrac{10^{-2} \, \cancel{m}}{1 \text{ cm}} \cdot \dfrac{1 \text{ km}}{10^{3} \, \cancel{m}} = \mathbf{1.61 \, km}$

(*S.F.*: Assume an integer **1** that is part of any equality or conversion is exact, with infinite *s.f.*)

6. $? \text{ g} = 7.7 \text{ lb.} \cdot \dfrac{1 \text{ kg}}{2.2 \text{ lb.}} \cdot \dfrac{10^{3} \text{ g}}{1 \text{ kg}} = \mathbf{3.5 \times 10^{3} \, g}$

(*S.F.*: 7.7 and 2.2 have two *s.f.* A 1 and metric-prefix *definitions* have infinite *s.f.* Round the answer to two *s.f.*)

7. $? \text{ fl. oz.} = 2.00 \text{ L} \cdot \dfrac{1 \text{ mL}}{10^{-3} \text{ L}} \cdot \dfrac{12.0 \text{ fl. oz.}}{355 \text{ mL}} = \mathbf{67.6 \, fl. \, oz.}$ (Check this answer on any 2-liter soda bottle.)

8a. C = **Carbon**

8b. Cl = **Chlorine**

8c. Ca = **Calcium**

Lesson 4.5

Practice A 1. $? \, \dfrac{\text{g}}{\text{dL}} = 355 \dfrac{\text{ⓖ}}{\text{L}} \cdot \dfrac{10^{-1} \text{ L}}{1 \, \text{ⓓⓛ}} = \mathbf{35.5} \, \dfrac{\text{g}}{\text{dL}}$

2. $? \, \dfrac{\text{m}}{\text{s}} = 4.2 \times 10^{5} \, \dfrac{\text{ⓜ}}{\text{hr}} \cdot \dfrac{1 \text{ hr}}{60 \text{ min}} \cdot \dfrac{1 \text{ min}}{60 \, \text{ⓢ}} = \mathbf{1.2 \times 10^{2}} \, \dfrac{\mathbf{m}}{\mathbf{s}}$

Practice B 1a. $? \, \dfrac{\text{mi.}}{\text{hr}} = 80.7 \, \dfrac{\text{ft.}}{\text{s}} \cdot \dfrac{1 \, \text{ⓜⓘ}}{5{,}280 \text{ ft.}} \cdot \dfrac{60 \text{ s}}{1 \text{ min}} \cdot \dfrac{60 \text{ min}}{1 \, \text{ⓗⓡ}} = \mathbf{55.0} \, \dfrac{\mathbf{mi.}}{\mathbf{hr}}$ (three *s.f.*)

1b. $? \, \dfrac{\text{m}}{\text{s}} = 250. \, \dfrac{\text{ft.}}{\text{min}} \cdot \dfrac{1 \text{ min}}{60 \, \text{ⓢ}} \cdot \dfrac{12 \text{ in.}}{1 \text{ ft.}} \cdot \dfrac{2.54 \text{ cm}}{1 \text{ in.}} \cdot \dfrac{10^{-2} \, \text{ⓜ}}{1 \text{ cm}} = \mathbf{1.27} \, \dfrac{\mathbf{m}}{\mathbf{s}}$

(*S.F.*: 250. due to the decimal has three *s.f.*, all other conversions are definitions, answer is rounded to three *s.f.*)

2a. $? \, \dfrac{\text{km}}{\text{hr}} = 1.17 \times 10^{4} \, \dfrac{\text{mm}}{\text{s}} \cdot \dfrac{10^{-3} \text{ m}}{1 \text{ mm}} \cdot \dfrac{1 \, \text{ⓚⓜ}}{10^{3} \text{ m}} \cdot \dfrac{60 \text{ s}}{1 \text{ min}} \cdot \dfrac{60 \text{ min}}{1 \, \text{ⓗⓡ}} = \mathbf{42.1} \, \dfrac{\mathbf{km}}{\mathbf{hr}}$

2b. $? \, \dfrac{\text{ng}}{\text{mL}} = 47 \times 10^{2} \, \dfrac{mg}{\text{dm}^{3}} \cdot \dfrac{1 \text{ dm}^{3}}{1 \text{ L}} \cdot \dfrac{10^{-3} \text{ L}}{1 \, \text{ⓜⓛ}} \cdot \dfrac{10^{-3} \text{ g}}{1 \, mg} \cdot \dfrac{1 \, \text{ⓝⓖ}}{10^{-9} \text{ g}} = \mathbf{4.7 \times 10^{6}} \, \dfrac{\mathbf{ng}}{\mathbf{mL}}$

2c. Hint: An English/metric bridge conversion for distance units is needed. Head for the bridge: first convert the given metric distance unit to the metric distance unit used in your known bridge conversion.

$? \, \dfrac{\text{ft.}}{\text{s}} = 95 \, \dfrac{\text{m}}{\text{s}} \cdot \dfrac{1 \text{ min}}{60 \, \text{ⓢ}} \cdot \dfrac{1 \text{ cm}}{10^{-2} \text{ m}} \cdot \dfrac{1 \, in.}{2.54 \, cm} \cdot \dfrac{1 \, \text{ⓕⓣ}}{12 \text{ in.}} = \mathbf{5.2} \, \dfrac{\mathbf{ft.}}{\mathbf{s}}$

5

Word Problems

Introduction

In this chapter you will learn to identify *given* quantities and equalities in word problems. You will then be able to solve nearly all of the initial problems assigned in chemistry with the same conversion method used in Chapter 4. In addition, you will learn a system to organize your data before you solve. Most students report that by using this structured approach, they have a better understanding of the steps to take to solve science calculations.

Lesson 5.1   Answer Units—Single or Ratio?

Types of Units

In these lessons, for our initial problem solving we will divide the units of measurements into three types:

- **Single units** have one kind of base unit in the numerator, but no denominator. Examples include meters, cubic centimeters, grams, and hours.
- **Ratio units** have one single unit in the numerator and a different single unit in the denominator. Examples include meters/second and g/cm^3.
- **Complex units** are all other units, such as 1/s or $(kg \cdot m^2)/s^2$.

Most of the calculations encountered initially in chemistry involve single units and ratios, but not complex units. Rules for single units will be covered in this chapter. Additional characteristics of single and ratio units will be covered in Chapter 11. Complex units will be addressed in Chapter 16.

Rule 1: First, Write the WANTED Unit

> To solve word problems, begin by writing "WANTED: ?" then the *unit* of the *answer*, and then an $=$ sign. The *first* time you read a word problem, look *only* for the *unit* of the answer.

Apply that rule to the following problem.

➤ TRY IT

(See "How to Use These Lessons," point 1, p. xv.)

   Q. At an average speed of 25 miles/hour, how many hours will it take to go 450 miles?

   Begin by writing:

Answer:

WANTED: ? **hr** $=$

Writing the answer unit first is essential to

- Help you choose the correct *given* to start your conversions
- Prompt you to write DATA conversions that you will need to solve
- Tell you when to stop conversions and do the math

Rules for Answer Units

When writing the WANTED unit, it is important to distinguish between single units and ratio units.

1. An answer unit is a *ratio* unit if a problem asks you to find

    a. "unit X/unit Y" or "unit $X \cdot$ unit Y^{-1}" or

    b. "unit X *per* unit Y" where there is no number after *per*

    All of those expressions are equivalent. All are ways to represent ratio units.
    Example:

    $$\frac{grams}{mL}$$

    also written grams/mL *or* $g \cdot mL^{-1}$, is a ratio unit.

2. For an answer unit, if there is no number in front of the denominator unit or after the word *per*, the number *one* is understood, and the WANTED unit is a ratio unit.
    Example: "Find the speed in miles/hour (or miles per hour)" is equivalent to "find the miles traveled per *one* hour."
    A ratio unit is one unit *per* ONE other unit.

3. An answer unit is a *single* unit if it has a one kind of base unit in the numerator (top term) but no denominator.
    Example: If a problem asks you to find miles, or cm^3, or dollars, it is asking for a single unit.

4. If a problem asks for a "unit per *more than one* other unit," it WANTS a *single* unit.
    Example: If a problem asks for "grams per 100 milliliters," or the "miles traveled in 27 hours," it is asking for a single unit.
    A ratio unit must be one unit per *one* other unit.

Writing Answer Units

1. If you WANT a ratio unit, write the unit as a *fraction* with a top and a bottom.
    Example: If an answer unit in a problem is miles/hour, to start,

    Write: WANTED: **? mi.** =
            hr

    Do *not* write: WANTED: ? miles/hour *or* ? mph

The slash mark (/), which is read as "per" or "over," is an easy way to *type* ratios and conversion factors. However, when solving with conversions, *writing* ratio answer units as a *fraction*, with a clear numerator and denominator, will help in arranging the conversions needed to solve.

2. If a problem WANTS a single unit, write the WANTED unit without a denominator.
   Write WANTED: **? mi. =**   or   WANTED: **? mL =**
   Single units have an understood 1 as denominator and are written without a denominator.

PRACTICE

For each problem, write "WANTED: ?" and the unit that the problem is asking you to find, using the rules above. After that WANTED unit, write an equals sign.
   Do not finish the problem. Write only the WANTED unit.

1. If 1.12 liters of a gas at standard temperature and pressure (STP) has a mass of 3.55 grams, what is the molar mass of the gas in grams/mole?

2. At an average speed of 25 mi./hr, how many minutes will it take to go 15 miles?

3. If a car travels 270 mi. in 6 hr, what is its average speed?

4. A student needs 420 special postage stamps. The stamps are sold with six stamps on a sheet, each stamp booklet has three sheets, and the cost is $14.40 per booklet. How much is the cost of all of the stamps?

5. How much is the cost per stamp in problem 4?

Lesson 5.2   Mining the DATA

The method we will use to *simplify* problems is to divide solving into three parts:

   WANTED

   DATA

   SOLVE

This method will break complex problems into pieces. You will always know what steps to take to solve a problem because we will solve all problems with the same three steps.

Rules for DATA

To solve word problems, get rid of the words.

By translating words into numbers, units, and labels, you can solve most of the initial word problems in chemistry by chaining conversions, as you did in Chapter 4.

To translate the words, write in the DATA section on your paper every *number* you encounter as you read the problem, followed by its *unit* and a *label* that describes the quantity being measured.

In the initial problems of chemistry, it is important to distinguish numbers and units that are parts of equalities from those that are not. To do so, we need to learn the many ways that quantities that are *equal* or *equivalent* can be expressed in words and symbols.

Rules for Listing DATA in Word Problems

1. Read the problem. Write "WANTED: ?" followed by the WANTED unit and an = sign.

2. On the next line down, write "DATA:"

3. Read the problem a second time.
   - Each time you find a number, *stop*. Write the number on a line under "DATA:"
   - After the number, write its *unit* plus a *label* that helps to identify the number.
   - Decide if that number, unit, and label is *paired* with another number, unit, and label as part of an equality.

4. In the DATA section, write each number and unit in the problem as part of an *equality* in these cases.

   a. If you see *per* or / (a slash), write *per* or / in DATA as an equal sign (=).
      - If a number is shown after *per* or /, write the number in the equality.
        Example: If you read "$8 *per* 3 lb.," write in the DATA: "$8 = 3 lb."
      - If *no* number is shown after *per* or /, write *per* as " = **1**"
        Example: If you see "25 km/hr," write "25 km = **1** hr"
      Per means / or =. A *per* statement can be used as a conversion factor.

   b. Treat *unit x · unit y*$^{-1}$ the same as *unit x/unit y*.
      Example: If you see "75 g · mL^{-1}," write "75 g = **1** mL"

   c. If the same *quantity* is measured using two different units, write the corresponding equality statement.
      Examples: If a problem says, "0.0350 moles of gas has a volume of 440 mL," write in your DATA: "0.0350 moles of gas = 440 mL"
      If a problem says a bottle is labeled "2 liters (67.6 fluid ounces)," write: "2 liters = 67.6 fluid ounces"
      In both examples, the same physical quantity is being measured using two different units.

   d. If the same *process* is measured using two different units, write out the equality statement.
      Example: If a problem says, "burning 0.25 grams of candle wax releases 1,700 calories of energy," write in your DATA section, "0.25 grams candle wax = 1,700 calories of energy"

Both sides are measures of what happened as this candle burned.

After each unit, if two different entities are being measured in the problem, write a label after the unit: additional words that identify what is being measured by the number and unit.

The labels "candle wax" and "energy" will help us to identify which numbers and units to use at points during problem solving.

5. Watch for words that mean *one*, such as *each* and *every*. *One* is a number, and you want *all* numbers in your DATA table.

Example: If you read, "*Each* student was given two sodas," write "1 student = 2 sodas."

6. Continue until *all* of the *numbers* in the problem are written in your DATA.

7. Note that in writing the WANTED unit, you write "per one" as a ratio unit and "per more than one" as a single unit.

In the DATA, however, "per one" and "per more than one" can be written in the same way in an equality.

PRACTICE

1. For each phrase below, write the equality that you will add to your DATA. On each side of the equals sign, include a number, a unit, and a label if available. After every few, check your answers.

   a. The car was traveling at a speed of 55 mi./hr.

   b. A bottle of designer water is labeled 0.50 L (16.9 fl. oz.).

   c. Every student was given 19 pages of homework.

   d. To melt 36 g of ice required 2,880 calories of heat.

   e. The molar mass is 18.0 g $H_2O \cdot (\text{mole } H_2O)^{-1}$.

   f. The dosage of the aspirin is 2.5 mg per kg of body mass.

   g. If 0.24 g of NaOH are dissolved to make 250 mL of solution, what is the concentration of the solution?

2. For problems 1–4 in the Practice for Lesson 5.1, write DATA: and then list the DATA *equalities* that are supplied in the problem.

Lesson 5.3   Solving for Single Units

Dimensional Homogeneity

By the law of **dimensional homogeneity**, the *units* on both sides of an equality must be the same at the *end* of a calculation. One implication of this law is: If a single unit amount is WANTED in a calculation, a single-unit amount must be supplied in the data as a *given* basis for the conversion.

When a single unit is WANTED, this rule will simplify using conversions. We will start each conversion calculation with an equality:

? WANTED single unit = number given single unit

and then convert the given unit to the WANTED unit.

DATA Formats If a Single Unit Is WANTED

If a problem WANTS a single unit, one number and unit in the DATA is likely to be

- Either a number and its unit that is not paired in an equality with other measurements, or
- A number and its unit that is paired with the WANTED unit in the format

"? unit WANTED = number unit given"

We will define the given as the term written to the right of the equal sign in the SOLVE step: the starting point for the terms that we multiply to solve conversion calculations.

If a problem WANTS a single-unit amount, by the laws of science and algebra, at least one item of DATA must be a single-unit amount. In problems that can be solved using conversions, often one measurement will be a single unit, and the rest of the DATA will be equalities.

If a single unit is WANTED, watch for one item of data that is a single-unit amount. In the DATA, write the single number, unit, and label on a line by itself.

It is a good practice to (circle) that single unit amount in the DATA, because it will be the given number and unit that is used to start your conversions.

TRY IT

For the following problem, in your notebook write only the WANTED and DATA steps.

Q. If a car's speed is 55 miles/hour, how many minutes are needed to travel 85 miles?

Answer:

Your paper should look like this.

WANTED: ? min =

DATA: 55 mi. = 1 hr

85 mi.

Variations on the above rules will apply when DATA includes two amounts that are equivalent in a problem. We address these cases in Chapter 11. However, for the problems you are initially assigned in first-year chemistry, the rules above will most often apply.

To SOLVE

After listing the DATA provided in a problem, below the DATA section, write "SOLVE." Then, if you WANT a single unit, write the WANTED and *given* measurements in the format of the *single-unit starting template*.

$$? \text{ unit WANTED } = \frac{\text{number and unit } given \text{ .}}{\text{unit } given}$$

The *given* measurement that is written after the $=$ sign will be the circled single unit listed in the DATA.

To convert to the WANTED unit, use the equalities in the DATA (and other fundamental equalities, such as metric prefix definitions, if needed).

In your notebook, finish the problem that you started above by adding the SOLVE step.

TRY IT

Q. If a car's speed is 55 miles/hour, how many minutes are needed to travel 85 miles?

Answer:

Your paper should look like this.

WANTED: ? min =

DATA: 55 mi. = 1 hr

　　　　　　(85 mi.)

SOLVE: $? \text{ min} = 85 \text{ mi.} \cdot \dfrac{1 \text{ hr}}{55 \text{ mi.}} \cdot \dfrac{60 \text{ min}}{1 \text{ hr}} = \mathbf{93 \text{ min}}$

You can solve simple problems without listing WANTED, DATA, SOLVE, but this three-part method works for *all* problems. It works especially well for the complex problems that soon you will encounter. By using the same three steps for every problem, you will know what to do to solve *all* problems. That's the goal.

SUMMARY

The Three-Step Method to Simplify Problem Solving

1. **WANTED:**
   When reading a problem for the first time, ask *one* question: What will be the *unit* of the answer? Then, write "WANTED: ?," the *unit* the problem is asking for, and a *label* that describes what the unit is measuring. Then add an $=$ sign.
   Write WANTED ratio units as $\dfrac{x}{y}$ fractions, and single units as single units.

2. **DATA:**
   Read the problem a second time.
   • Every time you encounter a *number*, in the DATA section write the number and its unit. Add a label after the unit if possible, identifying what is being measured.

- Next, decide whether that number and unit are equal to another number and unit.

  If a problem WANTS a single unit, most often *one* measurement will be a single unit and the rest will be equalities. Circle the *single* unit in the DATA.

3. **SOLVE:**

   Start each calculation with an *equality*:

   $$? \text{ WANTED unit} = \text{number and unit } given$$

   If you WANT a single unit, substitute the WANTED and *given* into this format.

   $$? \text{ unit WANTED} = \text{number and unit } given \cdot \frac{\rule{3cm}{0.4pt}}{\text{unit } given}$$

   Then, using equalities, convert to the WANTED unit.

PRACTICE

Many science problems are constructed in the following format: "Equality, equality," then, "? WANTED unit = a *given* number and unit."

The problems below are in that format. Using the rules above, solve on these pages *or* by writing the WANTED, DATA, and SOLVE sections in your notebook.

If you get stuck, read part of the answer at the end of this chapter, adjust your work, and try again. Do problems 1 and 3, and problem 2 if you need more practice.

1. If 2.20 pounds = 1 kilogram, what is the mass in grams of 12 pounds?

   WANTED: ?   (Write the **unit** you are looking for.)

   DATA: (Write every number and unit in the problem here. If solving for a single unit, often *one* number and unit is unpaired, and the rest are in equalities. Circle the unpaired single unit.)

   SOLVE: (Start with "? unit WANTED = number and unit *given* · $\frac{\rule{2cm}{0.4pt}}{\text{unit } given}$")

   ?

2. If there are 1.61 kilometers/mile, and one mile is 5,280 feet, how many feet are in 0.500 kilometers?

   WANTED: ?

   DATA:

   SOLVE:

   ?

(continued)

3. A bottle of drinking water is labeled "12 fl. oz. (355 mL)." What is the mass in centi-
    grams of 0.55 fluid ounces of the H_2O? (Assume a water density of 1.00 g/mL.)

    WANTED:

    DATA:

    SOLVE:

Lesson 5.4   Finding the *Given*

Single-Unit *Givens*

When solving for single units, the *given* quantity is not always clear.

Example: A student needs special postage stamps. The stamps are sold six per
sheet, each stamp booklet has three sheets, 420 stamps are needed, and the cost is
$43.20 per five booklets. What is the cost of the stamps?

Among all those numbers, which is the *given* that is needed as the first term
when you SOLVE?

For a single-unit answer, finding the *given* is often a process of elimination. If all
of the numbers and units are paired into equalities except one, that one is your *given*.

► TRY IT

Q. In your notebook, write the WANTED and DATA sections for the stamps problem
above (don't SOLVE yet). Then check your work below.

 Answer:

Your paper should look like this.

WANTED: ? $ =   *or* ? dollars =   (You could also solve in cents.)

DATA: 1 sheet = 6 stamps

3 sheets = 1 booklet

(420 stamps)

$43.20 = 5 booklets

Because you are looking for a *single* unit, dollars, your data has one number and
unit that did not pair up in an equality: 420 stamps. That is your *given*.

To SOLVE, the rule is

If you WANT a single unit, *start* with a single unit as your *given*.

⬛▬▬▬▬▬➤ **TRY IT**

Q. Apply this rule, assume all of these numbers are exact, and SOLVE the problem.

STOP **Answer:**

SOLVE: If you WANT a single unit, start with a single unit.

? $ = 420 stamps . _____
                              stamps

Putting the *given* unit where it *must* be to cancel in the next conversion will help you to pick the DATA for and arrange the DATA in the next conversion.

STOP   If you needed that hint, adjust your work and then finish.

$$? \$ = 420 \text{ stamps} \cdot \frac{1 \text{ sheet}}{6 \text{ stamps}} \cdot \frac{1 \text{ booklet}}{3 \text{ sheets}} \cdot \frac{\$43.20}{5 \text{ booklets}} = \mathbf{\$201.60}$$

Some Chemistry Practice

Many of the initial problems studied in chemistry can be solved using the conversion factor methods you have learned above. One additional rule that will be helpful in chemistry calculations is

> In measurements of chemical substances, always write the *chemical formula* of the substance being measured *after* the number and unit.

Examples:

Write single-unit data as: 9.00 g $\mathbf{H_2O}$

Write equalities in the format: 40.0 g $\mathbf{NaOH}$ = 1 mole $\mathbf{NaOH}$

In all measurements of chemical substances, write *number, unit, formula.*

The problems below supply the chemistry DATA needed for the conversion factors. In upcoming chapters, you will learn how to write these conversions automatically even when the problem does not supply them. That small amount of additional information will be all that you need to solve most initial chemistry calculations.

PRACTICE   A

Check (✓) and solve two of these problems in your notebook now. Do the remaining problems in your next study session. For each problem, use the WANTED, DATA, SOLVE method. If you get stuck, peek at the answer and try again.

1. Water has a molar mass of 18.0 grams H_2O per mole H_2O. How many moles of H_2O are in 450 milligrams of H_2O?

2. If 1 mole of Cl_2 gas has a volume of 22.4 liters at STP, and the molar mass of chlorine gas (Cl_2) is 71.0 grams Cl_2 per mole Cl_2, what is the volume, in liters, of 28.4 grams of Cl_2 gas at STP?

(continued)

3. If 1 mole of H_2SO_4 = 98.1 grams of H_2SO_4 and it takes 2 moles of NaOH per 1 mole of H_2SO_4 for neutralization, how many liters of a solution (soln.) that is 0.240 moles NaOH/liter are needed to neutralize 58.9 grams of H_2SO_4?

Listing Conversions and Equalities

Which is the best way to write DATA pairs: as *equalities* or in the *fraction* form as conversion-factor ratios? Mathematically, either form may be used.

In DATA: the equalities 1.61 km = 1 mi. and 3.0×10^8 m = 1 s can be listed as

$$\frac{1.61 \text{ km}}{1 \text{ mi.}}, \frac{3.0 \times 10^8 \text{ m}}{1 \text{ s}}$$

In these lessons, we will generally write *equalities* in the DATA section. This will emphasize that when solving problems using conversions, you need to focus on the relationship between two quantities. However, listing the data in the fraction format is equally valid. Data may be portrayed both ways in science texts.

Why "Want a Single Unit, Start with a Single Unit?"

Mathematically, the order in which you multiply conversions does not matter. You could solve with your single unit *given* written anywhere on top in your chain of conversions.

However, if you start with a *ratio* as your *given* when solving for a single unit, there is a 50% chance of starting with a ratio that is inverted. If this happens, the units will never cancel correctly, and you would eventually be forced to start the conversions over. *Starting* with the single unit is a method that automatically arranges your conversions "right side up."

PRACTICE B

The conversion factor methods learned in chemistry can be applied in other courses and careers. Problem 1 below is a type of calculation that you may encounter if managing a business or doing volunteer work. Problem 2 may be encountered in communications, physics, or engineering.

Use your WANTED, DATA, SOLVE method. On each of these, *before* you do the math, double-check each conversion, top and bottom, to make sure it is legal.

To be "quiz ready" for chemistry, be sure to complete problem 3.

1. You want to mail a large number of newsletters. The cost is 18.5 cents each at special non-profit rates. On the post office scale, the weight of exactly 12 newsletters is 10.2 ounces. The entire mailing weighs 125 pounds. There are exactly 16 ounces (oz.) in a pound (lb.).

   a. How many newsletters are being mailed?

   b. What is the cost of the mailing in dollars?

2. If the distance from an antenna on Earth to a geosynchronous communications satellite is 22,300 miles, given that there are 1.61 kilometers per mile and radio waves travel at the speed of light (3.0×10^8 m/s), how many seconds does it take for a signal from the antenna to reach the satellite?

3. On the following table, fill in the names and symbols for the atoms in the first three *rows* and the first two and last two *columns*.

PERIODIC TABLE

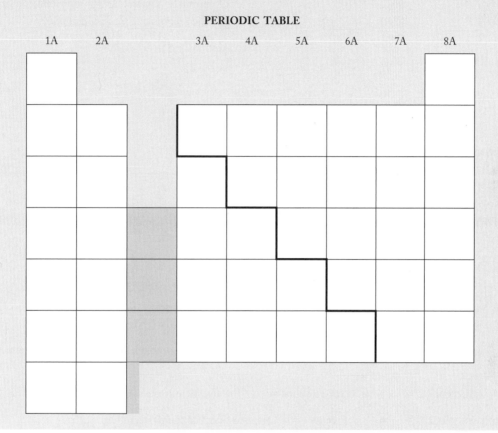

SUMMARY

1. To solve word problems, *get rid of the words.*

2. Organize your work into three parts: WANTED, DATA, and SOLVE.

3. First, under WANTED, write the unit you are looking for. As a part of the unit, include a label that describes what the unit is measuring.

4. If a *ratio* unit is WANTED, write the unit as a fraction with a top and a bottom.

5. Under DATA, to solve with conversions,
   - Write every number in the problem. Attach the units to the numbers. If the problem involves more than one substance, add a label after the unit that identifies the substance being measured.
   - If numbers and units are paired with other numbers and units, write the DATA as equalities.
   - Write *per* or a slash (/) in the data as $=$. Treat "$\cdot$ unit$^{-\text{number}}$" as "/ unit ^{number}." If no number is given after the *per* or /, write $= 1$.

(continued)

> • Write as equalities two different measurements of the same entity, or any units and labels that are equivalent or mathematically related in the problem.
>
> 6. To SOLVE, start each calculation with an *equality*:
>
> $$? \text{ WANTED unit } = \text{ number } given \text{ unit}$$
>
> If you WANT a single unit, start with a single number and unit as your *given*, use the format of the *single-unit starting template*
>
> $$? \text{ unit WANTED } = \text{ number and unit } given \cdot \underline{\qquad\qquad\qquad}$$
> $$\text{unit } given$$
>
> and chain conversions to solve.

ANSWERS

Lesson 5.1

Practice

1. Write WANTED: **? grams =**   (This is a ratio unit. Any unit that is in the form "unit *x* / unit *y*" is a ratio unit.)
       mole

2. Write WANTED: **? min =**   (This problem is asking for a single unit. *If* the problem asked for minutes per one mile, that would be a ratio unit, but minutes per *15* miles is asking for a single unit.)

3. In this problem, no unit is specified. However, because the data are in miles and hours, the easiest measure of speed is miles per hour, written

    WANTED: **? mi. =**   which is a familiar unit of speed. This problem is asking for a ratio unit.
         hr

4. WANTED: **? $ =**   *or* WANTED: **? dollars =**   (The answer unit is a single unit.)

5. WANTED: **? $ =**   *or* **? cents =**   (The cost per *one* stamp is a ratio unit.)
        stamp   **stamp**

Lesson 5.2

Practice   Terms that are equal may always be written in the reverse order. If there are two different entities in a problem, attach labels to the units that identify which entity the number and unit are measuring.

1a. 55 mi. = 1 hr (rule 4a)

1b. 0.50 L = 16.9 fl. oz. (rule 4c)

1c. 1 student = 19 pages (rule 5)

1d. 36 g ice = 2,880 calories heat (rule 4d: Equivalent)

1e. 18.0 g H_2O = 1 mole H_2O (rule 4b)

1f. 2.5 mg aspirin = 1 kg of body mass (rule 4a)

1g. 0.24 g NaOH = 250 mL of soln. (rule 4c)

2. Problem 1. DATA: 1.12 L gas STP = 3.55 g   (Two measures of the same gas.)

    Problem 2. DATA: 25 mi. = 1 hr   (Write / as = 1)

    Problem 3. DATA: 270 mi. = 6 hr   (Two measures of the same trip.)

    Problem 4. DATA: 6 stamps = 1 sheet

                  1 booklet = 3 sheets

                  $14.40 = 1 booklet

Lesson 5.3

Practice   1. WANTED: ? g =

   DATA: 2.20 lb. = 1 kg

   $\boxed{12 \text{ lb.}}$ (circled)

   SOLVE: $? \text{ g} = 12 \text{ lb.} \cdot \dfrac{1 \text{ kg}}{2.20 \text{ lb.}} \cdot \dfrac{10^3 \text{ g}}{1 \text{ kg}} = \dfrac{12 \cdot 10^3}{2.2} \text{ g} = \mathbf{5.5 \times 10^3 \text{ g}}$

   A single unit is WANTED, and the DATA has one single unit.
   Note that the SOLVE step begins with "How many grams equal 12 pounds?"
   Fundamental conversions such as kilograms to grams need not be written in your DATA section, but they will often be needed to solve.

2. WANTED: ? ft. =

   DATA: 1.61 km = 1 mi.

   1 mi. = 5,280 ft.

   $\boxed{0.500 \text{ km}}$ (circled)

   SOLVE: $? \textbf{ ft.} = \boxed{0.500 \text{ km}} \cdot \dfrac{1 \text{ mi.}}{1.61 \text{ km}} \cdot \dfrac{5,280 \textbf{ ft.}}{1 \text{ mi.}} = \dfrac{0.500 \cdot 5,280}{1.61} \text{ ft.} = \mathbf{1,640 \text{ ft.}}$

3. WANTED: ? cg =

   DATA: 12 fl. oz. = 355 mL

   $\boxed{0.55 \text{ fl. oz.}}$ (circled)

   $1.00 \text{ g } H_2O(\ell) = 1 \text{ mL } H_2O(\ell)$

   If needed, adjust your work and then complete the problem.

   SOLVE: $? \textbf{ cg} = 0.55 \text{ fl. oz.} \cdot \dfrac{355 \text{ mL}}{12 \text{ fl. oz.}} \cdot \dfrac{1.00 \text{ g } H_2O(\ell)}{1 \text{ mL } H_2O(\ell)} \cdot \dfrac{1 \text{ cg}}{10^{-2} \text{ g}} = \mathbf{1,600 \text{ cg}}$

Lesson 5.4

Practice A   1. WANTED: **? moles H_2O** =

   DATA: $18.0 \text{ g } H_2O = 1 \text{ mole } H_2O$

   $\boxed{450 \text{ mg } H_2O}$ (circled)

   SOLVE: $? \textbf{ moles } H_2O = 450 \text{ mg } H_2O \cdot \dfrac{10^{-3} \text{ g}}{1 \textbf{ mg}} \cdot \dfrac{1 \textbf{ mole } H_2O}{18.0 \text{ g } H_2O} = \mathbf{2.5 \times 10^{-2} \text{ moles } H_2O}$

   Write chemistry data in three parts: Number, unit, formula. Writing complete labels will make complex problems easier to solve. 450 has two *s.f.*

2. WANTED: ? L Cl_2 (Always attach chemical formulas, if known, to units.)

   DATA: 1 mole gas = 22.4 L gas

   $71.0 \text{ g } Cl_2 = 1 \text{ mole } Cl_2$

   $\boxed{28.4 \text{ g } Cl_2}$ (circled)

   SOLVE: $? \textbf{ L } Cl_2 = 28.4 \text{ g } Cl_2 \cdot \dfrac{1 \text{ mole } Cl_2}{71.0 \text{ g } Cl_2} \cdot \dfrac{22.4 \text{ L } Cl_2}{1 \text{ mole } Cl_2} = \mathbf{8.96 \text{ L } Cl_2}$

3. WANTED: ? L NaOH solution

    DATA: 1 mole H_2SO_4 = 98.1 g H_2SO_4

    2 moles NaOH = 1 mole H_2SO_4 (Assume whole numbers are exact.)

    0.240 moles NaOH = 1 L NaOH solution

    $\boxed{58.9 \text{ g } H_2SO_4}$

    SOLVE:

    ? L NaOH = 58.9 g $H_2SO_4 \cdot \dfrac{1 \text{ mole } H_2SO_4}{98.1 \text{ g } H_2SO_4} \cdot \dfrac{2 \text{ moles NaOH}}{1 \text{ mole } H_2SO_4} \cdot \dfrac{1 \text{ L NaOH soln.}}{0.240 \text{ moles NaOH}}$ = **5.00 L NaOH soln**.

Practice B   1a. WANTED: ? newsletters

    DATA: 18.5 cents = 1 newsletter

    12 exact newsletters = 10.2 oz.

    16 oz. = 1 lb.   (This is a definition that has infinite *s.f.*)

    $\boxed{125 \text{ lb.}}$

    SOLVE: ? newsletters = 125 lb. $\cdot \dfrac{16 \text{ oz.}}{1 \text{ lb.}} \cdot \dfrac{12 \text{ newsletters}}{10.2 \text{ oz.}}$ = **2,350 newsletters**

1b. WANTED: ? dollars

    (Strategy: Because you want a single unit, you can start over from your single *given* unit [125 lb.], repeat the conversions above, then add two more. *Or* you can start from your single unit answer in part a, and solve using the two additional conversions. In problems with multiple parts, to solve for a later part, using an answer from a previous part often saves time.)

    DATA: See part a.

    SOLVE: **? dollars** = 2,350 newsletters $\cdot \dfrac{18.5 \text{ cents}}{1 \text{ newsletters}} \cdot \dfrac{1 \text{ dollar}}{100 \text{ cents}}$ = **$435**

2. WANTED: ? s =

    DATA: $\boxed{22{,}300 \text{ mi.}}$

    1.61 km = 1 mi.

    3.0×10^8 m = 1 s

    SOLVE: ? s = 22,300 mi. $\cdot \dfrac{1.61 \text{ km}}{1 \text{ mi.}} \cdot \dfrac{10^3 \text{ m}}{1 \text{ km}} \cdot \dfrac{1 \text{ } s}{3.0 \times 10^8 \text{ m}}$ = $\dfrac{22{,}300 \cdot 1.61 \cdot 10^3}{3.0 \times 10^8}$ s = **0.12 s**

(This means that the time up *and* back for the signal is 0.24 seconds. You may have noticed this one-quarter-second delay during some live broadcasts that bounce video signals off satellites but use faster landlines for audio, or during overseas communications routed through satellites.)

6

Atoms, Ions, and Periodicity

For this chapter, you will need an alphabetical list of the atoms (provided at the back of the book) and a periodic table that closely resembles the type of table you will be allowed to consult during quizzes and tests in your course.

Lesson 6.1   Atoms

Terms and Definitions

The precise definition for some of the fundamental particles in chemistry is a matter of occasional debate, but the following simplified and somewhat arbitrary definitions will provide us with a starting point for discussing atoms.

1. **Matter.** Chemistry is primarily concerned with the measurement and description of the properties of matter and energy. Matter is anything that has mass and volume. In planetary environments, nearly all matter is composed of extremely small particles called atoms.

2. **Atoms.** In these lessons, we will define an atom as a particle with a single nucleus, plus the electrons that surround the nucleus.

    There are 91 different kinds of atoms that are found in Earth's crust. More than 20 additional atoms have been synthesized by scientists in nuclear reactors. All of the millions of different substances on Earth consist of only about 100 different kinds of atoms. It is how the atoms are grouped and arranged in space that results in so many different substances.

    A list of the atoms is found at the end of these lessons. Each atom is represented by a one- or two-letter **symbol**. The first letter of the symbol is always capitalized. The second letter, if any, is always lowercase.

3. **Electrical charges.** Some particles have a property known as electric charge.

    There are two types of charges: positive and negative. Particles with like electrical charges repel. Unlike charges attract.

4. **Atomic structure.** Atoms can be described as combinations of three **subatomic particles**: protons, neutrons, and electrons.

    a. **Protons** (abbreviated **p⁺**)
       - *Each proton has a **1+** electrical charge (one unit of positive charge).*
       - Protons have a mass of about 1.0 amu (**atomic mass units**).
       - Protons are found in the center of the atom, called the **nucleus**. The nucleus is extremely small and occupies very little volume in the atom.
       - *The number of protons in an atom is defined as the **atomic number** (symbol Z) of the atom.*
       - *The number of protons determines the **name** (and thus the symbol) of the atom.*
       - *The number of protons in an atom is never changed by chemical reactions.*

b. **Neutrons** (abbreviated n^0)
   * *A neutron has an electrical charge of zero.*
   * A neutron has about the same mass as a proton, 1.0 amu.
   * *Neutrons are located in the nucleus of an atom, along with the protons.*
   * Neutrons are thought to act as the glue of the nucleus: particles that play a role in keeping the repelling protons from flying apart.
   * Neutrons, like protons, are never gained or lost in chemical reactions.
   * The neutrons in an atom in *most* cases have very little influence on the chemical behavior of the atom.

c. **Electrons** (abbreviated e^-)
   * *Each electron has a **1−** electrical charge: one unit of negative charge, equal in magnitude but opposite the proton's charge.*
   * Electrons have very little mass, weighing about 0.00055 amu.
   * Electrons are found outside the nucleus of an atom, in regions of space called **orbitals**.
   * Nearly all of the *volume* of an atom is due to the space occupied by the electrons around the nucleus.
   * *Electrons are the only subatomic particles that can be gained or lost during chemical reactions.*

   Each of the above points will be addressed at various times in your course. For this lesson, the points in italics above must be *memorized*.

5. **Neutral atoms.** If an atom has an equal number of protons and electrons, the balance between positive and negative charges gives the atom a *net* charge of zero. The charges are said to "cancel" to produce an overall **electrically neutral** atom.

PRACTICE A

Commit to memory the eight italic points in section 4 above. Then, for the problems below, consult the alphabetical list of atoms at the back of the book.

1. Write the symbols for these atoms.

   a. Carbon   b. Oxygen   c. Osmium   d. Tungsten

2. Name the atoms represented by these symbols.

   a. K   b. F

   c. Fe   d. Pb

3. Fill in the blanks for the subatomic particles.

	Charge	Mass	Location
Proton			
Neutron			
Electron			

(continued)

4. Assume each atom below is electrically neutral. Fill in the blanks.

Atom Name	Symbol	Protons	Electrons	Atomic Number
Sodium				
	N			
		6		
			82	
				9

More Terms and Definitions

6. **Chemical reactions.** By definition, a chemical reaction cannot create or destroy atoms. A reaction cannot change an atom from one kind to another. However, during a chemical reaction, how atoms are bonded and arranged changes, and this alters the identity and behaviors of the substances that react.

7. **Physical changes.** When a substance undergoes a **physical change**, it does not change its identity. Melting ice to water is a physical change, because both ice and liquid water are composed of particles that internally have the same atoms in the same geometry. A physical change is not considered to be a chemical reaction.

8. **Ions.** During chemical reactions, the number of protons and neutrons in an atom never changes, but atoms can gain or lose one or more electrons. Any particle (atom or group of bonded atoms) that does not have an equal number of protons and electrons is termed an **ion**, which is a particle with a net electrical charge.
    *   Neutral particles that *lose electrons* become **positive ions**.
    *   Neutral particles that *gain electrons* become **negative ions**.
        An ion is *not* the same as a neutral particle with the same atom or atoms. The ion has a different number of electrons and different chemical behavior.
        The symbol or chemical formula for a particle that is not electrically neutral places the value of the net charge as a superscript to the right of the symbol.

9. **Examples of atoms and ions.**

    a. All atoms with a nucleus containing 16 protons are examples of **sulfur** (symbol S).
        An electrically neutral atom of sulfur has 16 protons and 16 electrons. The positive and negative charges balance to give a net charge of zero. The symbol for the neutral sulfur atom is written as S. The symbol S^0 may also be written to emphasize that the sulfur atom has a neutral charge.
        In substances, an ion of sulfur *may* be found that contains 16 protons and 18 electrons. The 16 protons cancel the charge of 16 electrons, leaving two un-cancelled electrons and an overall charge of **2 −**. The symbol for this particle is S^{2-} and it is named a **sulfide ion**.

b. All atoms with **19** protons are named **potassium** (symbol K). Potassium is a soft metal when it is in its elemental state. However, neutral potassium metal atoms react with many substances, including water and water vapor, and each potassium atom loses one electron in each of these reactions.

Because of this reactivity, in substances found in Earth's crust, potassium is always found as an *ion* with 18 electrons. The 18 electrons balance the charge of 18 protons. This leaves one positive charge uncancelled, so the ion has a net charge of **1+**. This particle is named **potassium ion** and its symbol is K^+. For the charges on ions, if no number in front of the sign is shown, a 1 is understood.

c. All atoms with 88 protons are named **radium** (symbol Ra).

Ra^{2+} ions must have two more positive protons than negative electrons.

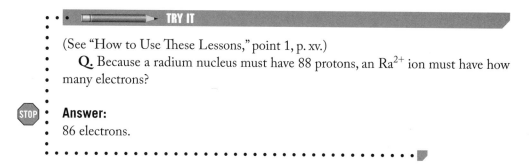

TRY IT

(See "How to Use These Lessons," point 1, p. xv.)

Q. Because a radium nucleus must have 88 protons, an Ra^{2+} ion must have how many electrons?

STOP

Answer:

86 electrons.

PRACTICE   B

For the problems below, use the alphabetical list of atoms at the back of this book.

1. Calcium has atomic number 20.

   a. A neutral Ca atom has how many protons?   How many electrons?

   b. How many protons are found in a Ca^{2+} ion?   How many electrons?

2. In their nucleus, during chemical reactions, atoms always keep a constant number of _____ , which have a positive charge. Atoms take on a charge and become ions by gaining or losing _____ , which have a _____ charge.

3. In terms of subatomic particles, an atom that is a positive ion will always have more _____ than _____ .

4. For these symbols, write the atom names from memory.

   a. Sr =   b. Si =   c. P =

5. For these atom names, write their symbols from memory.

   a. Bromine =   b. Boron =   c. Barium =

(continued)

6. For the particles below, fill in the blanks.

Symbol	Protons	Electrons
O		
O^{2-}		
Mg^{2+}		
	13	10
	79	79
	1	0
	35	36
	34	36
I^-		

Lesson 6.2   The Nucleus, Isotopes, and Atomic Mass

The Nucleus

At the center of an atom is the nucleus. The nucleus contains all of the protons and neutrons in the atom.

The nucleus is very small, with a diameter that is roughly 100,000 times smaller than the effective diameter of most atoms, yet the nucleus contains all of the atom's positive charge, and nearly all of its mass.

Because the nucleus contains nearly all of the atom's mass in a tiny volume, it is extremely dense. Outside of the nucleus, nearly all of the volume of an atom is occupied by its electrons. Because electrons have low mass but occupy a large volume compared to the nucleus, the region occupied by the electrons has a very low density. In terms of mass, an atom is mostly empty space.

However, an electron has a charge that is equal in magnitude (though opposite) to that of the much more massive proton.

Types of Nuclei Only certain combinations of protons and neutrons form a nucleus that is stable. In a nuclear reaction (such as occurs in radioactive decay or in a nuclear reactor), if a combination of protons and neutrons is formed that is unstable, the nucleus will decay.

The combinations of protons and neutrons in nuclei can be divided into three types.

- **Stable:** Stable nuclei contain combinations of protons and neutrons that do not change in a planetary environment such as Earth over many billions of years.
- **Radioactive:** Radioactive nuclei are *somewhat* stable. Once formed, they can exist for a time (from a few seconds to several billion years), but they fall apart (**decay**) at a constant, characteristic rate.
- **Unstable:** In nuclear reactions, if combinations of protons and neutrons form that are unstable, they decay within a few seconds.

Nuclei that exist in Earth's crust include all of the stable nuclei plus some radio-active nuclei.

For all atoms that have between 1 and 82 protons (except for technetium, with 43 protons and promethium, with 61), at least one stable nucleus exists. Atoms with 83–92 protons are also found in Earth's crust, but all are radioactive. Atoms with 93 or more protons exist on Earth only when they are created in nuclear reactions (such as in nuclear reactors).

Radioactive atoms comprise a very small percentage of the matter found on Earth. More than 99.99% of Earth's atoms have stable nuclei that have not changed since the atoms came together to form Earth billions of years ago.

Terminology Protons and neutrons are termed **nucleons** because they are found in the nucleus. A combination of a certain number of protons and neutrons is called a **nuclide**. A group of nuclides that have the same number of protons (so they are all the same kind of atom) but differing numbers of neutrons are called the **isotopes** of the atom.

Stable Nuclei Some atoms have only one stable nuclide; other atoms have as many as 10 stable isotopes.

Example: All atoms with 17 protons are called chlorine. Only two chlorine nuclei are stable: Those with

- 17 protons and **18** neutrons and
- 17 protons and **20** neutrons

Nuclei that have 17 protons and *other* numbers of neutrons can be made in nuclear reactions, but in all of those combinations, within a few seconds, the nucleus decays by emitting a particle from the nucleus.

Nuclide Symbols Each nuclide can be assigned a **mass number** that is the *sum* of its number of protons and neutrons.

$$\text{Mass number of a nucleus} = p^+ + n^0 = \text{protons} + \text{neutrons}$$

Example: A nucleus with $2\,p^+$ and $2\,n^0$ is helium with a mass number of **4**. A nuclide can be identified in two ways,

- Either by its number of protons and number of neutrons,
- Or by its **nuclide symbol** (also termed its **isotope symbol**).

A nuclide symbol has two required parts: the *atom symbol* and the *mass number*. The mass number is written as a superscript in front of the atom symbol.

Example: The two stable isotopes of chlorine can be represented as

- 17 protons + 18 neutrons *or as* ^{35}Cl (a nuclide named chlorine-35) and
- 17 protons + 20 neutrons *or as* ^{37}Cl (named chlorine-37)

Knowing one representation for the composition of a nucleus, you need to be able to write the other.

Using the table of atoms at the back of this book, write answers to these questions.

TRY IT

Q. A nuclide with 6 protons and 8 neutrons would have what nuclide symbol?

Answer:
Atoms with 6 protons are always named carbon, symbol **C**. The mass number of this nuclide is 6 protons + 8 neutrons = 14. This isotope of carbon, used in "radiocarbon dating," is named carbon-14 and its symbol is written ^{14}C.

Q. How many protons and neutrons are found in ^{20}Ne?

Answer:
All atoms called neon contain 10 protons. The mass number 20 is the total number of protons *plus* neutrons, so neon-20 contains **10 protons and 10 neutrons**.

Nuclide symbols may also include the **nuclear charge** of the particle written in front of and below the atom symbol. Including both the mass number and the nuclear charge is called the *A–Z* **notation** for a nuclide, illustrated below. *A* is the symbol for mass number and *Z* is the symbol for nuclear charge. In atoms, *Z* is also the number of *protons* in the nucleus.

$$^{37}_{17}\text{Cl}$$

Nuclide symbols can also be used to identify subatomic particles (particles smaller than atoms), and in those cases the nuclear charge may be zero or negative. Including the *Z* values is helpful when balancing nuclear reactions (a future topic), but the *Z* values are not required to identify an atom, because the *Z* repeats what the symbol identifies: the number of protons in the nucleus of the atom.

PRACTICE   A

Consult a table of atoms or a periodic table to fill in the blanks below.

1.

Protons	Neutrons	Atomic Number	Mass Number	Nuclide Symbol	Nuclide Name
	6	6			
7	7				
					Iodine-131
				^{235}U	
		2	4		

2. Which nuclides in problem 1 must be radioactive?

The Mass of Nuclides

The mass of a single nuclide is usually measured in **atomic mass units**, abbreviated **amu.**

Protons and neutrons have essentially the same mass, and both are much heavier than electrons. The mass of

- A proton is **1.0** amu
- A neutron is **1.0** amu
- An electron is 0.00055 amu

Based on those masses, you might expect that the mass of a ^{35}Cl atom would be just over 35.0 amu, because it is composed of 17 protons, 18 neutrons, and some very light electrons. In fact, for atoms of ^{35}Cl, the actual mass is 34.97 amu, slightly *lighter* than the combined mass of its protons, neutrons, and electrons.

Why do the masses of the three subatomic particles *not* add exactly to the mass of the atom? When protons and neutrons combine to form nuclei, a small amount of mass is either converted to, or created from, energy. This change is the relationship postulated by Albert Einstein:

Energy gained or lost = mass lost or gained times the speed of light squared. In equation form, this is written:

$$E = mc^2$$

In nuclear reactions, if a small amount of mass is lost, a very large amount of energy is created. In forming nuclei, however, because the gain or loss in mass is relatively small, the mass of a nuclide or atom in atomic mass units will *approximately* (but not exactly) *equal* its mass number.

The sum of the protons and neutrons of a nuclide approximately equals its mass in *amu*.

The Average Mass of Atoms (Atomic Mass)

For *most* atoms on Earth (those not formed by radioactive decay), one kind of atom may have several stable isotopes, but in all samples of that atom found in substances on Earth, the percentage of each isotope will be the same. Most atoms will therefore have the same *average mass* in any matter found on Earth.

This average mass of an atom, called its **atomic mass**, can be calculated from the **weighted average** of the mass of its isotopes. This means that for any mixture of isotopes, the average mass of the atoms can be found from the following equation:

Average mass = (Fraction that is isotope one) (mass of isotope one)
+ (Fraction that is isotope two) (mass of isotope two) + ...

In this equation, *fraction* means the *percentage*/100%
Example: If the percentage is 50.%, the fraction is 0.50
The sum of the fractions must be 1.00
Apply the equation to the following problem.

TRY IT

Q. In all samples of chlorine, 75.8% of the atoms have an atomic mass of 35.0 amu, and 24.2% have an atomic mass of 37.0 amu. Find the average mass of these particles.

Answer:

Average mass = (0.758) (35.0 amu) + (0.242) (37.0 amu) = **35.5 amu**

No single atom of chlorine will have this average mass, but in visible amounts of substances containing chlorine, all of the chlorine atoms can be assumed to have this *average* mass. Use of this average mass (the atomic mass) will simplify chemistry calculations involving mass.

Isotopes and Chemistry

The rules and the reactions for "standard chemistry" are very different from those of *nuclear* chemistry. For example,

- Chemical reactions can release substantial amounts of energy, such as seen in the burning of fuels or in conventional explosives. Nuclear reactions, however, can involve *much* larger amounts of energy, as in stars or nuclear weapons.
- An important rule in chemical reactions is that atoms can neither be created nor destroyed. In nuclear reactions, atoms are often created and destroyed.

Because the rules are very different, a clear distinction must be made between *chemistry* and *nuclear chemistry*. By convention, the rules that are cited as part of "chemistry" refer to processes that do *not* involve changes in nuclei (unless *nuclear* chemistry is specified). Processes that change the composition of a nucleus are termed *nuclear* reactions, which by definition are *not* chemical reactions.

The good news is that, except for experiments in nuclear chemistry, because all isotopes of an atom nearly always have the same chemical behavior, and because in visible amounts of substances, nearly all atoms have the same average mass, we can ignore the fact that atoms have isotopes as we investigate nearly all *chemical* reactions and processes.

We will return to the differences among isotopes when we consider nuclear chemistry, which includes reactions such as radioactive decay, fission, and fusion.

PRACTICE B

Fill in the blanks below.

Protons	Neutrons	Electrons	Atomic Number	Mass Number	Nuclide Symbol	Ion Symbol
	144	88	90			
	148					Pu^{2+}
		78			^{206}Pb	
		0				H^+
					^{3}H	H^-
		36			^{90}Sr	
11		10		23		
15	16	18				

Lesson 6.3   Elements, Compounds, and Formulas

> **PRETEST** Use the list of atoms at the back of the book. If you have a perfect Pretest score, skip to Lesson 6.4. Answers are at the end of the chapter.
>
> 1. In this list: A. H_2O; B. Cl_2; C. Au; D. S_8; E. CO_2; F. Co; G. H_2SO_4
>
>   a. Which formulas represent elements?
>
>   b. Which formulas represent a substance without ionic or covalent bonds?
>
>   c. Which formulas represent substances that are diatomic?
>
> 2. Write the number of oxygen atoms in each of these particles.
>
>   a. $Co(OH)_2$ b. CH_3COOH c. $Al_2(SO_4)_3$
>
> 3. Write the total number of atoms in each of the particles in question 2.

Substances

The definitions below are general and highly simplified, but they will give us a starting point for discussing how atoms may combine to yield different substances.

1. A **substance** is matter that contains *one* kind of chemical particle: All of its electrically neutral units are composed of the same number and kind of atoms, chemically bonded in the same manner and geometry. There are two types of substances: **elements** and **compounds**. **Chemical formulas** can be written to represent a substance.

   Substances have *characteristic* properties: Their melting points, color, and densities are some of the properties that will be the same for a substance no matter what steps are taken to form the substance. These properties can help in identifying the substance.

   A **mixture** is a combination of two or more substances.

2. In a substance, if the smallest particles that are stable independent units are neutral particles with two or more atoms, the particles are called **molecules**. If a substance consists of charged particles, the particles are called ions, and the smallest electrically neutral combination of ions is called a **formula unit**.

3. **Elements** are substances that contain only one kind of atom. Each atom has an **elemental state**. Its elemental state is the substance formula and phase (solid, liquid, or gas) that is the most stable form of the element that exists at room temperature and pressure.

   The basic particles for some elements, termed the **monatomic elements**, are individual atoms. The chemical formulas for monatomic elements are written as one instance of the atom's formula, reflecting the fact that the basic unit is a single atom that is not bonded to other atoms.

   Example: The basic particles of the **noble gases** (helium, neon, argon, krypton, xenon, and radon) are single atoms. Therefore, the chemical formulas for these elements are written as He for helium, Ne for neon, etc.

At typical room temperature and pressure, some substances that are elements consist of two or more atoms of the same kind that are chemically bonded to form a larger unit. Cl_2 and S_8 are formulas for elements because they are substances that contain only one kind of atom, and those formulas represent the most stable form in which a collection of those atoms will exist at room temperature and pressure.

4. **Bonds** are forces that hold particles together. Molecules of the **diatomic elements** consist of two atoms (*di-* means two), and their chemical formulas reflect the fact that each unit contains two atoms.

   Example: The elemental forms of oxygen, nitrogen, and chlorine are all diatomic. The chemical formula for chlorine is Cl_2, nitrogen is N_2, and oxygen is O_2.

   In chemical formulas, a **subscript** is the number written after a symbol that represents the number of that kind of atom or ion that is bonded within a particle.

   Polyatomic elements are molecules that contain two or more neutral atoms, but only one kind of atom.

   Example: The elemental formula for sulfur is S_8, indicating that it exists as eight atoms bonded together.

   In their elemental state, more than 70% of the atoms that can be found in Earth's crust are metals. Metals have a structure that is more complex than the monatomic or polyatomic elements. However, a metal in its elemental state is represented by formula written as a single atom, such as Ag for silver, and Al for aluminum.

5. A **compound** is a substance that consists of two or more different atoms that are chemically bonded. While there are just over 100 elements, there are millions of known compounds. In a given compound, the ratio of the atoms is always the same and is shown by their formulas. H_2O, NaCl, and H_2SO_4 are all formulas for compounds because they contain two or more different kinds of atoms. Compounds can be classified as either ionic or covalent, depending on their bonds.

6. The basic particles for **covalent compounds** (also known as molecular compounds) are molecules. Molecules are held together by **covalent bonds** in which electrons are shared between two neighboring atoms. Covalent bonds can be single bonds (involving two shared electrons), double bonds (four shared electrons), or triple bonds (six shared electrons). Covalent bonds hold atoms at predictable angles within a molecule.

7. **Molecular formulas** use atomic symbols and subscripts to represent the number and kind of atoms covalently bonded together to form a single molecule.
   • Water is a molecule that consists of two hydrogen atoms and one oxygen atom, represented by the molecular formula H_2O.
   • In chemical formulas, when there is no subscript written after a symbol, the subscript is understood to be *one*.
   • Carbon dioxide is composed of molecules that each consist of two oxygen atoms and one carbon atom. The molecular formula is CO_2.

PRACTICE   A

Use the atoms table at the back of the book to answer these questions.

1. Label the following formulas as representing elements or compounds.

   a. Ne

   b. H_2O

   c. NaCl

   d. S_8

   e. $C_6H_{12}O_6$

2. Which of these substances contain chemical bonds?

   a. H_2

   b. CO

   c. NH_3

   d. He

3. In problems 1 and 2, which formulas represent

   a. Diatomic elements?

   b. Monatomic elements?

   c. Four atoms?

Representing Substances

8. **Structural formulas** can be used to represent chemical particles that are held together by covalent bonds. These formulas show each of the atoms present along with information about their positions within the particle.

   Below is a structural formula for water. It shows that the oxygen atom is found in the middle of the molecule, and that water has two directional covalent single bonds and a *bent* shape.

   The structural formula for carbon dioxide, CO_2, is **O═C═O**. Carbon dioxide has two double bonds, and the molecule is linear in shape with the carbon atom in the middle.

   We generally write structural formulas when knowing the *shape* of the molecule is important, but we write the more compact molecular formulas when it is not.

9. Often, chemical formulas are written as a mixture of structural and molecular formulas.

   Example: The formula for ethyl alcohol can be written as C_2H_6O or as CH_3CH_2OH. The shorter formula, however, is also the molecular formula of dimethyl ether, which is usually written CH_3OCH_3 to show that the O is found in the middle in the ether, rather than toward one end as in the alcohol.

   Ethyl alcohol and dimethyl ether have the same number and kind of atoms, but the different arrangement of the atoms give the molecules very different properties. To predict chemical behavior, we often need to know a formula with structural information. In such cases, we write the longer formulas like those above.

10. **Ionic compounds** are substances consisting of a collection of positive and negative **ions** (particles with a net electrical charge). Ions can be **monatomic** (single atoms with an unequal number of protons and electrons) or **polyatomic** (a group of covalently bonded atoms that have an unequal number of protons and electrons). An **ionic bond** is the electrostatic attraction between oppositely charged ions.

11. **Ionic formulas** represent the ratio and kind of ions present in an ionic compound. The ions in an ionic compound are always present in a ratio that guarantees overall electrical neutrality.

    A **formula unit** is defined as the smallest combination of ions for which the sum of the electrical charges is zero. Chemical formulas for ionic compounds show the atom ratios in a single neutral formula unit. Examples:

    - Table salt consists of a 1:1 ratio of positively charged sodium ions (formula Na^+) and negatively charged chloride ions (Cl^-). The formal name of table salt is sodium chloride, and its ionic formula is written as **NaCl**. The formula unit NaCl contains two ions.
    - Parentheses are used if a formula unit contains more than one polyatomic ion. For example, calcium phosphate is an ionic compound composed of three monatomic Ca^{2+} ions for every two polyatomic PO_4^{3-} ions. The ionic formula is **Ca$_3$(PO$_4$)$_2$**, and one formula unit represents a total of five ions and 13 atoms.
    - Copper(II) nitrate is an ionic compound composed of one monatomic Cu^{2+} ion for every two polyatomic NO_3^- ions. The ionic formula is written as **Cu(NO$_3$)$_2$** to indicate that the compound contains two NO_3^- ions.

12. When you *write* formulas, be careful to distinguish between upper- and lowercase letter combinations such as CS and Cs, Co and CO, NO and No. Examples:

    - Co(OH)$_2$ has one cobalt atom, two oxygen atoms, and two hydrogen atoms.
    - CH$_3$COOH has two carbon, four hydrogen, and two oxygen atoms.

To summarize, although molecules of covalent substances and formula units of ionic compounds have different types of bonds, all substance formulas refer to a single, overall electrically neutral unit of the substance.

PRACTICE   B

Use the table of atoms at the back of the book to answer these questions.

1. Write the number of oxygen atoms in each of these formulas.

    a. Al(OH)$_3$ b. C$_2$H$_5$COOH   c. Co$_3$(PO$_4$)$_2$

2. Write the total number of atoms in each of the formulas in question 1.

3. Try every third letter below. Need more practice? Do a few more. Name each atom, and write the total number of those atoms, in each of the following chemical formulas.

    a. HCOOH

    b. $CoSO_4$

    c. $No(NO_3)_3$

    d. $Ca_3(PO_4)_2$

    e. $(NH_4)_3PO_4$

    f. $Pb(C_2H_5)_4$

4. If you need additional practice, redo the Pretest at the beginning of Lesson 6.3.

Lesson 6.4   The Periodic Table

Patterns of Chemical Behavior

Learning the behavior of more than 100 different atoms would be a formidable task. Fortunately, the atoms can be organized into **families**. The chemical behavior of one atom in a family helps to predict the behavior of other atoms in the family.

The grouping of atoms into families results in the **periodic table**. To build the table, the atoms are arranged in *rows* across (also called **periods**) in order of the number of protons in each atom. This order usually, but not always, matches the order of the increasing atomic mass of the atoms.

At certain points, the chemical properties of the atoms begin to repeat, somewhat like the octaves on a musical scale.

In the periodic table, under most graphic designs, when a noble gas atom is reached, it marks the end of a row. The next atom, with one more proton, starts a new row of the table. This convention places the atoms into vertical **columns** (called **families** or **groups**) with the noble gases in the last column on the right.

Within each *column*, the atoms tend to have similar chemical behavior.

Some Families in the Periodic Table

The noble gases (He, Ne, Ar, Kr, Xe, Rn) are monatomic (composed of single atoms) as elements. They can be liquefied by lowering temperature and/or increasing pressure, but in their elemental state at *room* temperature and pressure, all are gases.

These atoms are termed noble because they are chemically "content" with their status as single atoms: These atoms rarely bond with other atoms or with each other.

Although the noble gases take part in very few chemical reactions, they are important in predicting chemical behavior. Other atoms tend to react in ways that give them the same number of electrons as the nearest noble gas. The outer electrons of atoms are often said to react to attain a "cloak of nobility."

The alkali metals (Li, Na, K, Rb, Cs, and Fr) are in **column one** (also called **group 1A** or **group 1**) of the periodic table, at the far left. As elements, all are soft, shiny metals that tend to react with many substances, including the water vapor present in air.

In chemical reactions, alkali metal atoms tend to *lose* an electron to become a **1+ ion**. This ion has the same number of electrons as the noble gas that has one fewer proton. Once an alkali metal atom forms a 1+ ion, it becomes quite stable. Most chemical reactions do not change its 1+ charge.

The halogens (F, Cl, Br, I, and At) are in tall **column 7** (group 7A or 17) just to the left of the noble gas column. As neutral elements at room temperature, halogen atoms are stable only when they are found in the diatomic molecules F_2, Cl_2, Br_2, I_2, and At_2.

Like alkali metals, the halogens are very reactive. In many reactions, a neutral halogen atom tends to *gain* one electron to become a **halide ion** with a **1−** charge. This ion has the same number of electrons as the noble gas just to the right in the periodic table.

Halogen atoms can also share electrons with neutral nonmetal atoms. Shared electrons result in a covalent bond. Including the shared electrons, each neutral halogen atom will tend to be surrounded by the same number of electrons as the nearest noble gas.

Hydrogen is often placed in column one of the table, and the reactions of hydrogen are often like those of the alkali metals. However, other hydrogen reactions are like those of the halogens. Hydrogen is probably best portrayed as a unique family of one that has characteristics of both alkali metals and halogens.

The main group elements are those found in the *tall* column blocks on both sides of the table. They are termed either groups 1, 2, and 13 to 18, or groups 1A, 2A, and 3A–8A, depending on the version of the periodic table that you are using.

The transition metals are in the "middle dip" of the periodic table, in groups 3–12 or the "B" groups. There are 10 atoms in each row of the transition metals.

The inner transition atoms include the 14 **lanthanides** (or **rare earth metals**) that begin with lanthanum in the sixth row and the 14 **actinides** that begin with actinium in the seventh row. These atoms are usually listed below the rest of the periodic table in order to display a table that fits easily on a chart or page.

Predicting Behavior

For the following columns of the periodic table, the column numbers, family names, and likely ion charges should be memorized.

Group	1A	2A	3B → 2B	3A	4A	5A	6A	7A	8A
Family Name	(Except for H) alkali metals	Alkaline earth metals	Transition metals					Halogens	Noble gases
Charge If Monatomic Ion	1+	2+	Varies	3+ (or 1+)		3−	2−	1−	None

Example: Cesium (Cs) is in column one of the periodic table. Based on this placement, we can predict that it will:

- Behave like other alkali metal atoms and
- Exist as a Cs^+ ion in compounds

PRACTICE **A**

Use a copy of the periodic table and your memorized knowledge about the table (first learn the rules, then do the practice) to answer these questions.

1. Describe the location in the periodic table of the

    a. Noble gases

    b. Alkali metals

    c. Halogens

    d. Transition metals

2. Add a charge to these symbols to show the ion that a single atom of these elements tends to form.

    a. Br   b. Ra   c. Cs   d. In   e. Te

Metals, Metalloids, and Nonmetals

The atoms in the periodic table can be divided into metals, metalloids (also called semimetals), and nonmetals.

Metalloids Many periodic tables include a thick line, like a staircase, as shown in the section of the periodic table below. This line separates the metal and nonmetal atoms.

The six atoms bordering the line in **bold** below are the **metalloids**. They have chemical behaviors that are in between that of the metals and the nonmetals.

				(H)	He
B	C	N	O	F	Ne
	Si	P	S	Cl	Ar
	Ge	**As**	Se	Br	Kr
		Sb	**Te**	I	Xe
			(Po)	At	Rn

Unless you are allowed to use a periodic table that has the staircase and identifies the metalloids on tests, you should memorize the location of the staircase and the six metalloids.

If you memorize how the staircase looks at boron (B), the rest of the staircase is easy.

Some textbooks include polonium (Po) as a seventh metalloid, others do not. The halogen astatine (At) is usually, but not always, considered to be a nonmetal rather than a metalloid. For simplicity, in these lessons we will consider all of the halogens to be nonmetals.

Nonmetals Below are the 18 nonmetals. The nonmetals must be *memorized*: H, C, N, O, P, S, Se, plus the five halogens and six noble gases.

Note the shape of their positions: The nonmetals are all to the right of the staircase and to the right of the metalloids. All atoms in the last two columns are nonmetals.

				(H)	He
C	N	O		F	Ne
	P	S		Cl	Ar
		Se		Br	Kr
				I	Xe
				At	Rn

Note also that hydrogen, although it is often shown in column one, is considered to be a *non*metal. Hydrogen has unique properties, but it most often behaves as a nonmetal.

Metals The metals are *all* of the elements (except hydrogen) to the left of the thick line and the six metalloids. The metals include the inner transition elements that are usually listed below the rest of the chart.

Of the more than 100 elements, more than 75% are metals. To learn the atoms that are metals, memorize the six metalloids and 18 nonmetals. All of the remaining elements are metals.

PRACTICE   B

Use a copy of the periodic table and your memorized knowledge about the columns of the table to answer these questions.

1. How many atoms are nonmetals?

2. Without consulting a periodic table, add the metal/nonmetal dividing line to the portion of the periodic table below, then circle the metalloid atoms.

					(H)	He
	B	C	N	O	F	Ne
	Al	Si	P	S	Cl	Ar
Zn	Ga	Ge	As	Se	Br	Kr
Cd	In	Sn	Sb	Te	I	Xe
Hg	Tl	Pb	Bi	Po	At	Rn

Lesson 6.5 — A Flashcard Review System

Previous Flashcards

At this point, you may have a sizeable stack of flashcards, and we will soon add more. Before going further, let's organize the cards. Try this system.

Separate your existing flashcards into four stacks.

1. *Daily*: Those you have done until correct for less than three days.

2. *End-of-chapter/quiz*: Cards you have practiced for *more* than three days. Run these again before your next quiz on this material.

3. *Test*: Those you have done four or more times. Run these again before starting practice problems for your next major test.

4. *Final exam review*: Those you have retired until the final.

Add cards with those four *labels* to the top of each stack. Rubber-band each stack. You may want to carry the *daily* pack with you for practice during down time.

Chapter 6 Flashcards

If you have had a previous course in chemistry, you may recall much of the material in Chapter 6 after a brief review. Other points may be less familiar, and the material in Chapter 6 will need to be *firmly* in memory for the rest of the course.

For points that are not firmly in memory, make the flashcards. Use the method in Lesson 2.3 on the sample cards below: Cover the answers, put a check next to those that you can answer correctly and quickly. Make the flashcard if the answer is not automatic.

Run new cards for several days in a row. Run the two-way cards in both directions. Run the cards again before your next quiz, test, and final exam.

For Lesson 6.1

One-Way Cards (with Notch at Top Right)	Back Side—Answers
Like charges	Repel
Unlike charges	Attract
The particles in a nucleus	Protons and neutrons
Subatomic particle with lowest mass	Electron
Subatomic particles with charge	Protons and electrons
Mass of a proton in amu	1.0 amu
Mass of a proton in grams/mole	1.0 gram/mole
Protons minus electrons	Charge on particle
Number of protons determines	Atom name, symbol, and atomic number
Particles gained and lost in chemical reactions	Electrons
Zero charge on an atom means	Number of protons = number of electrons
Negative ions have	More electrons than protons
Subatomic particles with mass of 1.0 amu	Protons and neutrons

Two-Way Cards (with*out* Notch)

Ion	An atom or group of atoms with an electrical charge
Protons plus neutrons	Mass number

For Lesson 6.2

One-Way Cards (with Notch)	Back Side—Answers
Same number of p^+, different number of n^0	Isotopes
Different nuclides with same chemical behavior	Isotopes

Two-Way Cards (with*out* Notch)

1 proton and 1 neutron = ? nuclide symbol	^2H = contains what particles?
1 proton and 0 neutrons = ? nuclide symbol	^1H = contains what particles?
1 proton and 2 neutrons = ? nuclide symbol	^3H = contains what particles?
Protons plus neutrons $\approx$	Mass of nuclide in amu $\approx$

For Lesson 6.3

Two-Way Cards (with*out* Notch)

Define a substance	All particles have same chemical formula
A mixture	Contains two or more substances
Molecule	Neutral, independent particles with two or more atoms
Molecular formula	Shows the atoms inside a neutral particle
Structural formula	Shows atoms and positions in a particle
Elements	Stable neutral substances with one kind of atom
Compounds	Neutral substances with more than one kind of atom
Bonds	Forces holding atoms together

For Lesson 6.4

One-Way Cards (with Notch)	Back Side—Answers
Family that rarely bonds to other atoms or each other	Noble gases
Lightest nonmetal	Hydrogen (H)
Lightest metalloid	Boron (B)
Number of nonmetal atoms	18

Two-Way Cards (with*out* Notch)	
Position in periodic table of *alkali metals*	First column, below hydrogen
Position of *halogens*	Next-to-last column
Position of *noble gases*	Last column
Position of *rare earths (lanthanides)*	First row below body of table
Position of *transition metals*	In dip between tall columns 2 and 3
Tend to form 1− ions	Ions formed by *halogen* atoms
Tend to form 1+ ions	Ions formed by *alkali metals*
Tend to form 2+ ions	*Alkaline earth metals* ion charge
Name for halogen atoms with a 1− charge	Halide ions

Lesson 6.6   The Atoms (Part 4)

The following frequently encountered metals have symbols based on their Latin names. Test to see whether these are firmly in memory in both directions. If they are not, add them to your flashcards.

Two-Way cards (with*out* Notch)		Two-Way Cards (with*out* Notch)	
Copper	Cu	Iron	Fe
Tin	Sn	Lead	Pb
Mercury	Hg	Silver	Ag
Gold	Au	Sodium	Na
Potassium	K		

Lesson 6.7   Review Quiz for Chapters 4–6

You may use a calculator and a periodic table. Work out answers to calculations on your own paper. On multiple choice, it is recommended that you work out your own answer, then select the correct answer from the choices provided.

Set a 25-minute limit, then check your answers after the Summary that follows.

1. Lesson 4.4: Using 1 kg = 2.20 lb., solve

    ? mg = 4.0×10^{-2} lb.

    a. 8.8×10^{-7} mg   b. 8.8×10^{4} mg   c. 1.8×10^{-7} mg

    d. 1.8×10^{4} mg   e. 8.8×10^{1} mg

2. Lesson 4.5:

    $\dfrac{?\ kg}{mL} = \dfrac{2.4 \times 10^{5}\ \mu g}{dm^{3}}$

    a. $2.4 \times 10^{-7}\ \dfrac{kg}{mL}$ b. $2.4 \times 10^{5}\ \dfrac{kg}{mL}$ c. $2.4 \times 10^{-10}\ \dfrac{kg}{mL}$

    d. $2.4 \times 10^{-5}\ \dfrac{kg}{mL}$ e. $2.4 \times 10^{-4}\ \dfrac{kg}{mL}$

3. Lesson 5.4: If there are 96,500 coulombs per mole of electrons and 1 mole = 6.02×10^{23} electrons, what is the charge in coulombs on 100. electrons?

    a. 5.81×10^{30} coulombs   b. 1.72×10^{-27} coulombs

    c. 1.60×10^{-17} coulombs   d. 6.24×10^{20} coulombs

    e. 1.72×10^{-31} coulombs

4. Lesson 5.4: If each mole of iron (Fe) has a mass of 55.8 g, the density of iron is 7.87 g/mL, and there are 6.02×10^{23} atoms per mole of Fe, what would be the volume in mL of 3.01×10^{23} Fe atoms?

    a. 14.2 mL   b. 2.22 mL   c. 216 mL

    d. 0.0705 mL   e. 3.55 mL

5. Lesson 6.2: For a particle with atomic number 92 that contains 143 neutrons and 90 electrons, write the nuclide (isotope) symbol and then the symbol for the ion.

    a. ^{143}U and U^{2+} b. ^{235}U and U^{2+} c. ^{143}U and U^{2-}

    d. ^{235}U and U^{2-} e. ^{90}U and U^{2-}

6. Lesson 6.2: An atom contains the ^{107}Ag nucleus and is an Ag^{+} ion. How many protons, neutrons, and electrons does the atom contain?

   a. 47 p^{+}, 60 n^{0}, and 46 e^{-} b. 47 p^{+}, 107 n^{0}, and 48 e^{-}

   c. 46 p^{+}, 60 n^{0}, and 47 e^{-} d. 48 p^{+}, 107 n^{0}, and 47 e^{-}

   e. 47 p^{+}, 60 n^{0}, and 48 e^{-}

7. Lesson 6.4: Which of these lists contains all nonmetals?

   a. C, N, S, Na, O   b. H, I, He, P, C

   c. F, H, Ne, Si, S   d. Br, H, Al, N, C

8. Lesson 4.5: On the following table, fill in the names and symbols for the atoms in the first three *rows* and the first two and last two *columns*.

PERIODIC TABLE

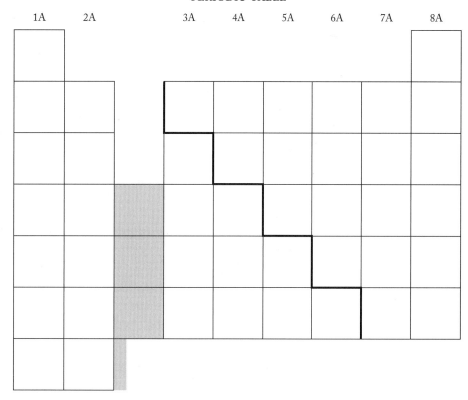

SUMMARY

1. Nearly all matter is made up of *atoms*. There are 91 different kinds of atoms that are found in Earth's crust, and others that can be produced by nuclear technology. Atoms are composed of *protons*, *neutrons*, and *electrons*. The properties of these subatomic particles include

	Charge	Mass	Location
Proton	1+	1.0 amu	Nucleus
Neutron	0	1.0 amu	Nucleus
Electron	1−	0.00055 amu	Outside nucleus

2. The number of protons in the nucleus determines the *name* and *symbol* of an atom. In an atom, the number of protons is also its *atomic number* and *nuclear charge*.

3. If an atom or group of atoms has the same number of protons and electrons, it is *electrically neutral*.

4. If an atom or group of atoms has an unequal number of protons and electrons, it is an *ion*.
     Charge on an ion = (number of protons) minus (number of electrons).
     The charge on an ion is identified by a following superscript, as in Fe^{3+} or SO_4^{2-}.

5. The identity of a *nucleus* is determined by its number of protons and neutrons.
     The *mass number* of a nucleus = (number of protons) plus (number of neutrons).
     The symbol for a nucleus is written with the mass number as a superscript in front of the atom symbol: 6 protons and 6 neutrons has the symbol ^{12}C.

6. The stability of a nucleus is determined by its number of protons and neutrons. Nuclei can be stable, radioactive (decaying gradually), or unstable.

7. The *isotopes* of an atom have the same number of protons but differing numbers of neutrons. Different isotopes of an atom nearly always have the same *chemical* behavior, but they often undergo different *nuclear* reactions.

8. One kind of atom may have several stable isotopes, but for most atoms, all samples of that atom will have the same *average mass*.

9. In a *substance*, all of the chemical particles are electrically neutral units with the same number and kind of atoms. *Chemical formulas* can be used to represent a substance.

10. A *mixture* is a combination of two or more substances.

11. Substances may be *elements* or *compounds*. Elements have only one kind of atom. Compounds have more than one kind of atom.

12. If the particles of a substance are composed of two or more neutral atoms, the particles are called *molecules*. If a substance is composed of ions, the smallest electrically neutral combination is called a *formula unit*.

13. Particles that contain more than one atom are held together by forces called *chemical bonds*.

14. In *covalent* bonds, atoms are held together by shared electrons. In *ionic* bonds, ions are held together by the attraction of opposite charges.

15. The periodic table has *rows* in which each atom increases in number of protons by one and *columns* in which atoms have similar chemical behavior. Atoms are classified as metals, metalloids, or nonmetals. More than 75% of the atoms are metals.

16. The families (columns) of the periodic table include
    • Hydrogen, which is a "family of one" with unique properties
    • The *alkali metals* below hydrogen in column one
    • The *alkaline earth metals* in column two
    • The *halogens* in the next-to-last table column
    • The *noble gases* in the last column

17. The 18 nonmetals are all of the noble gases and halogens, plus H, C, N, O, P, S, and Se.

ANSWERS

Lesson 6.1

Practice A 1a. C   1b. O   1c. Os   1d. W

2a. Potassium   2b. Fluorine   2c. Iron   2d. Lead

3.

	Charge	Mass	Location
Proton	1+	1.0 amu	Nucleus
Neutron	0	1.0 amu	Nucleus
Electron	1−	0.00055 amu	Outside nucleus

4.

Atom Name	Symbol	Protons	Electrons	Atomic Number
Sodium	Na	11	11	11
Nitrogen	N	7	7	7
Carbon	C	6	6	6
Lead	Pb	82	82	82
Fluorine	F	9	9	9

Practice B 1a. **20 protons, 20 electrons**   1b. **20 protons, 18 electrons**

2. In their nucleus, during chemical reactions, atoms always keep a constant number of **protons**, which have a positive charge. Atoms take on a charge, to become ions, by gaining or losing **electrons**, which have a **negative** charge.

3. In terms of subatomic particles, an atom that is a positive ion will always have more **protons** than **electrons**.

4a. Sr = **Strontium**   4b. Si = **Silicon**   4c. P = **Phosphorus**

5a. Bromine = **Br**   5b. Boron = **B**   5c. Barium = **Ba**

6.

Symbol	Protons	Electrons
O	8	8
O^{2-}	8	10
Mg^{2+}	12	10
Al^{3+}	13	10
Au	79	79

Symbol	Protons	Electrons
H^+	1	0
Br^-	35	36
Se^{2-}	34	36
I^-	53	54

Lesson 6.2

Practice A 1.

Protons	Neutrons	Atomic Number	Mass Number	Nuclide Symbol	Nuclide Name
6	6	6	12	^{12}C	Carbon-12
7	7	7	14	^{14}N	Nitrogen-14
53	78	53	131	^{131}I	Iodine-131
92	143	92	235	^{235}U	Uranium-235
2	2	2	4	^4He	Helium-4

2. **Uranium** must be radioactive, because no nuclei with more than 82 protons are stable.

Practice B

Protons	Neutrons	Electrons	Atomic Number	Mass Number	Nuclide Symbol	Ion Symbol
90	144	88	90	234	^{234}Th	Th^{2+}
94	148	92	94	242	^{242}Pu	Pu^{2+}
82	124	78	82	206	^{206}Pb	Pb^{4+}
1	0	0	1	1	^1H	H^+
1	2	2	1	3	^3H	H^-
38	52	36	38	90	^{90}Sr	Sr^{2+}
11	12	10	11	23	^{23}Na	Na^+
15	16	18	15	31	^{31}P	P^{3-}

Lesson 6.3

Pretest   1a. B, C, D, F   1b. C, F   1c. B

2a. 2   2b. 2   2c. 12

3a. 5   3b. 8   3c. 17

Practice A   1. 1a and 1d are elements because they have one kind of atom. The rest are compounds because they have more than one kind of atom.

2. 2a, 2b, and 2c have bonds because they have more than one atom. It takes bonds to hold two or more atoms together in particles.

3a. 2a, H_2 is the only diatomic element.

3b. 1a and 2d are the only monatomic elements.

3c. 2c, NH_3 is the only formula with four atoms.

Practice B   1a. 3 oxygen atoms   1b. 2   1c. 8
                 2a. 7 total atoms   2b. 11   2c. 13

3a. 2 hydrogen   3b. 1 cobalt   3c. 1 nobelium   3d. 3 calcium   3e. 3 nitrogen   3f. 1 lead
     1 carbon   1 sulfur   3 nitrogen   2 phosphorus   12 hydrogen   8 carbon
     2 oxygen   4 oxygen   9 oxygen   8 oxygen   1 phosphorus   20 hydrogen
                                                                          4 oxygen

Lesson 6.4

Practice A   1a. Noble gases—last column
                 1b. Alkali metals—column one (not including H)
                 1c. Halogens—group 7A (tall column 7), just before the noble gases
                 1d. Transition metals—the 10 columns in the middle dip
                 2a. Br^- 2b. Ra^{2+} 2c. Cs^+ 2d. In^{3+} 2e. Te^{2-}

Practice B   1. 18   2. See table in lesson.

Lesson 6.7

Some *partial* solutions are provided below. Your work on calculations should include WANTED, DATA, and SOLVE.

1. **d. 1.8×10^4 mg**   2. **a. 2.4×10^{-7} kg/mL** $L = dm^3$

3. **c. 1.60×10^{-17} coulombs**

$$? \text{ coulombs} = 100.\text{ electrons} \cdot \frac{1 \text{ mole of electrons}}{6.02 \times 10^{23} \text{ electrons}} \cdot \frac{96,500 \text{ coulombs}}{1 \text{ mole of electrons}} =$$

4. **e. 3.55 mL**

$$? \text{ mL} = 3.01 \times 10^{23} \text{ atoms Fe} \cdot \frac{1 \text{ mole Fe}}{6.02 \times 10^{23} \text{ atoms Fe}} \cdot \frac{55.8 \text{ g Fe}}{1 \text{ mole Fe}} \cdot \frac{1 \text{ mL Fe}}{7.87 \text{ g Fe}} =$$

5. **b. ^{235}U and U^{2+}**   6. **a. 47 p^+, 60 n^0, and 46 e^-**   7. **b. H, I, He, P, C**   8. See periodic table.

7

Writing Names and Formulas

Lesson 7.1 Naming Elements and Covalent Compounds

Systems for Naming Substances

Chemical substances are identified by both a unique name and a chemical formula. For names and formulas that both identify and differentiate substances, a *system* for writing formulas and names is required.

- Some compounds have names that are **nonsystematic** but familiar: Water (H_2O) and ammonia (NH_3) are examples.
- Historically, chemical substances have been divided into two broad categories. Compounds containing carbon and hydrogen are studied in **organic chemistry**, which has its own system for naming compounds. All other substances are part of **inorganic chemistry**, which is the focus of most first-year courses.

Different types of inorganic substances have different naming systems. We will begin with the rules for naming elements and binary covalent compounds.

Naming Elements An element is an electrically neutral substance that contains of only one kind of atom. The **name** of an element is simply the name of its **atoms**. Examples:

- The element composed of neutral atoms with 20 protons is called **calcium**. Calcium is a metal, and formulas of metals are written as if they were monatomic elements. The formula for the element calcium is therefore written as **Ca**.
- Neutral chlorine atoms, at room temperature, are most stable in diatomic molecules. For the element chlorine, the formula is written **Cl_2**.
- At room temperature, sulfur atoms tend to form molecules with eight bonded atoms. The formula for this elemental form of sulfur is written **S_8**.

When an element has more than one stable form, the forms are called **allotropes**. Example: Graphite, diamond, and fullerenes are all stable substances made from neutral, bonded carbon atoms, but the elemental state of carbon is generally designated to be graphite. Though graphite particles consist of many carbon atoms bonded together, in chemical reaction equations it is represented by the simplified substance formula **C**.

Note that for elements, the formula may distinguish between monatomic, diatomic, or polyatomic structures, but the name does not. This is only an issue for a few of the elements, but for the millions of chemical compounds, a more systematic **nomenclature** (naming system) is needed.

Naming Compounds In a compound, there is more than one *kind* of atom. Most compounds can be classified as either **ionic** or **covalent**.

Covalent compounds are molecules containing nonmetal atoms that are bonded together by electrons shared between the atoms. The attractive forces (bonds) *within* covalent molecules are strong compared to the attractions *between* neighboring molecules. Solids at room temperature may be ionic or covalent compounds, but compounds that are gases or liquids at room temperature are nearly always covalent compounds.

At room temperature, ionic compounds are nearly always solids. Ionic compounds are composed of an array of ions bonded strongly by electrostatic attractions.

Ionic and covalent compounds have different naming systems. To name a compound we must first identify it as ionic or covalent. To make that distinction, we must identify the types of bonds in the compound.

Types of Bonds

Chemical bonds can be separated into several categories, including **metallic** bonds found in metals, and the **hydrogen** bonds that are relatively weak but play an important role in the structure of proteins and DNA.

However, the two types of bonds that we encounter most often in substances are the relatively strong bonds termed **covalent** and **ionic bonds**.

1. In **covalent bonds**, electrons are *shared* between two atoms.

2. In **ionic bonds**, an atom (or group of atoms) has lost one or more electrons (compared to its electrically neutral form), and another neutral atom (or group of atoms) has gained one or more electrons. The loss and gain of electrons from neutral particles results in charged particles (ions). The ions are bonded by the attraction of their opposite charges.

3. For a bond between two atoms, the following rules will predict whether the bond is ionic or covalent in *most* cases.
    *   A bond between two *non*metal atoms is usually a *covalent* bond.
    *   A bond between a *metal* and a *non*metal atom is usually an *ionic* bond.

To identify the type of bond, begin by asking: Are both atoms nonmetals? If so, the bond is predicted to be covalent. The nonmetals are shown below. Recall that hydrogen is classified as a nonmetal, and that all atoms in the last two columns are nonmetals.

			(H)	He
C	N	O	F	Ne
	P	S	Cl	Ar
		Se	Br	Kr
			I	Xe
			At	Rn

The six noble gases rarely bond. The remaining 12 nonmetal atoms nearly always form covalent bonds when they bond with each other.

Is one of the atoms in the bond a metal and the other a nonmetal? If so, the bond is predicted to be ionic.

Using those rules and a periodic table, answer the following questions.

TRY IT

(See "How to Use These Lessons," point 1, p. xv.)

Q. Predict whether a bond between the following atoms will likely be ionic or covalent.

    1. C—H   2. Na—C   3. N—Cl   4. K—Cl   5. H—H

STOP

Answers:

  1. C—H   Both are nonmetals, so predict this to be a covalent bond.
  2. Na—C   A metal and a nonmetal; predict an ionic bond.
  3. N—Cl   Both are nonmetals; predict a covalent bond.
  4. K—Cl   A metal and nonmetal; predict an ionic bond.
  5. H—H   Both are nonmetals; predict a covalent bond.

Types of Compounds

1. If a compound contains *all* covalent bonds, it is classified as a **covalent compound**.

2. If a compound has *one* or more ionic bonds, even if it also has many covalent bonds, it will tend to have ionic behavior and is classified as an **ionic compound**.

These rules mean that in most cases,

- A compound with *all non*metal atoms is a *covalent compound*
- A compound that combines *metal* and *non*metal atoms is an *ionic compound*

The above general rules do not cover all types of bonds and compounds, and there are many exceptions. However, these rules will give us a starting point for both naming compounds and writing formulas that indicate the composition of compounds.

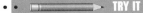 **TRY IT**

Q. Using the above rules and a periodic table, label these compounds as ionic or covalent.

    1. NaCl   2. CH_4 3. Cl_2 4. HCl

STOP

Answers:

1. NaCl: Na is a metal and Cl is a nonmetal; the compound is ionic.
2. CH_4: Both kinds of atoms are nonmetals; the compound is covalent.
3. Cl_2: Both atoms are nonmetals; the compound is covalent.
4. HCl: Both atoms are nonmetals; the compound is covalent.

Covalent Compounds The 12 nonmetals that tend to bond are a small percentage of the more than 100 atoms. However, because

- Covalent bonds are strong
- The nonmetal atoms are relatively abundant on our planet
- The molecules in living systems are based on a nonmetal (carbon)

a substantial percentage of the compounds studied in chemistry are covalent compounds.

PRACTICE   A

For the problems below, use the type of periodic table that you are permitted to view on tests in your course. You should not need to consult the metal versus nonmetal charts found in these lessons, because the locations of the metals and nonmetals in the chart should be committed to memory.

Check (✓) and do every other letter now. Do more in your next practice sessions.

1. Label these bonds as ionic or covalent.

   a. Na—I   b. C—Cl   c. S—O

   d. Ca—F   e. C—H   f. K—Br

2. Label these compounds as ionic or covalent.

   a. CF_4 b. KCl   c. CaH_2

   d. H_2O e. NF_3 f. CH_3ONa

3. Which two families in the periodic table are all nonmetals?

Rules for Naming Binary Covalent Compounds

Binary covalent compounds contain *two* different nonmetals (*bi-* means two). The naming of binary compounds uses the atom name or the *root* of the atom name.

Binary covalent compounds that include *hydrogen* are often given "common names" such as methane, water, and ammonia.

For the 11 remaining nonmetals that bond, the roots are C = carb-, N = nitr-, O = ox-, F = fluor-, P = phosph-, S = sulf-, Cl = chlor-, Se = selen-, Br = brom-, I = iod-, and At = astat-. Not all of those roots are "regular," but their use will become intuitive with practice.

For most compounds composed of two different *non*metal atoms, the rules for naming are

1. The name contains two words. The format is *prefix-atom name* then *prefix-root-***ide**.

   Example: The name of Br_2O_5 is dibromine pentoxide.

2. This rule takes precedence over the rules below. For covalent compounds that contain
   - O atoms, the second word is prefix-*oxide*
   - H atoms, the compound often has a name that does not follow these rules

3. The *first* word contains the name of the *atom* (of the two atoms in the formula) that is in a column farther to the *left* in the periodic table. If the two atoms are in the same column, the *lower* atom is named first.

4. The second word contains the *root* of the second atom name, with the suffix -*ide* added.

5. The *number* of atoms of each kind is represented by a Greek prefix from the following table.

mono- = 1 atom[1]	*penta-* = 5 atoms	*nona-* = 9 atoms
di- = 2 atoms	*hexa-* = 6 atoms	*deca-* = 10 atoms
tri- = 3 atoms	*hepta-* = 7 atoms	
tetra- = 4 atoms	*octa-* = 8 atoms	

[1]In the first word, *mono-* is left off and assumed if no prefix is given. *Mono-* is included if it applies to the second word.

If an *o* or *a* at the end of a prefix is followed by a first letter of an atom or root that is a vowel, the *o* or *a* in the prefix is *sometimes* omitted (both inclusion and omission of the *o* and *a* are allowed, and you may see such names both ways).

There are exceptions, but the rules above will accurately name binary covalent compounds in *most* cases.

Using a periodic table and the rules above, answer the following.

 TRY IT

Q. What is the name of CS_2?

 Answer:
Carbon is in the column farther to the left in the periodic table, so *carbon* is the first word. For one atom, the prefix would be *mono-*, but *mono-* is omitted if it applies to the first word. The name's first word is simply *carbon*.

For the root of the second word, sulfur becomes sulf*ide*. Because there are two sulfur atoms, the name of the compound is **carbon disulfide**.

Q. What is the name of the combination of four fluorine and two nitrogen atoms?

Answer:
Nitrogen is in the column more to the left in the periodic table, so the first word contains nitrogen. Because there are two nitrogen atoms, add the prefix *di-*. For the second word, the *root*–ide is fluoride, and the prefix for four atoms is *tetra-*. The name for the compound is **dinitrogen tetrafluoride**.

Flashcards

Cover the answers below, then check those that you can answer correctly and quickly. When you are done, make flashcards for the others. Run the new cards for several days in a row, then add them to the previous flashcards for quiz and test review.

One-Way Cards (with Notch)	Back Side—Answers
The formula for elemental oxygen	O_2
A bond between a metal and nonmetal is	Usually ionic
A bond between two nonmetals is	Usually covalent
A covalent compound has	Shared electrons and only covalent bonds
An ionic compound has	One or more ionic bonds
A compound with all nonmetal atoms is usually	A covalent compound

One-Way Cards (with Notch)	Back Side—Answers
Compounds with metal and nonmetal atoms are	Usually ionic compounds
Binary covalent name format	Prefix-atom prefix-root-ide
For binary covalent names: Which atom first?	Left column first, lower atom if in same column (H, O are exceptions)

Two-Way Cards (with*out* Notch)

Formula for ammonia = ?	Name of NH_3 = ?
Formula for carbon monoxide = ?	Name of CO = ?
Formula for dinitrogen tetrachloride = ?	Name of N_2Cl_4 = ?

PRACTICE B

Learn the rules, practice needed flashcards, then try every *other* problem. Wait a day, run the cards again, then try the remaining problems. If you need help in switching between an atom name and symbol, add its name and symbol to your two-way flashcards.

1. Write the name for these combinations of nonmetal atoms.

    a. Three chlorine plus one nitrogen

    b. One sulfur and six fluorine

    c. Four chlorines and one carbon

    d. Three chlorine and one iodine

    e. One oxygen and two chlorines

    f. One bromine and one iodine

2. Name these covalent compounds.

    a. SCl_2

    b. PI_3

    c. SO_2

    d. NO

3. Nonmetals often form several stable oxide combinations, including the combinations below. Name that compound!

    a. Five oxygen and two nitrogen

    b. Ten oxygen and four phosphorus

    c. NO_2

    d. N_2O

    e. SO_3

    f. Cl_2O_7

Lesson 7.2   Naming Ions

Ionic compounds are combinations of *ions*: particles with an electrical charge. In most first-year chemistry courses you will be asked to memorize the names and symbols for up to 50 frequently encountered ions. This task is simplified by the patterns for ion charges that are found in the periodic table. Learning these rules and patterns are a part of learning to speak the language of chemistry.

Categories of Ions

1. All ions are either positive or negative.
   - A positive ion is termed a **cation** (pronounced KAT-eye-un). The charges on cations are most often 1+, 2+, 3+, or 4+.
   - A negative ion is termed an **anion** (pronounced ANN-eye-un). The charges on anions are most often 1−, 2−, or 3−.

2. All ions are either **monatomic** or **polyatomic**.
   - A monatomic ion is composed of a single atom.
       Examples: Monatomic ions include Na^+, Al^{3+}, Cl^-, and S^{2-}.
   - A polyatomic ion is a particle that has two or more covalently bonded atoms and an overall electric charge.
       Examples: Polyatomic ions include OH^-, Hg_2^{2+}, NH_4^+, and SO_4^{2-}.

Ions of Hydrogen

Hydrogen has unique characteristics. It is classified as a nonmetal, and in many of its compounds hydrogen bonds covalently. However, in compounds classified as acids, one or more hydrogens form H^+ ions when the compound is dissolved in water. In addition, when bonded to metal atoms, hydrogen behaves as an anion: the hydride ion (H^-).

The Charge of Metal Ions

- More than 70% of the atoms in the periodic table are classified as metals.
- Geologically, in Earth's crust, *most* metals are found as metal *ions*. Exceptions to the "metals are found as ions" rule include the coinage metals: copper and silver, which may be found geologically both as ions or in their metallic, elemental form; and gold, which is always found in nature as a metal.
- In chemical *reactions*, neutral metal atoms tend to *lose* electrons to form *positive* ions.
- In compounds that contain both metal and nonmetal atoms, the metal atoms nearly always behave as ions with a *positive* charge. The charges are most often 1+, 2+, 3+, or 4+.
- With the exception of the mercurous (Hg_2^{2+}) ion, all frequently encountered metal ions are monatomic: The ions are *single* metal atoms that have lost one or more electrons.
    Examples: Metal ions include Na^+, Mg^{2+}, Al^{3+}, and Sn^{4+}.

In many cases, the charge (or possible charges) on a monatomic metal ion can be predicted from the position of the metal in the periodic table. Use a periodic table when learning the following rules for the charges on metal ions.

Predicting Metal Ion Charge

Metals in the *first two* columns of the periodic table form only *one* stable monatomic ion. The charge on that ion is easy to predict. Call this

Metal Ion Rule 1

Metal atoms in column *one* (the alkali metals) form only one stable ion: a single atom with a **1+** charge: Li^+, Na^+, K^+, Rb^+, Cs^+, and Fr^+ (H is not a metal).

Metals in column *two* form only one stable ion: a single atom with a **2+** charge: Be^{2+}, Mg^{2+}, Ca^{2+}, Sr^{2+}, Ba^{2+}, and Ra^{2+}.

The charges on metal ions in the remainder of the periodic table are more difficult to predict. Some metals form one stable cation, others form two. Call this

Metal Ion Rule 2

For metals to the right of the first two columns, the ions of metals that

- Form only *one* stable ion include Ag^+, Zn^{2+}, and Al^{3+}
- Form *two* stable ions include Cu^+ and Cu^{2+}, Fe^{2+} and Fe^{3+}, Sn^{2+} and Sn^{4+}, Pb^{2+} and Pb^{4+}, Hg^{2+} and Hg_2^{2+}

We will need to use rule 2, but given the many anions with unique names that you will need to commit to long-term memory in this course, adding rule 2 is a lot to remember. In these lessons, we will make this deal: Be aware of rule 2, but commit to memory rule 1 and the following.

Metal Ion Rule 3

Be able to add the charge when these atoms are ions: Ag^+, Al^{3+}, Hg^{2+} and Hg_2^{2+}.

The ions in rule 3 will be needed often. For the other ions covered by rule 2, when the charge on the ion is needed in problems, in these lessons it will be supplied.

PRACTICE   A

Use a periodic table. Memorize the rules, ion symbols, and names in the section above *before* doing the problems. On multi-part questions, save a few parts for your next study session.

1. For these atoms or groups, add the charge for the ion that the atom or group forms.

   a. Ba   b. Al   c. Na   d. Ag   e. Hg_2

(continued)

2. Write the symbols for the following.

   a. Potassium ion

   b. Lithium ion

   c. Hydride ion

   d. Calcium ion

3. Which ions in problems 1 and 2 are anions?

Naming Metal Ions

How a metal ion is named depends on whether the metal forms only one ion or forms two or more ions. To name metal ions, use these rules.

1. If a metal forms only *one* stable ion, the ion name is the atom name.
   Examples: Na^+ is sodium ion. Al^{3+} is aluminum ion.

2. For metals that form *two* different positive ions, the **systematic** name (or *modern* name) of the ion is the atom name followed by a Roman numeral in parentheses that states the ion's positive charge.
   Examples: Fe^{2+} is named an iron(II) ion and Fe^{3+} is named iron(III) ion.

3. Add the (Roman numeral) for ions of metal atoms that form *more* than one ion. Do *not* use (Roman numerals) in ion names for metals that can form only *one* ion.

4. The two ions of mercury are a special case.
   • One ion of mercury is a typical metal ion: Hg^{2+}, a single atom with a 2+ charge that is named mercury(II) ion by the standard naming rules.
   • Mercury(I) is the *only* frequently encountered metal ion that is polyatomic. It has two mercury atoms bonded together and has an overall 2+ charge, so its symbol is Hg_2^{2+}. It is given the name mercury(I) ion, based on the naming format of other metal ions, in part because it behaves in many respects as two loosely bonded 1+ ions.

SUMMARY

Metal Ion Rules

For metal ions, commit the following to memory.

• All metal ions are positive. Except for Hg_2^{2+}, nearly all metal ions are monatomic.
• In column one of the periodic table, all atoms form 1+ ions.
• In column two, all atoms form 2+ ions.
• When these symbols are ions, know their charges: Ag^+, Al^{3+}, Hg^{2+} and Hg_2^{2+}.
• If a metal forms only one ion, the ion name is the atom name followed by *ion*.
• If a metal forms **more** than one ion, the systematic ion name is the atom name followed by a Roman numeral in parentheses showing the positive charge of the ion.
• Hg_2^{2+} is named mercury(I) ion.

Flashcards

Using the flashcard steps in Lesson 2.3, make cards for any of these that you cannot answer from memory.

One-Way Cards (with Notch)	Back Side—Answers
Cation	Ion with one or more positive charges
Anion	Ion with one or more negative charges
Monatomic ion	One atom with a charge
Polyatomic ion	Two or more bonded atoms with an overall charge
All metal ions except mercury(I) are	Monatomic—contain only one atom
The charge on a metal ion is always	Positive
Column-one ions have what charge?	1+
Column-two ions have what charge?	2+
When are ( ) in an *ion* name needed?	If the metal forms more than one kind of positive ion

PRACTICE   B

Use a periodic table. Memorize the rules, ion symbols, and names in the section above *before* doing the problems. On multi-part questions, save a few parts for your next study session.

1. Write the name and symbol for the polyatomic metal ion often encountered.

2. Write the ion symbols for the following ions.

   a. Tin(II) ion
   b. Iron(III) ion

   c. Mercury(I) ion
   d. Mercury(II) ion

3. For the following metals, these are the ions that are stable. Name *each* ion.

   a. Ag^+
   b. Na^+

   c. Pb^{2+} and Pb^{4+}
   d. Cu^+ and Cu^{2+}

Monatomic Anions

Nine monatomic anions are often encountered in first-year chemistry. Their names and symbols should be memorized.

- One is H^- (hydride ion).
- Four are halides: fluoride, chloride, bromide, and iodide ions (F^-, Cl^-, Br^-, and I^-).
- Two are in tall column 6A: oxide (O^{2-}) and sulfide (S^{2-}) ions.
- Two are in tall column 5A: nitride (N^{3-}) and phosphide (P^{3-}) ions.

For monatomic anions, the name is the root of the atom name followed by -*ide*. For monatomic ions, the position of the atom in the periodic table predicts the charge.

Group	1A	2A		3A	4A	5A	6A	7A	8A
Family Name	Alkali metals		Transition metals			N family	O family	Halogens	Noble gases
Charge on Monatomic Ion	1+	2+		3+ (or 1+)		3−	2−	1−	None

Polyatomic Ions

A polyatomic ion is a particle that has two or more atoms held together by covalent bonds and has an overall electrical charge. In polyatomic ions, the total number of protons and electrons in the particle is not equal.

An example of a polyatomic ion is the hydroxide ion, OH^-. One way to form this ion is to start with a neutral water molecule $H-O-H$, which has $1 + 8 + 1 = 10$ protons and 10 balancing electrons, and take away an H^+ ion (which has one proton and no electrons).

The result is a particle composed of two atoms with a total of 9 protons and 10 electrons. Overall, the particle has a negative charge. The negative charge behaves as if it is attached to the oxygen. A structural formula for the hydroxide ion is

$$H-O^-$$

At this point, our interest is the *ratios* in which ions combine. For that purpose, it may help to think of a monatomic ion as a charge that has one atom attached, and a polyatomic ion as a charge with several atoms attached.

Polyatomic Cations

Two polyatomic cations with names and symbols that should be memorized are the NH_4^+ (ammonium) and Hg_2^{2+} [mercury(I)] ions.

Learning the Ion Names and Formulas

In most courses, you will be asked to memorize the names and formulas for a list of frequently encountered ions. In this book, that list is in the flashcards below. Being able to convert quickly and accurately between an ion name and its formula is essential to understanding chemistry.

You may want to use a unique card color to identify these as the *ion* cards.
Make these "two-way" flashcards following the procedure in Lesson 2.3:

- Cover the formula, and put a check if you are certain of the formula from the name.
- Then cover the names and put a check if you know the name from the formula.

You will need to be able to translate in *both* directions between the names and the ion formulas. Omit making flashcards for names and formulas that you already know well in both directions.

In nearly all situations, you will be allowed to consult a periodic table during quizzes and tests. For monatomic ions, that will help in assigning the ion charge. However, if you know the atoms and charges without looking at the table, it will reduce cognitive load and speed your work.

For a large number of new flashcards, allow yourself several days of practice. In the beginning, writing and saying the answers will assist in learning these fundamentals.

Two-Way Cards (with*out* Notch)

CH_3COO^-	acetate ion	SO_3^{2-}	sulfite ion	Hg_2^{2+}	mercury(I) ion
CN^-	cyanide ion	Na^+	sodium ion	Hg^{2+}	mercury(II) ion
OH^-	hydroxide ion	K^+	potassium ion	O^{2-}	oxide ion
NO_3^-	nitrate ion	Al^{3+}	aluminum ion	S^{2-}	sulfide ion
MnO_4^-	permanganate ion	F^-	fluoride ion	N^{3-}	nitride ion
CO_3^{2-}	carbonate ion	Cl^-	chloride ion	P^{3-}	phosphide ion
HCO_3^-	hydrogen carbonate (or bicarbonate) ion	Br^-	bromide ion	ClO_3^-	chlorate ion
CrO_4^{2-}	chromate ion	I^-	iodide ion	ClO_4^-	perchlorate ion
$Cr_2O_7^{2-}$	dichromate ion	Ca^{2+}	calcium ion	NH_4^+	ammonium ion
PO_4^{3-}	phosphate ion	Ba^{2+}	barium ion	H^+	hydrogen ion
SO_4^{2-}	sulfate ion	Mg^{2+}	magnesium ion	H^-	hydride ion
		Sr^{2+}	strontium ion		

PRACTICE **C**

Learn the rules and run the flashcards for the ion names and symbols in the section above, *then* try these problems.

1. In this chart of ions, from memory, add *charges*, *names*, and ion *formulas*.

Symbol	Ion Name	Symbol	Ion Name
	acetate ion	CO_3	
CN			radium ion
	silver ion	MnO_4	
	hydroxide ion	CrO_4	
Al		K	
ClO_4		?	dichromate ion
	nitrate ion	PO_4	
	sodium ion		sulfate ion
F			sulfide ion

2. Circle the polyatomic ion symbols in the first column (of four) above.

Lesson 7.3   Names and Formulas for Ionic Compounds

Ionic Compounds: Fundamentals

If ions have opposite charges, they attract. Ionic compounds contain positive ions (cations) combined with negative ions (anions).

The composition of an ionic compound can be expressed in three ways.

- By a **name**
    Example: ammonium phosphate
- As a **solid** formula (or *formula unit* formula)
    Example: $(NH_4)_3PO_4$
- And as balanced, **separated ions**
    Example: $3\ NH_4^+ + 1\ PO_4^{3-}$

As a part of solving many problems, given one type of identification, you will need to write the other two.

Ionic compounds can initially be confusing because their names and solid formulas do not clearly identify the charges on the ions. To solve problems that involve ionic compounds, a key step will be to translate the name or solid formula into the *separated-ions* format that *shows* the formulas of the ions, including their charges and their ratio in the compound.

In an ionic compound, the ions must be present in a *ratio* that balances the charges, resulting in electrical neutrality.

Balancing Separated Ions

It is a fundamental law of the universe that if matter has an electrical charge, it will tend to either arrange or react in ways that balance that charge, so that the overall number of positive and negative *charges* in a collection of particles is the same.

In the case of charged particles that are ions, the result is this rule:

> In any combination of ions, whether solid, melted, or dissolved in water, the total charges on the ions must *balance*. The total number of positive charges must equal the total number of negative charges, so that the overall net charge is *zero*.

When ions combine, only *one ratio* will result in electrical neutrality. In problems, you will often need to determine that ratio.

> When deciding the names and formulas for ionic compounds, the first steps are:
> - Write the separated-ion symbols, then
> - Write coefficients in front of each ion symbol that *balance* the *charges*

Let's learn how to do this with an example.

TRY IT

Q. Find the ratio that balances the charges when S^{2-} and Na^+ combine. In your notebook, apply the following steps, then check your answer below.

1. Write the two ion symbols separated by a $+$ sign. Writing the cation (positive ion) first is preferred. Leave space to write a number in front of each ion symbol.

2. **Coefficients** are numbers written in *front* of ion or particle symbols. Coefficients are a count of the number of particles that show the ratio in which the particles must exist or react.
   In all formulas for ionic compounds,

> (Coefficient *times* charge of cation) must balance (coefficient *times* charge of anion).

   When you are balancing, you *cannot* change the symbol or the charge of an ion.
   When balancing, the only change that you can make, and the one change that *you* must make, is to *write* whole-number *coefficients* in front of the particle symbols that balance the charges.
   Add these coefficients now.

3. Reduce the coefficients to the *lowest* whole-number ratios.

STOP   **Answer:**

1. $Na^+ +$ S^{2-}

2. **2** $Na^+ +$ **1** S^{2-} This is the *separated*-ions formula.
   There *must* be *two* sodium ions for every *one* sulfide ion. Why? For the charges, (2 times $1+ = 2+$) balances (1 times $2- = 2-$). In ion combinations, the ions are always present in ratios so that the total positive and negative *charges* balance.

3. 2 and 1 are the lowest whole-number ratios.

Try another problem using the steps above.

Q. Add coefficients so that the charges balance: _____ $Al^{3+} +$ _____ SO_4^{2-}

STOP   **Answer:**

One way to determine the coefficients is to make the *number* of charges on each ion equal to the coefficient of the *other* ion.

$$2\ Al^{3+} + 3\ SO_4^{2-}$$

For these ions, (2 times $+3 = +6$) balances (3 times $-2 = -6$). In an ionic compound, the total positives and total negatives must balance.

However, when balancing charges when using this method, you must often adjust the *final* coefficients to be the *lowest* whole-number ratios.

TRY IT

Q. Add proper coefficients: _____ Ba^{2+} + _____ SO_4^{2-}

Answer:

If balancing produces a ratio of **2** Ba^{2+} + **2** SO_4^{2-}, write the *final* coefficients as

$$1\, Ba^{2+} + 1\, SO_4^{2-}$$

To write ionic solid formulas, you will need the *lowest* whole-number ratio that results in electrical neutrality.

PRACTICE   A

Add lowest whole-number coefficients to make these separated ions balanced for charge. Start with the odd-numbered problems; save the even-numbered problems for your next practice session. After every two, check your answers at the end of the chapter.

1. _____ Na^+ + _____ Cl^- 5. NH_4^+ + CH_3COO^-

2. _____ Ca^{2+} + _____ Br^- 6. In^{3+} + CO_3^{2-}

3. Mg^{2+} + SO_4^{2-} 7. Al^{3+} + PO_4^{3-}

4. Cl^- + Al^{3+} 8. HPO_4^{2-} + In^{3+}

Writing the Separated Ions from Names   To write the separated ions from the *name* of an ionic compound, follow these steps.

1. The first word in the name of an ionic compound is always the positive ion.
   Write: positive ion symbol **+** negative ion symbol
   Leave a space in front of each symbol.

2. Add the lowest whole-number coefficients that balance the charges.

Apply those steps to the following problem.

TRY IT

Q. Write a balanced separated-ions formula for aluminum carbonate.

Answer:

1. Aluminum carbonate → Al^{3+} + CO_3^{2-}

2. Aluminum carbonate → **2** Al^{3+} + **3** CO_3^{2-}

The separated-ions formula shows clearly what the name does not: In aluminum carbonate, there must be two aluminum ions for every three carbonate ions.

When writing separated ions, write the charges *high*, any subscripts *low*, and the coefficients at the *same* level as the atom symbols.

PRACTICE B

If you have not done so today, run your ion flashcards. Then write balanced *separated-ion* formulas for the ionic compounds below. You may use a periodic table, but otherwise write the ion formulas from memory. Do odd-numbered problems now and evens later. Check answers as you go.

1. Sodium hydroxide →

2. Aluminum chloride →

3. Rubidium sulfite →

4. Iron(III) nitrate →

5. Lead(II) phosphate →

6. Magnesium chloride →

Writing Formulas of Solids from Names   In ionic solid formulas, charges are hidden, but charges must balance. The key to writing a correct solid formula is to write the balanced *separated* ions *first* so that you can see and balance the charges.

To write a *solid* formula from the name of an ionic compound, use these steps.

1. Based on the name, write the *separated* ions. Add lowest whole-number coefficients to balance charge. Then, to the right, draw an arrow →.

2. After the →, write the two ion symbols, positive ion first, with a small space between them. Include any *subscripts* that are part of the ion symbol, but *leave out* the charges and coefficients.

3. For the symbols after the arrow, **p**ut **p**arentheses ( ) around a **p**olyatomic ion *if* its coefficient in the separated-ions formula on the left is greater than 1.

4. Add *subscripts* after each symbol on the right. The subscript will be the same as the coefficient in front of that ion in the *separated-ions* formula.

   Omit subscripts of 1. For polyatomic ions, write the coefficients as subscripts *outside* and *after* the parentheses.

In your notebook, apply those steps to this example.

TRY IT

Q. Write the solid formula for potassium sulfide.

Answer:

1. Write the *separated*-ions formula first. For potassium sulfide:
   $2 K^+ + 1 S^{2-}$

2. Rewrite the symbols with*out* coefficients or charges: $2\,K^+ + 1\,S^{2-} \rightarrow K\,S$

3. Because both ions are monatomic, add no parentheses.

4. The K ion coefficient becomes its solid formula subscript:

$$2\,K^+ + 1\,S^{2-} \rightarrow K_2S$$

The sulfide subscript of one is omitted as understood. The *solid* formula for potassium sulfide is K_2S

Try another using the same steps.

TRY IT

Q. Write the solid formula for magnesium phosphate.

Answer:

1. Write the balanced separated ions.

$$\text{Magnesium phosphate} \rightarrow 3\,Mg^{2+} + 2\,PO_4^{3-}$$

2. Write symbols without coefficients or charges.

$$3\,Mg^{2+} + 2\,PO_4^{3-} \rightarrow Mg\,PO_4$$

3. Because Mg^{2+} is *mon*atomic (just one atom), it is not placed in parentheses. Phosphate is *both poly*atomic *and* we need more than **1**, so add **( )**.
$Mg\,(PO_4)$

4. The separated coefficient of Mg becomes its solid subscript. $Mg_3(PO_4)$
The phosphate coefficient becomes its solid subscript. $Mg_3(PO_4)_2$
$Mg_3(PO_4)_2$ is the *solid* formula for magnesium phosphate.

Recite the *three-Ps rule* until it is committed to memory. When writing ionic *solid* formulas,

Put *p*arentheses around ***p*olyatomic** ions—*if* you need more than one.

PRACTICE **C**

You may use a periodic table. Complete half of the lettered parts today and the rest during your next study session.

1. Circle the polyatomic ions.

    a. Na^+ b. NH_4^+ c. CH_3COO^- d. Ca^{2+} e. OH^-

2. When do you need parentheses? Write the rule from memory.

3. Write solid formulas for these ion combinations.

    a. $2 K^+ + 1 CrO_4^{2-} \rightarrow$

    b. $2 NH_4^+ + 1 S^{2-} \rightarrow$

    c. $1 SO_3^{2-} + 1 Sr^{2+} \rightarrow$

4. Balance these separated ions for charge, then write solid formulas.

    a. $Cs^+ +$ $N^{3-} \rightarrow$

    b. $Cr_2O_7^{2-} +$ $Ca^{2+} \rightarrow$

    c. $Sn^{4+} +$ $SO_4^{2-} \rightarrow$

5. From these names, write the separated-ions formula, then the solid formula.

    a. Ammonium sulfite $\rightarrow$

    b. Potassium permanganate $\rightarrow$

    c. Calcium chloride $\rightarrow$

    d. Sodium hydrogen carbonate $\rightarrow$

6. Write the solid formula.

    a. Tin(II) fluoride $\rightarrow$

    b. Calcium hydroxide $\rightarrow$

    c. Radium acetate $\rightarrow$

Writing Equations for Separated Ions from Solid Formulas   When placed in water, all ionic solids dissolve to *some* extent. In water, some ionic solids are highly soluble, but others are only slightly soluble. The ions that dissolve will separate and move about independently in the solution.

This dissolving process can be represented by a chemical equation that has a solid on the left and the separated ions on the right.

Example: When solid sodium phosphate dissolves in water, the equation is

$$Na_3PO_4(s) \xrightarrow{H_2O} 3 Na^+(aq) + 1 PO_4^{3-}(aq)$$

The (*s*) is an abbreviation for the *solid* phase. The (*aq*) is an abbreviation for the **aqueous** phase, which means "dissolved in water."

When an ionic compound separates into ions that can move about freely, the reaction is termed **dissociation**. If the reactant is an ionic solid, the ions are already present in the solid: Dissolving simply allows the ions to separate, move about, collide, and potentially react with other particles.

Every equation representing ion separation must balance atoms, balance charge, and result in correct formulas for the ions that are actually found in the solution.

In equations for an ionic solid separating into its ions, some subscripts in the solid formula become coefficients in the separated ions, but others do not. In the equation above, the subscript 3 became a coefficient, but the subscripts 1 and 4 did not. To correctly separate solid formulas into ions, you must be able to recognize the ions inside the solid formula. That's why the frequently encountered ion names and formulas must be memorized. For this question, read a part of the answer for a hint if needed, then try again.

 TRY IT

Q. Write the equation for the ionic solid Cu_2CO_3 separating into its ions.

Answer:

Follow these steps in going from a solid formula to separated ions.

1. Decide the *negative* ion's charge and coefficient first.

    The first ion in a solid formula is the positive ion, but many metal ions can have two possible positive charges. Because most negative ions only have one likely charge, and that charge is often needed to identify the positive ion's charge, we usually label the charge of the negative ion first.

    In Cu_2CO_3, the negative ion is CO_3, which always has a $2-$ charge. This step temporarily splits the solid formula into **Cu_2** and 1 CO_3^{2-}.

2. Decide the positive ion's charge and coefficients.

    Given Cu_2 and CO_3^{2-}, the positive ion or ions must include **2** copper atoms *and* must have a total **2+** charge to balance the charge of CO_3^{2-}.

    This means that **Cu_2**, in the separated-ions formula, must be *either* **1 Cu_2^{2+}** *or* **2 Cu^+**.

    Both possibilities balance atoms and charge. Which is correct?

    Recall that all *metal* ions are *mon*atomic [except mercury(I) ion].

    This means that Cu^+ must be the ion that forms, because Cu_2^{2+} is polyatomic.

    Because most metal ions are monatomic, a solid formula with a metal ion (M) will separate

    $$M_XAnion \rightarrow X\,M^{+?} + Anion \qquad (\textit{unless the metal ion is } Hg_2^{2+})$$

    The fact that copper ions are all monatomic leads us to predict that the equation for ion separation is

    $$Cu_2CO_3 \rightarrow 2\,Cu^+ + CO_3^{2-}$$

    Copper can also be a Cu^{2+} ion, but in the formula above, there is only one carbonate, and carbonate always has a $2-$ charge. Two Cu^{2+} ions cannot balance the single carbonate.

3. Check to make sure that the charges balance. Make sure that the number of atoms of each kind is the same on both sides. The equation must also make sense going backwards, from the separated to the solid formula.

Try another.

 TRY IT

Q. Write the equation for the ionic solid $(NH_4)_2S$ dissolving to form ions.

Answer:

* In an ionic solid formula, parentheses are placed around polyatomic ions. When you write the separated ions, a subscript *after* parentheses *always* becomes the polyatomic ion's *coefficient*.
    You would therefore split the formula $(NH_4)_2S \rightarrow 2\,NH_4 + 1\,S$
* Assign the charges that these ions prefer. $(NH_4)_2S \rightarrow 2\,NH_4^+ + 1\,S^{2-}$
* Check: In the separated formula, do the charges balance?
    Going backwards, do the separated ions combine to give the solid formula?

Keep up your practice for 15–20 minutes a day with your *ion* name and formula flashcards (Lesson 7.2). Identifying ions without consulting a table will be most helpful when solving the problems that lie ahead.

PRACTICE   D

If you have not done so today, run your ion flashcards in both directions, then try these problems. You may use a periodic table. To take advantage of the "spacing effect"(Lesson 2.3), do half of the lettered parts below today, and the rest during your next study session.

1. Finish balancing by adding ions, coefficients, and charges.

    a. Lead forms two different ions. Which ion must this be?

    $PbCO_3 \rightarrow \qquad Pb \qquad + 1\,CO_3^{2-}$

    b. $Hg_2SO_4 \rightarrow \qquad Hg_2 \qquad +$

2. Write equations for these ionic solids separating into ions.

    a. $KOH \rightarrow$

    b. For copper, two ions are possible. $CuNO_3 \rightarrow$

    c. Iron forms two ions. $Fe_3(PO_4)_2 \rightarrow$

    d. $Ag_2CO_3 \rightarrow$

    e. $NH_4Br \rightarrow$

    f. $Mg(OH)_2 \rightarrow$

Naming Ionic Compounds

From a solid or a separated-ions formula, writing the *name* is easy.

1. Write the *separated*-ions formula.

2. Write the *name* of the cation, then the name of the anion.

That's it! In ionic compounds, the name ignores the number of ions inside. Simply name the ions in the compound, with the positive ion first.

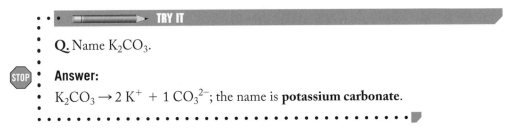

Q. Name K_2CO_3.

Answer:

$K_2CO_3 \rightarrow 2\ K^+ + 1\ CO_3^{2-}$; the name is **potassium carbonate**.

With time, you will be able to convert solid formulas to compound names without writing the separated ions, but to develop this accurate intuition requires practice.

PRACTICE   E

You may use a periodic table. If you are unsure of an answer, check it before continuing.

1. Return to Practice D and name each compound.

2. In Practice C, problems 3 and 4, name each compound.

3. Would CBr_4 be named carbon bromide or carbon tetrabromide? Why?

4. Name these ionic and covalent compounds. Try half today and half during your next study session.

    a. $CaBr_2$

    b. NCl_3

    c. NaH

    d. $CuCl_2$

    e. $RbClO_4$

    f. KI

    g. Li_3P

    h. MgO

    i. NH_4Cl

j. SO_2

k. $CaSO_3$

l. P_4S_3

Flashcards

Add these to your collection.

One-Way Cards (with Notch)	Back Side—Answers
What must be true for charges in all ionic substances?	Total + charges = total − charges Must be electrically neutral
Numbers you add to balance separated ions	Coefficients
To understand ionic compounds:	Write the *separated*-ion formulas
When are parentheses needed in formulas?	In *solid* formulas, put parentheses around polyatomic ions—*if* you need >1
In separated-ion formulas, what do the coefficients tell you?	The ratio in which the ions must be present to balance atoms and charge

PRACTICE   F

Fill in the blanks in the rows below. Complete half of the rows today and the rest during your next study session. Check answers after every few rows.

Ionic Compound Name	Separated Ions	Solid Formula
• Name by ion names	• Charges must show	• Positive ion first
• Must be two or more words	• Charges must balance	• Charges balance, but don't show
• Put name of +ion first	• Coefficients tell ratio of ions	• Put ( ) around polyatomic ions if you need >1
Sodium chloride	$1\,Na^+ + 1\,Cl^-$	NaCl
	$2\,Al^{3+} + 3\,SO_3^{2-}$	$Al_2(SO_3)_3$
Lithium carbonate		
Potassium hydroxide		
	____ Ag^+ + ____ NO_3^-	
	____ NH_4^+ + ____ SO_4^{2-}	
		$FeBr_2$
		$Fe_2(SO_4)_3$
Copper(I) chloride		

(continued)

Ionic Compound Name	Separated Ions	Solid Formula
Tin(II) fluoride		
	___ Al^{3+} + ___ $Cr_2O_7{}^{2-}$	
		K_2CrO_4
		$CaCO_3$
Aluminum phosphate		

SUMMARY

1. The name of an element is the name of its atoms.

2. In covalent bonds, electrons are shared. Two nonmetal atoms usually bond with a covalent bond.

3. An ionic bond exists between positive and negative ions. If a metal is bonded to a nonmetal, the bond is generally ionic. The metal is the positive ion.

4. Most compounds with all nonmetal atoms are covalent. Most compounds that have both metal and nonmetal atoms are ionic.

5. If a compound has only covalent bonds, predict it is covalent. If a compound has *one* or more ionic bonds, predict it is ionic.

6. The conventions for naming binary covalent compounds are

   a. Names have two words. Compounds with O end in (prefix)*oxide*. (This rule has precedence.) Compounds with H have many exceptions to the naming rules.

   b. The first word contains the name of the atom in the column farther to the left in the periodic table. For two atoms in the same column, the lower one is named first.

   c. The second word contains the root of the second atom name plus a suffix *-ide*.

   d. The number of atoms is shown by a prefix.
      Mono- = 1 atom. (For the first word of the name, *mono* is left off and is assumed if no prefix is given.)
      Di- = 2 atoms, *tri-* = 3, *tetra-* = 4, *penta-* = 5, *hexa-* = 6, *hepta-* = 7, *octa-* = 8.

7. Positive ions are cations (pronounced CAT-eye-uns). Negative ions are anions (pronounced ANN-eye-uns).

8. Metals can lose electrons to form positive ions. Column-one atoms form 1+ ions and column-two atoms form 2+ ions.

9. The name of a metal ion that forms only one ion is the name of the atom.

10. Metals to the right of column two often form two different cations. The name of these ions is the atom name followed by (I, II, III, or IV) stating the positive charge.

11. A polyatomic ion is composed of more than one atom.

12. To determine the names and formulas for ionic compounds,
    - Write the separated-ions formula first, and
    - Be certain that all names and formulas are electrically neutral combinations

13. To balance separated-ion formulas, add coefficients that balance charge. Coefficients are numbers written in front of the ion symbols that show the ratio of the ions in the compound. In balancing, you may not change the symbol or the stated charge of an ion.
    (Coefficient times charge of cation) must balance (coefficient times charge of anion). The overall charge for ionic compounds must equal zero.

14. To write solid formulas for ionic compounds from their names,
    - Write the separated ions with the lowest whole-number coefficient ratios
    - Write the two ion symbols, positive ion first, with*out* charges, a + sign, or coefficients
    - Put *p*arentheses ( ) around *p*olyatomic ions if you need more than one
    - Make the separated formula coefficients into solid formula subscripts. Omit subscripts of 1

15. To write separated ions from solid formulas,
    - Decide the *negative* ion's charge and coefficients first
    - Add the *positive* ion's charge based on what balances atoms and charges
    - Assume that metal atoms are monatomic (except Hg_2^{2+})

16. To name an ionic compound: Name the ions, positive first.

ANSWERS

Lesson 7.1

Practice A 1a. Na—I: **Ionic** 1b. C—Cl: **Covalent** 1c. S—O: **Covalent**

1d. Ca—F: **Ionic** 1e. C—H: **Covalent** 1f. K—Br: **Ionic**

2a. CF_4: **Covalent** 2b. KCl: **Ionic** 2c. CaH_2: **Ionic**

2d. H_2O: **Covalent** 2e. NF_3: **Covalent** 2f. CH_3ONa: **Ionic**

(All of the ionic compounds contain a metal atom.)

3. Halogens and noble gases.

Practice B 1a. Nitrogen is to the left, so it is the first word in the name. When the first word refers to a single atom, the prefix is omitted. For the second word, chlorine becomes chloride, and the prefix tri- is added. The name is **nitrogen trichloride**.

1b. Sulfur hexafluoride 1c. Carbon tetrachloride

1d. Iodine trichloride   (If in same column, name lower first.)

1e. Dichlorine monoxide   (If oxygen, end in oxide, drop last *o* in mono-.)

1f. Iodine monobromide   2a. Sulfur dichloride   2b. Phosphorus triiodide

2c. Sulfur dioxide   2d. Nitrogen monoxide   3a. Dinitrogen pentoxide (or pentaoxide)

3b. Tetraphosphorus decaoxide (or decoxide)   3c. Nitrogen dioxide   3d. Dinitrogen monoxide

3e. Sulfur trioxide   3f. Dichlorine heptaoxide (or heptoxide)

Lesson 7.2

Practice A 1a. Ba^{2+} 1b. Al^{3+} 1c. Na^+ 1d. Ag^+ 1e. Hg_2^{2+}

2a. K^+ 2b. Li^+ 2c. H^- 2d. Ca^{2+}

3. Only the hybride ion (H^-)

Practice B 1. Hg_2^{2+}

2a. Sn^{2+} 2b. Fe^{3+} 2c. Hg_2^{2+} 2d. Hg^{2+}

3a. Silver ion

3b. Sodium ion

3c. Lead(II) and lead(IV) ion

3d. Copper(I) and copper(II) ion

Practice C

Symbol	Ion Name	Symbol	Ion Name
CH_3COO^-	acetate ion	CO_3^{2-}	**carbonate ion**
CN^-	**cyanide ion**	Ra^{2+}	radium ion
Ag^+	silver ion	MnO_4^-	**permanganate ion**
OH^-	hydroxide ion	CrO_4^{2-}	**chromate ion**
Al^{3+}	**aluminum ion**	K^+	**potassium ion**
ClO_4^-	**perchlorate ion**	$Cr_2O_7^{2-}$	dichromate ion
NO_3^-	nitrate ion	PO_4^{3-}	**phosphate ion**
Na^+	sodium ion	SO_4^{2-}	sulfate ion
F^-	**fluoride ion**	S^{2-}	sulfide ion

Lesson 7.3

Practice A 1. $1\,Na^+ + 1\,Cl^-$ 2. $1\,Ca^{2+} + 2\,Br^-$ 3. $1\,Mg^{2+} + 1\,SO_4^{2-}$

4. $3\,Cl^- + 1\,Al^{3+}$ 5. $1\,NH_4^+ + 1\,CH_3COO^-$

6. $2\,In^{3+} + 3\,CO_3^{2-}$ 7. $1\,Al^{3+} + 1\,PO_4^{3-}$ 8. $3\,HPO_4^{2-} + 2\,In^{3+}$

Practice B 1. Sodium hydroxide $\rightarrow 1\,Na^+ + 1\,OH^-$

2. Aluminum chloride $\rightarrow 1\,Al^{3+} + 3\,Cl^-$

3. Rubidium sulfite $\rightarrow$ **2 Rb$^+$ + 1 SO$_3$$^{2-}$**

4. Iron(III) nitrate $\rightarrow$ **1 Fe^{3+} + 3 NO$_3$$^-$**

5. Lead(II) phosphate $\rightarrow$ **3 Pb^{2+} + 2 PO$_4$$^{3-}$**

6. Magnesium chloride $\rightarrow$ **1 Mg^{2+} + 2 Cl$^-$**

Practice C 1. The polyatomic ions: b. **NH$_4$$^+$**   c. **CH$_3COO^-$**   e. **OH$^-$**

2. For ionic solid formulas, put parentheses around polyatomic ions if you need more than one.

3a. **2 K$^+$ + 1 CrO$_4$$^{2-}$ $\rightarrow$ K$_2$CrO$_4$**   3b. **2 NH$_4$$^+$ + 1 S^{2-} $\rightarrow$ (NH$_4$)$_2$S**

3c. **1 SO$_3$$^{2-}$ + 1 Sr^{2+} $\rightarrow$ SrSO$_3$**   4a. **3 Cs$^+$ + 1 N^{3-} $\rightarrow$ Cs$_3$N**

4b. **1 Cr$_2$O$_7$$^{2-}$ + 1 Ca^{2+} $\rightarrow$ CaCr$_2$O$_7$**   4c. **1 Sn^{4+} + 2 SO$_4$$^{2-}$ $\rightarrow$ Sn(SO$_4$)$_2$**

5a. **2 NH$_4$$^+$ + 1 SO$_3$$^{2-}$ $\rightarrow$ (NH$_4$)$_2$SO$_3$**   5b. **1 K$^+$ + 1 MnO$_4$$^-$ $\rightarrow$ KMnO$_4$**

5c. **1 Ca^{2+} + 2 Cl$^-$ $\rightarrow$ CaCl$_2$**   5d. **1 Na$^+$ + 1 HCO$_3$$^-$ $\rightarrow$ NaHCO$_3$**

6. Write balanced, separated ions first to help with the solid formula:

6a. Tin(II) fluoride $\rightarrow$ 1 Sn^{2+} + 2 F$^-$ $\rightarrow$ **SnF$_2$**

6b. Calcium hydroxide $\rightarrow$ 1 Ca^{2+} + 2 OH$^-$ $\rightarrow$ **Ca(OH)$_2$**

6c. Radium acetate $\rightarrow$ 1 Ra^{2+} + 2 CH$_3$COO$^-$ $\rightarrow$ **Ra(CH$_3$COO)$_2$**

Practice D and E, 1

Practice D	Practice E, 1
1a. PbCO$_3$ $\rightarrow$ **1 Pb^{2+} + 1 CO$_3$$^{2-}$**	**Lead(II) carbonate**
1b. Hg$_2$SO$_4$ $\rightarrow$ 1 Hg$_2$$^{2+}$ + **1 SO$_4$$^{2-}$**	**Mercury(I) sulfate**
2a. KOH $\rightarrow$ **1 K$^+$ + 1 OH$^-$**	**Potassium hydroxide**
2b. CuNO$_3$ $\rightarrow$ **1 Cu$^+$ + 1 NO$_3$$^-$**	**Copper(I) nitrate**
2c. Fe$_3$(PO$_4$)$_2$ $\rightarrow$ **3 Fe^{2+} + 2 PO$_4$$^{3-}$**	**Iron(II) phosphate**
2d. Ag$_2$CO$_3$ $\rightarrow$ **2 Ag$^+$ + 1 CO$_3$$^{2-}$**	**Silver carbonate**
2e. NH$_4$Br $\rightarrow$ **1 NH$_4$$^+$ + 1 Br$^-$**	**Ammonium bromide**
2f. Mg(OH)$_2$ $\rightarrow$ **1 Mg^{2+} + 2 OH$^-$**	**Magnesium hydroxide**

Practice E, 2–4

2. (3a) Potassium chromate   (3b) Ammonium sulfide   (3c) Strontium sulfite

(4a) Cesium nitride   (4b) Calcium dichromate   (4c) Tin(IV) sulfate

3. Carbon tetrabromide. Carbon is a nonmetal, so the compound is covalent (see Lesson 7.1). Use prefixes in the names of *covalent* compounds. Practice recognizing the symbols of the nonmetals.

4a. Calcium bromide   4b. Nitrogen trichloride   4c. Sodium hydride

4d. Copper(II) chloride   4e. Rubidium perchlorate   4f. Potassium iodide

4g. Lithium phosphide   4h. Magnesium oxide   4i. Ammonium chloride

4j. Sulfur dioxide   4k. Calcium sulfite   4l. Tetraphosphorus trisulfide

Practice F

Ionic Compound Name	Separated Ions	Solid Formula
Sodium chloride	$1\,Na^+ + 1\,Cl^-$	NaCl
Aluminum sulfite	$2\,Al^{3+} + 3\,SO_3{}^{2-}$	$Al_2(SO_3)_3$
Lithium carbonate	$2\,Li^+ + CO_3{}^{2-}$	Li_2CO_3
Potassium hydroxide	$1\,K^+ + 1\,OH^-$	KOH
Silver nitrate	$1\,Ag^+ + 1\,NO_3{}^-$	$AgNO_3$
Ammonium sulfate	$2\,NH_4{}^+ + 1\,SO_4{}^{2-}$	$(NH_4)_2SO_4$
Iron(II) bromide (see next row)	$1\,Fe^{2+} + 2\,Br^-$	$FeBr_2$
Iron(III) sulfate (iron forms 2 ions)	$2\,Fe^{3+} + 3\,SO_4{}^{2-}$	$Fe_2(SO_4)_3$
Copper(I) chloride	$1\,Cu^+ + 1\,Cl^-$	CuCl
Tin(II) fluoride	$1\,Sn^{2+} + 2\,F^-$	SnF_2
Aluminum dichromate	$2\,Al^{3+} + 3\,Cr_2O_7{}^{2-}$	$Al_2(Cr_2O_7)_3$
Potassium chromate	$2\,K^+ + CrO_4{}^{2-}$	K_2CrO_4
Calcium carbonate	$1\,Ca^{2+} + 1\,CO_3{}^{2-}$	$CaCO_3$
Aluminum phosphate	$1\,Al^{3+} + 1\,PO_4{}^{3-}$	$AlPO_4$

8

Grams and Moles

Lesson 8.1   The Mole

Counting Particles

Atoms and molecules are extremely small. Visible quantities of a substance must therefore have a very large number of molecules, atoms, or ions.

Example: One drop of water contains about 1,500,000,000,000,000,000,000 (1.5×10^{21}) water molecules.

Rather than writing numbers of this size when solving calculations, chemists use a unit to count large numbers of particles. As we count eggs by the dozen, or buy printer paper by the ream (500 sheets), we count chemical particles such as molecules, atoms, and ions by the **mole**.

> A **mole** is 6.02×10^{23} particles.

The number 6.02×10^{23} is called **Avogadro's number**.

The definition of mole is based on the isotope carbon-12: Exactly 12 grams of C-12 contains exactly one mole of C-12 atoms. Using this definition simplifies the arithmetic in problems with other atoms, especially when we convert between *grams* and *moles*. That's our goal: to calculate a *count* of particles by measuring their mass on a balance or scale.

In chemical reactions, particles react and form in simple whole-number ratios if we *count* the particles. The ratios are usually *not* simple if we compare the masses of the particles. This means that in reaction calculations, we begin by converting the units we are supplied to the unit that is used to count visible numbers of particles: *moles*. If we do not know the moles supplied, the rule will be: *Calculate the moles first*.

Mole is abbreviated **mol** in the SI system. As with all metric abbreviations, *mol* is not followed by a period, and no distinction is made between singular and plural when the abbreviation is used.

Working with Moles

To simplify working with very large numbers, we use exponential notation. Recall that

- 10^{23} means a 1 followed by 23 zeros: 100,000,000,000,000,000,000,000
- When multiplying a *number* times a *number* times an *exponential*, the numbers multiply by the standard rules of arithmetic, but the exponential does not change Examples:

Half a mole $= 1/2 \times (6.02 \times 10^{23}) = 3.01 \times 10^{23}$ particles

Ten moles $= 10 \times (6.02 \times 10^{23}) = 60.2 \times 10^{23} = 6.02 \times 10^{24}$ particles

0.20 moles $= 0.20 \times (6.02 \times 10^{23}) = 1.2 \times 10^{23}$ particles

PRACTICE

How many particles are in the following? (Answer in scientific notation.)

1. 4.0 mol

2. 0.050 mol

Lesson 8.2   Grams per Mole (Molar Mass)

Atomic Mass

Each atom has a different average mass. The average atom of helium has a mass approximately four times that of the average hydrogen atom, because helium has more protons, neutrons, and electrons. On average, carbon atoms have *about* 12 times more mass than hydrogen atoms.

The average mass of an atom is its **atomic mass**. In most chemistry calculations, each kind of atom can be treated as if all atoms of that kind have a characteristic mass that is their atomic mass, measured in *atomic mass units* (amu; see Lesson 6.2). Atomic masses for the atoms are listed at the back of the book.

To encourage mental arithmetic, the atomic masses in these lessons use fewer significant figures than most textbooks. If you use a different table of atomic masses, your answers may differ slightly from the answers shown here.

Molar Mass

The **molar mass** of an atom is the mass of a mole of the atoms. The number that represents the atomic mass of an atom in amu is the *same* as the number that measures the molar mass of an atom in *grams per one mole*. The molar mass of the lightest atom, hydrogen, is 1.008 g/mol. A mole of uranium atoms has a mass of 238.0 grams.

For substances that contain more than one atom, the molar mass is easily determined. Simply add the molar masses of each of the atoms that make up the substance. In chemistry calculations, you will nearly always be supplied with atomic mass values *or* be allowed to consult a table of atomic masses.

Example: What is the molar mass of NaOH? Add these molar masses:

$$Na = 23.0$$
$$O = 16.0$$
$$H = \underline{1.008}$$
$$40.008 = \textbf{40.0}\,\text{g/mol NaOH}$$

S.F.: Recall from Lesson 3.1 that when *adding* significant figures, because the highest decimal *place* with doubt in the columns above is in the tenths place, the sum has doubt in the tenths place. Round the answer to that place.

The molar mass supplies an *equality*: 40.0 grams NaOH = 1 mole NaOH.

When solving problems, after calculating a molar mass, the *equality* format should be written in the DATA. Include the formula for the substance on *both* sides of the equality. This will greatly simplify the reaction calculations in upcoming lessons.

Molar Masses and Subscripts

To calculate the molar mass from chemical formulas containing *subscripts*, recall that subscripts are exact numbers. Multiplying by a subscript therefore does not change the doubtful digit's *place* in the result.

When calculating a molar mass, use the following *column* format to keep track of the numbers and the decimal place with doubt.

Example: Find the molar mass of phosphoric acid, H_3PO_4:

$$1 \text{ mol } H_3PO_4 = 3 \text{ mol } H = 3 \times 1.008 \text{ g/mol} = \ 3.024$$

$$1 \text{ mol } P \ = 1 \times 31.0 \text{ g/mol} \ = 31.0$$

$$4 \text{ mol } O = 4 \times 16.0 \text{ g/mol} \ = 64.0$$

$$98.024 \rightarrow \mathbf{98.0 \text{ g/mol}}$$

In your DATA, write

$$98.0 \text{ g } H_3PO_4 = 1 \text{ mol } H_3PO_4$$

If you know the chemical formula of a substance, you can calculate the molar mass of the substance.

The Importance of Molar Mass

The molar mass is the most frequently used conversion in chemistry. Why?

Grams of a substance are easy to measure using a balance or scale. Moles are difficult to measure directly because we don't have machines that count large numbers of particles. However, chemical processes are most easily understood by counting the particles. Using the molar mass, we can convert between the *grams* that we can measure and the particle *counts* that explain chemistry.

PRACTICE

Use the table of atomic masses at the back of the book. To speed your progress, try the last letter of each problem. If you have difficulty, try other letters of the same problem. Answers are at the end of this chapter.

1. Use your table of atoms to find the molar mass of these single atoms. Include the unit with your answer.

   a. Nitrogen   b. Au   c. Pb

2. How many *oxygens* are represented in each of these formulas?

   a. $Ca(OH)_2$ b. $Al_2(SO_4)_3$ c. $Co(NO_3)_2$

Do the next two problems in your notebook. Allow enough room on the paper for clear and careful work. Use the column format of the H_3PO_4 molar mass calculation above. If this is easy review, do a few problems, but be sure to do problems 3e and 4b.

3. Calculate the molar mass for these compounds. Include units and proper *s.f.* in your answers.

   a. H_2 b. NaH   c. KSCN   d. Na_3PO_4 e. Barium nitrate

4a. $1 \text{ mol } H_2S = ? \text{ g } H_2S$

4b. $? \text{ g } AgNO_3 = 1 \text{ mol } AgNO_3$

Lesson 8.3   Converting between Grams and Moles

Knowing how to calculate the grams per one mole, we now want to be able to calculate the mass of *any* number of moles of a substance.

The problem can be viewed as converting units, in this case from moles to grams. An equality, the molar mass, provides the conversion factor.

The Grams Prompt

A **prompt** is a word or two that reminds us of what to do next. In chemistry, certain words or conditions can prompt us to write relationships that will help in solving problems.

Commit to memory the following rule:

> ### The Grams Prompt
>
> In your WANTED *or* DATA, if you see *grams* or prefix-*grams* (such as k*g* or m*g*) of a substance with a known *formula*,
>
> - Calculate the molar mass of that formula, then
> - Write that molar mass as an *equality* in your DATA
>   Example: 1 mol H_2O = 18.0 g H_2O

The grams prompt will help to list in your DATA the conversions needed to SOLVE. If you see grams of a formula in a calculation problem, you will nearly always need the molar mass.

> To convert between grams and moles of a substance, use the molar mass as a conversion factor.

 TRY IT

(See "How to Use These Lessons," point 1, p. xv.)

Use WANTED, DATA, SOLVE, and the grams prompt to solve this problem in your notebook, then check your answer below.

Q. Find the mass in grams of 0.25 moles of O_2.

STOP

Answer:

WANTED: ? g O_2 =   (Write the unit and substance WANTED.)

DATA: 0.25 mol O_2

$\quad\quad\quad$ **32.0 g O_2 = 1 mol O_2**   (g O_2 in the WANTED is a grams prompt.)

SOLVE: ? g O_2 = 0.25 ~~mol O_2~~ · $\dfrac{32.0 \text{ g } O_2}{1 \text{ ~~mol O_2~~}}$ = **8.0 g O_2**

A single unit is WANTED, so the DATA contains a single unit to use as the *given* quantity. The remaining DATA will be in pairs, written as equalities or ratios.

S.F.: Because 0.25 has two *s.f.*, 32.0 has three *s.f.*, and 1 is exact, round the answer to two *s.f.*

In the WANTED, DATA, and conversions, you must write the *number*, *unit*, and chemical *formula* for all terms.

By writing the WANTED unit, you were prompted to write a conversion that was needed to solve. By listing the needed conversions first, you can focus on arranging your conversions when you SOLVE.

Writing out the WANTED, DATA, prompts, and labels takes time. However, this structured method of problem solving will greatly improve your success in the complex problems that soon will be encountered.

PRACTICE   A

Try the last letter on each numbered question. If you answer it correctly, go to the last letter of the next problem. If you need more practice to feel confident, do another letter of the problem. Molar masses for problems 2–4 are found in either problem 1 or the Lesson 8.2 Practice.

1. Working in your notebook, find the molar mass for the following compounds:

    a. H_2SO_4 b. Aluminum nitrate

2. Finish. ? g NaOH = 5.5 mol NaOH · $\dfrac{40.0\ \text{g NaOH}}{1\ \text{mol NaOH}}$ =

3. Supply the needed conversion and solve.

    a. ? g H_2SO_4 = 4.5 mol H_2SO_4 · _____ =

    b. ? g $AgNO_3$ = 0.050 mol $AgNO_3$

        (For molar mass, see Lesson 8.2, Practice problem 4b.)

4. Use WANTED, DATA, *prompt*, and SOLVE to do these problems in your notebook.

    a. 3.6 moles of H_2SO_4 would have a mass of how many grams?

    b. Find the mass in grams of 2.0×10^{-6} moles of $Al(NO_3)_3$. (Answer in scientific notation. See problem 1b.)

Converting Grams to Moles

If the *grams* of a substance with a known chemical formula are *given* in a problem, how do you find the *moles*? To convert between grams and moles, use the molar mass as a conversion.

 TRY IT

Solve the following problem in your notebook, then check the answer below.

   Q. How many moles are in 4.00 grams of O_2?

STOP   **Answer:**

   WANTED: ? mol O_2 =

DATA: 4.00 g O_2 (See grams of a formula? Write its molar mass.)

32.0 g O_2 = 1 mol O_2 (The grams prompt.)

SOLVE: ? mol O_2 = 4.00 $\cancel{g\ O_2}$ · $\dfrac{1\ \text{mol}\ O_2}{32.0\ \cancel{g\ O_2}}$ = **0.125 mol O_2**

S.F.: 4.00 has three *s.f.*, 1 is exact (infinite *s.f.*), 32.0 has three *s.f.*; your answer must be rounded to three *s.f.*

PRACTICE B

Start with the last letter on each numbered question. If you get it right, go to the next number. Need more practice? Do another part. Molar masses for these problems were calculated in the two prior sets of practice.

1. Supply conversions and solve. Answer in numbers without exponential terms.

   a. ? mol H_2SO_4 = 10.0 g H_2SO_4 · _____ =

   b. ? mol $Ba(NO_3)_2$ = 65.4 g $Ba(NO_3)_2$

2. Solve in your notebook. Answer in scientific notation.

   a. 19.6 kg of H_2SO_4 is how many moles?

   b. How many moles are in 51.0 mg of $AgNO_3$?

Lesson 8.4   Converting Particles, Moles, and Grams

Problems Involving a Large Number of Particles

We know that one mole of particles = 6.02×10^{23} particles. A mole is like a dozen, only bigger. You will likely need Avogadro's number in calculations that convert between a *count* of very small particles (such as molecules, atoms, or ions) and units used to measure visible amounts of particles, such as grams or liters.

Let's call this rule

The Avogadro Prompt

If the WANTED and/or *given*

- Includes *both* counts of *invisibly* small particles (such as atoms or molecules) and units used to measure *visible* amounts (such as grams or milliliters), *or*
- Contains any measure of a substance that includes a *two-digit* power of 10: (**10^{xx}** *or* **10^{-xx}**)

write in your DATA:

1 mol (substance formula) = 6.02×10^{23} (substance formula)

Add that rule to the previous

The Grams Prompt

In a calculation, if the WANTED or DATA includes *grams* (or *prefix*-grams) of a substance *formula*, write in your DATA the molar mass *equality* for that formula.
Example: 1 mol H_2O = 18.0 g H_2O

Using the prompts, you may not *always* need the prompt conversion to solve a problem, but you *usually* will. Problems often omit some conversions that you will need to calculate an answer. By applying the prompts, more of the conversions that you need will be in your DATA table for use in the SOLVE step.

In your notebook, apply WANTED, DATA, SOLVE and the two prompts to solve this problem.

TRY IT

Q. What is the mass in grams of 1.5×10^{22} molecules of H_2O?

Answer:

WANTED: ? g H_2O =

DATA: 1.5×10^{22} H_2O molecules

1 mol H_2O = 6.02×10^{23} H_2O molecules   (10^{xx} calls Avogadro prompt.)

1 mol H_2O = 18.0 g H_2O (WANTED unit calls the g prompt.)

SOLVE:

$$? \, \mathbf{g} \, \mathbf{H_2O} = 1.5 \times 10^{22} \text{ molecules } H_2O \cdot \frac{1 \text{ mole } H_2O}{6.02 \times 10^{23} \text{ molecules } H_2O} \cdot \frac{18.0 \, \mathbf{g} \, \mathbf{H_2O}}{1 \text{ mol } H_2O}$$

$$= \frac{1.5 \times 10^{22}}{6.02 \times 10^{23}} \cdot 18.0 \text{ g } H_2O = \mathbf{4.5 \times 10^{-1} \, g \, H_2O} = \mathbf{0.45 \, g \, H_2O}$$

There are several ways to do the arithmetic in the problem above. You may use any that work, but try doing the *exponential* math without a calculator (see Lesson 1.3).

TRY IT

In your notebook, try one more.
   Q. How many atoms are in 5.7 grams of F_2?

Answer:

WANTED: ? atoms **F**

DATA: 5.7 g F_2

1 mol F_2 *molecules* = 38.0 g F_2 *molecules*   (Grams prompt.)

$$1 \text{ mol } F_2 = 6.02 \times 10^{23} \text{ } F_2 \text{ molecules}$$

(Mix of *invisible* atoms and *visible* grams = Avogadro prompt.)

$$1 \text{ } F_2 \text{ } molecule = 2 \text{ F } atoms$$

Note the last equality and the *labels* that include *molecules* above. For most problems in which we work with molecules, the word *molecules* is left out of our labels as understood. However, if a calculation involves both counts of *atoms* and units measuring particles that have more than one atom, we need labels that distinguish between multi-atom particles and atoms.

Let's call this

The Atoms Prompt

If a calculation involves both *multi*-atom particles and a *count* of *atoms*,

- *Add labels* in the WANTED and DATA that distinguish atoms and multi-atom particles
- Write a conversion in the DATA relating the atoms and multi-atom particles

If needed, adjust your work and finish.

TRY IT

SOLVE:

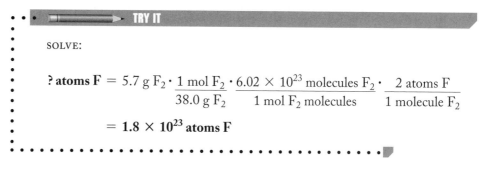

$$? \textbf{ atoms F} = 5.7 \text{ g } F_2 \cdot \frac{1 \text{ mol } F_2}{38.0 \text{ g } F_2} \cdot \frac{6.02 \times 10^{23} \text{ molecules } F_2}{1 \text{ mol } F_2 \text{ molecules}} \cdot \frac{2 \text{ atoms F}}{1 \text{ molecule } F_2}$$

$$= \textbf{1.8} \times \textbf{10}^{23} \textbf{ atoms F}$$

Flashcards

For the cards below, cover the answers, try the questions, and add questions that you cannot answer automatically to your collection. Run them once to perfection, then use them to do the problems below. Repeat for two more days, then put these cards in stack 2 (see Lesson 6.5).

One-Way Cards (with Notch)	Back Side—Answers
To find molar mass from a substance formula	Add the molar masses of its atoms
The units of molar mass	Grams per 1 mole
To convert between grams and moles	Use the molar mass equality
In DATA, write the molar mass as	**1 mol** formula = (molar mass) **g** formula
If you see *grams* or prefix-*grams* in WANTED or DATA	Write the molar mass *equality* in the DATA
If a calculation includes 10^{xx} of a substance	In the DATA, write 1 mol (formula) = 6.02×10^{23} (formula)
If a calculation mixes units measuring visible amounts (g, mol, mL, etc.) with units measuring invisibles (atoms, molecules, particles, etc.)	In the DATA, write 1 mol (formula) = 6.02×10^{23} (formula)

PRACTICE

Run the flashcards, then solve these problems in your notebook. Save one problem for your next study session.

1. 3.55 grams of $Cl_2(g)$ (chlorine gas) contain how many molecules of Cl_2?

2. 8.0×10^{24} atoms of aluminum have a mass of how many kilograms?

3. How many millimoles of oxygen atoms are in 6.40×10^{-2} g O_2?

4. 2.57 nanograms of S_8 contain how many sulfur atoms?

5. What is the mass in grams of exactly 25 molecules of water? Answer in both scientific and fixed decimal notation.

6. Check the conversions in your answers. In how many cases did you start by converting the *given* unit until moles of the *given* substance was on top?

7. Given a sample of dry crystals of a pure substance, what information would you need to know to find the moles of the substance in the sample?

Lesson 8.5   Review Quiz for Chapters 7–8

You may use a calculator and a periodic table. Work in your problem notebook and set a 20-minute limit, then check your answers at the end of the chapter.

1. Lesson 7.3: Write separated-ion formulas for these compounds.

    a. Ag_2SO_4

    b. $NaOH$

    c. K_2CrO_4

2. Lesson 7.3: Write ionic solid formulas for these compounds.

    a. Sodium dichromate

    b. Ammonium phosphate

    c. Aluminum chlorate

3. Lessons 7.2–7.3: Name these compounds.

    a. Br_2O_7

    b. $KClO_4$

    c. Na_2CO_3

4. Lesson 7.2: Which of the compounds in questions 2 and 3 are covalent?

5. Lesson 8.2: The molar mass of $Co(NO_3)_2$ is

    a. 104.9 g/mol   b. 134.9 g/mol   c. 150.9 g/mol

    d. 182.9 g/mol   e. 216.9 g/mol

6. Lesson 8.4: 1.80×10^{23} molecules of CO_2 have what mass in grams?

    a. 0.0840 g CO_2 b. 0.840 g CO_2 c. 8.40 g CO_2

    d. 13.2 g CO_2 e. 9.60 g CO_2

7. Lesson 8.4: 0.72 kg of water contains how many atoms?

    a. 2.4×10^{22} atoms   b. 2.4×10^{25} atoms   c. 7.2×10^{25} atoms

    d. 7.2×10^{22} atoms   e. 2.4×10^{26} atoms

SUMMARY

1. Chemical processes are easiest to understand if you count the particles involved. Large numbers of particles are counted using the mole.

2. 1 mole of particles = 6.02×10^{23} particles. That's Avogadro's number.

3. If you know the chemical formula for a substance, you can calculate the *grams per mole* of the substance: the molar mass.

4. To find the molar mass of a substance, add the molar masses of its atoms.

5. The units of molar mass are grams *per* one mole.

6. If molar mass is WANTED, write WANTED: ? $\dfrac{\text{g}}{\text{mol}}$

7. If a molar mass is DATA, write 1 mol formula = **XX g** formula

8. To *convert* between grams and moles, use the molar mass as a conversion factor.

9. The grams prompt: If a problem mentions *grams* or prefix-*grams* of a substance with a known *formula*, write in the DATA:

   (molar mass) grams of formula = 1 mol of formula

10. The Avogadro prompt: If the WANTED and/or *given* includes *both* counts of *invisibly* small particles and units used to measure *visible* amounts, *or* contains any measure of a substance that includes 10^{xx} or 10^{-xx}, write in your DATA:

    1 mol (formula) = 6.02×10^{23} (formula)

11. The atoms prompt: If a calculation involves both a *multi*-atom particle and a count of *atoms*, add labels in the WANTED and DATA that distinguish atoms and the *multi*-atom particle, and write a conversion in the DATA relating the atoms and the *multi*-atom particle.

12. To solve most chemistry calculations, *first convert to moles*.

ANSWERS

Lesson 8.1

Practice   1. 2.4×10^{24} particles

           2. ? particles = 0.050 mol $\times$ (6.02×10^{23} particles/mol) = **3.0×10^{22} particles**

Lesson 8.2

Practice   1a. Nitrogen: **$14.0 \dfrac{\text{g}}{\text{mol}}$**   1b. Au: **$197.0 \dfrac{\text{g}}{\text{mol}}$**   1c. Pb: **$207.2 \dfrac{\text{g}}{\text{mol}}$**

2a. $Ca(OH)_2$: **2 oxygens**   2b. $Al_2(SO_4)_3$: **12 oxygens**   2c. $Co(NO_3)_2$: **6 oxygens**

3a. $H_2 = 2 \times H = 2 \times 1.008 = $ **2.016 g/mol**

In the DATA, write **2.016 g H_2 = 1 mol H_2** (multiplying by an exact subscript does not change the *place* with doubt).

3b.   NaH =

Na = 23.0

H =   1.008

24.008 = **24.0 g/mol**

In the DATA, write **24.0 g NaH = 1 mol NaH**

3c.   KSCN =

K = 39.1

S = 32.1

C = 12.0

N = 14.0

97.2 g/mol

97.2 g KSCN = 1 mol KSCN

3d. Na_3PO_4 =

$3 \times Na = 3 \times 23.0 =$ 69.0

$1 \times P = 1 \times 31.0 =$ 31.0

$4 \times O = 4 \times 16.0 =$ 64.0

164.0 g/mol

164.0 g Na_3PO_4 = 1 mol Na_3PO_4

3e.   Barium nitrate = $Ba(NO_3)_2$ (Lesson 7.3) =

$1 \times Ba = 1 \times 137.3 = 137.3$

$2 \times N = 2 \times 14.0 =$ 28.0

$6 \times O = 6 \times 16.0 =$ 96.0

261.3 g/mol

261.3 g $Ba(NO_3)_2$ = 1 mol $Ba(NO_3)_2$

4.   This question asks for the grams per one mole. That's the molar mass.

4a. H_2S =

$2 \times H = 2 \times 1.008 =$ 2.016

$1 \times S = 1 \times 32.1 =$ 32.1

34.116 = **34.1** g/mol

1 mol H_2S = 34.1 g H_2S

4b. $AgNO_3$ =

$1 \times Ag = 1 \times 107.9 = 107.9$

$1 \times N = 1 \times 14.0 =$ 14.0

$3 \times O = 3 \times 16.0 =$ 48.0

169.9 g/mol

169.9 g $AgNO_3$ = 1 mol $AgNO_3$

Lesson 8.3

Practice A

1a. H_2SO_4 =

$2 \times H = 2 \times 1.008 =$ 2.016

$1 \times S = 1 \times 32.1 = 32.1$

$4 \times O = 4 \times 16.0 =$ 64.0

98.116 = **98.1** g/mol

98.1 g H_2SO_4 = 1 mol H_2SO_4

1b.   Aluminum nitrate = **$Al(NO_3)_3$**

$1 \times Al = 1 \times 27.0 =$ 27.0

$3 \times N = 3 \times 14.0 =$ 42.0

$9 \times O = 9 \times 16.0 = $ 144.0

213.0 g/mol

213.0 g $Al(NO_3)_3$ = 1 mol $Al(NO_3)_3$
(Note that multiplying by an exact 9—or any exact number—does not change the *place* with doubt.)

2.   ? g NaOH = 5.5 ~~mol NaOH~~ · $\dfrac{40.0 \text{ g NaOH}}{1 \text{ ~~mol NaOH~~}}$ = **220 g NaOH**

3a.   ? g H_2SO_4 = 4.5 mol H_2SO_4 · $\dfrac{98.1 \text{ g } H_2SO_4}{1 \text{ mol } H_2SO_4}$ = **440 g H_2SO_4**

3b. ? g $AgNO_3$ = 0.050 mol $AgNO_3 \cdot$ $\dfrac{169.9 \text{ g } AgNO_3}{1 \text{ mol } AgNO_3}$ = **8.5 g $AgNO_3$**

4a. ? g H_2SO_4 = 3.6 mol $H_2SO_4 \cdot$ $\dfrac{98.1 \text{ g } H_2SO_4}{1 \text{ mol } H_2SO_4}$ = **350 g H_2SO_4**

4b. ? g $Al(NO_3)_3$ = 2.0×10^{-6} mol $Al(NO_3)_3 \cdot$ $\dfrac{213.0 \text{ g } Al(NO_3)_3}{1 \text{ mol } Al(NO_3)_3}$ = **4.3×10^{-4} g $Al(NO_3)_3$**

Practice B

1a. ? mol H_2SO_4 = 10.0 ~~g H_2SO_4~~ $\cdot$ $\dfrac{1 \text{ mol } H_2SO_4}{98.1 \text{ ~~g H_2SO_4~~}}$ = **0.102 mol H_2SO_4**

1b. ? mol $Ba(NO_3)_2$ = 65.4 g $Ba(NO_3)_2 \cdot$ $\dfrac{1 \text{ mol } Ba(NO_3)_2}{261.3 \text{ g } Ba(NO_3)_2}$ =

Answer:

If you wrote **0.250 mol $Ba(NO_3)_2$**, go to the head of the class.

Always write a **0** in front of a decimal point if there is no number in front of the decimal point. This makes the decimal point *visible* when you need this answer for a later step of a lab report or test.

2a. ? mol H_2SO_4 = 19.6 kg $H_2SO_4 \cdot$ $\dfrac{10^3 \text{ g}}{1 \text{ kg}} \cdot \dfrac{1 \text{ mol } H_2SO_4}{98.1 \text{ g } H_2SO_4}$ = **2.00×10^2 mol H_2SO_4**

2b. ? mol $AgNO_3$ = 51.0 mg $AgNO_3 \cdot$ $\dfrac{10^{-3} \text{ g}}{1 \text{ mg}} \cdot \dfrac{1 \text{ mol } AgNO_3}{169.9 \text{ g } AgNO_3}$ = **3.00×10^{-4} mol $AgNO_3$**

Lesson 8.4

Practice   Your paper should look like this, but you may omit the comments in parentheses.

1. WANTED: ? *molecules* Cl_2

    DATA: 3.55 g Cl_2

    71.0 g Cl_2 = 1 mol Cl_2 (Grams prompt.)

    1 mol Cl_2 = 6.02×10^{23} molecules Cl_2 (Mix g and invisibles = Avogadro prompt.)

    SOLVE: ? molecules Cl_2 = 3.55 g $Cl_2 \cdot$ $\dfrac{1 \text{ mol } Cl_2}{71.0 \text{ g } Cl_2} \cdot \dfrac{6.02 \times 10^{23} \text{ molecules } Cl_2}{1 \text{ mol } Cl_2}$ = **3.01×10^{22} molecules Cl_2**

2. In metals, the particles in the "*molecular* formula" are individual atoms, so the metal "molecules" are the same as the metal *atoms*, and the molar mass is the mass of a mole of metal *atoms*.

    WANTED: ? kg Al

    DATA: 8.0×10^{24} Al atoms

    1 mol Al = 6.02×10^{23} Al atoms   (10^{xx} = Avogadro prompt.)

    1 mol Al = 27.0 g Al   (Any prefix-*grams* = grams prompt.)

    SOLVE: ? kg Al = 8.0×10^{24} Al atoms $\cdot$ $\dfrac{1 \text{ mol Al atoms}}{6.02 \times 10^{23} \text{ Al atoms}} \cdot \dfrac{27.0 \text{ g Al}}{1 \text{ mol Al}} \cdot \dfrac{1 \text{ kg}}{10^3 \text{ g}}$ = **0.36 kg Al**

3. WANTED: ? millimoles O atoms

    DATA: 6.40×10^{-2} g O_2 molecules

    32.0 g O_2 = 1 mol O_2 molecules   (Grams prompt.)

    1 molecule O_2 = exactly 2 atoms O   (Atoms prompt.)

SOLVE: ? mmol O atoms = 6.40×10^{-2} g $O_2 \cdot \dfrac{1 \text{ mol } O_2}{32.0 \text{ g } O_2} \cdot \dfrac{2 \text{ atoms O}}{1 \text{ mol } O_2} \cdot \dfrac{1 \text{ mmol}}{10^{-3} \text{ mol}}$ = **4.00 mmol O atoms**

If milliliters = mL, then millimoles = mmol.

Because the WANTED and *given* units were grams and moles, the molar mass that converts between grams and moles was needed, but the Avogadro conversion needed for invisibles was not.

4. WANTED: ? atoms S

   DATA: 2.57 nanograms S_8

   1 ng = 10^{-9} g   (Writing less frequently used prefix equalities is a good idea.)

   256.8 g S_8 = 1 mol S_8 molecules   (Any *prefix*-grams = grams prompt.)

   1 mol S_8 = 6.02×10^{23} molecules S_8 (*Invisible atoms* and visible *moles* = Avogadro prompt.)

   1 molecule S_8 = exactly 8 atoms S   (Atoms prompt.)

   (Note that the molar mass of S_8 is the mass of a mole of *molecules*. If needed, adjust your work and complete the problem.)

   SOLVE:

? atoms S = 2.57 ng $S_8 \cdot \dfrac{10^{-9} \text{ g}}{1 \text{ ng}} \cdot \dfrac{1 \text{ mol } S_8}{256.8 \text{ g } S_8} \cdot \dfrac{6.02 \times 10^{23} \text{ molecules } S_8}{1 \text{ mol } S_8} \cdot \dfrac{8 \text{ atoms S}}{1 \text{ molecule } S_8}$ = **4.82×10^{13} atoms S**

   S.F.: The 8 atoms per molecule is exact, and the 8 therefore does not restrict the *s.f.* in the answer. Exact numbers have infinite *s.f.*

5. WANTED: ? g H_2O

   DATA: 25 molecules H_2O (Exact.)

   1 mol H_2O = 6.02×10^{23} H_2O molecules   (Invisible molecules, visible g = Avogadro prompt.)

   1 mol H_2O = 18.0 g H_2O (WANTED unit = grams prompt.)

   SOLVE: ? g H_2O = 25 molecules $H_2O \cdot \dfrac{1 \text{ mol } H_2O \text{ molecules}}{6.02 \times 10^{23} \text{ } H_2O \text{ molecules}} \cdot \dfrac{18.0 \text{ g } H_2O}{1 \text{ mol } H_2O}$ = **7.48×10^{-22} g H_2O**

   7.48×10^{-22} g H_2O = **0.000 000 000 000 000 000 000 748 grams H_2O**

   Molecules have a very small mass.

6. All have moles of the *given* substance on top. In chemistry calculations, if you are not given moles, job 1 is nearly always to convert to moles of *given* substance, because nearly all relationships in chemistry are based on counts of particles, measured in moles of particles.

7. Methods to find moles of a pure substance include: knowing the mass of the sample in *grams* and its molar mass (*g/mol*), you can convert to its *moles*. Or, knowing the grams and the *formula* of a substance, from the formula you can calculate the g/mol and then the moles in the sample.

Lesson 8.5

1a. $2 \text{ Ag}^+ + 1 \text{ SO}_4^{2-}$ 1b. $1 \text{ Na}^+ + 1 \text{ OH}^-$ 1c. $2 \text{ K}^+ + 1 \text{ CrO}_4^{2-}$ (Coefficients of 1 may be omitted.)

2a. $Na_2Cr_2O_7$ 2b. $(NH_4)_3PO_4$ 2c. $Al(ClO_3)_3$

3a. Dibromine heptoxide (or heptaoxide)   3b. Potassium perchlorate   3c. Sodium carbonate

4. Only 3a   5. **d. 182.9 g/mol**   6. **d. 13.2 g CO_2**   7. **c. 7.2×10^{25} atoms**   (3 atoms per molecule.)

9

Stoichiometry

Lesson 9.1   Chemical Reactions and Equations

Chemical Reactions

An example of a **chemical reaction** is the burning of hydrogen gas (H_2) to produce steam (hot H_2O gas). In chemistry, to **burn** something is to react it with oxygen gas (O_2) to form one or more new substances.

Chemical equations are the language used to describe chemical reactions. In a chemical equation written using *molecular* formulas, the above reaction would be represented as

$$2\,H_2 + O_2 \rightarrow 2\,H_2O \qquad\qquad (9.1)$$

This equation can be read as either "two H two plus one O two react to form two H two O," or as "two molecules of hydrogen plus one molecule of oxygen react to form two molecules of water."

The substances on the left side of a reaction equation are termed the **reactants**. The substances on the right side of the arrow are the **products**.

> In chemical reactions, reactants are *used up* and products *form*.

Most chemical reactions are represented by equations using molecular formulas, as in equation 9.1 above. However, more information is supplied if the equation is written using structural formulas. An example is

$$
\begin{array}{c}
\text{H} \\
| \\
\text{H}
\end{array}
\text{O}=\text{O}
\begin{array}{c}
\text{H} \\
| \\
\text{H}
\end{array}
\rightarrow
\begin{array}{c}
\text{H} \\
\diagdown \\
\text{H} \diagup
\end{array}
\text{O} \quad \text{O}
\begin{array}{c}
\diagup \text{H} \\
\diagdown \text{H}
\end{array}
\qquad (9.2)
$$

 TRY IT

(See "How to Use These Lessons," point 1, p. xv.)

Q. Compare equation 9.1 to equation 9.2. Are they the same reaction?

STOP   **Answer:**

Yes.

However, by writing the structural formulas it is easier to see that in many respects, after the reaction, not much has changed. We began with four hydrogen atoms and two oxygen atoms; we end with the same.

> In chemical reactions, bonds break, and new bonds form, but the number and kinds of atoms stay the same.

The fact that chemical reactions can neither create nor destroy atoms is called the **law of conservation of atoms** or the **law of conservation of matter**. In this usage, *conservation* means that what you start with is conserved at the end.

Before, during, and after a reaction, there is also **conservation of mass**: the total mass of the reactants and products also does not change. Total mass is determined by the number and kind of atoms, which a chemical reaction does not change.

What does change? Because of the new positions of the bonds, after the reaction the products will have characteristics and behavior that are different from those of the reactants.

In the above reaction, the molecules on the left are explosive when ignited, but the water molecules on the right are quite stable. The oxygen molecules on the left cause many materials to burn. To stop burning, we often use the water on the right.

The position of the bonds can also change the *economic value* of atoms. The historic importance of chemistry to society has included the discovery of reactions that change

- Brittle rock into metals that can be molded and shaped
- Willow bark into aspirin; and fungus into antibiotics
- Animal waste into explosives; and sand into computer chips for electronic devices

Another outcome of chemical reactions is quite often the storage or release of *energy*. In burning hydrogen to form water, large amounts of stored energy are released. Including the energy term, the burning of hydrogen is represented as

$$2\,H_2 + O_2 \rightarrow 2\,H_2O + energy$$

It was the release of energy from this simple reaction that led to the explosive destruction of the airship *Hindenburg* in 1937. Currently, researchers are seeking ways to harness the energy released by burning hydrogen as an alternative to burning gasoline in cars.

Nearly all reactions either absorb or release energy; however, if energy is not the focus of a particular problem, the energy term is often omitted when writing a reaction equation.

Reaction Equation Terminology

1. In the substance formula H_2O, the 2 is a *subscript*. The omission of a subscript, such as after the O in H_2O, means the subscript is understood to be 1.

2. In a reaction equation, if **5 H_2O** is a term, the **5** is a called a **coefficient**. Coefficients are exact numbers that express the exact *particle ratios* in a reaction.

   It is important to distinguish between subscripts and coefficients.
   - Subscripts are numbers written *after* and *lower than* the atom symbols in a molecule or ion formula. Subscripts count the atoms of each type inside the particle.
   - Coefficients are numbers written in *front* of a particle formula. Coefficients represent the particle ratios in a reaction: a count of how many of one particle reacts with how many of another particle.

3. In a reaction equation involving substances, if the number and kind of atoms on each side of the arrow is the same, the equation is said to be **balanced**. The coefficients of a *balanced* equation show the exact ratios in which the particles react (are used up) and are formed.

In a balanced equation, writing a coefficient of 1 is optional. If no coefficient is written in front of an equation term, the coefficient is understood to be 1.

4. To balance equations, we will need to *count atoms* based on coefficients and subscripts.

> To count each kind of atom in a term in a reaction equation, multiply the coefficient times the subscript(s) for the atom.

Example: The term **5 H_2O** represents five molecules of water. Each molecule has three atoms. In those five molecules are $(5 \times 2) =$ **10 hydrogen** atoms *and* $(5 \times 1) =$ **5 oxygen** atoms.

TRY IT

Q. Count the **H** atoms in the following:

  a. **5 CH_4**

  b. **3 CH_3COOH**

  c. **2 $Pb(C_2H_5)_4$**

 Answer:

  a. Each molecule has four H atoms. Five molecules $= 5 \times 4\,H =$ **20 H** atoms

  b. Each molecule has four H atoms. Three molecules $= 3 \times 4\,H =$ **12 H** atoms

  c. The H in this case has *two* subscripts. Multiply $2 \times 4 \times 5 =$ **40 H** atoms

---

PRACTICE

If you are not sure that your answer is correct, check it before proceeding to the next question.

1. Label the *reactants* and *products* in this reaction equation. Circle the coefficients.

$$4\,Fe + 3\,O_2 \rightarrow 2\,Fe_2O_3$$

2. How many *oxygen* atoms are represented in the following?

  a. 7 Na_3PO_4 b. 3 $Co(OH)_2$

  c. 2 $No(NO_3)_2$ d. 5 $Al_2(SO_4)_3$

3. How many *total* atoms are represented in 2a and 2d above?

4. The following equation uses structural rather than molecular formulas.

$$C \; + \; C \; + \; H\!-\!H \; + \; H\!-\!H \; + \; H\!-\!H \; \rightarrow \; H\!-\!\underset{\underset{\displaystyle H}{|}}{\overset{\overset{\displaystyle H}{|}}{C}}\!-\!\underset{\underset{\displaystyle H}{|}}{\overset{\overset{\displaystyle H}{|}}{C}}\!-\!H$$

a. Is the equation balanced?

b. Write the reaction using *molecular* formulas (the type used in problem 1).

c. In going from reactants to products, what changed? What stayed the same?

Changes:

Stays the same:

Lesson 9.2   Balancing Equations

Balancing by Trial and Error

The coefficients that balance an equation are not always supplied with the equation. However, if you are given the chemical formulas for the reactants and products of a reaction, you can determine the coefficients that balance the equation by *trial and error.*

One consequence of the law of conservation of atoms is that *only one set of ratios* will balance a chemical equation. However, because coefficients are ratios, if you multiply all of the coefficients by the same number, you continue to have a balanced equation. This means that a balanced equation may be shown with different sets of coefficients, as long as the ratios among the coefficients are the same.

Example: $2 \; H_2 \; + \; O_2 \rightarrow 2 \; H_2O$ *and* $H_2 \; + \; 1/2 \; O_2 \rightarrow H_2O$ are the same balanced equation, because the coefficient *ratios* are the same. Note that if no coefficient is shown, a 1 is understood.

In a balanced equation, showing a coefficient that is *one* is not required, but it's not improper, either.

There are many ways to balance equations. In later lessons, we will learn methods that balance specific types of equations more quickly. However, trial-and-error balancing uses the same rules for all types of equations, and with perseverance works for all types of equations, so it's a good place to start.

Let's learn to balance with an example. Using the question below, apply the rules and steps below the question, then check your answers.

⟶ TRY IT

Q. Add the coefficients that balance this equation for the burning of *n*-propanol.

$$C_3H_7OH + O_2 \rightarrow CO_2 + H_2O$$

Steps in balancing:

1. Most important: During balancing, you cannot change a formula or a subscript. To balance, *you* must add numbers to the equation, but the *only* numbers that you can add are the *coefficients* that go in *front* of substance formulas.

2. To start, put a coefficient of 1 in front of the *most complex* formula (the one with the most atoms or the most different kinds of atoms). If two formulas seem complex, choose either one.

   Writing coefficients that are 1 is optional, but in an equation that you need to balance, including the 1s helps in tracking which coefficients have been determined.

STOP

Answer:

In this case, the first formula is the most complex, so start with

$$\mathbf{1}\,C_3H_7OH + O_2 \rightarrow CO_2 + H_2O$$

3. Now add coefficients to the other side that must be true if atoms are balanced.

STOP

   In chemical reactions, atoms cannot be created or destroyed. This means that in chemical equations, each side of the arrow must have the same kinds of atoms, and the same number of each kind of atom.

> Each **term** in a chemical equation is a coefficient followed by a substance formula.
> For a given atom, to count the number of atoms represented by a term, multiply the coefficient by the subscript(s) for that atom.
> To count each type of atom on a *side*, add the atoms in each term on that side.

   Above, the left side has three carbon atoms. Because only CO_2 on the right has C, the only way to have three carbon atoms on the right is to have the CO_2 coefficient be 3. The left has eight hydrogen atoms total; because only H_2O on the right has H, the only way to have eight H atoms on the right is to have the H_2O coefficient be 4. So far, this gives us

$$1\,C_3H_7OH + \underline{} O_2 \rightarrow \mathbf{3}\,CO_2 + \mathbf{4}\,H_2O$$

The right side is now finished, because each particle has a coefficient. Only the O_2 on the left side lacks a coefficient.

4. Add the coefficient that must be true for the oxygen to balance.

STOP

   We count the oxygens on the right side and get 10. On the left, we see one oxygen in propanol, which means we must have a total of **nine** oxygens from O_2. We can write

$$1\,C_3H_7OH + \mathbf{9/2}\,O_2 \rightarrow 3\,CO_2 + 4\,H_2O$$

or we can multiply *all* of the coefficients by 2 to avoid using fractions.

$$2\,C_3H_7OH + 9\,O_2 \rightarrow 6\,CO_2 + 8\,H_2O$$

Because coefficients are ratios, and both sets of ratios above are the same, both answers above are *equally correct*. We can multiply all the coefficients by the same number and still have the same ratios *and* a balanced equation.

Balancing Using Fractions as Coefficients

Initially, our primary use for coefficients will be as *ratios* in calculating amounts of substances involved in chemical reactions. The arithmetic in these calculations will be easier if the coefficients are converted to *whole numbers* at the end of balancing. Any whole numbers with proper ratios would be correct, but *lowest* whole-number ratios are preferred.

Fractions are permitted when adding coefficients to balance equations, and in some types of problems, including some energy calculations, the use of fractions to balance equations is required. In other situations, fractions may be inappropriate (you cannot have half a molecule). We will address these differences as we encounter them.

In a few problems below, we will include balancing with fractions to be familiar with their use. However, in *most* upcoming cases, if we encounter a fraction when balancing, it will simplify the arithmetic if at that point all of the coefficients are multiplied by the fraction's denominator. This will change coefficients that are fractions to lowest whole numbers.

Our rule will be

> Unless other coefficients are specified or required in a problem, convert to lowest-whole-number coefficients at the final step in balancing equations.

PRACTICE

Read each numbered step below, then do every other or every third lettered problem. As you go, check your answers. If you need more practice at a step, do a few more letters for that step. Save a few for your next practice session.

1. A balanced equation must have the same number of each kind of atom on both sides.
   To check for a balanced equation, count the total number of one kind of atom on one side, then count the number of that kind of atom on the other side. The left- and right-side counts must be equal.
   Repeat those counts for each kind of atom in the equation.
   Using this counting method, label each equation below as *balanced* or *unbalanced*.

   a. $2\,Cs + Cl_2 \rightarrow 2\,CsCl$

   b. $4\,HI + O_2 \rightarrow 2\,H_2O + I_2$

   c. $Pb(NO_3)_2 + 2\,LiBr \rightarrow PbBr_2 + LiNO_3$

   d. $BaCO_3 + 2\,NaCl \rightarrow Na_2CO_3 + BaCl_2$

(continued)

2. In the equations below, one coefficient has been supplied. Use that coefficient to decide one or more coefficients on the *other* side. Then use your added coefficient(s) to go back and forth, from side to side, filling the remaining blanks on both sides to balance the equation.

    In these, some coefficients may be fractions. Fractions are sometimes needed when balancing.

    Remember that balancing is trial and error. Do what works. If you need help, check your answer after each letter.

    Tip: It helps to balance *last* an atom that is used in two or more different formulas on the same side. Oxygen is the atom most frequently encountered in compounds, so "saving O until last" usually helps in balancing.

    a. $4 \, Al + \underline{\hspace{1cm}} O_2 \rightarrow \underline{\hspace{1cm}} Al_2O_3$

    b. $3 \, Ca + \underline{\hspace{1cm}} N_2 \rightarrow \underline{\hspace{1cm}} Ca_3N_2$

    c. $\underline{\hspace{1cm}} P_4 + \underline{\hspace{1cm}} O_2 \rightarrow 2 \, P_4O_6$

    d. $\underline{\hspace{1cm}} C_6H_{14} + \underline{\hspace{1cm}} O_2 \rightarrow 18 \, CO_2 + \underline{\hspace{1cm}} H_2O$

    e. $\underline{\hspace{1cm}} MgH_2 + \underline{\hspace{1cm}} H_2O \rightarrow 1 \, Mg(OH)_2 + \underline{\hspace{1cm}} H_2$

    f. $\underline{\hspace{1cm}} C_2H_6 + 7 \, O_2 \rightarrow \underline{\hspace{1cm}} CO_2 + \underline{\hspace{1cm}} H_2O$

3. Balance these equations. Start by placing a coefficient of **1** in front of the underlined substance.

    a. $K + \underline{F_2} \rightarrow KF$

    b. $Cs + \underline{O_2} \rightarrow Cs_2O$

    c. $PCl_3 \rightarrow \underline{P_4} + Cl_2$

    d. $\underline{C_2H_5OH} + O_2 \rightarrow CO_2 + H_2O$

    e. $FeS + O_2 \rightarrow \underline{Fe_2O_3} + SO_2$

4. When balancing equations without suggested ways to start,

    > *Begin* by putting a **1** in front of the most complex formula on *either* the left or right side of the equation (the one with the most atoms, or the most different kinds of atoms). If two formulas are complex, use either one.

    Then add as many coefficients as you are sure of to the side *opposite* the side where you put the **1**.

    On this problem, if you get a fraction as you balance, multiply *all* of the existing coefficients by the denominator of the fraction. Repeat this step if you get additional fractions while balancing. Fractions are permitted, but it will be easier in most calculations if you have whole-number coefficients.

Expect to need to start over on occasion, because balancing is trial and error. Be persistent! All of the equations below can be balanced.

a. $Mg + O_2 \rightarrow MgO$

b. $N_2 + O_2 \rightarrow NO$

c. $C_6H_6 + O_2 \rightarrow CO_2 + H_2O$

d. $P_4 + O_2 \rightarrow P_4O_6$

e. $Al + HBr \rightarrow AlBr_3 + H_2$

5. *Lowest*-whole-number coefficients are not required, but they are preferred when writing most balanced equations.

Example: If all of your coefficients are *even* numbers at the end of balancing, it is preferred to divide all the coefficients by 2.

This converts $4\,H_2 + 2\,O_2 \rightarrow 4\,H_2O$ to $2\,H_2 + O_2 \rightarrow 2\,H_2O$

The latter equation has *lowest-whole-number* coefficients.

On these, balance, then convert your answers to lowest-whole-number ratios.

a. $Al_2O_3 + HCl \rightarrow AlCl_3 + H_2O$

b. $Fe_3O_4 + H_2 \rightarrow Fe + H_2O$

c. $C + SiO_2 \rightarrow CO + SiC$

d. $N_2 + O_2 + H_2O \rightarrow HNO_3$

e. $Rb_2O + H_2O \rightarrow RbOH$

f. $Mg(NO_3)_2 + Na_3PO_4 \rightarrow Mg_3(PO_4)_2 + NaNO_3$

g. $Pb(C_2H_5)_4 + O_2 \rightarrow PbO + CO_2 + H_2O$

h. $Cd + HNO_3 \rightarrow Cd(NO_3)_2 + H_2O + NO$

6. Working in your notebook, write the formulas, then balance these. (Need formula help? See Lessons 7.2 and 7.3.)

a. Dinitrogen tetroxide $\rightarrow$ nitrogen dioxide

b. Barium carbonate + cesium chloride $\rightarrow$ cesium carbonate + barium chloride

c. Silver nitrate + calcium iodide $\rightarrow$ silver iodide + calcium nitrate

Lesson 9.3   Using Coefficients: Molecules to Molecules

Flashcards

Let's summarize the reaction fundamentals learned so far.

Cover the flashcard answers below, then put a check by the questions you can answer easily and quickly. Make flashcards (Lesson 2.3) for those you cannot. Consider using a different color of card for this new chapter. Run the new cards for three sessions in a row, then put them in stack 2 (the Run before End of Chapter/Next Quiz Stack).

One-Way Cards (with Notch)	Back Side—Answers
Cannot change during chemical reactions	Atoms present and total mass
In reaction equations, used up on left are the	Reactants
In reaction equations, formed on right are the	Products
Numbers inside a formula that count atoms	Subscripts
Numbers you add to balance equations	Coefficients
To balance an equation, start by	Putting 1 in front of most complex formula
The law of conservation of mass means	During a reaction, total mass does not change
What is balanced in a balanced equation that consists of neutral substance formulas?	The number and kind of atoms on each side

The Meaning of Coefficients

Balanced chemical equations tell the *ratios* in which molecules or formula units are used up and form. For example, the burning of paraffin, a hydrocarbon that is used as the fuel in some types of candles, can be represented as

$$C_{25}H_{52} + 38\ O_2 \rightarrow 25\ CO_2 + 26\ H_2O$$

The equation states that burning 1 molecule of paraffin uses up 38 molecules of oxygen. The products are 25 molecules of carbon dioxide and 26 molecules of water. Reactants are used up and products form in known, predictable ratios.

What would happen if you burned two molecules of paraffin? You would use up twice as much oxygen (76 molecules), and you would produce twice as much product: 50 molecules of CO_2 and 52 molecules of H_2O.

In a reaction, the amounts involved can be varied, but the *ratios* of the substances used up and formed must stay the same.

PRACTICE

Do these in your head and write your answers below. Check your answers after each problem as you go.

1. For the balanced equation: $4\ HBr + O_2 \rightarrow 2\ Br_2 + 2\ H_2O$

   a. If 16 molecules of HBr react, how many molecules of O_2 must react? How many molecules of Br_2 must form?

   b. If five molecules of O_2 react, how much HBr must react? How much Br_2 must form?

2. For the reaction $\quad\quad CS_2 + \quad\quad O_2 \rightarrow \quad\quad CO_2 + \quad\quad SO_2$

   a. Balance the equation.

   b. If 25 trillion molecules of CS_2 react, how many molecules of O_2 must also be used up? How many molecules of SO_2 must form?

Lesson 9.4 — Mole-to-Mole Conversions

Conversions Based on Coefficients

Coefficients are *exact* numbers because they are determined by counting atoms as you balance equations. In chemical reactions and processes, atoms are indivisible. You may have 1, 2, or 3 atoms, but you cannot have 3.1 atoms. Coefficients therefore have no uncertainty.

The balanced equation $2\,H_2 + O_2 \rightarrow 2\,H_2O$ means that

- If two molecules of H_2 are used up, exactly one molecule of O_2 must be used up
- Two molecules H_2 consumed equals exactly two molecules of H_2O formed
- If one molecule of O_2 reacts, exactly two molecules of H_2O must form

In this sense, any two terms in a balanced equation are "equal." Any two terms that are equal or equivalent can be made into conversion factors. From the equation above, we can write three equalities that can be used as conversions:

- Two molecules **H_2** used up = one molecule **O_2** also used up
- Two molecules **H_2** used up = two molecules **H_2O** formed
- One molecule **O_2** used up = two molecules **H_2O** formed

Mole-to-Mole Conversions

The units in the above equalities are molecules. Coefficients can also be read in *exact moles*, because once the coefficient ratios from the balanced equation are known, those ratios can be multiplied by any number and the equation is still exactly balanced. A mole is simply a large number.

This means that because the above three molecule-to-molecule equalities are true, the same equalities must be true if we substitute *moles* for molecules. In the ratios of reaction for

$$2\,H_2 + O_2 \rightarrow 2\,H_2O$$

- Two moles of H_2 are used up *per* one mole O_2 used up, and
- Two moles of H_2 are used up *per* two moles of H_2O formed

We can translate these words into equalities (as we did in Chapter 5):

$$2\,\text{mol}\ H_2 = 1\,\text{mol}\ O_2$$

$$2\,\text{mol}\ H_2 = 2\,\text{mol}\ H_2O$$

$$1\,\text{mol}\ O_2 = 2\,\text{mol}\ H_2O$$

We can use these equalities as conversions to solve calculations for this reaction.

By balancing an equation, we obtain coefficients that can be used to make simple whole-number conversion factors. In reaction calculations, these mole-to-mole ratios, using coefficients, will be our bridge conversions between the *given* and WANTED units.

We could use molecule-to-molecule coefficients to solve problems, and in some cases we do. However, most reaction calculations involve visible amounts of a substance, and for visible amounts the numbers are easier to work with if we count particles in *moles*.

The key rule in solving reaction calculations is

The Mole-to-Mole Bridge Prompt

In *reaction* calculations, after listing the WANTED and DATA, if the WANTED substance $\neq$ the *given* substance ($\neq$ means *does not equal*),

- Balance the equation using whole-number coefficients and
- Bridge by writing the *bridge* conversion: X **mol** WANTED $=$ Y **mol** *given* in which X and Y are the substance coefficients from the balanced equation
- Use the bridge conversion to SOLVE

 TRY IT

In your notebook, apply the bridge prompt to solve the problem below.

Q. Burning ammonia with appropriate catalysts can result in the formation of nitrogen monoxide by the reaction

$$4\,NH_3 + 5\,O_2 \rightarrow 4\,NO + 6\,H_2O$$

How many moles of NH_3 are needed to form 9.0 mol H_2O?

Answer:

WANTED: ? mol **NH_3**

DATA: 9.0 mol **H_2O**

Strategy: Because $NH_3 \neq H_2O$, balance the equation (done above) and then write the WANTED moles to *given* moles equality. We could write

4 mol NH_3 used up = 6 mol H_2O formed

and that would be a complete label, but in the interest of time we usually abbreviate the bridge conversion as simply

4 mol NH_3 = 6 mol H_2O   (Finish from here.)

SOLVE: Begin with the starting template.

? mol NH_3 = 9.0 mol H_2O · _____ ; then use the bridge.
                                     mol H_2O

? mol NH_3 = 9.0 ~~mol H_2O~~ · $\dfrac{\textbf{4 mol NH}_3}{\textbf{6 } \sim\sim\text{mol H}_2\text{O}}$ = 6.0 mol NH_3

S.F.: 9.0 has two *s.f.* Coefficients are exact with *infinite s.f.* Round the answer to two *s.f.*

PRACTICE

Learn the bridge prompt, then apply it from memory to solve these in your notebook.

1. In the production of steel, iron(III) oxide can be reduced by carbon. The *un*balanced equation is

$$Fe_2O_3 + \quad C \rightarrow \quad Fe + \quad CO$$

   a. How many moles of Fe_2O_3 are consumed in reacting with 4.80 mol carbon?

   b. To form 3.6 mol Fe, how many moles of carbon are needed?

2. Burning ammonia can result in the formation of water and elemental nitrogen:

$$4\,NH_3 + 3\,O_2 \rightarrow 2\,N_2 + 6\,H_2O$$

   a. In this reaction, how many moles of O_2 are required to burn 14.0 mol NH_3?

   b. How many moles of O_2 are used up if 20.0 mol N_2 form?

3. For the *un*balanced equation

$$NaClO_3 \rightarrow \quad NaCl + \quad O_2$$

how many moles of O_2 can be obtained from 2.50 mol $NaClO_3$?

Lesson 9.5 — Conversion Stoichiometry

Perhaps the most frequently encountered type of chemistry calculation is **stoichiometry** (stoy-kee-AHM-et-ree), a term derived from ancient Greek that means "measuring fundamental quantities." In stoichiometry, you are *given* a measured quantity of one substance and WANT to know *how much* of *another* substance will react or form.

Stoichiometry is a reaction calculation in which we

- Convert the units of the supplied DATA to moles
- Use the ratio supplied by the coefficients of the balanced equation to convert from the *given* moles to the WANTED moles
- Then convert to other WANTED units if needed

There are many types of stoichiometry. For now, in reaction calculations, we will limit our attention to reactions that *go to completion*: until at least one reactant is totally used up.

We will begin our study of reaction calculations with relatively simple cases: solving calculations that have a single amount WANTED and a single amount *given* that is completely used up in the reaction. Use the following system to structure these calculations.

The Seven Stoichiometry Steps

For a *reaction* calculation, if the WANTED substance $\neq$ *given* substance,

1. Write: "WANTED: ? *unit* and substance *formula*."

2. List the DATA, including prompts.

3. Balance the equation.

4. Bridge: Write the mole-to-mole ratio between the WANTED and *given* formulas, using the coefficients of the balanced equation.

5. To SOLVE, first write

    ? WANTED unit and substance = number and unit of *given* substance

    then convert to *moles* of the *given* substance.

6. Then convert to moles of WANTED substance, using the bridge mol–mol ratio.

7. Then convert to the unit WANTED.

With practice, these steps will become automatic. The short version is

The Stoichiometry Prompt

For reaction calculations, if WANTED substance $\neq$ *given* substance,

- Write four WDBB steps to gather tools: WANTED and DATA, **b**alance and **b**ridge
- Then three to solve: Convert "? WANTED unit *given*" to **mol** *given* to **mol** WANTED to unit WANTED

During initial problems, it can help to recite: "WDBB, then units to moles to moles to units."

The stoichiometry steps use conversions with a mole-to-mole bridge. The coefficients of the balanced equation provide the key bridge ratio.

TRY IT

In your notebook, solve the following problem by completing each of the seven stoichiometry steps, then check your answer below.

Q. Sodium burns to form sodium oxide (Na_2O). How many grams of Na_2O can be produced from 2.30 g of Na? The unbalanced equation is

$$Na + \quad O_2 \rightarrow \quad Na_2O$$

Answer:

1. WANTED: ? g *Na_2O*   (A single unit is WANTED.)

2. DATA: 2.30 g *Na*   (*Only* single unit in DATA—use as *given*.)
    1 mol Na = 23.0 g Na   (g Na in DATA = grams prompt.)
    1 mol Na_2O = 62.0 g Na_2O (g Na_2O WANTED = grams prompt.)

(Because this is a *reaction* and the WANTED and *given* substances differ, use the seven stoichiometry steps.)

3. Balance: $4\,Na + 1\,O_2 \rightarrow 2\,Na_2O$

4. Bridge: $4\,mol\,Na = 2\,mol\,Na_2O$ (Moles WANTED to moles *given*.)

If needed, adjust your work and finish.

5. SOLVE: Write "? WANTED = *given*":

   $?\,g\,Na_2O = 2.30\,g\,Na$

   When a single unit is WANTED, use single-unit DATA as a *given* (Lesson 5.3). Then convert to **moles** *given*. ("Grams to moles—use molar mass.")

   $?\,g\,Na_2O = 2.30\,g\,Na \cdot \dfrac{1\,mol\,Na}{23.0\,g\,Na} \cdot \dfrac{}{mol\,Na}$

6. Convert to **moles** WANTED using the bridge between *given* and WANTED.

   $?\,g\,Na_2O = 2.30\,g\,Na \cdot \dfrac{1\,mol\,Na}{23.0\,g\,Na} \cdot \dfrac{\mathbf{2}\,mol\,Na_2O}{\mathbf{4}\,mol\,Na} \cdot \dfrac{}{mol\,Na_2O}$

7. Convert to **units** WANTED.

   $\mathbf{?\,g\,Na_2O} = 2.30\,\cancel{g\,Na} \cdot \dfrac{1\,\cancel{mol\,Na}}{23.0\,\cancel{g\,Na}} \cdot \dfrac{\mathbf{2}\,\cancel{mol\,Na_2O}}{\mathbf{4}\,\cancel{mol\,Na}} \cdot \dfrac{62.0\,\mathbf{g\,Na_2O}}{1\,\cancel{mol\,Na_2O}}$

   $= \dfrac{2.30 \cdot 2 \cdot 62.0}{23.0 \cdot 4}\,\mathbf{g\,Na_2O} = \boxed{\mathbf{3.10\,g\,Na_2O}}$

   S.F.: Coefficients and 1 are always exact, 2.30 and 23.0 have three *s.f.*, round answer to three *s.f.*

Note the order of the units on the top in the multiplied conversions of the SOLVE step: units *given*, moles *given*, moles WANTED, units WANTED.

Note also the mole-to-mole bridge conversion that is needed in all reaction calculations when the WANTED substance formula differs from the *given* formula.

Units and Substance Formulas: Inseparable

In solving problems involving reactions, always write the *number*, the *unit*, and the substance *formula* in each term in your WANTED, DATA, and conversions. One reason to write *number, unit, formula* is to avoid errors in unit cancellation. The rule is

Units measuring one substance can*not* cancel the same units measuring a *different* substance.

TRY IT

Q. Is the following conversion *legal* or *illegal*? Write your answer, then check below.

$$? \text{ mol HCl} = 32 \text{ g NaOH} \cdot \frac{1 \text{ mol HCl}}{36.5 \text{ g HCl}} =$$

STOP **Answer:**

Illegal. Because grams of NaOH are not the same as grams of HCl, the units cannot cancel. The substance is an inseparable part of the unit.

Q. Is the following conversion *legal* or *illegal*?

$$? \text{ kg NaOH} = 32 \text{ g NaOH} \cdot \frac{1 \text{ kg}}{1000 \text{ g}} =$$

STOP **Answer:**

Legal, because different formulas are not cancelled. The ratio between grams and kilograms is true for all measurements.

When problems involve several formulas, writing substance formulas after units for all quantities in the WANTED, DATA, and conversions is essential to avoid errors.

> If a substance formula is known, it is a part of the *unit*, it must be written, and the unit cannot be cancelled in conversions unless the formula attached also cancels.

Writing the formulas also keeps straight which molar mass *numbers* to put where. In the previous problem, the 62.0 g applies to the Na_2O, *not* the Na.

PRACTICE **A**

Use the table of atomic masses at the back of the book.

1. Multiply and simplify. Your answers must include a number, unit, and formula.

    a. $5.1 \text{ g Al}_2O_3 \cdot \dfrac{1 \text{ mol Al}_2O_3}{102 \text{ g Al}_2O_3} \cdot \dfrac{3 \text{ mol O}_2}{4 \text{ mol Al}_2O_3} =$

    b. $1.27 \times 10^{-3} \text{ g Cu} \cdot \dfrac{1 \text{ mol Cu}}{63.5 \text{ g Cu}} \cdot \dfrac{2 \text{ mol Ag}}{1 \text{ mol Cu}} \cdot \dfrac{107.9 \text{ g Ag}}{1 \text{ mol Ag}} =$

2. Phosphorus (P_4) burns in air (O_2) to form the oxide P_4O_{10}.

    a. Write the balanced equation for the reaction.

    b. How many moles of O_2 are needed to burn 7.0 mol P_4? (In your notebook, solve with the seven stoichiometry steps.)

    c. How many grams of P_4O_{10} would be produced when 0.500 mol P_4 burns?

Why Not Go from Grams to Grams in One Conversion?

In reaction calculations, to go from grams *given* to grams WANTED, we first convert grams to moles. Why don't we solve in one conversion, using a gram–gram ratio? The reason is: gram-to-gram ratios are not easy to calculate.

In a reaction, if the substance *formulas* are known, the *particle* ratios are easy to determine. Simply balance the equation. The coefficients then supply exact and simple whole-number relationships for the numbers of particles that react and form, in units that count particles or *moles* of particles.

There is no similar way to calculate gram-to-gram ratios.

In reaction calculations, we do the steps we do, in the order we do, so that we can use the mole-to-mole bridge—the only bridge between the *given* and WANTED substances that is easy to determine.

To summarize: In reaction calculations involving visible amounts of reactants, the bridge that converts from the *given* to the WANTED substance is a *mole-to-mole* conversion.

THE STOICHIOMETRY BRIDGE: MOLES TO MOLES

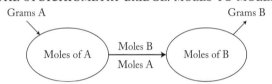

Flashcards

Put a check by the questions below that you can answer quickly and correctly. Make flashcards for those you cannot.

Run these cards until you can do them perfectly, then try the problems below. Save a few problems, then for your next study session, run the cards again and finish the practice.

Run the cards until you know each one for three sessions in a row, then add them to stack 2.

One-Way Cards (with Notch)	Back Side—Answers
Coefficients can be read as	Chemical particles, or moles of particles
When is a mol/mol conversion needed in conversions?	If WANTED formula ≠ *given* formula
When are the stoichiometry steps needed?	If WANTED formula ≠ *given* formula
What numbers go into a mol/mol conversion?	The coefficients of the balanced equation
In stoichiometry, convert the unit *given* to	mol *given* to mol *wanted* to unit *wanted*
Recite the seven steps to solve stoichiometry for a WANTED single unit	Wanted and Data, Balance and Bridge Convert *units* to *moles* to *moles* to *units*

PRACTICE B

Solve these in your notebook. For each calculation of an amount of a substance, write the seven stoichiometry steps. You will need the table of atomic masses at the back of the book.

Complete problems 1 and 2 now. Save problem 3 for your next practice session.

(continued)

1. The combustion of iron(II) sulfide can be represented as:

$$FeS + \qquad O_2 \rightarrow \qquad Fe_2O_3 + \qquad SO_2$$

   a. Balance the equation.

   b. Starting with 0.48 mol FeS and plenty of oxygen, how many grams of Fe_2O_3 can be formed?

2. Nitrogen dioxide can react with water to form nitric acid and nitrogen monoxide. The balanced equation is

$$3\,NO_2 + H_2O \rightarrow 2\,HNO_3 + NO$$

   a. How many moles of H_2O are required to use up 2.3 kg of NO_2?

   b. How many milligrams of NO can be formed from 2.4×10^{21} molecules of NO_2?

3. The metal silver reacts with concentrated nitric acid to produce the red-brown gas nitrogen dioxide. The balanced equation is

$$Ag + 2\,HNO_3 \rightarrow NO_2 + AgNO_3 + H_2O$$

   a. How many grams of nitric acid are required to use up 5.00 g Ag?

   b. How many grams of nitrogen dioxide would be formed?

Lesson 9.6

Finding the Limiting Reactant

Ratios of Mixing for Two Reactants

In a given chemical reaction, particles can react and form *only* in the particle ratios shown by the balanced equation. For the reactants, in any reaction between two substances that goes to completion, there are two possibilities.

- The particles are mixed in the exact ratio needed so that each of the reactants is completely used up, or
- One of the reactants is used up and some of the other reactant is left over

Let's take a closer look at these cases.

Stoichiometric Equivalents

Consider this reaction that when ignited, goes rapidly to completion:

$$2\,H_2 + O_2 \rightarrow 2\,H_2O$$

 **TRY IT**

Q. If four molecules of H_2 react, how many molecules of O_2 react?

STOP

Answer:

The ratio of H_2 molecules used up to O_2 used up must be 2 to 1. The four reacting H_2 must use up two molecules of O_2.

If one reactant is mixed with another, and both are exactly and completely consumed in the reaction, the reactants initially present are said to be in **stoichiometrically equivalent amounts** or **stoichiometric ratios**.

 TRY IT

Q. If 4.00 *mol* H_2 are to be reacted, how many moles of O_2 are needed for stoichiometric equivalence?

STOP

Answer:

2.00 mol. Moles is a unit that counts particles. For both reactants to be exactly used up, the initial particle ratio of H_2 to O_2 must be 2 to 1.

Q. If 4.00 mol H_2 are reacted with 2.00 mol O_2, how much H_2, O_2, and H_2O will be present in the reaction vessel after the reaction is over?

$H_2 =$

$O_2 =$

$H_2O =$

STOP

Answer:

Coefficients are ratios of reaction. If the coefficients for this reaction are doubled, the equation can be read as: four moles of H_2 react with two moles of O_2 to form four moles of H_2O. After the reaction, what is present is the **4.00 moles of H_2O** formed, but **no H_2** and **no O_2**. As products form, reactants are consumed. If the two reactants are mixed in stoichiometric amounts, both will be completely used up.

In many reaction calculations, you are asked to find a stoichiometric equivalent: the amount of one reactant that is needed to exactly use up another reactant. When the numbers are not as simple as those above, these amounts can be solved by our standard stoichiometry steps. Several problems of this type were solved in Lesson 9.5: one reactant amount was *given,* and the amount of another reactant needed to use up the *given* was WANTED.

Limiting Reactants

In many types of lab procedures (such as acid–base titration), the goal is to mix reactants in amounts that are as close to stoichiometric equivalence as possible.

However, in the reality of the laboratory, we cannot mix together two reactants in amounts that are *exactly* stoichiometrically equivalent. Let's explore what happens when reactant amounts are mixed that are not equivalent.

 TRY IT

Again considering this reaction that goes to completion:

$$2\,H_2 + O_2 \rightarrow 2\,H_2O$$

Q. If four molecules of H_2 are reacted with one molecule of O_2, how much H_2, O_2, and H_2O will be present after the reaction?

$H_2 =$

$O_2 =$

$H_2O =$

Answer:

The one molecule of O_2 reacts with exactly two molecules of H_2. At that point, there is no more O_2 to react, so the reaction must stop. After the reaction, what are present are the two molecules of H_2O that form, no O_2, and two molecules of H_2 that are left over because they had no O_2 with which to react. In this case, the amount of H_2O formed is determined not by the amount of H_2 originally supplied, but by the amount of the O_2, because it is used up first.

Limiting Reactant Terminology

Many reactions (and all reactions that we are concerned with at this point) go essentially to completion: until one reactant is very close to 100% used up. For those reactions, the following vocabulary, assumptions, and rules will simplify our problem solving.

Every reaction has at least *one* **limiting reactant** (or **limiting reagent**). The limiting reactant is the reactant used up first. This reactant limits and determines both how much of the other reactants are used up and how much of the products form.

A reactant that is *not* completely used up is said to be **in excess**, meaning

* Enough is present to use up all of the limiting reactant, and
* Some amount remains when the reaction stops

Which reactant is used up first depends on these conditions:

* The starting amounts of each reactant, and
* The ratios of reaction for the reactants (the reactant coefficients)

In calculations for reactions that go to completion, key rules include

* *One* reactant *must* be limiting (totally used up).
* The initial amount of the *limiting reactant* determines how much of the other reactants react and products form.
* The initial amount of the *limiting reactant* must be used as the *given* to calculate how much of the other reactants are consumed and products form.

- The initial amounts of *reactants in excess* do *not* determine how much of other reactants are used up or products form. The amount of a reactant *in excess* cannot be used as a *given* to calculate how much of the products form.
- If a reactant is *in excess*, its amount is more than enough to use up the limiting reactant, and measurements of its amount are not necessary (and can usually be ignored) in calculations.

In the special case of two reactants that are stoichiometrically equivalent, both are limiting, and either reactant amount can be used to calculate how much of the products will form.

Determining the Amounts of the Products

Using the logic of the rules above, in the space below, write your answers to the following questions based on "numbers in your head" rather than written conversions.

 **TRY IT**

Q. Given the balanced equation:

$$2\,H_2 + O_2 \rightarrow 2\,H_2O$$

if 10 molecules of H_2 are ignited with 6 molecules of O_2, and the reaction goes to completion,

  a. Which reactant is limiting?

  b. How much H_2O can be formed?

  c. How much of each reactant and product below is present in the mixture at the end of the reaction?

$$H_2 = \qquad O_2 = \qquad H_2O =$$

 Answer:

  a. For 6 molecules of O_2 to react, they need 12 molecules of H_2. We don't have that much H_2, so the H_2 is *limiting*. Though the H_2 molecules are initially present in a higher *count*, more H_2 molecules are needed per reaction. In this case, the 10 molecules of H_2 are used up first, *preventing* the O_2 molecules from being completely used up.

  Looked at another way: Reacting 10 molecules of H_2 requires 5 molecules O_2. We have more than that much O_2, so when we use up all of the H_2, some O_2 remains. The H_2 is therefore limiting (used up $\sim$100%), and there is *excess* O_2: some remains in the mixture at the end of the reaction.

  b. If 10 molecules of H_2 are used up, according to the balanced equation, 10 molecules of H_2O must form. Because some O_2 remains at the end of the reaction, the 6 molecules of O_2 initially present do *not* determine how much H_2O forms. Only amounts that *react* cause products to form. Only the initial amount of *limiting* reactant predicts the amount of a product that forms.

  c. In the mixture after the reaction *no* H_2 remains. Present are the 10 molecules of water formed, plus the 1 molecule of O_2 that did not react.

PRACTICE   A

Write answers to these using mental math (without a calculator). Based on the balanced equation:

$$4\ Fe + 3\ O_2 \rightarrow 2\ Fe_2O_3$$

1. To form four particles of Fe_2O_3,

    a. How many atoms of Fe are needed?

    b. How many molecules of O_2 are needed?

2. How much Fe_2O_3 is produced when nine molecules of O_2 are used up?

3. If 20 atoms of Fe are mixed with 20 molecules of O_2 and the reaction goes to completion,

    a. The Fe consumes how much O_2 as it reacts?

    b. How much O_2 is left over?

    c. How many particles of Fe_2O_3 are formed?

    d. Which reactant is in excess?

    e. Which reactant is limiting?

Identifying the Limiting Reactant

For reactions that go to completion, in prior lessons we have learned to solve three types of questions about reaction amounts (stoichiometry):

1. To form a *given* amount of product, what amount of one reactant must be used up?

2. Given a known amount of one reactant and *excess* amounts of the other reactants, what amount of a product can be formed?

3. Given a known amount of one reactant, what amount of a second reactant is the *stoichiometric equivalent* needed to exactly use up the first?

In each type of problem above,

- An amount of only *one* substance is identified, it is used as a *given*, and
- To solve, the preferred method is seven-step stoichiometry

In this lesson, we have considered problems in which *two* reactant amounts are supplied as DATA and the amount of a product formed is WANTED. In such cases, unless you are told which reactant is limiting (or that the reactants are stoichiometrically equivalent and both are limiting), you must determine which reactant is limiting *before* determining an amount of products that can form.

For reaction calculations, the rule is

> If an amount of *product* formed is WANTED and *two* reactant amounts are supplied, you must *first* determine the *limiting* reactant. The limiting reactant must be the *given* when you SOLVE.

If the particle counts and the reaction ratios are simple, you can often identify the limiting reactant by inspection, as has been done so far in this lesson. To find which reactant is limiting in more complex cases, one method is to use successive conversions.

Identifying the Limiting Reactant by Successive Conversions

In DATA, an *amount* of a substance will always be measured by a single unit, such as grams, moles, or molecules (see Lesson 4.5).

If single-unit amounts of *two* reactants are supplied in the DATA but the limiting reactant cannot be determined by inspection,

a. In two separate stoichiometry conversions, start with the same unit and product formula WANTED, but use each different reactant amount as a *given*.

b. The calculation that results in the *lowest* amount of the WANTED substance formed has the *limiting* reactant as its *given*. Label this reactant as a *limiting* reactant.

   The logic is: You cannot form more product than can be produced from the reaction ingredient that runs out first. The reactant that limits the reaction is the limiting reactant.

c. Once you determine the limiting reactant by the process above, if you are asked to calculate the amounts of *other* products that also form, use the supplied amount of the *limiting* reactant as *given*.

   The limiting reactant determines how much of *each* of the products forms.

 TRY IT

Apply the steps above to the following example, and then check your answer below.

   Q. If 2.00 g H_2 gas and 4.80 g O_2 gas are mixed and ignited, how many grams of water will form?

STOP

Answer:

In reaction calculations, start with WDBB.

1. WANTED: ? **g H_2O**   (A single unit is WANTED.)

2. DATA: 2.00 **g H_2**   (A single-unit amount supplied.)

               4.80 **g O_2**   (*Another* single-unit amount supplied.)

               1 mol H_2 = 2.016 g H_2

               1 mol O_2 = 32.0 g O_2

               1 mol H_2O = 18.0 g H_2O (There are three grams prompts.)

Strategy: If a single-unit amount of one or more products is WANTED, and the DATA includes single-unit amounts (such as grams or moles) of *two* reactants, you will need to identify a limiting reactant first. Calculate the WANTED amount in two separate stoichiometry conversions, each using a different reactant amount as a *given*.

3. Balance: $2\,H_2 + 1\,O_2 \rightarrow 2\,H_2O$

4. Bridge: (Moles WANTED to moles *given* will vary for the two *givens*.)

If needed, adjust your work and finish.

5. SOLVE:

$$? \, g\, H_2O = 2.00\, g\, H_2 \cdot \frac{1\, mol\, H_2}{2.016\, g\, H_2} \cdot \frac{2\, mol\, H_2O}{2\, mol\, H_2} \cdot \frac{18.0\, g\, H_2O}{1\, mol\, H_2O} = 17.9\, g\, H_2O$$

$$? \, g\, H_2O = \boxed{4.80\, g\, O_2} \cdot \frac{1\, mol\, O_2}{32.0\, g\, O_2} \cdot \frac{2\, mol\, H_2O}{1\, mol\, O_2} \cdot \frac{18.0\, g\, H_2O}{1\, mol\, H_2O} = 5.40\, g\, H_2O$$

At the SOLVE step in each calculation, for the *given* amount of one reactant used up, the amount of one product that would form is determined. Identify the *given* that produces the *lowest* amount of product as the *limiting* reactant.

The limiting reactant is the one that *most limits* the amount of products that can form.

For the above reaction, the limiting reactant must be O_2. The amount of water predicted to form is **5.40 g H_2O**. Though more grams of oxygen are supplied in the reaction mixture, it is the moles of the reactants and the ratios of reaction that decide the limiting reactant, and in this problem, the oxygen is used up first.

SUMMARY

1. If a reaction goes to completion, at least one reactant is *limiting*: it is used up ~100%.

2. To calculate how much of other reactants react or products form, you must start from an amount of a *limiting* reactant as *given*.

3. The amounts of *reactants in excess* (reactant substances not used up 100%) cannot be used to predict the amounts of products that form. In calculations, the amount of the reactant in excess can be ignored.

4. When two reactants exactly and completely use each other up in a reaction, both are limiting, and their amounts are said to be *stoichiometrically equivalent*.

5. If an amount of a product is WANTED, and two or more *reactant* amounts that are *not* stoichiometrically equivalent are in the DATA, you will need to identify a limiting reactant.

6. One way to identify the limiting reactant by *successive conversions*:
    *   In separate stoichiometry conversions, start with the same unit and product formula WANTED, but use each different reactant amount as a *given*.
    *   The maximum amount of product that can form is the *lowest* amount of product that results from those calculations.
    *   Identify the reactant that forms the lowest amount of product as the *limiting* reactant.

Learn the rules and terms above, then complete problems 1 and 3 in your notebook. If you need more practice, do problem 2.

1. For the reaction of hydrochloric acid and calcium carbonate, the balanced equation is

$$2\,HCl + CaCO_3 \rightarrow CO_2 + H_2O + CaCl_2$$

   If 0.400 mol HCl is reacted with 0.300 mol $CaCO_3$,

   a. Which reactant is limiting?

   b. How many moles of CO_2 can be formed?

2. For the reaction represented by the following unbalanced equation,

$$H_2 + \quad Cl_2 \rightarrow \quad HCl$$

   if 14.2 g chlorine gas is reacted with 0.300 mol hydrogen gas,

   a. How many moles of hydrogen chloride can be formed?

   b. How many moles of the "reactant in excess" must be used up to completely consume the limiting reactant?

3. For the following unbalanced reaction equation,

$$H_2SO_4 + \quad NaOH \rightarrow \quad H_2O + \quad Na_2SO_4$$

   if 2.00 g NaOH is reacted with 0.0100 mol H_2SO_4,

   a. How many grams of water can be formed?

   b. How many grams of the "reactant in excess" must be used up to completely consume the limiting reactant?

SUMMARY

1. In reaction equations, *reactants* on the left side are used up, and *products* on the right side form.

2. In chemical reactions, atoms and mass are neither created nor destroyed.

3. Balanced chemical equations have the same number and kind of atoms on both sides. The coefficients of a balanced equation show the exact ratios in which the particles react and form.

4. Coefficients can be read as molecules, particles, or any consistent multiple of molecules or particles. We most often read coefficients as exact *moles*.

5. Each **term** in a balanced equation is a coefficient followed by a substance formula.
    To count the number of each kind of atom represented by a term, multiply the coefficient by the subscript(s) for that atom.

(continued)

To count each type of atom on a side, add the atoms in each term on that side.

6. Only one set of ratios will balance a chemical equation. To balance equations, you must supply the coefficients by trial and error.

7. Start balancing by putting a 1 in front of the most complex formula. If you need to eliminate fractions, multiply all coefficients by the denominator.

8. **The seven stoichiometry steps:** If a reaction calculation WANTS a quantity of one substance and a quantity of a different substance is *given*, use these steps:

    1. Write the WANTED unit and substance.

    2. List the DATA.

    3. Balance the equation.

    4. Bridge: Write the mol WANTED to mol *given* equality, using coefficients.

    5. Convert "? WANTED = *given*" to moles *given*.

    6. To moles WANTED.

    7. To units WANTED.

        Summary: WDBB, *then* units → moles → moles → units.

9. **Limiting reactants:**

    a. A reaction that goes **to completion** continues until one reactant is ~100% used up. (For now, assume reactions go to completion unless otherwise noted.)

    b. A reactant 100% used up is a **limiting** reactant. Reactants not limiting are **in excess**. The amount of *limiting* particles used up determines the amounts of other reactants used up and products formed.

    c. If a reaction goes to completion, at least one reactant must be limiting. If two reactants are limiting (both exactly used up), the two reactants are *stoichiometrically equivalent*.

    d. If the *limiting* reactant is *known*, the amount of any other reactant used up or product formed can be calculated using seven-step conversion stoichiometry.

    e. If single-unit amounts of *two reactants* are supplied and the limiting reactant is *not* known, use seven-step stoichiometry in successive calculations to calculate the WANTED amount of one product. In each calculation, use as a *given* the known amounts for a *different* reactant. The calculation that forms the lowest amount of the WANTED substance has the limiting reactant as its *given*.

ANSWERS

Lesson 9.1

Practice   1. $\textcircled{4}\,Fe + \textcircled{3}\,O_2 \rightarrow \textcircled{2}\,Fe_2O_3$

Reactants   Products

2a. $7\,Na_3PO_4$: **28**   2b. $3\,Co(OH)_2$: **6**   2c. $2\,No(NO_3)_2$: **12**   2d. $5\,Al_2(SO_4)_3$: **60**

The total number of atoms is the *coefficient times each subscript* that applies to the atom.

3. 2a: **56**; 2d: **85**

4a. Yes. The same number and kinds of atoms are on each side.

4b. $2\,C + 3\,H_2 \rightarrow C_2H_6$

4c. Changes: The number of bonds, the bond locations, the molecules, the stored energy, and the appearance and characteristics of the substances involved.

Stays the same: The numbers of each kind of atom and the total mass.

Lesson 9.2

Practice   Coefficients of 1 may be omitted as understood.

1a. $2\,Cs + Cl_2 \rightarrow 2\,CsCl$: **balanced**

1b. $4\,HI + O_2 \rightarrow 2\,H_2O + I_2$: **not balanced**

1c. $Pb(NO_3)_2 + 2\,LiBr \rightarrow PbBr_2 + LiNO_3$: **not balanced**

1d. $BaCO_3 + 2\,NaCl \rightarrow Na_2CO_3 + BaCl_2$: **balanced**

2a. $4\,Al + \mathbf{3}\,O_2 \rightarrow \mathbf{2}\,Al_2O_3$ 2b. $3\,Ca + \mathbf{1}\,N_2 \rightarrow \mathbf{1}\,Ca_3N_2$

2c. $\mathbf{2}\,P_4 + \mathbf{6}\,O_2 \rightarrow \mathbf{2}\,P_4O_6$ (Reducing to lowest-whole-number ratios is usually preferred, but other ratios will be needed in some types of problems.)

2d. $\mathbf{3}\,C_6H_{14} + \mathbf{57/2}\,O_2 \rightarrow 18\,CO_2 + \mathbf{21}\,H_2O$

2e. $\mathbf{1}\,MgH_2 + \mathbf{2}\,H_2O \rightarrow \mathbf{1}\,Mg(OH)_2 + \mathbf{2}\,H_2$ (2e is tricky because H is used in more than one formula on each side, rather than the usual O. Save until last the atom that is in two compounds on one or both sides.)

2f. $\mathbf{2}\,C_2H_6 + 7\,O_2 \rightarrow 4\,CO_2 + \mathbf{6}\,H_2O$

3a. $\mathbf{2}\,K + F_2 \rightarrow \mathbf{2}\,KF$ 3b. $\mathbf{4}\,Cs + \underline{O_2} \rightarrow Cs_2O$

3c. $\mathbf{4}\,PCl_3 \rightarrow \underline{P_4} + 6\,Cl_2$ 3d. $\underline{C_2H_5OH} + 3\,O_2 \rightarrow 2\,CO_2 + 3\,H_2O$

3e. $2\,FeS + 7/2\,O_2 \rightarrow Fe_2O_3 + 2\,SO_2$ 4a. $\mathbf{2}\,Mg + \mathbf{1}\,O_2 \rightarrow \mathbf{2}\,MgO$

4b. $\mathbf{1}\,N_2 + \mathbf{1}\,O_2 \rightarrow 2\,NO$ 4c. $2\,C_6H_6 + \mathbf{15}\,O_2 \rightarrow 12\,CO_2 + 6\,H_2O$

4d. $\mathbf{1}\,P_4 + 3\,O_2 \rightarrow \mathbf{1}\,P_4O_6$ 4e. $\mathbf{2}\,Al + 6\,HCl \rightarrow 2\,AlCl_3 + 3\,H_2$

5. Lowest whole-number coefficients were requested, so your coefficients must match these exactly.

a. $\mathbf{1}\,Al_2O_3 + 6\,HCl \rightarrow 2\,AlCl_3 + 3\,H_2O$ b. $\mathbf{1}\,Fe_3O_4 + 4\,H_2 \rightarrow 3\,Fe + 4\,H_2O$

c. $\mathbf{3}\,C + \mathbf{1}\,SiO_2 \rightarrow 2\,CO + \mathbf{1}\,SiC$ d. $\mathbf{2}\,N_2 + 5\,O_2 + 2\,H_2O \rightarrow 4\,HNO_3$

e. $\mathbf{1}\,Rb_2O + \mathbf{1}\,H_2O \rightarrow 2\,RbOH$ f. $\mathbf{3}\,Mg(NO_3)_2 + 2\,Na_3PO_4 \rightarrow \mathbf{1}\,Mg_3(PO_4)_2 + 6\,NaNO_3$

g. $\mathbf{2}\,Pb(C_2H_5)_4 + 27\,O_2 \rightarrow 2\,PbO + \mathbf{16}\,CO_2 + 20\,H_2O$

h. $\mathbf{3}\,Cd + 8\,HNO_3 \rightarrow \mathbf{3}\,Cd(NO_3)_2 + 4\,H_2O + 2\,NO$

6a. $1\,N_2O_4 \rightarrow 2\,NO_2$

6b. $BaCO_3 + 2\,CsCl \rightarrow Cs_2CO_3 + BaCl_2$

6c. $2\,AgNO_3 + CaI_2 \rightarrow 2\,AgI + Ca(NO_3)_2$

Lesson 9.3

Practice   1a.   16 molecules HBr need **4** molecules O_2 (the ratio must be 4 HBr to 1 O_2); 16 molecules of HBr make **8** molecules Br_2 (the ratio must be 4 HBr to 2 Br_2).

               1b.   **20** molecules HBr react; **10** molecules Br_2 form.

               2a. $CS_2 + 3\,O_2 \rightarrow CO_2 + 2\,SO_2$ 2b.   75 trillion molecules O_2; 50. trillion molecules SO_2

Lesson 9.4

Practice   Your paper should look like this, but you may omit the comments in parentheses.

1a.   WANTED: ? mol **Fe_2O_3**

      DATA: 4.80 mol **C**   (Because WANTED $\neq$ *given*, balance the equation, then write the bridge.)

      Balance: **1** Fe_2O_3 + **3** C $\rightarrow$ **2** Fe + **3** CO

      Bridge: **1** mol **Fe_2O_3** = **3** mol C

      SOLVE: ? mol Fe_2O_3 = 4.80 mol C $\cdot$ $\dfrac{1\text{ mol }Fe_2O_3}{3\text{ mol C}}$ = $\boxed{\textbf{1.60 mol Fe}_2\textbf{O}_3}$

1b.   WANTED: ? mol C

      DATA: 3.6 mol Fe

      (Strategy: For a reaction, if formula WANTED $\neq$ formula *given*, balance and bridge.)

      Balance: **1** Fe_2O_3 + **3** C $\rightarrow$ **2** Fe + **3** CO

      Bridge: 3 mol C = 2 mol Fe

      SOLVE: ? mol C = 3.6 mol Fe $\cdot$ $\dfrac{3\text{ mol C}}{2\text{ mol Fe}}$ = $\boxed{\textbf{5.4 mol C}}$

2a.   WANTED: ? mol O_2

      DATA: 14.0 mol NH_3

      (Strategy: Because the formula WANTED $\neq$ formula *given*, balance and bridge.)

      Balance: See problem.

      Bridge: **3** mol **O_2** = **4** mol **NH_3**

      SOLVE: (Order your conversions in the usual way, beginning with the starting template.)

                    ? mol O_2 = 14.0 mol NH_3 $\cdot$ $\dfrac{3\text{ mol }O_2}{4\text{ mol }NH_3}$ = $\boxed{\textbf{10.5 mol O}_2}$

2b.   WANTED: **? mol O_2**

      DATA: 20.0 mol N_2

      (Strategy: Because the formula WANTED $\neq$ formula *given*, write the mole-to-mole conversion.)

      Balance: See problem.

      Bridge: **3** mol **O_2** = **2** mol **N_2**

      SOLVE: ? mol O_2 = 20.0 mol N_2 $\cdot$ $\dfrac{3\text{ mol }O_2}{2\text{ mol }N_2}$ = $\boxed{\textbf{30.0 mol O}_2}$

3. WANTED: ? mol O_2

DATA: 2.50 mol of $NaClO_3$ (Because $O_2 \neq NaClO_3$, balance and bridge.)

Balance: **2** $NaClO_3 \rightarrow$ **2** $NaCl$ + **3** O_2

Bridge: 3 mol O_2 = 2 mol $NaClO_3$ (Always a mol WANTED to mol *given* ratio.)

SOLVE: ? mol O_2 = 2.50 mol $NaClO_3 \cdot \dfrac{3 \text{ mol } O_2}{2 \text{ mol } NaClO_3}$ = $\boxed{\textbf{3.75 mol } O_2}$

Lesson 9.5

Practice A 1a. **0.038 mol O_2**   1b. **4.32 $\times$ 10^{-3} g Ag**

2a. **1 P_4 + 5 $O_2 \rightarrow$ 1 P_4O_{10}**

2b. 1. WANTED: **? mol O_2**

2. DATA: 7.0 mol **P_4**

(For reaction calculations, if WANTED $\neq$ *given*, use the seven stoichiometry steps.)

3. Balance: (See part a.)

4. Bridge: 1 mol P_4 = 5 mol O_2 (Bridge must be *mol* WANTED = mol *given*.)

5. SOLVE: Begin with ? mol O_2 = 7.0 mol $P_4 \cdot \dfrac{}{\text{mol } P_4}$

6–7. (Convert unit *given* to **moles** *given* to **moles** WANTED to unit WANTED. Here, the *given* and WANTED units are moles. Use the one conversion in the DATA to SOLVE.)

$$? \text{ mol } O_2 = 7.0 \text{ mol } P_4 \cdot \frac{5 \text{ mol } O_2}{1 \text{ mol } P_4} = \boxed{\textbf{35 mol } O_2}$$

2c. 1. WANTED: **? g P_4O_{10}**   (WANT single unit.)

2. DATA: 0.500 mol **P_4**   (Single unit *given*.)

1 mol P_4O_{10} = 284.0 g P_4O_{10} (The grams prompt.)

(For a reaction calculation, if WANTED $\neq$ *given*, use the seven stoichiometry steps.)

3. Balance: 1 P_4 + 5 $O_2 \rightarrow$ 1 P_4O_{10}

4. Bridge: 1 mol P_4 = 1 mol P_4O_{10}

5–7. SOLVE: (? WANTED = Unit *given* to **moles** *given* to **moles** WANTED to unit WANTED. Here, moles are *given*. Chain the two DATA conversions to SOLVE.)

$$? \text{ g } P_4O_{10} = 0.500 \text{ mol } P_4 \cdot \frac{1 \text{ mol } P_4O_{10}}{1 \text{ mol } P_4} \cdot \frac{284.0 \text{ g } P_4O_{10}}{1 \text{ mol } P_4O_{10}} = \boxed{\textbf{142 g } P_4O_{10}}$$

(*S.F.*: 0.500 has three *s.f.* and 1 is exact, so the answer must be rounded to three *s.f.*)

Practice B 1a. 4 FeS + 7 $O_2 \rightarrow$ 2 Fe_2O_3 + 4 SO_2

1b. 1. WANTED: ? *g* Fe_2O_3

2. DATA: 0.48 mol FeS

1 mol Fe_2O_3 = 159.6 g Fe_2O_3 (Grams prompt.)

(A reaction calculation and WANTED $\neq$ *given*. Use the seven stoichiometry steps.)

3. Balance: (See part a.)

4. Bridge: 4 mol FeS = 2 mol Fe_2O_3

5. SOLVE: (Write ? WANTED = *given*. Convert to *given* moles is not needed. Moles was *given*.)

6. To moles WANTED, via the bridge.

7. To units WANTED.

$$? \text{ g Fe}_2\text{O}_3 = 0.48 \text{ mol FeS} \cdot \underbrace{\frac{2 \text{ mol Fe}_2\text{O}_3}{4 \text{ mol FeS}}}_{\text{(step 6)}} \cdot \underbrace{\frac{159.6 \text{ g Fe}_2\text{O}_3}{1 \text{ mol Fe}_2\text{O}_3}}_{\text{(step 7)}} = \boxed{\textbf{38 g Fe}_2\textbf{O}_3}$$
$$\text{(step 5)}$$

2a. 1. WANTED: ? mol H_2O

　　2. DATA:　　2.3 kg NO_2

　　　　　　　1 mol NO_2 = 46.0 g NO_2　　(**kg** = grams prompt.)

　　　　　　　(A reaction calculation. WANTED substance ≠ *given* substance. Use the seven stoichiometry steps.)

　　3. Balance: 3 NO_2 + H_2O → 2 HNO_3 + NO

　　4. Bridge:　3 mol NO_2 = 1 mol H_2O

　5–7. SOLVE:　(? WANTED = unit *given* to mole *given* to mole WANTED to unit WANTED.)

$$? \text{ mol H}_2\text{O} = 2.3 \text{ kg NO}_2 \cdot \frac{10^3 \text{ g}}{1 \text{ kg}} \cdot \frac{1 \text{ mol NO}_2}{46.0 \text{ g NO}_2} \cdot \frac{1 \text{ mol H}_2\text{O}}{3 \text{ mol NO}_2} = \boxed{\textbf{17 mol H}_2\textbf{O}}$$

2b. 1. WANTED: ? mg NO

　　2. DATA:　　2.4×10^{21} molecules of NO_2　　(Only single unit in data = *given*.)

　　　　　　　1 mol = 6.02×10^{23} molecules　　(Avogadro prompt.)

　　　　　　　1 mol NO = 30.0 g NO　　(**g**, **mg**, **kg** of formula = *grams* prompt.)

　　　　　　　(WANTED substance ≠ *given* substance. A reaction calculation. Use stoichiometry steps.)

　　3. Balance: 3 NO_2 + H_2O → 2 HNO_3 + NO

　　4. Bridge:　1 *mol* NO = 4 *mol* NO_2

　5–7. SOLVE:　(? WANTED = Unit *given* to **mol** *given* to **mol** WANTED to unit WANTED.)

$$? \textbf{ mg NO} = 2.4 \times 10^{21} \text{ NO}_2 \cdot \frac{1 \text{ mol NO}_2}{6.02 \times 10^{23} \text{ NO}_2} \cdot \frac{1 \text{ mol NO}}{\textbf{3} \text{ mol NO}_2} \cdot \frac{30.0 \textbf{ g NO}}{1 \text{ mol NO}} \cdot \frac{1 \text{ mg}}{10^{-3} \text{g}} = \boxed{\textbf{40. mg NO}}$$

3a. 1. WANTED: ? g HNO_3　　(Nitric acid is what Ag reacts with. Reactants are on the left.)

　　2. DATA:　　5.00 g Ag　　(A single unit is WANTED. This is your single unit given.)

　　　　　　　1 mol Ag = 107.9 g Ag

　　　　　　　1 mol HNO_3 = 63.0 g HNO_3

　　3. Balance: (supplied)

　　4. Bridge:　1 mol Ag = 2 mol HNO_3

　5–7. SOLVE:　$? \text{ g HNO}_3 = 5.00 \text{ g Ag} \cdot \dfrac{1 \text{ mol Ag}}{107.9 \text{ g Ag}} \cdot \dfrac{2 \text{ mol HNO}_3}{1 \text{ mol Ag}} \cdot \dfrac{63.0 \text{ g HNO}_3}{1 \text{ mol HNO}_3} = \boxed{\textbf{5.84 g HNO}_3}$

3b. 1. WANTED: **? g NO_2**

　　2. DATA:　　5.00 **g Ag**; $\dfrac{1 \text{ mol Ag}}{107.9 \text{ g Ag}}$; $\dfrac{1 \text{ mol NO}_2}{46.0 \text{ g NO}_2}$

　　　　　　　(List DATA as equalities or conversions.)

　　4. Bridge:　1 mol Ag = 1 mol NO_2

5–7. SOLVE: $? \text{ g NO}_2 = 5.00 \text{ g Ag} \cdot \dfrac{1 \text{ mol Ag}}{107.9 \text{ g Ag}} \cdot \dfrac{1 \text{ mol NO}_2}{1 \text{ mol Ag}} \cdot \dfrac{46.0 \text{ g NO}_2}{1 \text{ mol NO}_2} = \boxed{\textbf{2.13 g NO}_2}$

Lesson 9.6

Practice A   1. Because coefficients are ratios, they can be doubled to match this problem: $8 \text{ Fe} + 6 \text{ O}_2 \rightarrow \textbf{4 Fe}_2\textbf{O}_3$

a. 8 atoms Fe needed.   b. 6 molecules O_2 needed.

2. The ratio is 3 O_2 produce $2 \text{ Fe}_2\text{O}_3$, so 9 molecules of O_2 produce **6 particles of Fe_2O_3**.

3a. 15 molecules O_2.   3b. 5 molecules O_2 are left over.

3c. 10 particles of Fe_2O_3 are formed.   3d. O_2 is in excess.   3e. Fe is limiting.

Practice B   1a. WANTED: Which reactant is limiting?

To find the limiting reactant, find the amount of one product that can be formed by *each* reactant using seven-step stoichiometry.

Because we WANT moles of CO_2 in part b, it is easiest to solve for moles of CO_2 in part a.

1. WANTED: $? \text{ mol CO}_2$ (A single unit is WANTED.)

2. DATA: 0.400 mol HCl

0.300 mol $CaCO_3$

3. Balance: $2 \text{ HCl} + 1 \text{ CaCO}_3 \rightarrow 1 \text{ CO}_2 + 1 \text{ H}_2\text{O} + 1 \text{ CaCl}_2$

4. Bridge: (Moles WANTED to moles *given* will vary for the two *givens*.)

5. SOLVE: $? \text{ mol CO}_2 = 0.400 \text{ mol HCl} \cdot \dfrac{1 \text{ mol CO}_2}{2 \text{ mol HCl}} = \boxed{\textbf{0.200 mol CO}_2 \text{ forms}}$

$? \text{ mol CO}_2 = 0.300 \text{ mol CaCO}_3 \cdot \dfrac{1 \text{ mol CO}_2}{1 \text{ CaCO}_3} = \textbf{0.300 mol CO}_2 \text{ forms}$

The limiting reactant must be **HCl**. The limiting reactant is the one that *most limits* the amount of products that can form.

1b. WANTED: $? \text{ mol CO}_2$

Use the limiting reactant as *given* to calculate the amount of product that can form. HCl is limiting. This calculation is done in the first SOLVE above: **0.200 mol CO_2** forms, then the reaction stops.

2a. For a reaction amount calculation, if WANTED substance $\neq$ *given* substance, the calculation is stoichiometry. Start with WDBB.

1. WANTED: $? \text{ mol HCl}$ (A single unit is WANTED.)

2. DATA: **14.2 g Cl_2**

0.300 mol H_2

$1 \text{ mol Cl}_2 = 71.0 \text{ g Cl}_2$

(Strategy: Because the DATA include single-unit amounts of *two* reactants, you must first identify the limiting reactant. Find the WANTED amount using the seven stoichiometry steps for each reactant amount.)

3. Balance: $1 \text{ H}_2 + 1 \text{ Cl}_2 \rightarrow 2 \text{ HCl}$

4. Bridge: (Moles WANTED to moles *given* will vary for the two *givens*.)

5. SOLVE: $? \text{ mol HCl} = 14.2 \text{ g Cl}_2 \cdot \dfrac{1 \text{ mol Cl}_2}{71.0 \text{ g Cl}_2} \cdot \dfrac{2 \text{ mol HCl}}{1 \text{ mol Cl}_2} = \boxed{\textbf{0.400 mol HCl}}$

$? \text{ mol HCl} = 0.300 \text{ mol H}_2 \cdot \dfrac{2 \text{ mol HCl}}{1 \text{ mol H}_2} = \textbf{0.600 mol HCl}$

The limiting reactant is the one that *most limits* the amount of products that can form. The reactant that is limiting determines how much product can form.

The limiting reactant must be **Cl₂**. When **0.400 mol HCl** forms, the reaction stops.

2b. In reaction amount calculations (stoichiometry), start with WDBB.

    1. WANTED: ? mol H_2 (From part a, H_2 is the *reactant* in excess.)

    2. DATA: 14.2 g Cl_2 (The *given* is the known amount of the limiting reactant.)

               1 mol Cl_2 = 71.0 g Cl_2

               To calculate the amount of a stoichiometric equivalent, use standard conversion stoichiometry.

    3. Balance: 1 H_2 + 1 Cl_2 → 2 HCl

    4. Bridge: 1 mol H_2 = 1 mol Cl_2

    5. SOLVE: ? mol H_2 = 14.2 g Cl_2 · $\dfrac{1\ mol\ Cl_2}{71.0\ g\ Cl_2}$ · $\dfrac{1\ mol\ H_2}{1\ mol\ Cl_2}$ = $\boxed{\textbf{0.200 mol H}_2}$

This answer makes sense. You started with 0.300 mol H_2. You use up 0.200 mol H_2 in the reaction with the Cl_2. This confirms that H_2 is *in excess*: some is left over when the reaction stops.

3a. In reaction calculations, start with WDBB.

    1. WANTED: ? **g** H_2O

    2. DATA: 2.00 **g** NaOH

              0.0100 mol H_2SO_4

              1 mol NaOH = 40.0 g NaOH

              1 mol H_2O = 18.0 g H_2O (Two grams prompts.)

             (Strategy: Because the data include single-unit grams or counts of *two* reactants, you must identify the limiting reactant. Find the WANTED amount using stoichiometry with *each* reactant amount as *given*.)

    3. Balance: **1** H_2SO_4 + **2** NaOH → **2** H_2O + **1** Na_2SO_4

    4. Bridge: (Moles WANTED to moles *given* will vary for the two *givens*.)

    5. SOLVE: ? g H_2O = **2.00 g NaOH** · $\dfrac{1\ mol\ NaOH}{40.0\ g\ NaOH}$ · $\dfrac{\textbf{2}\ mol\ H_2O}{\textbf{2}\ mol\ NaOH}$ · $\dfrac{18.0\ g\ H_2O}{1\ mol\ H_2O}$ = **0.900** g H_2O

           ? g H_2O = **0.0100 mol H₂SO₄** · $\dfrac{\textbf{2}\ mol\ H_2O}{\textbf{1}\ mol\ H_2SO_4}$ · $\dfrac{18.0\ g\ H_2O}{1\ mol\ H_2O}$ = **0.360** g H_2O

The limiting reactant *most limits* the amount of product that can form. In this reaction, the limiting reactant is **H₂SO₄**. The mass of water formed is decided based on the limiting reactant: $\boxed{\textbf{0.360 g H}_2\textbf{O}}$.

3b. The calculation of a stoichiometrically equivalent amount is standard seven-step stoichiometry.

    1. WANTED: ? g NaOH   (NaOH is the reactant in excess.)

    2. DATA: 0.0100 mol H_2SO_4 (The *given* is the amount of the limiting reactant.)

              1 mol NaOH = 40.0 g NaOH

    3. Balance: **1** H_2SO_4 + **2** NaOH → **2** H_2O + **1** Na_2SO_4

    4. Bridge: **1** mol H_2SO_4 = **2** mol NaOH

    5. SOLVE: ? g NaOH = 0.0100 mol H_2SO_4 · $\dfrac{\textbf{2}\ mol\ NaOH}{1\ mol\ H_2SO_4}$ · $\dfrac{40.0\ g\ NaOH}{1\ mol\ NaOH}$ = $\boxed{\textbf{0.800 g NaOH}}$

This answer makes sense. You started with **2.00 g** NaOH. You use up **0.800** g NaOH in the reaction with the H_2SO_4. This confirms that NaOH is *in excess*: some is left over after the reaction.

10

Solution Concentration

Introduction

Most problems we have done so far have asked for a single-unit answer, such as grams, moles, or molecules. Problems seeking to find the concentration of a solution or the density of a substance require finding a *ratio* unit such as "moles per liter of solution" or "grams per liter of gas."

In Chapters 10 and 11, you will learn two ways to solve for ratio units: (1) starting with a ratio, and (2) solving in two parts.

Lesson 10.1   Ratio Unit Review

Solving for ratio units using conversions was covered in Lesson 4.5. To briefly review:

- The *order* in which conversions are multiplied does not affect the answer.
- When solving for a ratio unit, either the top or the bottom *given* unit may be converted first.
- If a unit and label to the right of the equals sign *matches* a unit and label WANTED, in both what it is and where it is (top or bottom), circle it and leave it alone.
- If a unit and label *in* or *after* the *given* unit is *not* what you WANT and where you want it, put it where it will cancel, and convert until it matches what is WANTED.

PRACTICE

Complete every other part, and do more if you need more practice. For additional review, see Lesson 4.5.

1. Complete the conversions, cancel units, do the math, check your answers.

   a. $? \dfrac{m}{s} = 95 \dfrac{km}{hr} \cdot \dfrac{m}{} \cdot \dfrac{}{min} \cdot \dfrac{}{} =$

   b. $? \dfrac{mol\ H_2O}{L\ of\ solution} = \dfrac{0.36\ g\ H_2O}{mL\ of\ solution} \cdot \dfrac{}{L} \cdot \dfrac{1\ mol\ H_2O}{} =$

2. Add legal conversions in any order and solve.

   a. $? \dfrac{mg}{L} = \dfrac{0.025\ g}{mL} \cdot$

   b. $? \dfrac{joules}{g\ H_2O} = \dfrac{6.02\ kilojoules}{mol\ H_2O}$

Lesson 10.2   Word Problems with Ratio Answers

Recall these rules (from Chapter 5) for solving word problems using conversions:

1. Write the WANTED unit first.

   a. If the WANTED unit is a single unit *or* a single unit per *more* than *one* of another unit, write the WANTED unit as a single unit.

   b. If the WANTED unit is one unit *per one* other unit, write the WANTED unit as a *ratio*: a fraction with a top and a bottom.

2. Write DATA as *single* units *or* as *equalities*.

   a. In problems that can be solved using conversions, most of the DATA will be two measurements that are *paired* because they are equivalent in some way.

   b. Paired measurements are written in the DATA as equalities.

3. If you WANT a *single* unit, a single-unit amount from the DATA will be used as the *given*. The rest of the DATA will be paired measurements.

   If you WANT a *ratio* unit, all of the DATA will be paired measurements.

Solving for a Ratio

There are several ways to solve for ratio units using conversions. To learn the first method, we will add to our list above the following new rule (which must be committed to memory):

4. To SOLVE, in choosing a *given*, if you WANT a single unit, start with a single unit; if you WANT a ratio, start with a ratio.

Rule 4 is based on dimensional homogeneity, which is a law that requires the following:

- If a single-unit amount is alone on one side of an equal sign, a single-unit amount must result on the other.
- If a ratio unit is on one side of an equal sign, a ratio must start and result on the other.

Otherwise, the two sides cannot be *equal*.

If you WANT a ratio unit, it does not matter which DATA ratio you choose as your *given*, because in multiplication the order of the terms does not matter. However, it is possible to pick a ratio as your *given* that will never cancel to give the WANTED unit, because your *given* is algebraically upside down from the answer unit.

If this happens, it will become apparent as you do your conversions. The solution is to simply start over with your *given* inverted (flipped over). However, the following methods will help to pick a *given* ratio that is right-side up the first time.

Method 1: Start with One *Given* Unit Where It Is in the WANTED Unit

To solve for a ratio unit, pick as your *given* an equality from your DATA that *includes* one of the units that is WANTED. Write that unit in your *given* ratio where it needs to be (on the top or bottom) to match the WANTED unit. This way, the other unit of your *given* is the only one that needs converting, *and* the conversions will be right-side up.

▷ **TRY IT**

(See "How to Use These Lessons," point 1, p. xv.)

For the following question, write only the WANTED and DATA (do not write the SOLVE step).

Q. If a car is traveling 60. feet per second, what is its speed in miles per hour (5,280 feet = 1 mile)?

Answer:

WANTED: $? \dfrac{\text{mi.}}{\text{hr}}$

DATA: 60. ft. = 1 s

5,280 ft. = 1 mi.

Now, using method 1, write just your initial *given ratio* after the = sign below.

SOLVE: $? \dfrac{\text{mi.}}{\text{hr}} =$

SOLVE: $? \dfrac{\textbf{mi.}}{\text{hr}} = \dfrac{\textbf{1 mi.}}{5{,}280 \text{ ft.}}$

The only WANTED unit that is in the DATA is *miles*, so start with a *given* that has *miles* where you want it in the answer. Once any one *given* unit, on the top or bottom, is where it should be, the other conversions will lead to a "right-side up" answer. Finish solving for the WANTED unit.

SOLVE: $? \dfrac{\textbf{mi.}}{\text{hr}} = \dfrac{\textbf{1 mi.}}{5{,}280 \text{ ft.}} \cdot \dfrac{60 \text{ ft.}}{\text{s}} \cdot \dfrac{60 \text{ s}}{1 \text{ min}} \cdot \dfrac{60 \text{ min}}{1 \text{ hr}} = \textbf{41} \dfrac{\textbf{mi.}}{\textbf{hr}}$

PRACTICE   **A**

Using method 1 above, do *both* questions below.

1. In 1988, Florence Griffith-Joyner set a new women's world record in the 200-meter dash with a time of 21.34 seconds. What was her average speed in miles per hour? (Use 1.609 km = 1 mi. Assume the distance is 200.**0** m: very carefully measured in a world-record certified race.)

2. For a given substance, 0.0500 moles has a mass of 9.01 grams.

    a. To find the molar mass, what unit is WANTED?

    b. Calculate the molar mass of the substance.

Method 2: Arrange the *Given* Based on Complete Labels

A second way to pick a *given* ratio that is right-side up is, in your WANTED and DATA, to include complete and descriptive *labels* after the units: words that describe what the unit is measuring. Then pick and arrange the *given* to match the *labels* in the WANTED unit.

Using method 2, *start* the following problem and pick as your *given* an equality which has something about the NaCl *solution* on the bottom.

TRY IT

Q. If a 250 mL solution contains 0.020 lb. of dissolved NaCl, how many moles of NaCl are dissolved per liter of solution? (Use 1 kg = 2.20 lb. and 58.5 g NaCl = 1 mol NaCl.)

STOP

Answer:

WANTED: $\dfrac{? \text{ mol NaCl}}{\text{L } solution}$

DATA: 250 mL *solution* = 0.020 lb. NaCl   (Two measures of same object.)

$1 \text{ kg} = 2.20 \text{ lb.}$

$58.5 \text{ g NaCl} = 1 \text{ mol NaCl}$

SOLVE: $\dfrac{? \text{ mol NaCl}}{\text{L } solution} = \dfrac{0.020 \text{ lb. NaCl}}{250 \text{ mL } solution}$

Now finish.

STOP

$$\frac{? \text{ mol NaCl}}{\text{L } solution} = \frac{0.020 \text{ lb. NaCl}}{250 \text{ mL } solution} \cdot \frac{1 \text{ kg}}{2.20 \text{ lb.}} \cdot \frac{10^3 \text{g}}{1 \text{ kg}} \cdot \frac{1 \text{ mol NaCl}}{58.5 \text{ g NaCl}} \cdot \frac{1 \text{ mL}}{10^{-3} \text{ L}}$$

$$= \mathbf{\frac{0.62 \text{ mol NaCl}}{\text{L solution}}}$$

By using the label *solution* to arrange the *given*, the final units are right-side up.

PRACTICE   B

Use method 2 above on these problems. Start with problem 1. Solve problem 2 if you need more practice.

1. 10.0 g of NaOH is dissolved to make 1,250 mL of solution. Calculate the concentration of the solution in moles of NaOH per liter of solution (40.0 g NaOH/mol).

2. A water bath absorbs 24 calories of heat from a reaction that forms 0.88 g of carbon dioxide. What is the amount of heat released by the reaction, in kilocalories of heat per mole of CO_2?

Method 3: Arrange the *Given* Based on Dimensions

A third way to pick a *given* ratio that is right-side up is to arrange the *given dimensions* to match the WANTED *dimensions*.

Example: If km/s is WANTED, and 24 miles per hour is in the DATA, because you want *distance* over *time*, to start, use as your *given* distance over time.

WANTED: ? $\dfrac{km}{s}$ = $\dfrac{24 \text{ mi.}}{hr}$

The above three methods may lead you to pick different ratios as your *given*. That's OK. In solving for a ratio, if the *given* ratio is *any* ratio in the DATA that is written right-side up relative to the WANTED unit, you will be able to convert to the answer unit.

From the above, you do not need to remember which method is 1, 2, or 3, but you do need to be able to recall and apply each of these three strategies for arranging a *given* ratio right-side up.

Ratios Represented by Negative Powers

A ratio unit can be represented as "A per B" *or* A/B *or* $A \cdot B^{-1}$. The algebraic rule is $A \cdot B^{-1} = A/B$. When solving with conversions, units in the form $A \cdot B^{-1}$ must be treated as a *ratio* unit.

Example: Treat $g \cdot mol^{-1}$ in the same manner as g/mol; as a ratio unit.
If DATA includes $18.0 \text{ g} \cdot mol^{-1}$, write DATA: 18.0 g = **1** mol
If $g \cdot mol^{-1}$ is WANTED, write:

WANTED = ? $\dfrac{g}{mol}$

The general rule is $A \cdot B^{-x} = A/(B^x)$.
Example: In conversion calculations, treat $g \cdot cm^{-3}$ as g/cm³; as a ratio unit.

PRACTICE C

1. Rewrite these in the slash (/) format.

   a. $14 \text{ m} \cdot s^{-1}$

   b. $0.47 \text{ mg} \cdot mol^{-1}$

   c. $9.2 \text{ kg} \cdot dm^{-3}$

2. If a sample of Rn gas at a given temperature and pressure has a density of $8.60 \text{ mg} \cdot mL^{-1}$, what is its density in $kg \cdot L^{-1}$? (Solve using method 3 above.)

3. A typical 16 fluid ounce (fl. oz.) can of soft drink contains 52 g of sugar. Convert the concentration of sugar to $mg \cdot cm^{-3}$ (12 fl. oz. = 355 mL). (Use method 3.)

Lesson 10.3   Molarity

Solvents and Solutions

Many substances dissolve to some extent in other substances. When a larger amount of one substance has smaller amounts of other substances completely dissolved and uniformly distributed within it, the result is termed a **homogeneous solution**. A mixture of gases can be considered a solution: air can be considered a solution in which other gases are dissolved in the predominant gas, nitrogen. However, the solutions encountered most often in chemistry contain substances dissolved in liquids.

A liquid used to dissolve a substance is termed the **solvent**, and the substance dissolved is the **solute**. When two liquids are mixed and dissolve in each other, the liquid with the larger original volume is the solvent. When a substance dissolves to a substantial extent in a solvent, it is said to be **soluble** in that solvent. A substance that does not dissolve or dissolves only slightly in a liquid is **insoluble** or **slightly soluble** in that solvent.

Any liquid can be a solvent, but water is the solvent most commonly employed in chemistry and everyday use. When substances are dissolved in water, the result is termed an **aqueous** solution (from the Latin *aqua*, for water). In these lessons, if the solvent for a solution is not specified, assume the solvent is water.

Aqueous solutions are of special interest in the biological sciences. Water is the most abundant component of all cells, and most biochemical processes are the result of chemical reactions in the aqueous solution of a cell or other parts of organisms.

Calibrated glassware such as burets and pipettes can measure solution volumes with high accuracy. If a solution has a known concentration of a dissolved substance, by measuring the volume of a sample, we can convert to a count of the particles in the sample. Counting the particles is the key to solving chemistry calculations and understanding chemical processes.

Preparing Solutions

The *volume* and the **concentration** of a solution are the dimensions most often used to describe solutions. The **molar concentration** (or **molarity**) of a solution is defined as *moles* of dissolved solute *per liter* of *solution*.

In chemistry calculations, concentration can be measured by other units, but unless otherwise specified, you should assume that the units used to measure concentration are moles per liter (mol/L).

To make an aqueous solution of a known *molarity*, a substance is weighed, placed in a volumetric flask, and completely dissolved in distilled water (water with minerals and impurities removed). The quantity of water is then increased, with mixing, until a precisely marked volume is reached.

In preparing an aqueous solution, the amount of water *added* is neither measured nor used to calculate the solution concentration. What *is* measured carefully and used in calculations is the volume of the *mixture* of the dissolved solute and water: the volume of the *solution*.

Rules for Concentration Calculations

The concentration of a solution, measured in *moles per liter*, is termed the **molarity** of the solution. Moles per liter can be written as *mol/L* or as $mol \cdot L^{-1}$, or it can be abbreviated as a capital **M**.

A solution labeled "0.50 M HCl" is termed a "0.50 **molar** HCl solution" and these terms mean that the solution contains 0.50 moles of HCl per liter of solution volume.

Brackets [ ] are used as shorthand for "the molar concentration of a solution." [NaCl] is read as "the molar concentration of NaCl" or as "the molarity of the NaCl."

Solution concentration is a *ratio* unit: moles of dissolved substance *per* liter of solution. Molarity calculations can be solved with conversions if you use this rule:

The M Prompt (the Molarity Ratio)

In *conversions*, treat *concentration* or a capital *M* or *molar* or *molarity* or *mol · L⁻¹* or [ ] *brackets* as **"moles *per* 1 liter of solution"** in the WANTED and DATA.

$$M = [\ ] = molar = mol/L = mol \cdot L^{-1} = \textbf{moles \textit{per} 1 liter} = a \textit{ ratio unit}$$

The abbreviation **M** for mol/L can*not* be used when writing the WANTED or DATA or conversions. Concentration is a ratio, and ratio units must be written as a *fraction* when solving with conversions.

Examples:

- If a problem WANTS the "*concentration* of NaCl" or "*molarity* of NaCl" or [NaCl], write

    WANTED: ? $\dfrac{\textbf{mol NaCl}}{\textbf{L of NaCl soln.}}$ Concentration is a ratio unit.

- If the DATA in a problem includes *0.25 M NaOH* or *0.25 molar NaOH*, write this *equality* (*or* conversion) in the DATA:

    DATA: 0.25 mol NaOH = **1** L NaOH soln. $\left(or \dfrac{0.25 \text{ mol NaOH}}{1 \text{ L NaOH soln.}} \right)$

As always, when solving with conversions,

- If a *ratio* unit is WANTED, write it as a *fraction* with a top and bottom, but
- You may write a ratio in DATA in either the equality format or in the fraction/ratio/conversion format. (In these lessons, we usually list equalities.)

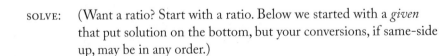
TRY IT

Apply the rules from above.

Q. If 2.98 g KCl is dissolved to make 250. mL of solution, what is the [KCl]?

 Answer:

WANTED: ? $\dfrac{\text{mol KCl}}{\text{L of KCl soln.}}$ (You want concentration. Its *units* are mol solute *per* L soln.)

DATA: *2.98 g KCl* = 250. mL KCl soln.   (Two measures of the *same* solution.)

    74.6 g KCl = 1 mol KCl   (Grams prompt.)

When using conversions to solve for a *ratio* unit, all of the DATA will consist of paired measurements, listed as equalities or ratios. There should be no single-unit DATA. If needed, complete the problem.

SOLVE: (Want a ratio? Start with a ratio. Below we started with a *given* that put solution on the bottom, but your conversions, if same-side up, may be in any order.)

$$? \ \frac{\text{mol KCl}}{\text{L KCl soln.}} = \frac{2.98 \text{ g KCl}}{250 \text{ mL soln.}} \cdot \frac{1 \text{ mol KCl}}{74.6 \text{ g KCl}} \cdot \frac{1 \text{ mL}}{10^{-3} \text{ L}} = 0.160 \ \frac{\text{mol KCl}}{\text{L KCl soln.}}$$

The molarity equality can also be used in problems that solve for single units. Apply the fundamental rules.

> • If you WANT a single unit, *start* with a single unit as your *given*.
> • If you WANT a ratio, start with a ratio.

 TRY IT

Q. How many grams of NaOH are required to make 150. mL of a 0.300 M NaOH solution?

Answer:

WANTED: ? g NaOH   (You want a single unit.)

DATA: 150. mL NaOH soln.   (Single-unit data.)

0.300 mol NaOH = 1 L NaOH soln.   (M prompt in DATA.)

40.0 g NaOH = 1 mol NaOH   (WANTED unit = grams prompt.)

If needed, adjust your work and finish.

SOLVE: Want a single unit? Start with a single unit.

$$? \text{ g NaOH} = 150. \text{ mL NaOH soln.} \cdot \frac{10^{-3}L}{1\ mL} \cdot \frac{0.300 \text{ mol NaOH}}{1 \text{ L NaOH soln.}} \cdot \frac{40.0 \text{ g NaOH}}{1 \text{ mol NaOH}}$$

$$= \textbf{1.80 g NaOH}$$

Labeling Solution Conversions

In the problem above, ratios that are not always true were given a *complete* label: number, unit, and substance. Because we rely on units and labels to solve problems, complete labels are important.

Often, however, some parts of labels can be omitted as "understood." For example, in the problem above, all of the *volumes* are of NaOH solution. In such cases, the fact that each volume is of "NaOH solution" should be indicated once in the problem, but after that, volume units may omit the label "NaOH solution" as understood.

However, this shortcut needs to be used carefully. When we encounter problems that involve two different solutions, or problems that involve both volumes of gas and volumes of a solution, we will need full labels that clearly identify what each unit is measuring.

> To solve complex problems, write complete labels.

Flashcards

Add any flashcards below that you cannot answer quickly to your collection. Practice until you can answer each card correctly, then do the following problems. Repeat this process for two more days, then put these cards in stack 2 (see Lesson 6.5).

One-Way Cards (with Notch)	Back Side—Answers
If you WANT a single unit	Start with a single unit
If you WANT a ratio	Start with a ratio
If unit X/unit Y is WANTED, write WANTED $=$	$\dfrac{\text{Unit } X}{\text{Unit } Y}$
If unit X/unit Y is DATA, write in DATA	Unit X = Unit Y
$X \cdot Y^{-1} =$	X/Y (a ratio unit)
If 0.50 M of X is in the DATA, write	DATA: 0.50 mol X = 1 L X
If 0.50 mol $\cdot$ L^{-1} X is DATA in problem, write	DATA: 0.50 mol X = 1 L X
If 0.50 M $=$ $[X]$ is DATA in problem, write	DATA: 0.50 mol X = 1 L X
If $[X]$ is WANTED, write WANTED $=$	$\dfrac{\text{mol } X}{\text{L } X \text{ soln.}}$
Treat M $=$ [ ] $=$ molar $=$ mol/L $=$ mol $\cdot$ L^{-1} as	Moles *per* 1 liter, a ratio unit
If you get stuck on a complex problem	Add detail to WANTED and DATA labels

PRACTICE

Do every other problem. Need more practice? Do more. Problem 9 must be done and checked; it has important information discussed in its answer.

1. Find the moles of solute in 100. mL of 0.40 M KBr.

2. In problem 1, name the solute and the solvent.

3. How many moles of dissolved potassium hydroxide are present in 0.20 L of 0.60 mol $\cdot$ L^{-1} KOH?

4. How many grams of HCl are needed to prepare 500. mL of 0.200 M HCl?

5. How many milliliters of 0.100 molar KCl contain 4.48 g of dissolved KCl?

6. In 400. mL of solution is dissolved 2.34 g of NaCl. What is the molarity of this NaCl solution?

7. If 0.020 lb. of NaOH is dissolved to make 1 quart (qt.) of aqueous solution, find the [NaOH] (2.2 lb. = 1 kg, 12 fl. oz. = 355 mL, 1 qt. = 32 fl. oz.).

8. Given a solution that contains one dissolved substance, what information would you need to know to find the moles of the substance in the sample?

9. At typical room temperatures, liquid water has a density of 1.00 g/mL. What is the concentration of this liquid water in moles per liter?

Lesson 10.4   Conversions and Careers

The methods used in Lessons 10.2 and 10.3 to solve for single and ratio units can be useful in a variety of science courses and careers.

Students with an interest in health and medical careers should try problem 1 below, a type of problem you might encounter in the *dosage lab* of a pharmacy or veterinary medicine program.

Students preparing for engineering, physics, or applied math will likely encounter a variation on problem 2 during introductory courses.

Problems similar to 3–5 are often included on chemistry and engineering final exams.

Try at least three of the problems below.

PRACTICE

Apply the conversion rules in Lessons 10.2 and 10.3. If you get stuck, peek at a *part* of the answer, adjust your work, and try again.

1. (Don't let the vocabulary intimidate you. Apply the rules for writing the WANTED and DATA.)
   A 9.0 lb., 7-year-old cat examined at your veterinary clinic has glaucoma. You prescribe acetazolamide for treatment. The dosage recommended for acetazolamide is 15 mg/kg to be given orally twice daily (1 kg = 2.20 lb.).

   a. What is the appropriate milligram dosage?

   b. If acetazolamide is supplied in 125 mg quarter-scored tablets, how should your client be instructed?

2. The speed of light is the highest speed at which energy or matter can travel. This "speed limit of the universe" in SI units is 3.0×10^8 m/s, but medievalists prefer more traditional measures. If there are 14 days in a fortnight, 8 furlongs is a mile, and a mile is 1.61 km, calculate the speed of light in furlongs per fortnight.

3. A drop of water has a volume of about 0.050 mL. If a drop takes 2.0 hr to evaporate, how many molecules evaporate per second? [Use 1 mL $H_2O(\ell)$ = 1.00 g $H_2O(\ell)$.]

4. An atom has a mass of 6.6×10^{-24} g. What is the molar mass of the atom? Which atom is it likely to be?

5. The density of iron (Fe) is 7.87 g/mL. What is the volume in milliliters occupied per Fe atom?

Lesson 10.5   Fractions and Percentages

Fractions and Decimal Equivalents

A **fraction** is a ratio: one quantity divided by another. In math, a fraction can be any ratio, but in science, "fraction" often (but not always) refers to a *part* of a larger total: a smaller quantity over a larger quantity. In dealing with percentages and fractions in chemistry, use the following rule:

Rule 1

$$\text{Fraction} = \frac{\text{quantity A}}{\text{quantity B}}$$

and often equals

$$\frac{\text{part}}{\text{total}} = \frac{\text{smaller number}}{\text{larger number}}$$

The **decimal equivalent** of a fraction is a number that results by dividing the top number of the fraction (the **numerator**) by the bottom number (the **denominator**). Example: For the fraction 1/2, its decimal equivalent is 0.50.

Rule 2

To find the decimal equivalent of a fraction, divide the numerator by the denominator.

(In these lessons, if a numeric fraction is written in the *x/y* format, such as 1/2 or 2/5, assume for purposes of significant figures that it is exact.)

TRY IT

Q. (Use a calculator if needed.) The decimal equivalent of 5/8 =

Answer:

0.625

In chemistry calculations, a *fraction* generally can be any number that has a value between zero and one (that can be written as $0.X\dots$).

Rule 3

In chemistry vocabulary:

Fraction = decimal equivalent = any value between 0 and 1

Example: An equation that we will use in calculating radioactive half-life is

$$\ln(fraction) = -kt$$

The fraction could be written either in an *x/y* format (such as 1/4) *or* in a decimal equivalent format (such as 0.834).

Percentages and Decimal Equivalents

A **percentage** is a fraction or its decimal equivalent multiplied by 100%.
   A familiar example is 1/2 = 0.50; 0.50 × 100% = 50%.
   A percentage is a way to express fractions that is often used in scientific measurements, but a percentage must be converted to its *decimal equivalent* before it can be used in most conversions or equations.

If you are able to convert between a percentage and its decimal equivalent by mental arithmetic, it will simplify calculations. To do so, use the following rule.

> ### Rule 4: Converting between Percentage and Decimal Equivalents
>
> A. To convert a decimal equivalent to its percentage, multiply the decimal equivalent by 100% (moving the decimal twice to the right).
>
> B. To convert a percentage to its decimal equivalent (its fraction), divide the percentage by 100% (moving the decimal twice to the left).
>
> C. *Check*: The number in the percentage is always 100 times *larger* than its decimal equivalent.

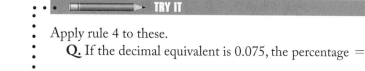 **TRY IT**

Apply rule 4 to these.
   Q. If the decimal equivalent is 0.075, the percentage =

STOP **Answer:**

$$0.075 \times 100\% = \mathbf{7.5\%}$$

   Q. If the percentage = 85%, its decimal equivalent (its fraction) =

STOP **Answer:**

$$85\%/100\% = \mathbf{0.85}$$

 PRACTICE **A**

Practice writing the fraction definition (rule 1) until you can do so from memory. Learn rules 2–4 if they are not already in memory. Then complete these problems.

1. 4.8% has what decimal equivalent?

2. Write 9.5/100,000 as a decimal equivalent and a percentage.

Percentages, Fractions, and Ratios

In chemistry calculations, we will define a percentage in three ways.

> ### Rule 5: Three Ways to Define Percentage
>
> A. **Percent** = $\dfrac{\text{part}}{\text{total}} \times 100\%$ = fraction × 100%
>
>   = (decimal equivalent) × 100%
>
> B. *X*% means: *X* parts *per* **100 parts** total
>
> C. *X*% means: (decimal equivalent) *parts* *per* **one** part *total*

All of those statements are mathematically equivalent. A percentage is always based on a ratio (a fraction), and a ratio can be written as an equality or a *per* statement or a conversion factor.

In percentage calculations, the 100 in "100 parts" and "100%" is an exact number that does not affect the uncertainty (and significant figures) in a calculated answer.

When a Percentage Is WANTED Note in rule 5.A that a percentage is based on a fraction. For most calculations in which a percentage is WANTED, our work is simplified if we use the following rule.

> **Rule 6**
>
> If a percent is WANTED, first write the *fraction* WANTED, then find the decimal equivalent value of the fraction, then convert to its percent.

The short version of rule 6 is: If a percentage is WANTED, write the fraction, then decimal equivalent, then percent.

➤ TRY IT

Apply rule 6 to the following questions.
  Q. 1/8 is what percentage?

 Answer:

WANTED: Percent   (To find percentage, write the fraction, then decimal, then percentage.)

? = fraction = **1/8** = 0.125; % = decimal × 100% = 0.125 × 100% = **12.5%**

  Q. 25 is what percent of 400? (Assume exact numbers.)

 Answer:

WANTED: Percent   (To find percentage, write the fraction, then decimal, then percentage.)

$$? = \text{fraction} = \frac{\text{part}}{\text{total}} = \frac{\text{smaller number}}{\text{larger number}} = \frac{25}{400} = \mathbf{0.0625}$$

Percentage = 6.25%

With practice, you can solve many percent calculations using mental arithmetic by writing the *fraction*, then *decimal*, then percentage.

➤ TRY IT

Q. (Try without a calculator.) 2/5 is what percent?

 Answer:

2/5 = 0.40 = 40% (exactly)

PRACTICE   B

Write the percent definition (rule 5) until you can do so from memory. Then complete each of these problems.

1. 3/5 is what decimal equivalent and what percent?

2. What percentage of 25 is 7?

3. Twelve is what percent of 24,000?

When Percentage Is Supplied as DATA When a percentage is supplied as DATA, it may be a "number percent," such as 25%, or a percent with a label attached, as in "2.5% KCl by mass." Let's consider these cases one at a time.

When a Number Percent Is DATA If a "number percent" (a percentage without a label) is supplied as DATA, in most problems you can calculate using the rule

X% *of* Y means   (X/100%) *times* Y *or*   (decimal equivalent of X%) *times* Y

This can be remembered as

Rule 7

X% *of* Y = (decimal equivalent) *times* Y

 TRY IT

Apply rule 7 to the following problem.
   Q. 3.5% of 12,000 g is how many grams?

STOP **Answer:**

$$\textbf{3.5\% of } 12{,}000 \text{ g} = \textbf{0.035} \times 12{,}000 \text{ g} = \textbf{420 g}$$

When a Percent in the DATA Has a Label Attached If a supplied percentage has a unit and substance attached, we can simplify calculations by

• Methodically listing the WANTED and DATA, and
• Converting the percentage to its *fraction* (which can then be used as a conversion factor)

Most equation and conversion calculations require that a percentage be converted to its fraction: the ratio that the percentage represents. Unless an equation specifically calls for a percentage, use the following rule:

Rule 8

Convert a *supplied* percentage to a *fraction* by writing in the DATA *either*

- *X*% means: *X* units of the part *per* **100 units** of the *total* (rule 5.B), or
- *X*% means: (decimal equivalent) units of *part* *per* **one** unit of *total* (rule 5.C)

Per means / or =. *Per* statements can be used as conversion factors. Note that

- In rule 5.B, one way to define a percentage is "parts *per* 100 parts total"
- Rule 5.C is simply 5.B with both sides divided by 100

Both definitions are mathematically equivalent and you may write either one.

TRY IT

Apply rule 8 and the WANTED, DATA, SOLVE steps to this problem.

Q. In a solution of KCl dissolved in water, 2.0% of the total grams are KCl. How many grams of KCl are in 150 g of the solution?

Answer:

WANTED: ? g KCl

DATA: **2.0** g KCl = **100** g soln.   (Using rule 5.B.)

150 g soln.

SOLVE: ? g KCl = 150 g soln. · $\dfrac{2.0 \text{ g KCl}}{100 \text{ g soln.}}$ = **3.0 g KCl**

SUMMARY

Percentage and Fraction Calculations

You don't need to know the numbers of the rules above, but you need to be able to apply the rules automatically. Being able to write the fraction (rule 1) and percent (rule 5) definitions from memory is especially helpful. If you are not already fluent in fraction calculations, practice writing the unfamiliar rules until you can apply them from memory.

PRACTICE   C

1. How much is 25% of 180?

2. How much is 0.0450% of 7,500?

3. If 5.0% of the mass of an aqueous solution is NaCl, how many grams of the solution contain 12.5 g of NaCl?

4. 3/8 has what decimal equivalent and what percent?

5. 16.7% has what decimal equivalent?

6. 12% of 750 = ?

7. 1.8 is what percent of 45?

8. Write the decimal equivalent of 0.6%.

9. 45/10,000 has what decimal equivalent and what percent?

10. 0.25% of 12,400 = ?

11. In a solution of glucose and water, if 4.5% of the mass is glucose, how many grams of the solution contain 2.25 g of glucose?

Lesson 10.6   Solution Concentration in Mass Percent

Mass Percent

In chemistry, the concentration of a solution is most often measured as a molarity, but concentration may also be expressed as a **mass percent** (also called **percent by mass**).

Example: In a *5% by mass* NaCl solution, 5 g of NaCl are dissolved per 100 g of solution.

To assist with calculations, we will define a mass percent in three ways. These three definitions apply the general definitions for percentages in Lesson 10.5 to the specific case of a substance that is dissolved in a solution.

Definition A

Solution concentration in *mass percent* ≡ $\dfrac{grams \text{ of } solute \text{ } dissolved}{total \text{ } grams \text{ of } solution} \times$ **100%**

The logic of this definition is

- A percent is (part/total) times 100%
- A *mass* percent is (mass of part/mass of total) times 100%: a mass *fraction* times 100%
- The mass percent of a *solution* is (mass of solute/mass of solution) times 100%

Definition A is useful if a percentage by mass is WANTED. When a mass percent is DATA, use

---

Definition B

X% by mass means: *X* grams of solute *per 100* grams of solution, or

---

Definition C

X% by mass means: (*decimal equivalent*) grams of solute *per 1* gram of solution.

Definitions A, B, and C are all mathematically equivalent.

Mass Percentage Calculations

Percent by mass solution calculations can be solved using conversions. Specific rules for these calculations are:

1. If a mass *percent* is WANTED: solve for the mass *fraction first.*
   If a *percent by mass* is WANTED, write

   WANTED: percentage by mass   (To find percentage, write the fraction, then decimal, then percentage.)

   Write $\dfrac{\text{grams of } solute}{\text{grams of } soln}$, then solve for decimal, then percentage.

2. If a *X% by mass* is supplied, write

   DATA: *X% by mass* means: *X grams of solute* = **100** grams of *solution*
      (We could write either definition B or C, but we will generally use B.)

3. *Label* each number and unit as measuring either the *solute* formula or the *solution*.

TRY IT

Q. If a problem includes "3.0% HCl by mass," in your DATA what would you write?

Answer:

3.0% by mass HCl means: 3.0 **g** *HCl* = 100 **g** *solution*   (Rules 2 and 3.)

---

4. If the *moles* of a substance are mentioned in the WANTED or DATA, write its molar mass in the DATA. In percentage concentration calculations, if moles are not mentioned, you will not need the molar mass.

   Most chemistry relationships are defined in terms of moles, but percentage concentration is based on masses rather than moles. You will need a g/mol ratio only if *moles* appear as a unit in the WANTED or DATA.

5. In chemistry, if a concentration is expressed as a percent, assume it is a *mass* percent unless otherwise noted.

   Example: If *3% HCl* is supplied as DATA, write: 3 **g** HCl = 100 **g** soln.

To find a percentage, we *first* find the fraction on which the percentage is based. Finding the fraction means solving for a ratio using conversions (which we know how to do). That answer will result in a decimal equivalent that is easily converted to a percent.

TRY IT

Apply rules 1–5 to this problem:

Q. If the density of a 3.31 M KOH solution is 1.14 g/mL, what is its percentage concentration?

Answer:

WANTED: ? percentage concentration = mass percent = percent by mass

(Rule 5 above.)

The fundamental rule: if a percentage is WANTED, solve for the *fraction first*. If a mass percent is wanted, find the mass fraction first. Write

WANTED: ? $\frac{\text{g KOH}}{\text{g soln.}}$, then decimal, then percentage.   (Rule 1.)

DATA: 3.31 mol KOH = 1 L solution   (M prompt.)

1.14 g solution = 1 mL solution   (Solution density.)

56.1 g KOH = 1 mol KOH   (Moles is in DATA.)

(Rule 3: Write labels distinguishing numbers that measure the dissolved solute from numbers that measure the complete solution.)

SOLVE: (Want a ratio? Start with a ratio.)

$$\frac{?\ \text{g KOH}}{\text{g soln.}} = \frac{56.1\ \text{g KOH}}{1\ \text{mol KOH}} \cdot \frac{3.31\ \text{mol KOH}}{1\ \text{L soln.}} \cdot \frac{10^{-3}\ \text{L soln.}}{1\ \text{mL soln.}} \cdot \frac{1\ \text{mL soln.}}{1.14\ \text{g soln.}}$$

$$= 0.163\ \frac{\text{g KOH}}{\text{g soln.}} = \text{decimal fraction; } \% = \textbf{16.3\% by mass KOH}$$

(Your conversions may be in a different order as long as they are "same side up.")

PRACTICE

First learn the rules, then do the problems.

1. If 6.0 g of glucose is dissolved in 150 g of solution, find the percentage glucose by mass.

2. A 44% aqueous solution of H_2SO_4 has a density of 1.34 g/mL. How many moles of H_2SO_4 are in 80. mL of this solution?

Dilute and Non-Dilute Solutions

In these lessons, if no solvent for a solution is identified, assume the solution is aqueous: The solute is dissolved in water.

A *dilute* aqueous solution can generally be considered to be one in which the solute concentration is *less* than either 1.0% or 0.2 molar.

When calculating with percentage concentrations, we often need to know the density of a solution. Because a dilute (~1%) aqueous solution is 99% water by mass, and water has a density of about 1.00 g/mL at or near room temperature, we can add the following *approximation* to our rules for percentage concentration calculations.

6. The *dilute* solution prompt: In a percentage concentration calculation about a *dilute* aqueous solution, if no density is supplied, add to your DATA:

1 mL dilute soln. ≈ 1.00 mL H_2O solvent ≈ 1.00 g H_2O solvent ≈ 1.00 g soln.

In a solution that is *not* dilute, the mass of the solute becomes significant and we use rule 7.

7. In higher concentration solutions (all those that are not dilute),

  a. Total mass of solution = mass of solute(s) *plus* mass of solvent

  b. Attach labels that distinguish the *solute*, the *solvent*, and the *solution*

➤ TRY IT

Apply the above rules to the following.

  Q. In 100. g of 14% by mass KCl,

  a. What is the mass of the KCl in the solution?

  b. What is the mass of the water in the solution?

Answer:

  a. The mass of the KCl is 14% of 100. g = $0.14 \times 100.\,g = \mathbf{14\,g}$

  b. The solution is KCl plus water, so the H_2O mass must be
     $100.\,g - 14\,g = \mathbf{86\,g}$

PRACTICE **B**

Write a quick summary of rules 6 and 7, add it to your prior "quick summary," learn your rules, then complete this practice.

1. How many milliliters of dilute 0.50% KCl solution contain 1.6 g of KCl?

2. Express the concentration of a relatively dilute 0.080 M HCl solution as a mass percent.

3. To make 500. g of an aqueous solution that is 6.0% NaCl by mass, how many grams of water would be needed?

Weight Percent Solutions

In the health sciences, solution concentration may be expressed in different terms and by different measures than those employed in chemistry. Two such measures are **percent (w/w)** and **percent (w/v)**.

The expression **percent (weight/weight)**, abbreviated as **% (w/w)** or as **percent by weight**, has the same meaning as a *mass percent* or *percentage by mass* solution concentration in chemistry.

Though mass and weight are not the same, at constant gravity they are proportional and can be measured with the same instruments. In percentage solution calculations, "weight" is generally measured in mass units. Any consistent unit can be used to measure mass or weight, but *grams* should be assumed if no unit is specified.

The term **percent (weight/volume)**, abbreviated as **% (w/v)**, can be defined as

  • (*grams* of solute/*milliliters* of solution) times 100%, or as

  • *X* g of solute per **100** *mL* of solution

Rules for % (w/w) and % (w/v)

To cover the specific case of % (w/w) and % (w/v) concentrations, let's add the following rules to our list of rules above.

8. Unless otherwise specified, assume the unit of mass and weight (**w**) is grams and unit of volume (**v**) is milliliters.

9. If a problem expresses solution concentration as **percentage (weight/weight)** or **percent (w/w)** or **percent by weight**, assume weight means *mass* and use the *mass percent* rules to solve.

10. If a %(w/v) is WANTED, solve for the *fraction first*.
    If a *%(w/v)* is WANTED, write

    WANT: %(w/v); *first* find $\dfrac{\text{g of } solute}{\text{mL of } soln.}$, then decimal, then percentage.

11. If any *X% concentration* is supplied, write

    DATA: *X*% means: *X* units of *solute* = **100** units of *solution*

TRY IT

Q. If a problem includes "1.5% (w/v) NaCl," in your DATA what would you write?

 Answer:

1.5% (w/v) NaCl means: 1.5 **g** *NaCl* = 100 **mL** *solution*   (Rules 3, 8, and 11.)

One implication of the dilution approximation is that in a *dilute* solution, the *grams* of the water and solution and the *milliliters* of the water and solution all have the same *numeric value*. From this we can derive rule 12.

12. In *dilute* solutions: percentage by mass = % (w/**w**) = % (w/**v**).

PRACTICE C

Write and learn a "quick summary" of the rules above, then do these problems.

1. If 0.050 mol of NaOH is dissolved to make 200. g of solution, find the % (w/w) NaOH.

2. The **normal saline** solution employed in intravenous drips is 0.90% (w/v) NaCl dissolved in water. How many grams of NaCl would be contained in 125 mL of normal saline?

3. The *normal saline* solution of 0.90% (w/v) NaCl is a relatively dilute aqueous solution. What is the molarity of normal saline?

Lesson 10.7   Review Quiz for Chapters 9–10

You may use a calculator and a periodic table. Work on your own paper.

To answer *multiple-choice* questions, it is suggested that you solve as if the question is *not* multiple choice, then circle your answer among the choices provided. Set a 30-minute limit. Answers are at the end of the chapter.

1. For the reaction

$$Fe_2O_3 + \quad CO \rightarrow \quad Fe + \quad CO_2$$

   a. Lesson 9.2: Balance the equation.

   b. Lesson 9.5: If 33.5 g Fe is produced, how many moles of carbon monoxide are used up?

      a. 0.400 mol CO   b. 0.601 mol CO   c. 1.20 mol CO

      d. 1.81 mol CO   e. 0.901 mol CO

2. Lesson 9.5: How many grams of H_2O gas are produced when 64 mg of CH_3OH is burned?
   The unbalanced equation is

$$CH_3OH + \quad O_2 \rightarrow \quad CO_2 + \quad H_2O$$

   a. 0.018 g H_2O b. 0.036 g H_2O c. 0.072 g H_2O

   d. 0.14 g H_2O e. 0.36 g H_2O

3. Lesson 9.6: If 40.2 g H_2 gas is burned with 128 g O_2 gas, how many moles of water form?

   a. 1.00 mol H_2O b. 2.00 mol H_2O c. 4.00 mol H_2O

   d. 8.00 mol H_2O e. 16.0 mol H_2O

4. Lesson 10.3: How many grams of NaOH are required to make 500. mL of 0.150 M NaOH solution?

   a. 0.550 g NaOH   b. 133 g NaOH   c. 0.834 g NaOH

   d. 1.35 g NaOH   e. 3.00 g NaOH

5. Lesson 10.3: If 150 mg NaOH is dissolved to form 50.0 mL of NaOH solution, find the [NaOH].

   a. 0.33 M NaOH   b. 3.3 M NaOH   c. 0.30 M NaOH

   d. 0.075 M NaOH   e. 0.75 M NaOH

6. Lesson 10.4: If a drop of water contains 0.050 mL and requires 6.0 hr to evaporate, how many molecules evaporate per minute?

   a. 4.6×10^{18} molecules/min   b. 8.3×10^{21} molecules/min

   c. 2.8×10^{20} molecules/min   d. 1.8×10^{19} molecules/min

   e. 7.7×10^{17} molecules/min

7. Lesson 10.6: A 10.0% by mass aqueous solution of H_2SO_4 has a density of 1.07 g/mL. How many milliliters of this solution would be needed in order to supply 26.8 g H_2SO_4?

    a. 0.400 mL soln.   b. 250. mL soln.   c. 287 mL soln.

    d. 28.7. mL soln.   e. 25.0 mL soln.

SUMMARY

1. In choosing a *given*,
   - If you WANT a *single-unit amount*, start with a single-unit amount
   - If you WANT a ratio, start with a ratio

2. When using conversion factors,

   a. The order in which you multiply conversions does not affect the answer

   b. In conversions after a ratio-unit that is *given*, whether you convert the top or bottom unit first makes no difference

   c. When you WANT a ratio, it does not matter which ratio you pick as your *given*. However, you need to start with a ratio that is right-side up

3. To pick a *given* ratio that is right-side up the first time,
   - Pick as your *given* a ratio that includes one of the units and labels WANTED, and put that unit where it is WANTED (on top or bottom), or
   - Arrange the *given* ratio based on descriptive labels in your WANTED unit, or
   - Arrange the given ratio based on the dimensions in the WANTED unit

4. During conversions, to SOLVE,
   - If a unit and label on the right of the $=$ sign *matches* a unit and label wanted on the left, in both what it is and where it is (top or bottom), circle it and leave it alone
   - If a unit and label in or after the given is not what you want where you want it, put it where it will cancel. Convert until it matches what you WANT

5. $A \cdot B^{-x} = A/(B^x)$ When solving with conversions, treat units in the form $A \cdot B^{-x}$ in the same way as other *ratio* units.

6. In solutions, substances are dissolved in fluids. The fluid used to dissolve a substance is the solvent, and the substance dissolved is the solute. Substances that dissolve are soluble in the solvent.

7. In problems involving molar concentration (molarity),

   a. Concentration is a ratio unit: moles per liter

   b. Brackets are used as shorthand for "the concentration of a solution." [NaCl] is read as "the concentration of NaCl"

   c. The molarity equality (M prompt): Treat *concentration* or [ ] or *molar* or *molarity* or capital M or mol · L^{-1} as "moles *per* 1 liter" in conversions.

(continued)

$$\mathbf{M} = [\ ] = molar = mol/L = mol \cdot L^{-1} = \textbf{moles \textit{per} 1 liter} = \text{a \textit{ratio} unit}$$

8. In conversions, treat *ratios* as ratios.
   - You may write *ratio* unit data in your DATA as *equalities* (or *fractions* or *conversions*).
   - You must write WANTED *ratio* units as a *fraction* with a top and bottom.

9. Rules for fractions and percents:

   a. **Fraction** $= \dfrac{\text{quantity A}}{\text{quantity B}}$ and often equals $\dfrac{\text{part}}{\text{total}} = \dfrac{\text{smaller number}}{\text{larger number}}$

   b. Fraction = decimal equivalent = numerator divided by denominator

   c. To convert a decimal equivalent to its percent, multiply the decimal by 100% (moving the decimal twice to the right).
      To convert a percentage to its decimal equivalent (its fraction), divide the percent by 100% (moving the decimal twice to the left).

   d. Three ways to define percentage:

   $$\textbf{Percent} = \dfrac{\text{part}}{\text{total}} \times 100\% = \text{fraction} \times 100\% = (\text{decimal equivalent}) \times 100\%$$

      X% means this fraction: (decimal equivalent) *parts **per** one* part *total*, or

      X% means: *X* parts ***per* 100 total** parts

   e. If a percent is WANTED, write the fraction (if needed), then the decimal equivalent, then the percent.

   f. In DATA, write a supplied percent as one of these:

      X% means this fraction: (decimal equivalent) *parts **per** one* part *total*, or

      X% means: *X* parts ***per* 100** total parts

   g. Calculate "*X*% of *Y*" using "(decimal equivalent) times *Y*"

10. For calculations that include a percentage concentration,

   a. Unless otherwise stated, assume masses and weights (w) are in grams and volumes (v) are in milliliters.

   b. In the fraction and percentage definitions, the solute is the *part* and the solution is the *total*.

   c. Write labels that distinguish the *solute* and *solution* units and numbers.

   d. If a percentage is WANTED, solve for its part/total *fraction first*, then find percentage.

   e. If a percentage is supplied, write

      DATA: *X*% soln. means *X* units solute= 100 units soln.

   f. For a percentage solution, if no (*x/x*) term is given, assume the percentage is a mass percent (which is the same as a % (w/w) notation in the health sciences).

g. Dilute solution prompt: If a calculation is stated to be about a *dilute* aqueous solution and if no density is supplied, add to your DATA:

$$1 \text{ mL } dilute\ soln. \approx 1.00 \text{ mL } H_2O \text{ solvent} \approx 1.00 \text{ g } H_2O \approx 1.00 \text{ g soln.}$$

h. For non-dilute solutions,

mass of soln. = mass of solute + mass of solvent

i. Unless *moles* is a unit in the WANTED or DATA, you will probably not need the molar mass (g/mol) of the solute or solvent in the calculation.

ANSWERS

Lesson 10.1

Practice 1a. $\ \dfrac{? \text{ m}}{\text{s}} = 95 \dfrac{\text{km}}{\text{hr}} \cdot \dfrac{10^3 \text{ m}}{1 \text{ km}} \cdot \dfrac{1 \text{ hr}}{60 \text{ min}} \cdot \dfrac{1 \text{ min}}{60 \text{ s}} = \mathbf{26} \dfrac{\mathbf{m}}{\mathbf{s}}$

(In the *given*, because kilometers was not the unit WANTED on top, it is put where it cancels and converted to the unit WANTED on top. Next, as hours is on the bottom, but seconds is WANTED, hours is converted to the seconds WANTED on the bottom.)

1b. $\ \dfrac{? \text{ mol } H_2O}{\text{L of soln.}} = \dfrac{0.36 \text{ g } H_2O}{\text{mL of soln.}} \cdot \dfrac{1 \text{ mL}}{10^{-3} \text{ L}} \cdot \dfrac{1 \text{ mol } H_2O}{18.0 \text{ g } H_2O} = \mathbf{20.} \dfrac{\mathbf{mol\ H_2O}}{\mathbf{L\ of\ soln.}}$

(Note the grams prompt. If you see grams of a formula, you will likely need the molar mass.)

2. Your conversions should match these, but they may be in any order.

a. $\ \dfrac{? \text{ mg}}{\text{L}} = \dfrac{0.025 \text{ g}}{\text{mL}} \cdot \dfrac{1 \text{ mg}}{10^{-3} \text{ g}} \cdot \dfrac{1 \text{ mL}}{10^{-3} \text{ L}} = \mathbf{2.5 \times 10^4} \dfrac{\mathbf{mg}}{\mathbf{L}}$

b. $\ \dfrac{? \text{ joules}}{\text{g } H_2O} = \dfrac{6.02 \text{ kilojoules}}{\text{mol } H_2O} \cdot \dfrac{10^3 \text{ joules}}{1 \text{ kilojoule}} \cdot \dfrac{1 \text{ mol } H_2O}{18.0 \text{ g } H_2O} = \mathbf{334} \dfrac{\mathbf{joules}}{\mathbf{g\ H_2O}}$

(Note the grams prompt in the WANTED unit.)

Lesson 10.2

Practice A In both questions 1 and 2 below, a *given* was chosen that matched the WANTED *top* unit.

1. WANTED: $\dfrac{? \text{ mi.}}{\text{hr}}$ (Write WANTED *X per Y* units as a *ratio*.)

DATA: 200.0 m = 21.34 s   (Equivalent: two measures of one event. See Lesson 5.2.)

1.609 km = 1 mi.

(When you solve for a *ratio* unit, all of the data will be in *pairs*.)

SOLVE: (Your conversions may be in a different order, but should all be right-side up compared to these. Want a ratio? Start with a ratio.)

$$\dfrac{? \text{ mi.}}{\text{hr}} = \dfrac{1 \text{ mi.}}{1.609 \text{ km}} \cdot \dfrac{1 \text{ km}}{10^3 \text{ m}} \cdot \dfrac{200.0 \text{ m}}{21.34 \text{ s}} \cdot \dfrac{60 \text{ s}}{1 \text{ min}} \cdot \dfrac{60 \text{ min}}{1 \text{ hr}} = \mathbf{20.97} \dfrac{\mathbf{mi.}}{\mathbf{hr}}$$

(Strategy: If you start your *given* by putting the *miles* data on top, all that remains is to convert *kilometers* on the bottom to *hours*. The **arrows** above show a path through the data from kilometers to hours, giving the needed conversions in order. Using arrows may be a useful technique.)

S.F.: 1 and definitions are exact and have infinite *s.f.*

2. WANTED: $\frac{?\ g}{mol}$ (The unit of molar mass is g/mol.)

DATA: $0.0500\ mol = 9.01\ g$

(If a ratio is WANTED, all of the DATA will be in equalities, i.e., conversions or ratios.)

SOLVE: (Your conversion may be in a different order, but should be "right-side up" compared to these. Want a ratio? Start with a ratio.

You WANT the ratio g/mol. You know the grams and moles. The fundamental rule is: *Let the units tell you what to do.*)

$$\frac{?\ g}{mol} = \frac{9.01\ g}{0.0500\ mol} = \mathbf{180.}\ \frac{\mathbf{g}}{\mathbf{mol}}$$ (When the units on the two sides match, stop conversions and do the math.)

Practice B   Data labels are used in solving.

1. WANTED: $\frac{?\ mol\ NaOH}{L\ of\ soln.}$ (Write wanted *X per Y* units as a *ratio.*)

DATA: $10.0\ g\ NaOH = 1{,}250\ mL\ soln.$ (Equivalent: two measures of the same solution.)

$40.0\ g\ NaOH = 1\ mol\ NaOH$

SOLVE: (Below, the label "solution" is placed in the bottom of the *given* to match the WANTED unit, but your conversions may be in a different *order*. Want a ratio? Start with a ratio.)

$$\frac{?\ mol\ NaOH}{L\ soln.} = \frac{10.0\ g\ NaOH}{1{,}250\ mL\ soln.} \cdot \frac{1\ mL}{10^{-3}\ L} \cdot \frac{1\ mol\ NaOH}{40.0\ g\ NaOH} = \mathbf{0.200}\ \frac{\mathbf{mol\ NaOH}}{\mathbf{L\ soln.}}$$

S.F.: All of the non-exact numbers have three *s.f.*; the answer is rounded to three *s.f.*

Note that the second conversion is true for all substances and processes. The first and last conversions need a substance formula after the unit because that ratio of numbers and units is not always true.

2. WANTED: $\frac{?\ kilocalories\ heat}{mol\ CO_2}$ (Write WANTED *X per Y* units as a *ratio.*)

DATA: $24\ calories\ heat = 0.88\ g\ CO_2$ (Equivalent: see Lesson 5.2.)

$44.0\ g\ CO_2 = 1\ mol\ CO_2$ (Grams prompt.)

$1\ kilo\text{-}anything = 10^3\ anythings$ (Prefix conversions are optional in DATA.)

SOLVE: (Pick as your *given* any data equality that puts a heat term on top or a CO_2 term on the bottom. Your conversions may be in a different order, but must all be right-side up compared to these and must result in the same answer. Want a ratio? Start with a ratio.)

$$\frac{?\ kilocalories\ heat}{mol\ CO_2} = \frac{24\ calories\ heat}{0.88\ g\ CO_2} \cdot \frac{1\ kilocalorie}{10^3\ calories} \cdot \frac{44.0\ g\ CO_2}{1\ mol\ CO_2} = \mathbf{1.2}\ \frac{\mathbf{kilocalories\ heat}}{\mathbf{mol\ CO_2}}$$

Practice C   1a. 14 m/s   1b. 0.47 mg/mol   1c. $9.2\ kg/dm^3$

2. WANTED: $\frac{?\ kg\ Rn}{L}$ $(kg \cdot L^{-1} = kg/L$; write a WANTED ratio unit as a *ratio.*)

DATA: $8.60\ mg = 1\ mL$ $(mg \cdot mL^{-1} = mg/mL$; write ratio DATA as an equality.)

SOLVE: $\dfrac{?\ kg\ Rn}{L} = \dfrac{8.60\ mg}{1\ mL} \cdot \dfrac{10^{-3}\ g}{1\ mg} \cdot \dfrac{1\ kg}{10^3\ g} \cdot \dfrac{1\ mL}{10^{-3}\ L} = \mathbf{8.60 \times 10^{-3}}\ \dfrac{\mathbf{kg\ Rn}}{\mathbf{L}}$

3. WANTED: $\frac{?\ mg\ sugar}{mL\ drink}$ $(mg \cdot mL^{-1} = mg/mL = $ a ratio unit is WANTED$)$

DATA: 16 fl. oz. drink = 52 g sugar   (Two measures of the same drink.)

355 mL = 12 fl. oz.

SOLVE: [You want mass over volume. The first DATA equality has a mass unit (g) and a volume unit (fluid ounces). The second has two volume units. To use method 3, start with mass on top. However, you could use a different method and order as long as you get the same answer.]

$$? \frac{\text{mg sugar}}{\text{cm}^3 \text{ drink}} = \frac{52 \, g \, \text{sugar}}{16 \, \text{fl. oz.}} \cdot \frac{12 \, \text{fl. oz.}}{355 \, \text{mL}} \cdot \frac{1 \, \text{mL}}{1 \, \text{cm}^3} \cdot \frac{1 \, \text{mg}}{10^{-3} g} = \mathbf{110} \, \frac{\textbf{mg sugar}}{\textbf{cm}^3 \textbf{ drink}}$$

Lesson 10.3

Practice   Your answers should look like these, with the comments omitted.

1. WANTED: ? mol KBr

    DATA: ⟨100. mL of KBr solution⟩

    0.40 mol KBr = 1 L KBr solution   (M prompt.)

    SOLVE: (WANT a single unit? Start with the single unit in your data.)

    $$? \, \textbf{mol KBr} = 100 \, \text{mL KBr soln.} \cdot \frac{10^{-3} \, \text{L}}{1 \, \textbf{mL}} \cdot \frac{0.40 \, \textbf{mol KBr}}{1 \, \text{L KBr soln.}} = \mathbf{0.040 \, mol \, KBr}$$

2. The solute is KBr and, if the solvent is not specified, assume it is water.

3. WANTED: ? mol KOH

    DATA: ⟨0.20 L KOH soln.⟩

    0.60 mol KOH = 1 L KOH soln.   (mol · L^{-1} = mol/L = M prompt)

    SOLVE: (WANT a single unit? Start with a single unit.)

    $$? \, \text{mol KOH} = 0.20 \, \text{L KOH soln.} \cdot \frac{0.60 \, \text{mol KOH}}{1 \, \text{L KOH soln.}} = \mathbf{0.12 \, mol \, KOH}$$

4. WANTED: ? g HCl

    DATA: ⟨500. mL HCl solution⟩

    0.200 mol HCl = 1 L HCl solution   (M prompt.)

    36.5 g HCl = 1 mol HCl   (g prompt—see Lesson 8.3.)

    SOLVE: (WANT a single unit? Start with a single unit.)

    $$? \, \text{g HCl} = 500. \, \textbf{mL HCl} \cdot \frac{10^{-3} \, \text{L}}{1 \, \textit{mL}} \cdot \frac{0.200 \, \text{mol HCl}}{1 \, \text{L}} \cdot \frac{36.5 \, \text{g HCl}}{1 \, \text{mol HCl}} = \mathbf{3.65 \, g \, HCl}$$

    (The "soln." and "HCl soln." labels are omitted from some of the volume units as "understood." When a problem is about one solution, volumes can be assumed to be for that solution.)

5. WANTED: ? mL soln.

    DATA: 4.48 **g** KCl (Because you WANT a single unit, start with this as your *given.*)

    0.200 moles KCl = 1 liter of solution   (M prompt.)

    74.6 g KCl = 1 mol KCl   (Grams prompt.)

    SOLVE: (The arrows trace the path from the *given* to the WANTED unit.)

    $$? \, \text{mL KCl soln.} = 4.48 \, g \, \textit{KCl} \cdot \frac{1 \, \text{mol KCl}}{74.6 \, g \, \textit{KCl}} \cdot \frac{1 \, \text{L soln.}}{0.100 \, \text{mol KCl}} \cdot \frac{1 \, \text{mL}}{10^{-3} \, \text{L}} = \mathbf{601 \, mL \, KCl}$$

6. WANTED: $\dfrac{?\ \text{mol NaCl}}{\text{L soln.}}$ (You want molarity. Its **units** are moles *per* liter.)

DATA: 2.34 *g NaCl* = 400. mL soln.   (Two measures of *same* solution.)

58.5 g NaCl = 1 mol NaCl   (Grams prompt.)

(If you want a ratio unit, all of the DATA will be in equalities.)

SOLVE: (You want a ratio. Start with a ratio. Your conversions may be in any order if they are the "same side up" as these.)

$$\dfrac{?\ \text{mol NaCl}}{\text{L soln.}} = \dfrac{2.34\ \text{g NaCl}}{400.\ \text{mL soln.}} \cdot \dfrac{1\ \text{mol NaCl}}{58.5\ \text{g NaCl}} \cdot \dfrac{1\ \text{mL}}{10^{-3}\ \text{L}} = \dfrac{\mathbf{0.100\ mol\ NaCl}}{\textbf{L soln.}}$$

7. WANTED: $\dfrac{?\ \text{mol NaOH}}{\text{L soln.}}$ (Brackets mean concentration: *units* are moles *per liter*.)

DATA: 0.020 lb. NaOH = 1 qt. soln.   (Two measures of *same* solution.)

40.0 g NaOH = 1 mol NaOH   (Convert lb. to k**g** = grams prompt.)

(Plus the conversions listed in the problem.)

SOLVE: (Want a ratio? Start with a ratio. Your conversions may be in any order "right-side up" compared to these.)

$$\dfrac{?\ \text{mol NaOH}}{\text{L soln.}} = \dfrac{0.020\ \text{lb. NaOH}}{1\ \textit{qt.}\ \text{soln.}} \cdot \dfrac{1\ \text{kg}}{2.2\ \text{lb.}} \cdot \dfrac{10^3\ \text{g}}{\textbf{1 kg}} \cdot \dfrac{\textbf{1 mol}\ \text{NaOH}}{40.0\ \text{g NaOH}} \cdot \dfrac{1\ \textit{qt.}}{32\ \text{fl. oz.}} \cdot \dfrac{12\ \text{fl. oz.}}{355\ \text{mL}} \cdot \dfrac{1\ \text{mL}}{10^{-3}\ \text{L}} = \dfrac{\mathbf{0.24\ mol\ NaOH}}{\textbf{L soln.}}$$

8. If you measure the volume of the sample in *liters*, and if you know the concentration of the sample in *mol/L*, you could convert to the *moles* dissolved. If you knew the grams of substance added and its formula, you could find its molar mass and then its moles.

9. WANTED: $\dfrac{?\ \text{mol H}_2\text{O}}{\text{L H}_2\text{O}}$

DATA: 1.00 *g* liquid H_2O = 1 mL liquid H_2O

18.0 g H_2O = 1 mol H_2O (Grams prompt.)

SOLVE: $\dfrac{?\ \text{mol H}_2\text{O}}{\text{L H}_2\text{O}} = \dfrac{1\ \text{mol H}_2\text{O}}{18.0\ \text{g H}_2\text{O}} \cdot \dfrac{1.00\ \text{g H}_2\text{O}(\ell)}{1\ \text{mL H}_2\text{O}(\ell)} \cdot \dfrac{1\ \text{mL}}{10^{-3}\ \text{L}} = \dfrac{\mathbf{55.6\ mol\ H_2O}}{\textbf{L H}_2\textbf{O}}$

As calculated in this answer, pure liquid water has a concentration of *about* **55** moles per liter. For substances dissolved in water, even for those that are very soluble in water, the highest concentration solutions usually have a limit of about **20** moles per liter.

If you calculate a concentration for an aqueous solution that is higher than 20 mol/L, check your work.

Lesson 10.4

Practice   1a. Hint 1: Write complete *labels*, especially as both of the answer units are mass units. The dosage is "milligrams of *medicine* per kilogram of *cat*."

Hint 2: This problem is asking for a single-unit answer. If you want a single unit, start with a single unit as your *given*.

WANTED: ? mg acetaz.

DATA: 9.0 lb. of cat

7 yr old

15 mg acetaz. = 1 kg **cat**

SOLVE: $?\ \text{mg acetaz.} = \mathbf{9.0\ lb.\ cat} \cdot \dfrac{1\ \text{kg}}{2.20\ \textbf{lb.}} \cdot \dfrac{15\ \text{mg}\ \text{acetaz.}}{1\ \text{kg}\ \textit{cat}} = \mathbf{61\ mg\ acetaz.}$

(This is a problem where you are given a number that you don't use, the cat's age, but complete labels will help you to choose the DATA you need.)

1b. Your prescription might be "Give 1/2 of the 125 mg tablet twice daily." This would be a dose of 62.5 mg. For medicines, if precise dosages are not critical, a small additional amount should be safe.

2. WANTED: ? $\frac{\text{furlongs}}{\text{fortnight}}$

    DATA: 3.0×10^8 m = 1 s   (You may list the DATA as equalities.)

    14 days = 1 fortnight

    8 furlongs = 1 mi.

    1 mi. = 1.61 km

    (**OR** you may list DATA that is two related units in a *conversion* or *ratio* format.)

    $$\frac{3.0 \times 10^8 \text{ m}}{1 \text{ s}} \qquad \frac{14 \text{ days}}{1 \text{ fortnight}} \qquad \frac{8 \text{ furlongs}}{1 \text{ mi.}} \qquad \frac{1 \text{ mi.}}{1.61 \text{ km}}$$

    (Both forms are equivalent. In DATA, use whichever you wish.)

    SOLVE: (Your conversions may be in a different order, provided that they are right-side up compared to these and arrive at the same answer.)

$$? \; \frac{\textbf{furlongs}}{\textbf{fortnight}} = 3.0 \times 10^8 \; \frac{\text{m}}{s} \cdot \frac{1 \text{ km}}{10^3 \text{m}} \cdot \frac{1 \text{ mi.}}{1.61 \text{ km}} \cdot \frac{8 \text{ furlongs}}{1 \text{ mi.}} \cdot \frac{60 \; s}{1 \text{ min}} \cdot \frac{60 \text{ min}}{1 \text{ hr}} \cdot \frac{24 \text{ hr}}{1 \text{ day}} \cdot \frac{14 \text{ days}}{1 \text{ fortnight}} = \textbf{1.8} \times \textbf{10}^{\textbf{12}} \; \frac{\textbf{furlongs}}{\textbf{fortnight}}$$

3. WANTED: ? $\frac{\text{molecules } H_2O}{s}$ (A ratio unit.)

    DATA: 0.050 mL H_2O = 1 drop

    1 drop = 2.0 hr   (Two measures of the same process: evaporation of drop.)

    1 mL $H_2O(\ell)$ = **1.00 g $H_2O(\ell)$**

    18.0 **g H_2O** = 1 mol H_2O (Grams of formula = grams prompt.)

    1 mol H_2O = 6.02×10^{23} molecules H_2O (Molecules = Avogadro prompt.)

    SOLVE: (Want a ratio? Start with a ratio. Because you WANT H_2O on top, you may want to pick a *given* ratio with H_2O on top, but any order is OK for multiplied conversions.)

$$? \; \frac{\text{molecules } H_2O}{s} = \frac{0.050 \text{ mL } H_2O}{\text{drop}} \cdot \frac{1 \text{ drop}}{2 \text{ hr}} \cdot \frac{1 \text{ hr}}{60 \text{ min}} \cdot \frac{1 \text{ min}}{60 \text{ s}} \cdot \frac{1.00 \text{ g } H_2O}{1 \text{ mL } H_2O} \cdot \frac{1 \text{ mol } H_2O}{18.0 \text{ g } H_2O} \cdot \frac{6.02 \times 10^{23} \text{ molecules}}{1 \text{ mol } H_2O}$$

$$= \textbf{2.3} \times \textbf{10}^{\textbf{17}} \; \frac{\textbf{molecules } H_2O}{s} \; (\; = 230{,}000 \; \textit{trillion} \text{ molecules/second! Molecules are } \textit{small.})$$

4. WANTED: ? $\frac{\text{g}}{\text{mol}}$ (The unit wanted for molar mass.)

    DATA: 6.6×10^{-24} g = 1 atom

    1 mol atoms = 6.02×10^{23} atoms   (Any 10^{xx} or 10^{-xx} DATA = Avogadro prompt.)

    SOLVE: (Want a ratio? Start with a ratio. You may want to pick a *given* ratio with grams on top, or moles on the bottom, but any order is permitted for multiplied conversions.)

$$? \; \frac{\text{g}}{\text{mol}} = \frac{6.6 \times 10^{-24} \text{ g}}{1 \text{ atom}} \cdot \frac{6.02 \times 10^{23} \text{ atoms}}{1 \text{ mol}} = 39.7 \times 10^{-1} = \textbf{4.0} \; \frac{\textbf{g}}{\textbf{mol}}$$

The atom with a mass of 4.0 g/mol is helium.

5. WANTED: ? $\dfrac{\text{mL}}{\text{Fe atom}}$

   DATA: 7.87 g Fe = 1 mL Fe

   55.8 g Fe = 1 mol Fe atoms   (Grams prompt in density.)

   1 mol atoms = 6.02 × 10²³ atoms   (Visible grams, invisible atoms = Avogadro prompt.)

   SOLVE: (Want a ratio? Start with a ratio. Your conversions may be in a different order.)

$$\dfrac{?\ \ \text{mL}}{\text{atom Fe}} = \dfrac{1\ \text{mL Fe}}{7.87\ \text{g Fe}} \cdot \dfrac{55.8\ \text{g Fe}}{1\ \text{mol Fe}} \cdot \dfrac{1\ \text{mol atoms}}{6.02 \times 10^{23}\ \text{atoms}} = \mathbf{1.18 \times 10^{-23}}\ \dfrac{\mathbf{mL}}{\mathbf{Fe\ atom}}$$

Lesson 10.5

Practice A 1. Decimal equivalent = percent/100% = 4.8%/100% = **0.048**

2. To divide by 100,000, move the decimal five times. Decimal equivalent = **0.000095 = 9.5 × 10⁻⁵**

   Percent = decimal equivalent × 100% = 0.000095 × 100% = **0.0095% = 9.5 × 10⁻³%**

Practice B 1. Decimal equivalent of 3/5 = **0.60 (exact)**; percent = decimal equivalent × 100% = **60% (exact)**

2. Write fraction, then decimal, then %.

$$\text{Fraction} = \dfrac{\text{smaller}}{\text{larger}} = \dfrac{7}{25} = 0.28;\ \% = \text{decimal equivalent} \times 100 = \mathbf{28\%}$$

3. Write fraction, then decimal, then %.

$$\text{Fraction} = \dfrac{\text{part}}{\text{total}} = \dfrac{12}{24{,}000} = \dfrac{12}{24 \times 10^3} = 0.5 \times 10^{-3};\ \% = \text{decimal} \times 10^2\% = \mathbf{0.050\%}$$

A decimal (decimal equivalent) is a number between zero and one derived from a fraction.

Practice C 1. The rule is: X% of Y = (decimal equivalent) times Y; 25% of 180 = 0.25 × 180 = **45**

2. X% of Y = (decimal equivalent) times Y; 0.0450% of 7,500 = 0.00045 × 7,500 = **3.4**

3. WANTED: ? g soln.

   DATA: 5.0% NaCl means: **5.0 g NaCl = 100 g soln.**   (Using rule 5.B.)

   12.5 g NaCl

   SOLVE: ? g soln. = 12.5 g NaCl · $\dfrac{100\ \text{g soln.}}{5.0\ \text{g NaCl}}$ = **250 g soln.**

4. 3/8 = 0.375 = **37.5%**

5. 16.7% = **0.167**

6. X% of Y = (decimal equivalent) times Y; 12% of 750 = 0.12 × 750 = **90.**

7. To find %: Write fraction, then decimal, then %.

$$\text{Fraction} = \dfrac{\text{smaller}}{\text{larger}} = \dfrac{1.8}{45} = 0.040 = \mathbf{4.0\%}$$

8. Decimal equivalent of percent = percent/100% = 0.6%/100% = **0.006**

9. To find the decimal equivalent of a fraction, divide.

   45/10,000 = **0.0045**; percent = decimal equivalent × 100% = 0.0045 × 100% = **0.45%**

10. *X*% of *Y* = (decimal equivalent) times; 0.25% of 12,400 = 0.0025 × 12,400 = **31**

11. WANTED: ? g soln.

DATA: **4.5** g glucose = **100** g soln.   (Using rule 5.B.)

2.25 g glucose

SOLVE: ? g soln. = 2.25 g glucose · $\dfrac{100 \text{ g soln.}}{4.5 \text{ g glucose}}$ = **50. g soln.**

Lesson 10.6

Practice A   1. WANTED: ? mass % glucose; write $\dfrac{? \text{ g glucose}}{\text{g soln.}}$ fraction, then decimal, then %   (Rule 1.)

DATA: 6.0 g glucose = 150 g soln.   (Two measures of one solution.)

SOLVE: $\dfrac{? \text{ g glucose}}{\text{g soln.}}$ = $\dfrac{6.0 \text{ g glucose}}{150 \text{ g soln.}}$ = **0.040** = fraction = **4.0%** glucose by mass

2. WANTED: ? mol H_2SO_4

DATA: 1.34 g soln. = 1 mL soln.   (What a solution density of 1.34 g/mL means.)

44% H_2SO_4 means: 44 g H_2SO_4 = 100 g soln.   (Rules 2 and 5.)

98.1 g H_2SO_4 = 1 mol H_2SO_4 (Rule 4: Moles is WANTED.)

80. mL soln.

SOLVE: ? mol H_2SO_4 = 80. mL soln. · $\dfrac{1.34 \text{ g soln.}}{1 \text{ mL soln.}}$ · $\dfrac{44 \text{ g } H_2SO_4}{100 \text{ g soln.}}$ · $\dfrac{1 \text{ mol } H_2SO_4}{98.1 \text{ g } H_2SO_4}$ = **0.48 mol H_2SO_4**

Practice B   1. WANTED: **? mL KCl soln.**

DATA: 0.50% KCl means 0.50 **g KCl** = **100 g soln.**   (Assume mass percent by rule 5.)

1.6 g KCl

1 mL dilute soln. ≈ 1.00 mL water solvent ≈ 1.00 g H_2O ≈ 1.00 g soln.   (Dilute prompt.)

SOLVE: **? mL KCl soln.** = 1.6 g KCl · $\dfrac{100 \text{ g soln.}}{0.50 \text{ g KCl}}$ · $\dfrac{1 \text{ mL soln.}}{1.00 \text{ g soln.}}$ = **320 mL KCl soln.**

2. WANTED: ? mass % HCl. Find decimal value for $\dfrac{? \text{ g HCl}}{\text{g soln.}}$ fraction, then write %.   (Rule 1.)

DATA: 0.080 *mol* HCl = 1 L soln.   (M prompt.)

1 mL dilute soln. ≈ 1.00 mL water solvent ≈ 1.00 g H_2O ≈ 1.00 g soln.   (Dilute prompt.)

36.5 g HCl = 1 mol HCl   (Needed as moles is in DATA.)

SOLVE: $\dfrac{? \text{ g HCl}}{\text{g soln.}}$ = $\dfrac{36.5 \text{ g HCl}}{1 \text{ mol HCl}}$ · $\dfrac{0.080 \text{ mol HCl}}{1 \text{ L soln.}}$ · $\dfrac{10^{-3} \text{ L soln.}}{1 \text{ mL soln.}}$ · $\dfrac{1 \text{ mL soln.}}{1.00 \text{ g soln.}}$

= **0.0029** $\dfrac{\text{g HCl}}{\text{g soln.}}$ = fraction; **%** = **0.29%** HCl by mass

(When solving for a ratio, your conversions may be in any order so long as they are "right-side up" compared to these.)

3. WANTED: ? g water =   (Hint: Find the grams of NaCl in the solution first.)

DATA: 6.0 g NaCl = **100 g total** soln.   (Rules 2 and 3.)

500. g soln.

SOLVE: $?\,\mathbf{g\,NaCl}$ = 500 g soln. · $\dfrac{6.0\ \text{g NaCl}}{100\ \text{g total soln.}}$ = 30. g NaCl   What must be the mass of the *water*?

$?\,\mathbf{g\,water}$ = (grams total solution) − (30. g NaCl) = 500. g total − 30. g NaCl = **470. g water**

Practice C   1. WANTED: ? % (w/w) NaOH; find $\dfrac{?\ \text{g NaOH}}{\text{g soln.}}$ fraction, then %   (Rule 9.)

DATA: 0.050 mol NaOH = 200. g soln.   (Two measures of one solution.)

40.0 g NaOH = 1 mol NaOH   (Rule 4: moles in data.)

SOLVE: $\dfrac{?\ \text{g NaOH}}{\text{g soln.}}$ = $\dfrac{0.050\ \text{mol NaOH}}{200\ \text{g soln.}}$ · $\dfrac{40.0\ \text{g NaOH}}{1\ \text{mol NaOH}}$ = **0.010** = fraction = **1.0% (w/w) NaOH**

(Your conversions may be in any order so long as they are "right-side up" compared to these.)

2. WANTED: $?\,\mathbf{g\,NaCl}$

DATA: 0.90% (w/v) NaCl means: 0.90 **g NaCl** = **100 mL** soln.   (Rule 11.)

125 mL soln.

SOLVE: (Want a single unit?) $?\,\mathbf{g\,NaCl}$ = 125 mL soln. · $\dfrac{0.90\ \text{g NaCl}}{100\ \text{mL soln.}}$ = **1.1 g NaCl**

3. WANTED: $\dfrac{?\ \text{mol NaCl}}{\text{L soln.}}$ =   (Molarity is mol/L, a ratio unit.)

DATA: 0.90% (w/v) NaCl means: 0.90 **g NaCl** = 100 **mL** soln.   (Rule 11.)

125 mL soln.

58.5 g NaCl = 1 mol NaCl   (Needed because moles is in WANTED unit.)

1 mL dilute soln. ≈ 1.00 mL water solvent ≈ 1.00 g H_2O ≈ 1.00 g soln.   (Dilute prompt.)

SOLVE: (Want a ratio?) $\dfrac{?\ \text{mol NaCl}}{\text{L soln.}}$ = $\dfrac{1\ \text{mol NaCl}}{58.5\ \text{g NaCl}}$ · $\dfrac{0.90\ \text{g NaCl}}{100\ \text{mL soln.}}$ · $\dfrac{1\ \text{mL soln.}}{10^{-3}\ \text{L soln.}}$ = **0.15 mol NaCl / L soln.**

Lesson 10.7

1a. **1 Fe_2O_3 + 3 CO → 2 Fe + 3 CO_2**

1b. **e. 0.901 mol CO**

2. **c. 0.072 g H_2O**

3. **d. 8.00 mol H_2O**   (Two reactant amounts are supplied. Find the limiting reactant first.)

4. **e. 3.00 g NaOH**

5. **d. 0.075 M NaOH**

6. **a. 4.6 × 10^{18} molecules/min**   (Data include: 1 drop = 6.0 hr, 1 drop = 0.050 mL, 1.00 g H_2O = 1 mL H_2O liquid, 18.0 g H_2O = 1 mol H_2O, 1 mol = 6.02 × 10^{23} molecules.)

7. **b. 250. mL soln.**   (Data include: 10.0 g H_2SO_4 = 100 g soln., 1.07 g soln. = 1 mL soln.)

11

Dimensions

Lesson 11.1 Units and Dimensions

Measurement Fundamentals

So far, we have focused on calculations that answer one question based on one set of data. We also need to be able to solve calculations that have several questions about one set of data. Answering these questions will require distinctions between two types of *equalities* that we have been listing in our data. To make these distinctions, let's take a closer look at quantities, measurements, and units.

The Fundamental Quantities

Quantities that can be measured are termed **physical quantities** or **dimensions**. The physical properties of the universe are measured based on the **fundamental quantities**.

The fundamental quantities most often studied in chemistry are distance, mass, time, and electrical charge. Two additional dimensions, *temperature* and the *numbers* used to count quantities, differ in some respects from the other fundamental quantities, but it will simplify our work in chemistry at most points if we add them to the list.

A **measurement** must have a **magnitude** (a number) and a **unit**. Any system of measurement must begin with an arbitrary definition for each of the **base units** used to measure the fundamental quantities. For the quantities most often studied in chemistry, the base units of the SI system are

Fundamental Quantities	SI Base Units
distance	meter
mass	gram or kilogram[1]
time	second
electric charge	coulomb
count of particles	mole
temperature	kelvin

[1]In the SI system, the kilogram is the base unit for mass, but when using metric *prefixes*, the gram is treated as the base unit for mass.

Adding a metric prefix (such as *kilo-*, *centi-*, or *nano-*) in front of a base unit does not change which quantity is being measured. A metric prefix is best thought of as an abbreviation for an exponential term that is part of the magnitude (the number in front of the unit in a measurement) rather than part of the unit.

Example: The units mm, km, and cm are all measures of distance.

The fundamental quantities are always measured by a *single base unit* (with or without a prefix) to the *first* power. The units meters, centimeters, grams, kilograms, and milliseconds are all measures of fundamental quantities.

PRACTICE **A**

Before doing the following problems, learn the *fundamental quantities* and their *base units* in the table above. Then cement your knowledge by completing these problems.

1. Which of these is not a measure of a fundamental quantity: kilograms, millimeters, or cubic centimeters?

2. Write the fundamental quantity that is measured by each of these units.

   a. kilograms

   b. nanoseconds

   c. millimoles

   d. decimeters

Derived Quantities

Derived quantities have definitions that either use more than one fundamental quantity or take one fundamental quantity to a power that is not equal to one.

Derived quantities are measured in base units that include either two or more of the base units for fundamental quantities or one base unit raised to a power that is not equal to one.

Examples: Meters, kilograms, and milliseconds measure fundamental quantities. Cubic meters (m^3), seconds^{-1}, and g/mol measure derived quantities.

The table below lists some derived quantities that are frequently measured in chemistry.

Derived Quantity	Definition	SI Base Units	Units Often Used in Chemistry
area	**distance squared**	$meter^2$	m^2, cm^2
volume	**distance cubed**	$meter^3$	mL ($= cm^3$), liter ($= dm^3$)
speed (velocity)	**distance over time**	meters/second	m/s
density	**mass over volume**	kg/m^3	g/cm^3, g/mL
concentration	**particles per volume**	mol/m^3	mol/L
molar mass	**mass per particle**	kg/mol	g/mol
acceleration	velocity over time	m/s^2	m/s^2
energy	mass $\times$ acceleration $\times$ distance	$(kg \cdot m^2)/s^2$	joules, calories
pressure	(mass $\times$ accel.)/area	$kg/(m \cdot s^2)$	pascal, bar, torr, mm Hg, atm
frequency	1/time	1/s	1/s $=$ hertz

The units used to measure a quantity must be based on the definition of the quantity.

Examples:

- Volume is the three-dimensional space an object occupies. All geometric formulas for volume are based on a *distance* multiplied three times.
  Volume of a cube = (**side**)³; volume of a cylinder = $\pi r^2 h$
  Units for volume can always be related to a distance unit *cubed*.
  Milliliters are cubic centi*meter*s; liters are cubic deci*meter*s.

- Speed (or velocity) is always a ratio of the *distance* an object is moving per unit of *time*. The units of speed must be *distance* units over *time* units, such as *meters per second* or *miles per hour*.

Some units for derived quantities are abbreviations: units that are equivalent to a number of base units or to a combination of base units that are used to measure the fundamental quantities. For example, most chemistry textbooks use

- *Liter* (L) as an abbreviation for *cubic decimeter* (dm^3)
- *Milliliter* (mL) as an abbreviation for *cubic centimeter* (cm^3)
- *Molar* as an abbreviation for *moles/liter* which is *moles per cubic decimeter*
- *Hour* as an abbreviation for *3,600 seconds*
- *Joule* (which measures energy) as an abbreviation for *$(kg \cdot m^2)/s^2$*

Some calculations require the translation of unit abbreviations into their base units for units to cancel properly in calculations. With practice, you will know when this is necessary and when it is not.

The Three Types of Units

In these lessons, to solve problems we will divide units into three categories:

1. **Single units** are those with *one kind* of *base* unit. The base unit may
   - Have a prefix. Examples: mm, km, kg
   - Be a multiple of a base unit. Example: minute = 60 seconds
   - Have a positive exponent. Examples: dm^3 (liters), cm^3 (mL), m^2

   However, a single unit can*not* have a *unit* in the denominator.

2. **Ratio units** have *one base unit* in the *numerator* and *one different base unit* in the *denominator*. Each base unit may include prefixes and/or powers.
   Examples: km/hr, mg/cm³, g/mol, and mol · L⁻¹ are all ratio units.
   In the table of derived quantities above, *speed, density, molar mass,* and *concentration* are defined as ratios and they are measured by ratio units.

3. **Complex units** are combinations of base metric units that either have *two* or more different base units in the numerator or denominator, or have *no* term in the numerator.
   Examples: Joules, an energy unit, is an abbreviation for ($kg \cdot m^2)/s^2$.
   Wave frequency is measured in hertz, an abbreviation for 1/seconds.
   We will address complex units in later lessons.

PRACTICE B

Before doing the following problems, learn the *fundamental quantities* and their *base units* in the table above; the information in **bold** in the *derived quantities* table above; and the difference between single, ratio, and complex units.

Then cement your knowledge by completing the problems below.

1. Write two metric units that can be used to measure

   a. time

   b. mass

   c. distance

   d. volume

2. Which unit measures the molar concentration of solutions?

3. Label each of these units as measuring the quantity of: distance, area, volume, mass, density, time, speed, solution concentration, or molar mass.

   a. liters

   b. cubic meters

   c. kilograms

   d. decimeters

   e. millimeters

   f. deciliters

   g. fluid ounces

   h. acres

   i. nanoseconds

   j. grams/milliliter

   k. feet/second

   l. square feet

   m. moles dissolved per liter

   n. grams/mole

4. Which units in problem 3 measure fundamental quantities?

5. Which quantities listed in the first sentence of problem 3 are *derived* quantities?

6. Which units in problem 3 are ratio units?

7. Which quantities listed in the first sentence of problem 3 are defined as *ratios* of two quantities?

8. How do complex units differ from ratio units?

Intensive and Extensive Quantities

Physical quantities can also be classified as **intensive** and **extensive**.

An **extensive** quantity is a physical quantity whose value is proportional to the *size* of the system it describes. For extensive quantities, the *amount* matters. Examples of extensive quantities include all of the fundamental quantities, as well as area, volume, and energy.

An **intensive** quantity is a quantity whose value does *not* depend on the amount of substance being measured. Pressure, density, molar mass, melting points, temperature, and concentration are examples of intensive quantities. In general, if a ratio is composed of two extensive quantities, it is an intensive quantity.

Dimensional Analysis

The fundamental and derived quantities together are termed **dimensions**. The technique of using dimensions and their units to solve problems is called **dimensional analysis**.

Dimensional analysis is a powerful tool for solving science problems. Using conversion rules, you can often solve a problem even if you are not sure what the units measure. However, in complex calculations, a careful analysis of dimensions must often be done to correctly choose which data to use in each part. This will require a clear understanding of which units measure which dimensions.

To develop that understanding, it is important to learn which quantities are fundamental, the definitions of derived quantities, and the units that are used to measure both types of quantities. These fundamentals are a foundation for higher-level work.

Flashcards

Add any of the following that you cannot quickly answer to your collection. Practice until you can do each correctly. Repeat this practice for two more days, then put them in your weekly stack.

One-Way Cards (with Notch)	Back Side—Answers
Name six fundamental quantities	Distance, mass, time, temperature, particles, charge
What are the two parts of a measurement?	A magnitude and a unit
What quantity is based on distance cubed?	Volume
Name four quantities defined as ratios of two fundamental quantities	Density, speed, concentration, molar mass
Define an extensive quantity	A quantity based on an amount
Define an intensive quantity	A quantity in which amount does not matter
Name three intensive quantities	Concentration, pressure, density, temperature, or molar mass
What quantity is measured by distance/time?	Speed or velocity
What quantity is mass per unit of volume?	Density
What is the SI base unit for distance?	meters
What is the SI base unit for mass?	kilograms

PRACTICE   C

1. Label each of these equalities as *always true* (**AT**) or *not always true* (**Not AT**). Some of these will be challenging yet educational. They will help to get you thinking about *units* and *dimensions*. Check your answers frequently.

   a. 60 s = 1 min

   b. 60 mi. = 1 hr

   c. 18.0 g = 1 mol

   d. 18.0 g H_2O = 1 mol H_2O

    e. 6.02×10^{23} molecules $= 1$ mol of molecules

    f. 2.0 mol HCl $= 750$ mL soln.

    g. 355 mL $= 12$ fl. oz.

2. The following two quantities are defined as ratios. Which parts of problem 1 could be measurements of these quantities?

    a. speed   b. concentration

3. Which of the two quantities in problem 2 are intensive quantities?

Lesson 11.2   Ratios versus Two Related Amounts

Types of DATA

So far, in solving with conversions, we have divided DATA into two types: single unit and ratio unit. To solve conversion calculations that ask one question about one set of data, those are often the only distinctions that we need to solve problems.

In some calculations, however, you will need to *select* the correct data to use in each part of a problem. To do so, it will be necessary to distinguish between *ratios* and *two related amounts*. This will result in three types of DATA: single-unit amounts, ratios, and two *related* single-unit amounts.

Single-Unit Amounts

Single-unit data measures an *amount* of an object or process.

A single unit may measure a fundamental or derived quantity. It may have a prefix (such as *kilo*grams), may be a multiple of a base unit (1 minute $= 60$ seconds), and may be raised to a positive power, but it must be defined using *one base* unit in the numerator and *no unit* in the denominator.

Examples: mm, km, g, kg, dm^3 (liters), cm^3 (mL), m^2, s, min, mol.

This represents *no change* in our past definition of single-unit data.

Ratios

Ratios do *not* represent one amount or two amounts of a physical quantity. Ratios represent a relationship between two units.

Ratios must have *two* measurements. Ratios in DATA can be written either as an equality (as we have usually been doing for ratio units) or as a *fraction* or *ratio* or *conversion* (all of those terms have the same meaning).

Examples of *ratio* data include the following:

    a. Metric-prefix definitions are ratios. Example: 1 kilometer $= 10^3$ meters.

b. Conversions relating two units that measure the same dimension are ratios. Example: 1 minute = 60 seconds *or* 1 min/60 s.

c. Some derived quantities are *defined* as ratios. Examples: *Concentration, speed, density,* and *molar mass* are ratios by definition.

Quantities defined as ratios must have ratio units. Note in the examples of data below: each of the quantities being measured is defined as a ratio, the units that measure the quantity are ratios, and the units are those that measure the two quantities in the definition.

- The speed of the car was 55 miles/hour. (Speed is distance/time.)
- The density of the gold alloy is 15.5 g/cm^3. (Density is mass/volume.)
- The concentration of the solution is 0.25 moles per liter.
- The molar mass of water is 18.0 g H_2O = 1 mol H_2O.

d. The *coefficients* of a balanced equation are exact particle-to-particle ratios.

Two Related Amounts

A third type of DATA is two amounts that are related because they both measure the same object or process. If this relationship involves two different dimensions or two different substances or entities, the two related amounts are termed an **equivalency**.

Two related amounts are especially important in solving problems because they supply three items of DATA: two single-unit amounts and a ratio.

Example: If a problem reads, "the car traveled 84 miles in 2.0 hours," two *amounts* have been measured.

- The *distance* the car traveled was 84 miles. We can convert this distance amount to other distance units, such as kilometers or feet.
- The *time* the car traveled was 2.0 hours. We can convert this amount of time to minutes (120 minutes), seconds, or any other time unit.

With this data, we can also calculate a *ratio,* the average speed of the car, in distance over time units.

$$\frac{?\ miles}{hour} = \frac{84\ miles}{2.0\ hours} = 42\ \frac{miles}{hour}$$

Ratios versus Two Related Amounts

If a problem (let's call it Problem A) reads, "the car traveled 42 miles in 1 hour," we have been writing in our DATA,

$$\textbf{42 miles = 1 hour} \qquad\qquad (A)$$

The problem A data supplies two amounts that are equivalent because they measure, in two different dimensions, the same object or process: the trip.

$$42\ miles = \textbf{the trip} = 1\ hour$$

so "42 miles = 1 hour" is true in this problem.

The wording of problem A supplies two single-unit amounts *and* a ratio that you can use to solve problems based on that data.

If problem B reads, "the car was traveling 42 miles *per* hour," we also write in our data,

$$\textbf{42 miles} = \textbf{1 hour} \tag{B}$$

This is the same equality that we wrote for problem A. However, equalities A and B are different.

- From the wording of problem A, we know both the amount of distance the car traveled and the amount of time required. From that data, we can find also the average speed of the car. We know two amounts and one ratio.
- The wording of problem B data tells us the speed of the car, but *not* how far it went nor how long it traveled. Problem B supplies one ratio, but *no amounts*.

The ratio that is the speed of the car is the same in both problems A and B. However, problem A gives us more information.

We have been writing ratios and two related amounts in the same way to this point, because both can be used in the same way as ratios and/or conversions. However, when solving for *single* units, we often need to make a distinction between measurements that are ratios (which we will label **R**) and those that are two *related* amounts that can be used as amounts *or* used as a ratio (which we will label **2A + R**).

Distinguishing Ratios from Two Related Amounts

How do you make the distinction between a ratio and two related amounts? It often requires a careful analysis of the wording of a problem.

To practice making this distinction between ratios and two related amounts, answer these questions.

 TRY IT

(See "How to Use These Lessons," point 1, p. xv.)

Q1. If 0.24 g NaOH are dissolved to make 250 mL of solution, find the [NaOH]. Write only the DATA, then *label* the DATA equality as a *ratio* (**R**) or as two single-unit amounts *and* a ratio (**2A + R**).

 Answer:

DATA: 0.24 g NaOH = 250 mL soln. (**2A + R**)

This data represents two measures of the *amount* of the material in the solution. These two amounts are related because they are both measurements of the same solution.

Q2. In Q1, how many *liters* are present in the solution?

Answer:

$$? \text{ L soln.} = 250 \text{ mL soln.} \cdot \frac{10^{-3} \text{ L}}{1 \text{ mL}} = \textbf{0.25 L soln.}$$

You can calculate that single unit *amount* because you have a single unit *amount* to start from.

Q3. If the conversion between the English unit, fluid ounces, and milliliters is 12 fl. oz. = 355 mL, what is the number of milliliters per fluid ounce?

Write only the DATA, then *label* the DATA equality as a *ratio* (**R**) or as two single-unit amounts *and* a ratio (**2A + R**).

 Answer:

12 fl. oz. = 355 mL (**R**) The conversion given is a ratio that applies in all cases. In this problem, the two sides of the equality are not measurements of an object.

Q4. In Q3, how many *liters* are present in the solution?

 Answer:

You cannot answer. The data in this problem includes no amounts: it is only a ratio. Try

Q5. A typical can of soda is labeled "12 fl. oz. = 355 mL." What is the number of milliliters per fluid ounce?

Write only the DATA, then *label* the DATA equality as either a *ratio* (**R**) or as two single-unit amounts *and* a ratio (**2A + R**).

 Answer:

$$12 \text{ fl. oz.} = 355 \text{ mL } (\mathbf{2A + R})$$

The relationship can be written in the same way as in Q3, but in this case, these are two related amounts *and* a ratio. Both measurements represent the *amount* of liquid in the can.

Q6. In Q5, how many liters of soda does the can hold?

Answer:

0.355 L. The equality in Q5 is *both* two separate amounts and a *ratio* between the two units.

In some problems, and especially for problems in which multiple questions are asked about a single set of data, you will need to be able to distinguish between data that is *two related amounts* and data that is a *ratio*. Practice will help in making these distinctions.

PRACTICE **A**

For each "part of a problem" below, write the DATA only. Then label each DATA equality as a ratio (**R**) or as two single-unit amounts *and* a ratio (**2A + R**). Check your answers at the end of this chapter as you go.

1. The car was traveling at a speed of 55 mi./hr.

2. A bottle of designer water is labeled 0.50 L (16.9 fl. oz.).

3. To melt 36 grams of ice required 2,880 calories of heat.

4. The dosage of the aspirin is 2.5 mg per kg of body mass.

5. The concentration of the acid solution is 2.0 M HCl.

6. 2.0 mol HCl were dissolved in water to make 1 L of solution.

7. Given that there are 1.61 kilometers per mile, and radio waves travel at the speed of light (3.0×10^8 m/s), . . .

Labeling Amounts and Ratios

For calculations in which you need to distinguish ratios from amounts, in the DATA we will label

- **Single-unit amounts** as (**SUA**)
- **Ratios** as (**R**)
- **Two single-unit amounts** *and* a ratio as (**2A + R**)

A helpful rule is that when a measurement in a problem is written in ratio units [with the word *per* or with a slash mark (/) meaning *per one*], it nearly always represents a *ratio* (**R**) rather than two related amounts.

PRACTICE B

For each part below, write the DATA section *only*. Include prompts (such as the grams prompt and M prompt), but you may omit metric-prefix conversions. Then *label* each line in the DATA as a single-unit amount (**SUA**); a ratio (**R**); or two single-unit amounts and a ratio (**2A + R**).

1. 10.0 g NaOH is dissolved to make 1,250 mL of solution. Calculate the concentration of the solution in moles of NaOH per liter of solution. (Molar mass of NaOH: 40.0 g/mol.)

2. A water bath absorbs 24 calories of heat from a reaction that forms 0.88 g CO_2. What is the heat released by the reaction, in kilocalories per mole of CO_2?

3. How many milliliters of 0.100 molar KCl contain 1.49 g KCl?

4. A 0.50 L (16.9 fl. oz.) soft drink contains an artificial sweetener. The concentration is 2.5 mg per fluid ounce. Express this concentration in mg/mL.

Lesson 11.3   Solving Problems with Parts

Solving in Parts

Until this point, for all calculations in which a single unit was WANTED, one item of data had a single unit, and the rest of the data was in ratios/equalities/conversions. In some problems, however, you will need to solve for a single unit when *all* of the DATA is in ratios.

To learn a method to solve these calculations, let us begin with a variation on a problem we have done previously.

TRY IT

Q. In 1988, Florence Griffith-Joyner set a new women's world record in the 200.0 meter dash (a carefully measured distance in a world-record certified race) with a time of 21.34 s (1.609 km = 1 mi.).

  A. What was the length of the race in miles?

  B. What was the time of the race in hours?

  C. What was her average speed in miles per hour?

In a problem with one set of data but *multiple* questions, it is helpful to write the DATA section *first*, then write the WANTED unit for each part.

In your notebook, first write the DATA section, then solve for *just* parts A and B above.

STOP

Answer:

Your paper should include

  DATA: 200.0 m = 21.34 s

     1.609 km = 1 mi.

  A. WANTED: ? mi. =

  B. WANTED: ? hr =

Which of the measurements in the DATA should be chosen as the *given* to solve each part?

Because the above units are familiar, you may know by intuition which quantity to use as the *given* in each part. Science problems, however, often involve less familiar units. Let's use this track and field example to develop rules for picking a *given* from several possibilities.

Rules for Selecting the *Given*

In the current problem, you WANT single units in parts A and B, but the DATA has two equalities. To handle such cases, let's expand our conversion rules.

> **When All DATA Are Equalities, But a Single Unit Is WANTED**
>
> 1. Label each *equality* in the DATA as a ratio (**R**) or as two related single-unit amounts *and* a ratio (**2A + R**).
>
> 2. As the *given*, choose from the DATA *one* single-unit amount of *two amounts* that are related.
>
> 3. Choose as your *given* the single-unit amount that converts more easily to the WANTED unit.

When solving for a single-unit amount, you must start from a *given* single-unit amount. You cannot choose as a *given* one part of a ratio (which we write as an equality or conversion) *unless* the ratio is *also* two related *amounts*.

> ●   ● ▭━━━━━▶ **TRY IT**

Q. For the track problem, use rules 1–3 above to label each line in your DATA table, then SOLVE parts A and B.

Answer:

Your paper should include

DATA: 200.0 m = 21.34 s (**2A + R**)

  1.609 km = 1 mi. (**R**)

The top equality is two related amounts because both are measurements of the same object (this race).

The second equality "1.609 km = 1 mi." is a *ratio* relating two distance units. Those two numbers and their units are not measured amounts in this race. Those two numbers and their units cannot be used as single-unit *givens*.

If needed, adjust your work and use step 3 to SOLVE parts A and B.

As a *given* in each part, you must choose an *amount* that you can convert to the WANTED unit. The 2A + R data above supplies two amounts.

To choose a *given* for each part, begin by identifying the *dimension* of each WANTED unit, and then search for a unit in a DATA *amount* that measures that dimension.

$$\text{SOLVE: A. ? mi.} = \mathbf{200.0\ m} \cdot \frac{1\ km}{10^3\ m} \cdot \frac{1\ mi.}{1.609\ km} = \boxed{0.12430\ \mathbf{mi.}}$$

$$\text{B. ? hr} = \mathbf{21.34\ s} \cdot \frac{1\ min}{60\ s} \cdot \frac{1\ hr}{60\ min} = \boxed{0.0059277\ \mathbf{hr}}$$

In part A, you WANT miles, a measure of the dimension *distance*. The DATA supplies an amount for meters, which is also a distance unit.

For part B, the dimension *time* is WANTED and seconds is an *amount* in the DATA.

> When solving for a single unit, eliminating ratios that are not two amounts simplifies the selection of the *given* amount.

In Problems with Multiple Steps, Carry an Extra Significant Figure

Note the extra digit, after the doubtful digit, in the above two answers. When solving a problem or calculation that has *multiple steps* or *multiple parts*, it is a good practice to carry an extra digit until the *final* answer. Then, in the *final* part, round to the correct number of *s.f.* based on the *s.f.* in the original DATA in the problem. This method limits variations in the final answer owing to rounding in the parts.

Solving for a Ratio in Two Parts

Note also that the answers above were boxed for each part, A and B. When a problem has parts, it is a good idea to box answers from early parts. Those answers are often helpful as DATA for a later part.

▬▬▬▬▬▶ **TRY IT**

Q. With that tip, SOLVE part C of the track problem in one conversion.

For part C of the track problem, the answer unit is a *ratio* unit.

When we solved this problem in Lesson 10.2, we started from the original data and used the rule: "If you want a ratio, start with a ratio." Solving required five conversions. But in many problems that have multiple parts, a faster way to solve later parts is to use answers from *previous parts* as DATA.

Part C asks for the ratio of the *miles* run in the race to the *hours* of the race. *Miles* was solved in part A, and *hours* was solved in part B. Apply the fundamental rule of conversions: *Let the units tell you what to do.*

The units say to divide miles by hours.

$$\text{C. WANTED: } ? \; \frac{\textbf{mi.}}{\textbf{hr}} = \frac{\text{part A answer}}{\text{part B answer}} = \frac{0.12430 \textbf{ mi.}}{0.0059277 \textbf{ hr}} = 20.97 \frac{\text{mi.}}{\text{hr}}$$

S.F.: In this *final* step, the answer is rounded to four *s.f.* based on the data supplied in the problem.

SUMMARY

Problems with Multiple Questions about One Set of Data

1. List the DATA common to all of the parts first, then list the WANTED unit for each part.

2. Box the answer to each part. When a problem has parts, watch to use an answer for an early part as DATA for a later part.

3. When a part WANTS a single-unit amount, at least one measurement in the DATA will be a single-unit amount. Label the lines in your DATA table as **SUA, R,** or **2A + R.** Choose as your *given* the single-unit amount that most easily converts to the WANTED single unit.
   Ratio (**R**) data can*not* be used as a *given* to solve for an single-unit amount.

4. If a later part WANTS a ratio unit, see if you can use answers for the earlier parts as a *given* to find the WANTED ratio.

The Three Methods to Solve for a Ratio Unit

To solve for a ratio unit, there are three methods you can use:

1. Apply the conversion rule, "if you want a ratio, start with a ratio."

2. Solve in two parts: Solve *separately* for the values of the top and bottom amounts, then divide the top by the bottom amount.

3. Use a mathematical equation.

For many complex relationships, solving with equations is necessary. However, most initial relationships in first-year chemistry can be solved with conversions, and this avoids having to memorize dozens of equations.

> With conversions, the same rules apply to each problem.

To solve for a *ratio* unit using conversions, should you choose method 1 or method 2? In many cases, both methods work, but for *less complex* calculations, "if you want a ratio, start with a ratio" is usually quicker.

Method 2, solving for the top and bottom units separately, is necessary in calculations that require both conversions and mathematical equations. For calculations that have multiple questions about common data, method 2 may not be necessary, but it is often quicker.

For now, a good strategy is to first try method 1. If you get stuck, try method 2. If neither method 1 nor 2 works, try method 3 (look for an equation that will solve the problem).

Flashcards

Add these to your collection. Practice for three days, then put them in stack 2.

One-Way Cards (with Notch)	Back Side—Answers
What types of units always measure amounts?	Single units
State three ways to solve for a ratio unit	1. Want a ratio, start with a ratio. 2. Solve for top and bottom unit, then divide. 3. Use an equation.
In calculations with parts, watch for	Answers to one part that can be used as data for later parts.

PRACTICE

Use the rules in the Summary section and the flashcards above. Save one or two numbered problems for your next study session.

1. 11.7 g NaCl is dissolved to make 250. mL of aqueous solution.
   a. How many moles of NaCl were used?
   b. How many liters of solution were prepared?
   c. What is the molar concentration of the NaCl in the solution?

2. 0.0250 mol KCl are used to make 750. mL of solution.
   a. How many grams of KCl are needed?
   b. How many liters of solution were prepared?
   c. What is the [KCl]?

(continued)

3. If 0.050 lb. NaCl are dissolved in water to make 32 fl. oz. of solution (2.2 lb. = 1 kg, 12.0 fl. oz. = 355 mL),

    a. How many moles of NaCl were used?

    b. How many liters of solution were prepared?

    c. What is the [NaCl]?

4. An 18-carat brick of gold measures 2.50 cm × 5.00 cm × 10.0 cm and weighs 4.27 lb. (1 kg = 2.20 lb.). Find

    a. The mass of the brick in grams.

    b. The volume of the brick in cubic centimeters, then milliliters.

    c. The density of the brick in grams/milliliter.

SUMMARY

1. When solving for a ratio, an alternative to starting with a ratio is to solve in two parts.
   - Solve for the numeric value of the top answer unit as a single unit.
   - Solve for the value of the bottom answer unit as a single unit.
   - Divide the top answer by the bottom answer to obtain the final answer.

2. If you WANT a single-unit amount, pick as your *given* a single-unit amount.
   When solving for a single-unit amount, you may choose as a *given*
   - A single-unit amount (SUA data), or
   - One of the single-unit amounts from two that are related (2A+R)

   but you cannot use a measurement from a *ratio* (R) as a single-unit *given*.

3. When solving in parts, or solving for any single unit when all of the data are in equalities/ratios/conversions,
   - Label the data as single-unit amounts (SUA), ratios (R), or two amounts that are equivalent because they measure the same object or process (2A+R)
   - Use a single unit of the SUA or 2A+R DATA as your "single-unit *given*"
   - For each part, pick a single-unit amount as your *given* that can be most easily converted to the WANTED unit
   - In problems that have one set of DATA but multiple questions, box answers from early parts and use them as DATA for later parts

ANSWERS

Lesson 11.1

Practice A   1. **cubic centimeters**; a fundamental quantity must be measured by a base unit to the *first* power.

2a. **mass**   2b. **time**   2c. **count of particles**   2d. **distance**

Practice B   1. These are a few of the possibilities:

a. time: **seconds, minutes, hours, days, years, centuries**

b. mass: **grams, kilograms, milligrams**

c. distance: **meters, kilometers, centimeters**

d. volume: **liters, milliliters, deciliters, cubic meters, cubic centimeters, decimeters cubed**

2. Unit used to measure solution concentration: **moles/liter**

3. The fundamental quantities, their base units, and the definitions of derived quantities must be memorized. A metric prefix does not change the dimension being measured.

a. liters: **volume**

b. cubic meters: **volume** (Volume by definition must be a distance cubed.)

c. kilograms: **mass** (Grams is the base unit for mass.)

d. decimeters: **distance** (Deci- is a prefix meaning one-tenth, meters is distance.)

e. millimeters: **distance**

f. deciliters: **volume** (Liter is a unit of volume.)

g. fluid ounces: **volume** (An English-system unit for volume.)

h. acres: **area** (An English-system unit for area.)

i. nanoseconds: **time**

j. grams/milliliter: **density** (Density is defined as mass over volume.)

k. feet/second: **speed** or **velocity** (Speed is defined as distance over time.)

l. square feet: **area** (Area by definition is distance squared.)

m. moles/liter: **concentration**

n. grams/mole: **molar mass**

4. kilograms, decimeters, millimeters, nanoseconds

5. Those that are *not* fundamental quantities: **area, volume, density, speed, concentration, molar mass**

6. Feet/second, moles/liter, grams/milliliter, grams/mole

7. Density, speed, solution concentration, molar mass

8. Ratio units have *one* base unit on the top and *one* on the bottom. Complex units have *more* than one base unit on the top or bottom, *or* none on top.

Practice C   1. Don't worry if you miss a few. It takes experience and context to make these judgments in many cases. This is practice.

a. 60 s = 1 min   **AT**—always true.

b. 60 mi. = 1 hr   **Not AT**—60 miles per hour is a speed; speed varies.

c. 18.0 g = 1 mol   **Not AT**—g/mol (molar mass) ratios vary depending on substance formulas.

d. 18.0 g H_2O = 1 mol H_2O **AT**—always true for H_2O, even if ice or steam.

e. 6.0 × 10^{23} molecules = 1 mol   **AT**—the mole definition: true for particles of everything.

f. 2.0 mol HCl = 750 mL soln.   **Not A**—this is a concentration; concentration can vary.

g. 355 mL = 12 fl. oz.   **AT**—two volume units will always be related by the same ratio.

2a. Speed **b**   (Distance units over time units = speed.)

2b. Concentration **f**   ("Particles per volume" is a concentration by definition.)

3. **Both speed and concentration** are intensive quantities. Both are a ratio of two extensive quantities.

Lesson 11.2

Practice A   Consider these as "best guess" answers. Taking a line out of context can *sometimes* change its meaning.

1. **55 mi. = 1 hr (R)** You don't know the *amount* of the distance or time.

2. **0.50 L = 16.9 fl. oz. (2A + R)** Two amounts. Both measure the volume of the bottle.

3. **36 g ice = 2,880 calories heat (2A + R)** *Equivalency:* the amounts of ice and heat in melting.

4. **2.5 mg = 1 kg (R)** You are not told the *amount* of the aspirin or the body mass.

5. **2.0 mol HCl = 1 L soln. (R)** You don't know the number of moles or liters, only the ratio of moles to liters in the solution.

6. **2.0 mol HCl = 1 L soln. (2A + R)** An equivalency. You know the single-unit amount that is the moles of HCl; so you could solve for grams HCl. You could solve for the milliliters of solution present. This is the same equality as problem 5, but the wording of the problem makes this data an equivalency.

7. **1.61 km = 1 mi. (R)** Unit conversions and/or "per one" data nearly always means a ratio.

    3.0×10^8 m = 1 s (R) This data is called a speed in the problem, and speed is defined as a ratio.

Practice B   1. DATA: 10.0 g NaOH = 1,250 mL soln.   **(2A + R**—two amounts for same solution.)

    40.0 g NaOH = 1 mol NaOH   (g prompt.) **(R**—the ratio of grams to moles for NaOH.)

2. DATA: 24 calories heat = 0.88 g CO_2 **(2A + R**—two amounts, same reaction.)

    44.0 g CO_2 = 1 mol CO_2 (g prompt.) **(R**—the ratio of grams to moles for CO_2.)

3. DATA: **1.49 g** KCl   **(SUA)**

    0.100 mol KCl = 1 L soln.   **(R**—molarity is a ratio: mole *per* liter.)

    74.6 g KCl = 1 mol KCl   (g prompt.) **(R**—molar mass is a ratio: grams *per* mole.)

4. DATA: 0.50 L = 16.9 fl. oz.   **(2A + R**—two amounts for container volume.)

    2.5 mg sweetener = 1 fl. oz.   **(R**—concentration is a ratio by definition. Neither measurement is an amount of an object in the problem.)

Lesson 11.3

Practice   1. DATA: 11.7 **g** *NaCl* = 250. mL NaCl soln. **(2A + R)**   (Two amounts—same soln.)

    58.5 **g** *NaCl* = 1 mol NaCl **(R)**   (Molar mass is a ratio by definition.)

a. WANTED: ? mol NaCl

    SOLVE: (When a problem has multiple questions about one set of data, label the data as **SUA, R,** or **2A + R.** Then, on the *parts* of the problem where you WANT a single unit, pick a single unit from the SUA or 2A+R data.

    To find moles NaCl, which side of the 2A+R data above do you pick as the given? Grams can be converted to moles using molar mass.)

$$? \text{ mol NaCl} = 11.7 \text{ g NaCl} \cdot \frac{1 \text{ mol NaCl}}{58.5 \text{ g NaCl}} = \boxed{0.200\textit{0} \text{ mol NaCl}}$$

b. WANTED: **?** L soln.

    DATA: Above

    SOLVE: (Pick a single-unit-*given amount* that easily converts to the L WANTED.)

$$? \text{ L soln.} = 250. \text{ mL soln.} \cdot \frac{10^{-3} \text{ L}}{1 \text{ mL}} = \boxed{0.2500 \text{ L soln.}}$$

c. WANTED: $? \dfrac{\text{mol NaCl}}{\text{L soln.}}$ (Units of molarity = mol/L.)

    SOLVE: (You found moles and liters in parts A and B.)

$$\frac{? \text{ mol NaCl}}{\text{L soln.}} = \frac{\text{part A answer}}{\text{part B answer}} = \frac{0.2000 \text{ mol NaCl}}{0.2500 \text{ L soln.}} = \mathbf{0.800 \frac{mol\ NaCl}{L\ soln.}}$$

(*S.F.*: Both original measurements were three *s.f.*; round *final* answer to three *s.f.*)

2. DATA: 0.0250 mol KCl = 750. mL soln.   **(2A + R)**   (Two related amounts for soln.)

a. More DATA: 74.6 *g* KCl = 1 mol KCl   **(R)**   (Molar mass is a ratio, not amounts.)

    WANTED: ? g KCl

    SOLVE: [If you WANT a single unit, start with a single unit. All of the data is in pairs, but two *amounts* related in a pair (2A+R) may be used as single unit data.

                You know moles and want grams. "Grams and moles, use molar mass."]

$$? \text{ g KCl} = 0.0250 \text{ mol KCl} \cdot \frac{74.6 \text{ g KCl}}{1 \text{ mol KCl}} = \boxed{\mathbf{1.865} \text{ g KCl}}$$

(Carry an extra *s.f.* until the final part.)

b. WANTED: ? L soln.

    SOLVE: (If you WANT a single unit, start with a single unit. One of the 2A+R related amounts is **mL** soln.; convert it to the **liters** WANTED.)

$$? \text{ L soln.} = 750. \text{ mL soln.} \cdot \frac{10^{-3} \text{ L}}{1 \text{ mL}} = \boxed{\mathbf{0.7500} \text{ L soln.}}$$

c. WANTED: $? \dfrac{\text{mol KCl}}{\text{L soln.}}$ (Write the unit that you WANT.)

    SOLVE: [When solving for a ratio in parts, let the units tell you what to do. The answer divides moles (supplied in 2A+R data) by liters (found in part B).]

$$\frac{? \text{ mol KCl}}{\text{L soln.}} = \frac{0.0250 \text{ } \mathbf{mol\ KCl}}{0.750 \text{ L soln.}} = \mathbf{0.0333 \frac{mol\ KCl}{L\ soln.}}$$

3. DATA: 0.050 lb. NaCl = 32 fl. oz. soln.   **(2A + R)**   (Two amounts measuring one solution.)

            2.2 lb. = 1 kg   **(R)**   (An English-metric unit conversion.)

            12.0 fl. oz. = 355 mL   **(R)**   (An English-metric volume conversion.)

a. WANTED: ? mol NaCl

(To solve for moles, use one of your 2A+R amounts. Which one? *Pounds NaCl* can be converted to kilograms, kilograms to *grams NaCl*, grams to moles using the molar mass of NaCl.)

    DATA: 58.5 *g NaCl* = 1 mol NaCl   **(R)**   (Molar mass is a ratio, not an amount.)

SOLVE: ? mol NaCl = 0.050 lb. NaCl · $\dfrac{1 \text{ kg}}{2.2 \text{ lb.}}$ · $\dfrac{10^3 \text{ g}}{1 \text{ kg}}$ · $\dfrac{1 \text{ mol NaCl}}{58.5 \text{ g NaCl}}$ = $\boxed{0.389 \text{ mol NaCl}}$

(Carry an extra *s.f.* until the final part.)

b.   WANTED: ? L soln.

(The other single-unit amount measured is a volume. You have a conversion from fluid ounces to milliliters, and milliliters to liters you can do.)

SOLVE: ? L soln. = 32 fl. oz. soln. · $\dfrac{355 \text{ mL}}{12.0 \text{ fl. oz.}}$ · $\dfrac{10^{-3} \text{ L}}{1 \text{ mL}}$ = $\boxed{0.947 \text{ L soln.}}$

c.   WANTED: $? \dfrac{\text{mol NaCl}}{\text{L soln.}}$

SOLVE: (When solving for a ratio in parts, let the units tell you what to do. The units say to divide moles—found in part A—by liters—found in part B.)

$$? \dfrac{\text{mol NaCl}}{\text{L soln.}} = \dfrac{0.389 \text{ mol NaCl}}{0.947 \text{ L soln.}} = \mathbf{0.41 \dfrac{\text{mol NaCl}}{\text{L soln.}}}$$

(*S.F.*: Both original measurements were two *s.f.*; round *final* answer to two *s.f.*)

4.   (For problems with multiple parts, list the data first.)

DATA: 4.27 lb. Au = (2.50 cm × 5.00 cm × 10.0 cm) Au   (**2A + R**—weight and volume of one brick.)

        1 kg = 2.20 lb.   (**R**—a conversion always true at earth's surface gravity.)

a.   WANTED: ? g Au

(You WANT a single unit, but all of the DATA are in equalities. Label each equality as R or 2A+R. Pick the SUA measurement that can convert to grams.)

$$? \, \mathbf{g \, Au} = 4.27 \text{ lb. Au} \cdot \dfrac{1 \text{ kg}}{2.20 \text{ lb.}} \cdot \dfrac{10^3 \text{ g}}{1 \text{ kg}} = \boxed{1,941 \text{ g Au}}$$

(Carry an extra *s.f.* until the final part.)

b.   (Use volume = length × width × height to calculate the volume in cubic centimeters, which equal milliliters.)

$$? \, \mathbf{cm^3 \, brick} = 2.50 \text{ cm} \times 5.00 \text{ cm} \times 10.0 \text{ cm} = 125.0 \, \mathbf{cm^3} = \mathbf{125.0 \, mL}$$

c.   WANTED: $\mathbf{?} \dfrac{\text{g Au}}{\text{1 mL}}$

(The answer units tell us to divide the grams of the gold sample by its volume in **milliliters**. Both of those quantities were found in the parts above.)

$$? \dfrac{\text{g Au}}{\text{mL}} = \dfrac{1,941 \text{ g Au}}{125.0 \text{ mL}} = \mathbf{15.5 \dfrac{\text{g Au}}{\text{mL}}} \qquad \text{(The original data had three } s.f.)$$

12

Molarity Applications

Lesson 12.1   Dilution

Converting between mL and L

When a simple operation is needed often, it is best to practice until you can perform the conversion "in your head." During calculations involving molarity, it is especially helpful to be able to convert quickly between milliliters (mL) and liters (L). Let's review the logic of a "quick rule" for these conversions.

  1. Milli- means 10^{-3}, so $\times \, 10^{-3}$ can be *substituted* for **milli-**
      Example: $? \, \text{L} = 47 \, \textbf{mL} = 47 \times 10^{-3} \, \textbf{L}$ or $\textbf{0.047 L}$

  2. Recall that 1 L = 1000 mL, so 4.2 L = 4,200 mL. The rule is this:

  > When converting between mL and L in fixed decimal notation,
  >
  > • Move the decimal three times
  > • Make the number of **milliliters** 1000 times *larger* than the number of **liters**

  3. *Check*: Compare the mL and L. The number of mL must be 1000 times larger.

PRACTICE   A

Learn the rules, then apply them to these problems.

  1. Using rule 1, fill in the blanks. Check your answer using rule 3.

      a. 35 mL = 35 × 10____ L = _____ L   b. 125 mL = 125 × ____ L = _____ L

          Notation: exponential fixed decimal   exponential fixed decimal

  2. Apply rule 2, answering without exponentials. Check your answers using rule 3.

      a. 2.500 L = _____ mL   b. 0.77 L = _____ mL   c. 15 mL = _____ L

  3. For volumes written in *fixed* notation, when converting between mL and L, the decimal point in the L should always be ____ places to the _____.

  4. Given milliliters or liters, write the other in fixed notation.

      a. 150 mL =   b. 33 L =   c. 9.21 mL =

      d. 0.45 L =   e. 0.833 mL =   f. 0.0655 L =

Dilution Fundamentals

To mix a solution, a *solute* is dissolved in a *solvent* in which the solute is soluble.

A solution is **diluted** when more solvent is added. As it is diluted, the solution becomes less **concentrated**: The average distance between the dissolved particles has increased.

At a given temperature, reactions involving a solute in a solution occur more slowly in more dilute solutions. To react, nearly always the solute must collide with another particle, and in diluted solutions collisions occur less often. By studying the impact of dilution on the speed of a chemical reaction, we can explore the steps by which the reaction takes place.

When a solution is diluted, its volume increases, and concentration of its solute decreases, but the number of particles of the solute stays the same. That's the key relationship in dilution: the *moles* of dissolved solute are the *same* in both the concentrated and the diluted solutions.

In dilution: *moles* solute in **C**oncentrated soln. = *moles* solute in **D**iluted soln.

Solving Dilution by Inspection

If the numbers in a dilution involve simple multiples, calculations can often be solved "by inspection" (in your head). For example,

- If the *volume* of a solution is *doubled* (**2×**) by adding more solvent, the *concentration* of the solute is cut in *half* (**1/2**).
- If a solution concentration is WANTED that is **1/10th** the [original], add solvent to increase the volume of the solution by a factor of **10**.

Dilution Calculations for Easy Multiples

In dilution calculations, if either the *volume* or the *molarity* is changed by a *multiple*, multiply the other supplied value by *1/multiple*.

Example: If the volume of a solution is quadrupled by adding solvent, the concentration becomes 1/4 its original value.

 TRY IT

(See "How to Use These Lessons," point 1, p. xv.)

Based on the logic above, solve this problem "in your head." Write your answer, then check below.

Q. To 250 mL of a 0.45 M aqueous glucose solution, distilled water is added until the volume is 750 mL. What is the new [glucose]?

To solve dilution by inspection, compare the two items in the supplied DATA that have the same *unit*. Figure out which easy multiple takes you from the initial to the final value for that unit. Multiply the stated value for the *other* unit times 1/multiple. That's the answer.

 Answer:

The total volume increased from 250 **mL** to 750 **mL**: it *tripled*. So the final concentration of the glucose will be *1/3* the original 0.45 M = **0.15 M**

Learn the logic above, then treat this practice as a quiz: solve without looking back at the rules. Write your answers by inspection in the space below.

1. If 100. mL of 2.0 M KCl is diluted to 400. mL total volume, what is the final [KCl]?

2. To dilute 250. mL of 1.00 M HCl to a concentration of 0.200 M, what must be the final volume of the solution?

The Dilution Equation

When a dilution problem cannot be done by inspection, you can solve using conversions or an equation. Dilution is a case where the equation method is generally faster.

In dilution, because the moles of solute stays constant, we can write:

$$\text{liters } \mathbf{C} \cdot \frac{\text{moles } \mathbf{C}}{\text{liter } \mathbf{C}} = \text{moles Concentrated} = \text{moles Diluted} = \text{liters } \mathbf{D} \cdot \frac{\text{moles } \mathbf{D}}{\text{liter } \mathbf{D}}$$

A way to rewrite this equation is

$$\text{Volume}_C \times \text{Molarity}_C = \text{Volume}_D \times \text{Molarity}_D$$

which is written in symbols as the **dilution equation**:

In dilution: $\mathbf{V_C} \times \mathbf{M_C} = \mathbf{V_D} \times \mathbf{M_D}$

and memorized by recitation:

"In dilution, volume times molarity equals volume times molarity."

A restriction on the use of this and other equations is that the units must be consistent, meaning in this equation that the volume units must be the same. The equation works in any volume units (such as mL, L, or gallons), provided that all of the volumes are converted to the *same unit*.

Let's learn to use the dilution equation by solving an example.

TRY IT

Q. To 225 mL of an aqueous 0.200 M KOH solution, water is added until the total volume is 4.00 L. What is the resulting [KOH]?

Complete the following steps in your notebook.

Steps for Solving with the Dilution Equation

1. List the WANTED unit and DATA in the usual manner.

2. If the problem involves dilution, write

SOLVE: In dilution: $V_C \times M_C = V_D \times M_D$

3. Label each item of WANTED and DATA with a *symbol* from the equation.

    a. First, label each item in the WANTED and DATA as being a volume (**V**) or a concentration (**M**), based on its unit. The molarities (M) will be written as ratios in the WANTED and equalities in the DATA.

    b. Then mark each **V** as being V_C or V_D. Mark each **M** as being M_C or M_D.
        The solution with the higher molarity *or* lower volume is the more concentrated (**C**). Once you have one solution identified as **C** or **D**, the other must be the opposite. *Each* of the four symbols in the equation must be used *once* in labeling the WANTED and DATA.
        It often helps to mark the WANTED unit as **C** or **D** *last*, by process of elimination.

Do those steps, then check below.

Your paper should look like this:

WANTED: ? $\dfrac{\text{mol KOH}}{\text{L KOH soln.}}$ M_D

DATA: 225 mL KOH soln. V_C

    0.200 mol KOH = 1 L KOH soln. M_C

    4.00 L KOH soln. V_D

SOLVE: In dilution: $V_C \times M_C = V_D \times M_D$

4. Convert the DATA to consistent units. (If the two volume units differ, either convert both to milliliters *or* both to liters.)
    Do that step, then check below.
    You can work in liters: ~~225 mL KOH soln.~~ V_C 0.225 **L** KOH soln.
    Or in milliliters: ~~4.00 L KOH soln.~~ V_D 4.00×10^3 *mL* KOH soln.
    One volume must be converted so that both volumes have the *same* unit, but which unit you choose will not affect the final answer.

5. Using algebra, solve the equation in *symbols* for the *symbol* WANTED.

    WANTED: ? $M_D = \dfrac{V_C \times M_C}{V_D}$

    Do not plug numbers into an equation until *after* you have solved for the WANTED symbol. When numbers include units, symbols move faster and with fewer mistakes.

6. Now plug in the numbers with their units and solve.
    When using *conversions*, you must write **M** as moles *per* liter because you use **M** as a ratio to solve. With the dilution *equation*, however, use **M** to abbreviate moles/liter so that the equation will solve more quickly.

    $? M_D = \dfrac{V_C \times M_C}{V_D} = \dfrac{225 \text{ mL KOH } \mathbf{C} \times 0.200 \text{ M KOH } \mathbf{C}}{4.00 \times 10^3 \text{ mL KOH } \mathbf{D}}$

    = **0.0113 M KOH soln. diluted**

7. *Check* your answer. Round the numbers in the problem to make the supplied change an easy multiple. Then estimate the answer by inspection (in your head). Compare your estimate to the calculated answer. They should be close.

Try that step, then check below.

If 200 mL is increased to 4,000 mL, the increase is by a factor of 20. The concentration should therefore be cut by 1/20. 0.200 M × 1/20 = 0.010 M. This is close to the answer above. Check!

A Caution about Equations, Units, and Labels

In step 6 above, the *units* cancelled properly to give the WANTED unit, but the C and D *labels* with each quantity seem not to cancel. Actually, units and labels must cancel in equations. If instead of abbreviating molarity for M_C, we had written out the full "mol KOH/L KOH C soln.," the units and labels would have cancelled properly, but the equation does not solve quickly. If you abbreviate mol/L as **M**, the equation solves quickly, but the unit-label cancellation may not work.

The bottom line: To solve quickly, our rule will be: abbreviate mol/L as **M** when you substitute into the dilution equation, but do so carefully. You will not have label cancellation as a check on your work.

The Dilution Prompt: Two Volumes, Two Concentrations, Same Formula

How do you recognize that a problem is *about* dilution? Dilution calculations often do not include the word dilution, but they can be recognized by using the following:

The Dilution Prompt

- If a problem involves *two* solutions that contain the *same* dissolved substance, *or*
- If the WANTED and DATA contain *two volumes* (usually in mL or liters) *and two concentrations* (M) of the *same* substance,

Try a dilution method (solving by inspection or the dilution equation) to solve.

SUMMARY

Dilution Calculations

1. A solution is *diluted* by adding additional solvent. When solvent is added, the volume of the solution increases and the concentration of the dissolved substance (solute) decreases.

2. In dilution calculations, the fundamental relationship is: the *moles* of dissolved solute are the *same* before and after dilution.

In dilution: *moles* solute **C**oncentrated = *moles* solute **D**iluted

3. In dilution calculations, if the volume or molarity is changed by an *easy multiple*, try to solve by inspection: If the *volume or* the *molarity* is changed by a *multiple X*, multiply the other supplied value by *1/X*.

    Example: In dilution, if the volume is doubled, the concentration is cut in half.

4. The dilution equation is $V_C \times M_C = V_D \times M_D$, memorized by recitation as

    "In dilution, volume times molarity equals volume times molarity."

5. If the numbers in a dilution calculation do not involve easy multiples, solve using the dilution equation. List the WANTED and DATA in the usual manner. To solve with an equation,
    *   Write the needed equation: In dilution: $V_C \times M_C = V_D \times M_D$
    *   Label each WANTED and DATA item with a *symbol* from the equation
    *   Convert the DATA to consistent units (such as mL *or* L)
    *   Solve the equation for the WANTED symbol using the symbols
    *   Plug in the numbers and units and solve
    *   *Check* your answer: Round so that the supplied change is an easy multiple, estimate an answer in your head, and compare to the calculated answer.

6. When solving a dilution equation, use M to abbreviate mol/L. The units will cancel properly (but the labels *diluted* and *concentrated* may not).

7. **The dilution prompt:**
    *   If a problem involves *two* solutions that contain the *same* dissolved substance, *or*
    *   If the WANTED and DATA contain *two volumes* (usually in milliliters or liters) *and two concentrations* (M) of the *same* substance, try a dilution method (solving by inspection or the dilution equation) to solve.

Flashcards

Add these to your collection. Run them to perfection three days in a row.

One-Way Cards (with Notch)	Back Side—Answers
The fundamental rule of dilution	*Moles of solute* is not changed by dilution
In dilution calculations, if *volume* or *concentration* is changed by a multiple,	The other value is multiplied by 1/multiple
The dilution equation in words	Volume times molarity = volume times molarity
The dilution equation in symbols	$V_C \times M_C = V_D \times M_D$
When solving with equations, label the WANTED and DATA with	The *symbols* used in the equations

PRACTICE   C

1. Given liters, write milliliters; given milliliters, write liters. Answer in fixed notation.

    a. 250 mL =   b. 0.62 L =   c. 3.5 mL =

Below, complete problems 2 and 5. Save 3 and 4 for your next practice session.

2. If 50.0 mL of 3.00 M NaOH is diluted

    a. To 500. mL total volume, what is the [NaOH]?

    b. To 0.750 L total volume, what is the [NaOH]?

3. To what volume must 20.0 mL of 2.5 M KCl be diluted to make 0.24 M KCl?

4. To make 250 mL of 0.65 M NaCl solution starting from a 2.00 M NaCl solution, how many milliliters of the concentrated solution are needed?

5. If 6.2 mL of an HCl solution is added to water to make 0.250 L of 0.15 mol · L^{-1} HCl, what was the concentration of the initial HCl?

Lesson 12.2 | Ion Concentrations

Ion Separation in Water

Ionic compounds have relatively high melting points: they are solids at room temperature. In the solid, the ions are locked in place, held together by electrostatic attraction. However, when an ionic solid is melted or is dissolved in water, the ions **dissociate**: they separate and move about. The electrical charges on the ions can "flow". This ability for the charges to move means that when ions are melted or dissolved in water, they can conduct electricity. A substance that separates into ions when dissolved in water is termed an **electrolyte**.

Some ionic compounds dissolve only slightly in water, while others dissolve 100% in dilute solutions. In either case, dissolving an ionic solid in water can be written as if it were a simple chemical reaction. As the solid dissolves, the original formula units of the solid are *used up* as the *separated* ions form.

For the portion of an ionic substance that dissolves in water, unless otherwise specified, you should assume that *all* of the ions separate: that in water, the ions do not stick together despite the attraction of their opposite charges. In dilute solutions, this "ideal solution behavior" is generally close to true. Our rule will be this approximation:

> For the portion of an ionic solid that dissolves in water, to find [ions] in the solution, assume the ions separate ~100%.

In the case of an ionic solid that dissolves 100% in water, the relationship between the number of formula units added to the solution and the number of ions formed are the simple whole-number ratios of the balanced equation.

Example: Na_3PO_4 is highly soluble in water—in dilute solutions, the ionic solid dissolves 100%. In dilute solutions, ions also separate by very close to 100%. This reaction can be expressed as

$$1\,Na_3PO_4(s) \rightarrow 3\,Na^+(aq) + 1\,PO_4^{3-}(aq)$$

The (*s*) is an abbreviation for *solid* and the (*aq*) abbreviates *aqueous*, which means dissolved in water. The products of the ionization of an ionic solid can be predicted using the rules for *separated-ion* formulas (see Lesson 7.3).

Note that *one* particle was used up and *four* particles form. One formula unit of an ionic compound, when its ions separate, will always produce *two or more* ions.

Note also that the reaction is balanced for both atoms and charge. In chemical reactions, the total number of particles on each side will often differ. However, the number of each kind of atom, and the overall charge, must be the *same* on each side of any balanced equation.

To simplify problem solving, use this rule:

> For any substance that forms ions when it dissolves in water, write the balanced equation showing the formation of its separated ions.

Calculating Moles of Ions

If an ionic solid dissolves 100% in water and the moles of solid dissolved are known, then the moles of ions formed can often be "solved by inspection" (done in your head), because the calculation is based on coefficients: simple whole-number mole ratios.

<hr/>

TRY IT

For this problem, see if you can "solve by inspection," doing any needed math by mental arithmetic.

Q. The ionic solid $BaCl_2$ dissolves 100% in dilute aqueous solutions. If 0.40 mol solid $BaCl_2$ dissolves when mixed into water,

a. Write the balanced equation for ion separation.

b. How many moles of Cl^- ions are present in the solution?

c. How many moles of $BaCl_2$ particles are present in the solution?

Answers:

a. The balanced equation for dissolving and separation is

$$1\,BaCl(s) \rightarrow 1\,Ba^{2+}(aq) + 2\,Cl^-(aq)$$

For every *one mole* of $BaCl_2$ formula units used up as it dissolves and separates, *two moles* of Cl^- ions form.

b. Dissolving **0.40** mol $BaCl_2$ will therefore result in **0.80 mol** Cl^- in the solution.

c. **0 mol** $BaCl_2$ particles are present in the solution. All of the $BaCl_2$ formula units are used up as $BaCl_2$ dissociates (separates into ions).

PRACTICE   **A**

A solution is made by adding 0.50 mol Na_2SO_4 to water. The Na_2SO_4 dissolves and dissociates 100%. In the resulting solution, what will be the number of moles of

1. Na^+ ions?
2. SO_4^{2-} ions?
3. Na_2SO_4 particles?
4. Ions in the solution?

Coefficients and Ion Concentrations

When ionic solids dissolve 100% in water, the simple coefficient ratios that predict *moles* used up and formed also predict the *concentrations* (mol/L) of ions formed. Why?

- The coefficients of balanced equations can be used as *ratios* between *numbers* of reacting particles.
- Because coefficients can be used as ratios, coefficients can be multiplied by any number and still be true. This means that coefficients can be read as either counts of particles or as *moles* of particles, since a mole is simply a large number.
- Because coefficients can be mole ratios, they can also be mole *per liter* (concentration) ratios *if* all of the substances in the balanced equation are dissolved in the *same* solution, because they will all be in the same *volume* (the same liters).

For a reaction of dissolved substances that takes place in a solution that keeps a constant volume during the reaction (which is almost always the case—and you should assume is true unless otherwise noted), the mole *ratios* for the balanced equation will remain the *same* when *divided* by the constant volume, so the coefficients can be read as mole *per liter* ratios.

To summarize,

For the Reaction of an Ionic Solid Dissolving in Water

1. Coefficients of a balanced equation can be read as
   - Particles, or moles of particles, or
   - Moles *per liter* if all reactants and products are dissolved in the same solution

2. In calculations involving ions dissolved in water, assume the ions separate ~100% unless otherwise noted, and write the balanced equation for ion separation.

Ion Concentration Calculations

Describing the concentration of particles in a solution is a bit tricky because chemists often use a shortcut when writing solution concentrations.

TRY IT

Using the rules above, solve this problem in your notebook.

Q. In a bottle of solution labeled 0.15 M K_2CrO_4, what is the

a. $[K^+]$?   b. $[CrO_4^{2-}]$?   c. $[K_2CrO_4]$?

STOP   **Answer:**

To find moles or [ions] when ionic solids dissolve, begin by writing the balanced equation for the separation of the solid into ions (see Lesson 7.3).

$$1\ K_2CrO_4(s) \rightarrow 2\ K^+(aq) + 1\ CrO_4^{2-}(aq)$$

Solving each part of this problem by inspection,

a. $[K^+] = \mathbf{0.30\ M}$

b. $[CrO_4^{2-}] = \mathbf{0.15\ M}$, based on the balanced equation

Because all moles are in the same liters of solution, the coefficients show the ratios in which the moles *per liter* are used up and formed. For every 0.15 moles per liter of K_2CrO_4 formula units used up, *0.30 mol/L* of K^+ and *0.15 mol/L* of CrO_4^{2-} forms. The answer to part c is a bit more complex.

c. The $[K_2CrO_4]$ is *labeled* as "**0.15 M**," but in the solution are *zero* particles of K_2CrO_4.

Why? As the solid K_2CrO_4 dissolves in water, its formula units are used up.

The moles that are added per liter to *mix* a solution is a legitimate quantity to measure. For the above solution, the label "0.15 M K_2CrO_4" represents the moles of K_2CrO_4, per liter that were dissolved to *make* the solution, so we call this the "$[K_2CrO_4]$ **as mixed**." However, in most solutions of soluble ionic compounds, as the particles of the solid dissolve they immediately dissociate essentially 100% to form separated ions.

Describing the solution as "0.15 M K_2CrO_4" is a shortcut: it is quicker than writing "0.30 M K^+ *and* 0.15 M CrO_4^{2-}," so this "as mixed" concentration is usually written to describe the solution. However, we will need to remember that in most solutions of soluble ionic compounds, the actual concentration of the undissociated particles is zero. For most calculations, we will need to know the concentrations of the *separated ions* that are actually present in the solution.

The REC Steps

For a reaction in which a substance separates 100% into ions, a fast method to calculate the ion concentrations is to solve by inspection by *writing* what in these lessons we will call the **REC** steps.

For Substances Separating 100% into Ions

To find the [ions] for an ionic solid dissolving 100%, first convert all units to mol/L, then write the **REC** steps. Write

- **R**: The balanced **R**eaction equation. After the equation, write
- **E**: The **E**xtent of the reaction (such as "goes 100%"). Below each particle, write
- **C**: The **C**oncentration of each particle based on the coefficient ratios of the balanced equation.

TRY IT

In your notebook, apply the REC steps to answer the following problem.

Q. Calcium nitrate is dissolved in water to form a solution labeled 0.25 M $Ca(NO_3)_2$. What are the following concentrations?

a. $[Ca(NO_3)_2]_{\text{as mixed}}$ b. $[Ca^{2+}]$ c. $[NO_3^-]$

Answer:

Because this reaction is a solid separating 100% into ions and the particle concentrations are WANTED, solve by inspection using the REC steps. Write the

Reaction and **E**xtent: $1\ Ca(NO_3)_2 \rightarrow 1\ Ca^{2+}(aq) + 2\ NO_3^-(aq)$ (Goes 100%.)

$\qquad\qquad\qquad\qquad\qquad \wedge \qquad\qquad\quad \wedge \qquad\qquad\quad \wedge$

Concentrations: **0.25 M** ($\rightarrow$ 0 M)   **0.25 M**   **0.50 M**

First write the *given* concentration, then write the others based on the coefficient ratios. The REC notation above shows the mol/L of the initial ionic solid *both* as mixed (0.25 M) and after it separates (0 M).

Flashcards

Add these to your collection. Run each until perfect for three days, then weekly and monthly.

One-Way Cards (with Notch)	Back Side—Answers
Coefficients can be read as	Particles or moles—or mol/L if all reactants and products are in same volume
To find solution concentrations for substances that separate 100% into ions	Write the REC steps
The REC steps are	Write the balanced **R**eaction equation, **E**xtent, and **C**oncentrations

Two-Way Cards (with*out* Notch)	
Electrolyte	A substance or mixture that is composed of ions and conducts electricity

PRACTICE   B

Assume these solids dissolve 100% and form ions 100%. Solve using the REC steps. Save a problem or two for later review. Check your answers as you go.

1. In a 0.30 M solution of radium nitrate,

    a. $[Ra^{2+}]$ = ?

    b. $[NO_3^-]$ = ?

2. In a 0.60 molar solution of sodium carbonate,

    a. $[Na^+]$ = ?

    b. $[CO_3^{2-}]$ = ?

3. If a solution of potassium phosphate has a $[K^+]$ of 0.45 M,

    a. $[PO_4^{3-}]$ = ?

    b. $[K_3PO_4]_{as\ mixed}$ = ?

4. In a solution of aluminum sulfate, if the $[SO_4^{2-}]$ = 0.036 M,

    a. $[Al^{3+}]$ = ?

    b. $[Al_2(SO_4)_3]_{as\ mixed}$ = ?

5. If 10.62 g solid K_3PO_4 are completely dissolved to make 0.250 L of solution ($212.3\ g\ K_3PO_4$/mol),

    a. What is the concentration of the K_3PO_4 solution, as mixed?

    b. What is the concentration of K^+ ions in the solution?

Lesson 12.3   Solution Stoichiometry

Calculations for Solution Reactions

For the reaction calculations in Lessons 12.3, we will limit our attention to reactions that *go to completion* (that go until at least one reactant is totally used up) and at least one limiting reactant is known.

Chemical reactions are often carried out in aqueous solutions. A calculation of *how much* of the reactants and products are involved in a reaction is stoichiometry. For reactions carried out in solutions, calculations can be solved using the same stoichiometry methods that we have studied previously.

For Single-Unit WANTED, *Given* Quantity Known

In solution stoichiometry, if a single-unit amount is WANTED and the limiting reactant is supplied in the DATA, that amount is your *given* quantity. When a single-unit *given* is known, solve by seven-step conversion stoichiometry.

For reactions, if WANTED substance ≠ *given* substance,

- Write four steps to gather tools: WANTED and DATA, **b**alance, and **b**ridge (WDBB), and
- Three to solve: Convert "? WANTED = unit *given*" to **mol** *given* to **mol** *wanted* to unit *wanted*.

The quick version of the stoichiometry steps is this: WDBB, then units to *moles* to *moles* to units.

In solution reactions, an additional factor is

The M Prompt

In the WANTED and DATA, translate *M, molar, molarity, concentration,* and [ ] *brackets* into *moles per 1 liter.* Write moles per liter as a ratio when it is WANTED and as an equality or conversion in the DATA.

The Importance of Complete Labels

By definition, stoichiometry involves two or more substances. In any problem dealing with two or more entities, a careful *labeling* of measurements is essential.

- In the WANTED, DATA, and conversions, if a number and unit apply to only *one* entity in the problem, write after the unit a label: a substance formula and/or words that identify what the unit is measuring.
- The substance, its state, and other parts of its descriptive label are an inseparable part of the unit.
- A unit that measures *one* substance cannot be used to cancel the same unit measuring a *different* substance.

Solution Stoichiometry at an Equivalence Point

For calculations in which two reactants "react exactly" in some way, such as being exactly "neutralized," you should assume the reactants are stoichiometrically equivalent: that the moles of the two reactants have been mixed in a ratio that matches the coefficients of the balanced equation. For calculations at an **equivalence point**, both reactants are limiting (completely and exactly used up), and we can use conversion stoichiometry to solve.

▷ **TRY IT**

Using the rules above, try this problem in your notebook.

Q. How many milliliters of a 0.0500 M H_2SO_4 solution are required to exactly neutralize 25.0 mL of 0.220 M KOH? The unbalanced equation is

$$H_2SO_4 + \quad KOH \rightarrow \quad H_2O + \quad K_2SO_4$$

Answer:

For stoichiometry, start with WDBB.

WANTED: ? mL H_2SO_4 soln.   (You want a single unit.)

DATA: 0.0500 mol H_2SO_4 = 1 L H_2SO_4 (M prompt.)

25.0 mL KOH soln.   (The single unit *given*.)

0.220 mol KOH = 1 L KOH   (M prompt.)

Because you have solutions of two different substances, include substance formulas after all units.

In a calculation that includes volume measurements for both gases and solutions with liquid solvents, you would need to label each volume as being for the *gas* or the *solution*. However, in this problem and in *most* calculations for reactions in solutions, all of the volumes are measurements of *solutions*. If that is the case, you may label one volume for each substance as "soln." but leave off "solution" as understood for the other volume labels in the DATA and conversions.

Balance: **1** H_2SO_4 + **2** KOH → **2** H_2O + **1** K_2SO_4

Bridge: **1 mol** H_2SO_4 = **2 mol** KOH

Now apply stoichiometry steps: convert units to moles to moles to units.

SOLVE:

$$? \text{ mL } H_2SO_4 = 25.0 \text{ mL KOH} \cdot \frac{\cancel{10^{-3} \text{ L}}}{1 \text{ mL}} \cdot \frac{0.220 \text{ mol KOH}}{1 \text{ L KOH}} \cdot \frac{1 \text{ mol } H_2SO_4}{2 \text{ mol KOH}}$$

$$\cdot \frac{1 \text{ L } H_2SO_4}{0.0500 \text{ mol } H_2SO_4} \cdot \frac{1 \text{ mL}}{\cancel{10^{-3} \text{ L}}}$$

$$= 55.0 \text{ mL } H_2SO_4 \text{ soln.}$$

PRACTICE

1. Potassium hydrogen phthalate ($KHC_8H_4O_4$, abbreviated as KHPht) is a solid organic acid with a molar mass of 204.2 g/mol. How many milliliters of 0.0750 M KOH solution is required to neutralize 0.300 g of this acid? The balanced equation can be written as:

    1 KHPht + 1 KOH → 1 H_2O + 1 K_2Pht

2. A 0.0250 M solution of lead(IV) oxide is reacted with 0.268 g manganese(II) nitrate. The balanced equation is

    2 $Mn(NO_3)_2$ + 5 PbO_2 + 6 HNO_3 → 5 $Pb(NO_3)_2$ + 2 $HMnO_4$ + 2 H_2O

    How many milliliters of the acidic PbO_2 solution is needed to use up all of the manganese(II) nitrate [$Mn(NO_3)_2$ = 179 g/mol]?

3. How many grams of silver metal will react with 1.50 L of 6.0 M HNO_3?
    The balanced equation is

    Ag + 2 HNO_3 → $AgNO_3$ + NO_2 + H_2O

Take the Paper You Need

In many problems, a methodical approach will require using an entire sheet of paper to solve each question. That's OK. Paper recycles. Trees grow. Your understanding will also grow if you take the time and paper that is needed for careful work.

If you are solving in a graph-paper notebook, try working with the paper in the landscape mode (turned sideways). This provides more room for the long string of conversions that stoichiometry often requires.

Lesson 12.4   Review Quiz for Chapters 11–12

You may use a calculator and a periodic table. Work on your own paper. On the multiple-choice problems, solve—*then* find your answer in the choices provided.
Set a 25-minute limit, then check your answers at the end of the chapter.

1. Lesson 11.4: A 2.0 L (67.6 fl. oz.) bottle of a soft drink contains an artificial sweetener that has a concentration of 4.5 mg per fluid ounce.

   a. Write an equality from this data that represents two related amounts.

   b. Write an equality from this data that represents a ratio that is not two measured amounts.

   c. Calculate the milligrams of sweetener that is found in 0.25 L of the drink.

      a. 38. mg   b. 76 mg   c. 1.88 mg   d. 0.033 mg   e. 19 mg

2. Lesson 12.1: To what volume must 50.0 mL of 2.0 M KOH be adjusted to make 0.24 mol $\cdot$ L^{-1} KOH?

   a. 600. mL   b. 420 mL   c. 60. mL   d. 104 mL   e. 24 mL

3. Lesson 12.2: In a 0.20 M solution of ammonium sulfate, the $[NH_4^+]$ is

   a. 0.10 M   b. 0.20 M   c. 0.30 M   d. 0.40 M   e. 0.60 M

4. Lesson 12.2: If 4.25 g K_3PO_4 (212.3 g/mol) is dissolved to make 400. mL solution, what is the resulting $[K^+]$?

   a. 0.0500 M   b. 0.150 M   c. 0.166 M   d. 0.667 M   e. 0.100 M

5. Lesson 12.3: How many milliliters of a 0.250 M HCl solution are required to exactly neutralize 0.741 g calcium hydroxide? The unbalanced equation is

   $$HCl + \quad Ca(OH)_2 \rightarrow \quad H_2O + \quad CaCl_2$$

   a. 20.0 mL HCl   b. 40.0 mL HCl   c. 80.0 mL HCl
   d. 100. mL HCl   e. 120. mL HCl

SUMMARY

1. For quick conversions between L and mL:
   - Milli- means 10^{-3}, so $\times\ 10^{-3}$ can be *substituted* for **milli-**
   - The number of milliliters is always 1000 times larger than the number of liters
   - When converting between mL and L, move the decimal three times

2. **In dilution**, the amount of *solvent* changes. This changes the volume of the solution but *not* the number of solute particles dissolved. The key relationship:

   moles solute concentrated (**C**) = moles solute diluted (**D**)

3. **Dilution prompt:** If a problem has two volumes and two concentrations for *one* substance formula, use a dilution method to solve.
   - If **V** or **M** changes by an easy multiple, multiply the other by 1/multiple.
   - If not easy multiples, solve with the dilution equation:

   $$V_C \times M_C = V_D \times M_D$$

4. **Coefficients** can be read in units of particles, or moles of particles. They can also be read as moles *per liter* of particles *if* all the particles are evenly distributed and measured in the same volume at the beginning and end of the reaction or process.

5. Assume (unless otherwise noted) that ions dissolved in water separate 100%.

6. **To find [ions in solution]** for ionic compounds that dissolve completely in water, solve by inspection by writing the **REC** steps.
   - *R*: The balanced *R*eaction equation
   - *E*: The *E*xtent of the reaction ("goes 100%"), and on the next line,
   - *C*: Each *C*oncentration based on the coefficient ratios

7. **Steps for solution stoichiometry:** If a limiting reactant is known and an amount of another substance is WANTED, write WDBB and then convert:

   given unit → *given* moles → WANTED moles → WANTED units

ANSWERS

Lesson 12.1

Practice A 1a. 35 mL $= 35 \times 10^{-3}$ L $= $ **0.035 L**   1b. 125 mL $= 125 \times 10^{-3}$ L $= $ **0.125 L**

2a. 2.500 L $= $ **2,500.** mL   2b. 0.77 L $= $ **770** mL   2c. 15 mL $= $ **0.015 L**

3. **Three** places to the **left**.

4a. 150 mL $= $ **0.15 L**   4b. 33 L $= $ **33,000 mL**   4c. 9.21 mL $= $ **0.00921 L**

4d. 0.45 L $= $ **450 mL**   4e. 0.833 mL $= $ **0.000833 L**   4f. 0.0655 L $= $ **65.5 mL**

Practice B 1. Because the volume has been *quadrupled*, the concentration must be cut to 1/4 the original amount. The final [KCl] $= $ **0.50 M**.

2. Because the final concentration has been cut to **1/5** of the original value, the volume must become *five times* higher. Final volume $= 5 \times 250.$ mL $= $ **1,250 mL**.

Practice C 1a. 250 mL = **0.25 L**   1b. 0.62 L = **620 mL**   1c. 3.5 mL = **0.0035 L**

2a. **0.300 M NaOH**   If the volume increases by a factor of 10, the solution becomes 1/10th as concentrated.

2b. WANTED: [NaOH] = ? M NaOH = ? $\dfrac{\text{mol NaOH}}{\text{L NaOH soln.}}$ M_D

DATA: 50.0 **mL** V_C (C, because lower of two volumes.)

3.00 mol NaOH = 1 L soln. M_C (In wording, goes with 50.0 mL, which is C.)

~~0.750 L~~ 750. mL soln. V_D (Higher of two volumes = diluted.)

(Work in milliliters or liters, but not both. In WANTED and DATA, two must be C and two D. To label this WANTED unit as C or D, label the DATA first.)

(Strategy: Write the needed equation. Label the DATA with the equation's symbols. Make the units consistent. Solve the equation for the missing symbol, *then* plug in numbers and units and solve. Check the answer *unit* to make sure it makes sense.)

SOLVE: Dilution: $V_C \times M_C = V_D \times M_D$ Solve the equation for the WANTED symbol.

$$? M_D = \frac{V_C \times M_C}{V_D} = \frac{50.0\ \text{mL} \times 3.00\ \text{M NaOH}}{750.\ \text{mL}} = \textbf{0.200 M NaOH diluted}$$

Check: For the variable (volume or molarity) known *both* before and after dilution, round one of the numbers to make it an easy multiple of the other. Then do the dilution estimate by mental arithmetic, write an answer, and compare it to the calculated answer. Try that step.

50 to 800 mL makes the volume 16 times larger, so the new concentration will be about 1/16 × original = 1/16 × 3 = about 3/15 = about 1/5 = 0.20 M. Check!

3. WANTED: ? mL KCl soln. V_D (Choose mL for consistent volume units.)

DATA: 20.0 mL KCl V_C (Goes with 2.5 M, which is C.)

2.5 mol KCl = 1 L KCl M_C (Higher of the two concentrations.)

0.24 mol KCl = 1 L KCl M_D

(Strategy: Labeling the DATA as C or D *first* helps in labeling the WANTED as C or D. Using the dilution equation, use the M abbreviation for mol/L.)

SOLVE: $V_C \times M_C = V_D \times M_D$ Solve the equation for the WANTED symbol.

$$\text{WANTED: } ? \textbf{ mL } V_D = \frac{V_C \times M_C}{M_D} = \frac{20.0\ \text{mL} \cdot 2.5\ \text{M}}{0.24\ \text{M}} = \textbf{210 mL KCl soln. D}$$

Check: From 2.5 M **C** to 0.24 M **D** is *about* 1/10th. The **D** volume must increase the original **C** volume by *about* 10 times. 10 × 20 mL = 200 mL. Pretty close.

4. WANTED: ? **mL** NaCl soln. V_C (Given a choice of a WANTED unit, make it consistent with the DATA.)

DATA: 250 mL soln. V_D (D because it is paired with 0.65 M, which is D below.)

0.65 mol NaCl = 1 L M_D (Lower concentration.)

2.00 mol NaCl = 1 L M_C (Higher concentration.)

SOLVE: $V_C \times M_C = V_D \times M_D$

$$? \text{ mL NaCl soln.} = V_C = \frac{V_D \times M_D}{M_C} = \frac{250\ \text{mL} \cdot 0.65\ \text{M}}{2.00\ \text{M}} = \textbf{81 mL NaCl soln.}$$

Check: If the 0.65 M is rounded to 0.50 M, that number must be quadrupled to reach 2.00 M, so the volume must be cut to about 1/4 of the original 250 mL = about 60 mL, in the ballpark of the answer.

5. WANTED: [HCl] = ? M HCl = ? $\dfrac{\text{mol HCl}}{\text{L HCl soln.}}$ M_C

DATA: 6.2 mL HCl soln. V_C (Before dilution, soln. is concentrated.)

0.15 mol HCl = 1 L M_D (mol · L^{-1} = mol / L = *ratio* unit data.)

~~0.250 L soln.~~ V_D = **250.** *mL* HCl soln.   (mL is always 1000 × liters.)

(To use an equation, make the units consistent: convert both volumes to liters *or* to milliliters.)

(Strategy: Labeling the DATA as C and D first can help label the WANTED amount as C or D.)

SOLVE: In dilution: $V_C \times M_C = V_D \times M_D$

$$[HCl] = ? \, M_C = \frac{V_D \times M_D}{V_C} = \frac{250 \text{ mL} \cdot 0.15 \text{ M}}{6.2 \text{ mL}} = \textbf{6.0 M HCl}$$

Check: Increasing the volume from 6 mL to 250 mL is about a $40\times$ increase, so the new M_C volume is the original multiplied by *1/40*. Because 1/40th of 6 M is about 0.15 M, the answer checks.

Lesson 12.2

Practice A   First, balance: $1 \, Na_2SO_4(s) \qquad \rightarrow \qquad 2 \, Na^+(aq) \quad + \quad 1 \, SO_4^{2-}(aq)$

$\qquad\qquad\qquad\qquad\qquad\qquad\qquad \wedge \qquad\qquad\qquad\qquad\quad \wedge \qquad\qquad\qquad \wedge$

Then, based on coefficients: 0.50 moles *used up* →   **1.0 mol**   +   **0.50 mol** *formed*

After the reaction, there will be (1) **1.0 mol Na$^+$** ions, (2) **0.50 mol SO$_4^{2-}$** ions, (3) **no Na$_2$SO$_4$ particles**. The un-separated Na$_2$SO$_4$ formula units are all used up in the reaction. (4) **1.5 total** moles of ions.

Practice B   1. Radium nitrate = $1 \, Ra^{2+}$ + $2 \, NO_3^-$ = $1 \, Ra(NO_3)_2$

*R*eaction and *E*xtent: $1 \, Ra(NO_3)_2 \quad \rightarrow 1 \, Ra^{2+} \quad + \quad 2 \, NO_3^- \qquad$ (Goes 100%.)

$\qquad\qquad\qquad\qquad\qquad\qquad \wedge \qquad\qquad\quad \wedge \qquad\qquad \wedge$

*C*oncentrations: 0.30 M (→ **0 M)** → 0.30 M + **0.60 M** formed

2. *R*eaction and *E*xtent: $1 \, Na_2CO_3 \quad \rightarrow 2 \, Na^+ \quad + \quad 1 \, CO_3^{2-} \qquad$ (Goes 100%.)

$\qquad\qquad\qquad\qquad\qquad\qquad \wedge \qquad\qquad \wedge \qquad\qquad \wedge$

*C*oncentrations: 0.60 M (→ **0 M)** → **1.2 M** + **0.60 M**

3. *R*eaction and *E*xtent: $1 \, K_3PO_4(aq) \quad \rightarrow \quad 3 \, K^+ \quad + \quad 1 \, PO_4^{3-} \qquad$ (Goes 100%.)

$\qquad\qquad\qquad\qquad\qquad\qquad\qquad \wedge \qquad\qquad \wedge \qquad\qquad \wedge$

*C*oncentrations: **0.15 M** (→ **0 M)** → 0.45 M + **0.15 M** formed

  a. Based on the coefficients, 3 K$^+$ are formed for every **one** PO$_4^{3-}$. **[PO$_4^{3-}$] = 0.15 M**

  b. [K$_3$PO$_4$]$_{as \, mixed}$? The ratio of K$_3$PO$_4$ to K$^+$ is one to three, so **mol/L K$_3$PO$_4$** used up = **0.15 M**. That is the mol/L of K$_3$PO$_4$ *mixed*. However, there are no K$_3$PO$_4$ particles in the solution.

4. *R*eaction and *E*xtent: $1 \, Al_2(SO_4)_3 \quad \rightarrow \quad 2 \, Al^{3+} + 3 \, SO_4^{2-} \qquad$ (Goes 100%.)

$\qquad\qquad\qquad\qquad\qquad\qquad\qquad \wedge \qquad\qquad\quad \wedge \qquad\qquad \wedge$

*C*oncentrations: **0.012 M** (→ **0 M)** → **0.024 M** + 0.036 M formed

  a. **[Al^{3+}] = 0.024 M**

  b. [Al$_2$(SO$_4$)$_3$]$_{as \, mixed}$? One-third of the given [SO$_4^{2-}$] = 1/3 (0.036 M) = **0.012 mol/L Al$_2$(SO$_4$)$_3$** is used to make the solution, and this may be used as a label for the solution, but there are no particles of **Al$_2$(SO$_4$)$_3$** in the solution.

5a. WANTED: $\dfrac{\text{mol } K_3PO_4}{\text{L soln.}}$ (Write the *unit* WANTED as mixed.)

   DATA: 10.62 g K$_3$PO$_4$ = 0.250 L soln.   (Equivalent: two measures of same solution.)

   212.3 g K$_3$PO$_4$ = 1 mol K$_3$PO$_4$ (g prompt.)

   SOLVE: (Want a ratio? Start with a ratio.)

$$? = \frac{\text{mol } K_3PO_4}{\text{L soln.}} = \frac{10.62 \, g \, K_3PO_4}{0.250 \, \text{L soln.}} \cdot \frac{1 \text{ mol } K_3PO_4}{212.3 \, g \, K_3PO_4} = \textbf{0.200} \, \frac{\textbf{mol } K_3PO_4}{\textbf{L soln.}}$$

   The [K$_3$PO$_4$] is 0.200 M *as mixed*, which answers the question.

5b. *R*eaction and *E*xtent: $1 \, K_3PO_4 \qquad \rightarrow \quad 3 \, K^+ \quad + \quad 1 \, PO_4^{3-} \qquad$ (Goes 100%.)

$\qquad\qquad\qquad\qquad\qquad\qquad\qquad \wedge \qquad\qquad \wedge \qquad\qquad \wedge$

*C*oncentrations: 0.200 M (→ 0 M) → **0.600 M** + 0.200 M

Lesson 12.3

Practice 1. WANTED: ? mL KOH solution   (Single unit WANTED.)

  DATA: 0.300 g KHPht (the acid)   (Single unit *given*.)

  204.2 g KHPht $=$ 1 mol KHPht   (g prompt.)

  0.0750 mol KOH $=$ 1 L KOH soln.   (M prompt, a ratio.)

Balance: See problem.

Bridge: **1** mol KHPht $=$ **1** mol KOH

  Because we have *two* substances, DATA and conversions must be labeled with substance formulas.
  Finish the remaining steps and then check below.

  SOLVE: Because a single unit is WANTED, start with a single unit, and solve in one string of conversions.

$$? \text{ mL KOH} = 0.300 \text{ g KHPht} \cdot \frac{1 \text{ mol KHPht}}{204.2 \text{ g KHPht}} \cdot \frac{1 \text{ mol KOH}}{1 \text{ mol KHPht}} \cdot \frac{1 \text{ L KOH}}{0.0750 \text{ mol KOH}} \cdot \frac{1 \text{ mL}}{10^{-3} \text{ L}}$$

$$= \textbf{19.6 mL KOH soln.}$$

  2. WANTED: ? mL PbO_2 soln.   (Want a *single* unit.)

  DATA: 0.268 g $Mn(NO_3)_2$ (One single unit in DATA.)

  179 g $Mn(NO_3)_2$ $=$ 1 mol $Mn(NO_3)_2$

  0.0250 mol PbO_2 $=$ 1 L PbO_2 soln.

(For reaction calculations with measurements of *two* substances, use stoichiometry steps.)

Balance: (See problem.)

Bridge: **5 mol** PbO_2 $=$ **2 mol** $Mn(NO_3)_2$ (Relate the WANTED and *given* moles.)

  SOLVE: (WANT a single unit? Start with a single-unit.)

$$? \textbf{ mL } PbO_2 = 0.268 \text{ g } Mn(NO_3)_2 \cdot \frac{1 \textbf{ mol } Mg(NO_3)_2}{179 \text{ g } Mn(NO_3)_2} \cdot \frac{5 \textbf{ mol } PbO_2}{2 \textbf{ mol } Mn(NO_3)_2} \cdot \frac{1 \text{ L } \textbf{PbO}_2}{0.0250 \text{ mol } PbO_2} \cdot \frac{1 \text{ mL}}{10^{-3} \text{ L}}$$

$$= \textbf{150. mL } PbO_2 \textbf{ soln.}$$

  3. WANTED: ? *g Ag*

  DATA: 1.50 L ***HNO₃*** soln.

  107.9 *g* Ag $=$ 1 mol Ag   (Grams prompt.)

  6.0 mol HNO_3 $=$ 1 L HNO_3 soln.   (M prompt.)

(Note that by listing the WANTED unit and formula first, it prompts for a conversion needed to solve.)

Balance: See problem.

Bridge: 1 mol Ag $=$ 2 mol HNO_3

  SOLVE: (WANT a single unit? Start with the single-unit amount in the DATA.)

$$? \text{ g Ag} = 1.50 \text{ L } HNO_3 \cdot \frac{6.0 \text{ mol } HNO_3}{1 \text{ L } HNO_3} \cdot \frac{1 \text{ mol Ag}}{2 \text{ mol } HNO_3} \cdot \frac{107.9 \text{ g Ag}}{1 \text{ mol Ag}} = \textbf{490 g Ag}$$

Lesson 12.4

Some *partial* solutions are provided below. Your work on calculations should include WANTED, DATA, and SOLVE.

1a. **2.0 L $=$ 67.7 fl. oz.**

1b. **4.5 mg $=$ 1 fl. oz.**

1c. **a. 38 mg sweetener**

  2. **b. 420 mL** (dilution)

  3. **d. 0.40 M** $[(NH_4)_2SO_4 \rightarrow 2 NH_4^+ + SO_4^{2-}]$

  4. **b. 0.150 M K^+** (0.0600 mol K^+ / 0.400. L soln.)

  5. **c. 80.0 mL HCl**

13

Ionic Equations and Precipitates

Prerequisites

During Chapter 13, if you have any problems translating between the names, solid formulas, and separated-ion formulas for ionic compounds, review your ion name and formula flashcards from Lessons 7.2 and 7.3.

Lesson 13.1   Predicting Solubility for Ionic Compounds

Solubility Terminology

All *ionic* compounds dissolve to *some* extent in water. Some dissolve only slightly, some dissolve to a substantial extent, and some ionic compounds have borderline solubility at room temperature. In addition, the solubility of ionic compounds is temperature dependent: some dissolve to a greater extent in warmer solvents and some dissolve less.

Those messy realities aside, *most* ion combinations can generally be classified as **soluble** or **insoluble** in water. The following definitions are commonly accepted:

- If 0.10 moles or more of a solid dissolve per liter of solution at room temperature, the solid is termed **soluble**.
- If the solid dissolves less than 0.10 moles per liter, it is considered either **slightly soluble** or **insoluble**.

A Solubility Scheme

Most introductory chemistry courses ask that you memorize a set of solubility rules that will allow you to predict the solubility of many ion combinations. The solubility scheme that follows is limited, but it will cover many of the ions customarily encountered in first-year chemistry.

The scheme below is hierarchical: higher rules take *precedence* over those beneath them. To predict the solubility of ion combinations, apply the rules *in order* from the top.

If only one ion of the two in a compound is in this table, you may presume that the compound will follow the rule for the ion that is in the table. (This will not always be accurate, but is a best guess. With solubility, there are exceptions.)

Commit this chart to memory:

	Positive Ions	**Negative** Ions	**Solubility**
1.	(alkali metals)$^+$, NH_4^+	NO_3^-, CH_3COO^-, ClO_3^-, ClO_4^- (nitrate, acetate, chlorate, perchlorate)	**Soluble**
2.	Pb^{2+}, Hg_2^{2+}, Ag^+	CO_3^{2-}, PO_4^{3-}, S^{2-}, CrO_4^{2-}	**Insoluble**
3.		Cl^-, Br^-, I^- (except when with Cu^+)	**Soluble**

Exceptions include: Column 2 sulfides and aluminum sulfide decompose in water; $AgCH_3COO$ is moderately soluble.

Devote special attention to rule 1. In a hierarchical table, the higher the rule, the more likely it is to be used.

Using the Scheme to Make Predictions

When using any solubility scheme, if you are unsure of which *ions* are in a compound, you should write out the *separated-ion* formula that *shows* the ion charges. Atoms that have two possible ion charges, such as Cu^+ and Cu^{2+}, can have differing solubilities.

 TRY IT

(See "How to Use These Lessons," point 1, p. xv.)

Q. For the questions below, use the solubility scheme. Write your answer and your reasoning.

  a. Is $Ba(NO_3)_2$ soluble?

  b. Is $PbCl_2$ soluble or insoluble?

  c. Is $Hg(NO_3)_2$ soluble or insoluble?

STOP

Answers:

  a. Yes. $Ba(NO_3)_2 \rightarrow Ba^{2+} + 2\ \mathbf{NO_3^-}$. All nitrates are soluble by rule 1.

  b. $PbCl_2 \rightarrow \mathbf{Pb^{2+}} + 2\ Cl^-$. Compounds containing Pb^{2+} ion by rule 2 are **in**soluble. Compounds containing chloride ion by rule 3 are soluble, but rule 2 takes precedence. $PbCl_2$ is **insoluble**.

  c. $Hg(NO_3)_2 \rightarrow Hg^{2+} + 2\ NO_3^-$. The above solubility scheme makes a specific prediction for mercury(I) ion, but not for Hg^{2+}, the mercury(II) ion. However, based on rule 1 that all nitrates are soluble, predict **soluble**. If only one ion in a pair is in the table, base your prediction on the rule for that ion.

PRACTICE

Practice writing the solubility chart until you can write the chart from memory, then use your "written from memory" chart to solve the problems below.

  1. Label each ion combination as soluble or insoluble and state a reason for your prediction. Check your answers after each one or two parts.

    a. $K^+ + S^{2-}$

    b. $Sr^{2+} + Cl^-$

    c. $Ca^{2+} + CO_3^{2-}$

    d. $Ag^+ + CrO_4^{2-}$

  2. Write the separated ions for these combinations, then label the ion combination as soluble or insoluble and state a reason for your prediction.

    Do every other part, and check your answers frequently. Do more if you need more practice. If you have trouble writing the separated ions, redo Lessons 7.2 and 7.3.

(continued)

a. Lead(II) bromide $\rightarrow Pb^{2+} + 2\ Br^-$; **insoluble**—rule 2 for Pb^{2+} (example)

b. Barium carbonate $\rightarrow$

c. Sodium hydroxide $\rightarrow$

d. $SrBr_2 \rightarrow$

e. Silver nitrate $\rightarrow$

f. Ammonium hydroxide $\rightarrow$

g. $Fe_3(PO_4)_2 \rightarrow$

h. $Pb(CH_3COO)_2 \rightarrow$

i. $Ni(ClO_3)_2 \rightarrow$

j. $RbBr \rightarrow$

k. $Fe_2S_3 \rightarrow$

3. Name the ions that in combination with Pb^{2+} form soluble compounds.

4. Are all compounds containing nitrate ions soluble?

5. Are all compounds containing phosphate ions insoluble?

Lesson 13.2 Total and Net Ionic Equations

Mixing Ions

When solutions containing different ions are mixed, chemical reactions can occur. One type of reaction is **precipitation**. When aqueous solutions of different soluble ionic compounds are mixed, the ions can *trade partners*: new combinations of positive and negative ions are possible. If a new combination is possible that is insoluble in water, it will form a **precipitate**: a cloud of very small but *solid* particles that will slowly fall to the bottom of the reaction vessel.

Example: When a test tube containing a 0.2 M of dissolved sodium chloride (table salt) is added to a test tube containing 0.2 M silver nitrate, a white cloud forms immediately as the solutions are mixed. Over several minutes, the particles from the cloud settle, leaving a clear solution with a layer of solid on the bottom.

The precipitate is silver chloride (AgCl). This reaction can be written as

$$NaCl(aq) + AgNO_3(aq) \rightarrow AgCl(s) + NaNO_3(aq) \qquad (13.1)$$

However, for understanding, a better format to see what is happening in the reaction is the **total ionic equation**:

$$Na^+(aq) + Cl^-(aq) + Ag^+(aq) + NO_3^-(aq) \rightarrow$$
$$AgCl(s) + Na^+(aq) + NO_3^-(aq) \qquad (13.2)$$

The (*aq*) means that a particle is in the aqueous phase, which means dissolved in water. The (*s*) after AgCl abbreviates *solid*: the state of a precipitate.

Note the difference in how the equations above are written. The total ionic equation (13.2) shows that

- Before mixing, in both solutions, all of the particles are dissolved, separated ions
- To form the solid, a positive silver ion combines with a negative chloride ion

An important general rule is this:

To understand the reactions of ionic compounds that are dissolved in aqueous solutions, write the reaction equation with those compounds in their *separated ions* format.

In the *total* ionic equation (13.2), it is clear that the sodium and nitrate ions do not change in the reaction: they are the *same* dissolved and separated ions before and after the reaction. Ions that are present during a reaction but do not change their formula and state are termed **spectator ions**. In *total* ionic equations, spectator ions are *included* on each side and they have the *same* formula on each side. Including the spectators helps us to see all of the reactants and products.

However, as in mathematical equations, terms that are the same on each side of an equation can be cancelled. Canceling the spectators gives the **net ionic equation**.

 TRY IT

Q. In the total ionic equation above, *cancel* the spectators, then rewrite the equation including only the particle formulas that did not cancel.

STOP **Answer:**

Net ionic equation:

$$Ag^+(aq) + Cl^-(aq) \rightarrow AgCl(s) \tag{13.3}$$

A net ionic equation shows only the particles that *change* their state or formula (those that react) during the reaction. In a *net* ionic equation, the spectator ions are left out.

PRACTICE   A

Answer question 1. If you need more practice, do question 2.

In these lessons, for reactions in aqueous solutions, if an equation shows the *state* (or *phase*) after some particle formulas but not others, assume the omitted phases are (*aq*).

For these total ionic equations, circle the precipitate, cross out the spectators, then write the *net* ionic equation.

1. $Pb^{2+}(aq) + 2\,NO_3^-(aq) + Cu^{2+}(aq) + 2\,Cl^-(aq) \rightarrow Cu^{2+}(aq) + 2\,NO_3^-(aq) + PbCl_2(s)$

2. $6\,Na^+ + 2\,PO_4^{3-} + 3\,Mg^{2+} + 3\,SO_4^{2-} \rightarrow Mg_3(PO_4)_2(s) + 6\,Na^+ + 3\,SO_4^{2-}$

Balancing Total Ionic Equations

For precipitation reactions, you will often be asked to balance a total ionic equation. To balance properly, you must first balance each of the reactant and product formulas for *charge*, then balance again to account for *ratios of reaction*.

⯈ TRY IT

Q. In the following equation, the brackets show the original ion combinations in the reactants and the new combinations in the products.

$$[\quad Ca^{2+} + \quad NO_3^-] + \quad [\quad K^+ + \quad PO_4^{3-}] \rightarrow$$
$$1\, Ca_3(PO_4)_2(s) + \quad [\quad K^+ + \quad NO_3^-]$$

To balance, first add the lowest whole-number coefficients *inside* the brackets so that the charges are balanced for each substance formula. Do that step now.

STOP

$$[\mathbf{1}\, Ca^{2+} + \mathbf{2}\, NO_3^-] + \quad [\mathbf{3}\, K^+ + \mathbf{1}\, PO_4^{3-}] \rightarrow$$
$$1\, Ca_3(PO_4)_2(s) + \quad [\mathbf{1}\, K^+ + \mathbf{1}\, NO_3^-]$$

Now add coefficients in front of the *brackets* so that *all* of the atoms balance.

STOP

To form the precipitate, three calcium ions and two phosphate ions are needed, so

$$\mathbf{3}\, [\mathbf{1}\, Ca^{2+} + 2\, NO_3^-] + \mathbf{2}\, [3\, K^+ + 1\, PO_4^{3-}] \rightarrow$$
$$1\, Ca_3(PO_4)_2(s) + \mathbf{6}\, [1\, K^+ + 1\, NO_3^-]$$

Now rewrite the equation, taking out the brackets in the same way you would take out parentheses in algebra: multiply each coefficient inside by the coefficient outside. The result is the *total* ionic equation.

When done, check that your equation is balanced for atoms and charge.

Start with 3 Ca^{2+} +

STOP

$$3\, Ca^{2+} + 6\, NO_3^- + 6\, K^+ + 2\, PO_4^{3-} \rightarrow 1\, Ca_3(PO_4)_2(s) + 6\, K^+ + 6\, NO_3^-$$

Check: 3 Ca, 6 N, 6 K, 2 P, and 26 O on each side; zero net charge on each side; balanced.

PRACTICE **B**

Put a ✓ by and do problems 1 and 2. Save problem 3 for your next practice session.
   On problems 1, 2, and 3, first balance inside the brackets, then in front of the brackets.
Then rewrite the equation without the brackets as the *total ionic* equation.

1. $[\quad K^+ + \quad SO_4^{2-}] + \quad [\quad Sr^{2+} + \quad NO_3^-] \rightarrow$
    $$1\, SrSO_4(s) + \quad [\quad K^+ + \quad NO_3^-]$$

2. $[\quad Fe^{2+} + \quad Br^-] + \quad [\quad Na^+ + \quad PO_4^{3-}] \rightarrow$
    $$[\quad Na^+ + \quad Br^-] + 1\, Fe_3(PO_4)_2(s)$$

3. $[\quad Ag^+ + \quad NO_3^-] + \quad [\quad K^+ + \quad CrO_4^{2-}] \rightarrow$
    $$[\quad K^+ + \quad NO_3^-] + 1\, Ag_2CrO_4(s)$$

Converting Ionic Solid Formats to Total and Net Ionic Equations

In equations for reactions that form precipitates, ionic compounds that are dissolved in water are often written as ionic *solid* formulas in the (*aq*) phase. Including the states or phases does not affect the steps taken to balance the equation.

 TRY IT

Q. Balance:

$$BaCl_2(aq) + \qquad K_2SO_4(aq) \rightarrow \qquad BaSO_4(s) + \qquad KCl(aq)$$

Answer:

$$\textbf{1}\,BaCl_2(aq) + \textbf{1}\,K_2SO_4(aq) \rightarrow \textbf{1}\,BaSO_4(s) + \textbf{2}\,KCl(aq)$$

During balancing, writing coefficients that are **1** is helpful in keeping track of which particles have been balanced. In a final balanced equation, however, coefficients of 1 may be omitted as understood.

It is relatively quick to write and to balance equations in which ionic compounds are represented by ionic solid formulas, as in the problem above. However, for a reaction of dissolved ionic compounds, those solid formulas are a simplification. When an ionic solid is dissolved in water, its ions are dissociated (separated).

 TRY IT

Q. To show the ions that actually exist in the solution, rewrite the balanced equation above. In place of each ionic solid formula that is (*aq*), write its *separated*-ion formulas inside *brackets*. For the solid, write the ionic solid formula as shown, followed by (*s*). In front of the brackets and solids, put the coefficients that you found above by balancing.

The phase of an ion dissolved in water is (*aq*), but in these lessons, in the case of reactions that are carried out in aqueous solutions, if no state or phase is shown, assume (*aq*).

Answer:

Start with $\textbf{1}\,[\textbf{1}\,Ba^{2+} + \textbf{2}\,Cl^-] + \ldots$

$$\textbf{1}\,[\textbf{1}\,Ba^{2+} + \textbf{2}\,Cl^-] + \textbf{1}\,[\textbf{2}\,K^+ + \textbf{1}\,SO_4^{2-}] \rightarrow \textbf{1}\,BaSO_4(s) + \textbf{2}\,[\textbf{1}\,K^+ + \textbf{1}\,Cl^-]$$

Now rewrite the equation taking out the brackets. Label this the *total* ionic equation. Check that your equation is balanced for atoms and charge.

Total ionic equation:

$$\textbf{1}\,Ba^{2+} + \textbf{2}\,Cl^- + \textbf{2}\,K^+ + \textbf{1}\,SO_4^{2-} \rightarrow \textbf{1}\,BaSO_4(s) + \textbf{2}\,K^+ + \textbf{2}\,Cl^-$$

Now write the *net* ionic equation.

To find the net ionic equation, cross out the spectators in the total ionic equation. *Net* ionic equation:

$$1\,Ba^{2+}(aq) + 1\,SO_4^{2-}(aq) \rightarrow 1\,BaSO_4(s)$$

Do problems 1, 2, and 3 for part 1a. If you need more practice, do the same for part 1b.

1. Balance these equations.

    a. $Fe(NO_3)_3(aq) +$ $NaOH(aq) \rightarrow$ $Fe(OH)_3(s) +$ $NaNO_3(aq)$

    b. $Ca(NO_3)_2(aq) +$ $K_3PO_4(aq) \rightarrow$ $KNO_3(aq) +$ $Ca_3(PO_4)_2(s)$

2. Rewrite the balanced equations for problems 1a and 1b using *separated*-ion formulas in brackets *if* the ionic solid formulas are (*aq*). For solids, write the ionic *solid* formula as shown. Then write the *total* ionic equation by taking out the brackets.

3. Write the *net* ionic equations for each reaction.

Balancing Equations That Omit Some Spectator Ions

Until this point, most of the equations we have encountered have contained either

- Formulas for neutral compounds, or
- Separated ions with a total charge of *zero* on each side of the equation

For reactions that involve ions, equations may be written that omit *all* of the spectator ions (as in net ionic equations) or omit *some* of the spectator ions.

Example: The reaction of silver nitrate and magnesium chloride solutions can be represented by the total ionic equation

$$2\,Ag^+ + 2\,NO_3^- + Mg^{2+} + 2\,Cl^- \rightarrow 2\,AgCl(s) + Mg^{2+} + 2\,NO_3^- \quad (13.4)$$

However, it is common in ionic reactions to leave out *some* of the spectators. If we leave out the magnesium ions above, one way to write the reaction is

$$AgNO_3(aq) + Cl^-(aq) \rightarrow AgCl(s) + NO_3^-(aq) \quad (13.5)$$

In this format, the equation emphasizes that a silver nitrate solution will form a precipitate when mixed with *any* solution that contains chloride ions.

Note that *partial* ionic equation 13.5, unlike total ionic equation 13.4, does not have a zero charge on each side. However, it does have a *balanced* charge: the total charge on each side is negative one. That's an important rule that we will state in two ways:

> Equations are considered to be balanced if they have the same number and kind of atoms *and* the *same net charge* on each side. The net charge is *not* required to be zero.
>
>   An equation is considered to be *balanced* even if some of the ions that must be present to balance charge have been left out.

In the *partial* ionic equations that are frequently written, the atom counts and total charge must simply be the *same* on each side.

Leaving out the Spectators

Writing *net* or *partial* ionic equations (leaving out all or some spectators) results in an equation that is quicker to write than the total ionic equation. Without spectator ions, the equation focuses on the most important particles: those that change in the reaction.

In a similar fashion, in the laboratory a test solution may be labeled Ag^+ or OH^-. However, when a label shows a *single* ion, spectator ions *must* also be present in the substance or solution so that the net charge of the substance is zero. "Leaving out the spectators" on container labels and in equations is a *shortcut* that emphasizes the ion that is likely to react.

When these shortcuts are taken, the presence of other ions that balance the charge is "understood." A fundamental principle in science is that

> In all stable matter containing charged particles, the charges must add up to zero.

Positive or negative charges can accumulate temporarily on particles, but processes will have a strong tendency to occur that rebalance the charges. A small spark of static electricity is one example of this charge-equalizing process. Bolts of lightning are another.

PRACTICE   D

Balance these partial ionic equations. Do every other number now, then complete the rest in your next practice session.

1. $Co(NO_3)_2 + \quad OH^- \rightarrow \quad Co(OH)_2 + \quad NO_3^-$

2. $AgNO_3 + \quad S^{2-} \rightarrow \quad Ag_2S + \quad NO_3^-$

3. $CO_3^{2-} + \quad HCl \rightarrow \quad H_2CO_3 + \quad Cl^-$

4. $Al^{3+} + \quad NO + \quad H_2O \rightarrow \quad Al + \quad NO_3^- + \quad H^+$

Lesson 13.3   Precipitation

Predicting Precipitation

There are many different types of chemical reactions, but a factor that many reactions have in common is that when two different reactants are *mixed*, ions in the original reactants can attract new partners that have opposite charges. In some cases this attraction results in *reactions* to form new compounds, but in other cases it does not.

When solutions of ionic compounds are mixed, formation of a precipitate is one reaction that can occur. Whether precipitation occurs depends on the solubility of the *new* combinations that are possible.

In typical precipitation experiments, we mix approximately equal volumes of 0.2 M aqueous solutions of two different *soluble* ionic compounds. The result is one solution in which, because of the dilution that occurs with mixing, all of the initial compounds before any reaction would have a 0.1 M concentration.

The mixing makes new ion combinations possible. Knowing the formulas for the two reactants, we can predict formulas for the new possible products. Knowing those formulas, because our solubility rules are based on 0.1 M concentrations, we can predict whether new products covered by the rules will be *soluble* or *insoluble*.

Knowing the product solubilities, we can predict whether precipitation will occur by applying this rule:

> When solutions of two soluble ionic compounds are *mixed*, if a new combination is *possible* that is **in**soluble, it **will** precipitate.

To begin investigating reactions that form precipitates, we begin by asking four questions:

1. If two 0.2 M solutions, each containing one known and dissolved ionic compound, are mixed in roughly equal volumes, will a precipitate form?

2. If a precipitate forms, what is its formula?

3. What is the *total* ionic equation for the possible precipitation reaction between the two solutions?

4. If a precipitate forms, what is the *net* ionic equation for the reaction?

To answer these questions, apply the steps below to the following problem.

TRY IT

Q. Aqueous solutions of 0.2 M $Ca(NO_3)_2$ and Na_2CO_3 are mixed. Will a precipitate form? If so, what is its formula?

Predicting Precipitate Formulas: Chart Steps and Rules

1. In row 2 below, to the left of the arrow, write the ionic *solid* formulas above for the two reactants. In row 3, inside the parentheses, write balanced *separated* ion formulas for each of those reactants. Based on row 3, in row 1 write the *names* of the two reactants.

1	Names	_____ _____ + _____ _____ →
2	Balanced solid formulas	+ →
3	Balanced separated formulas	(+) + (+) →
4	Solubility	

Complete that step, then check your answers below.

Your chart should match the one below. If you need help with this step, review Lesson 7.3.

1	Names	calcium nitrate + sodium carbonate →
2	Balanced solid formulas	$Ca(NO_3)_2$ + Na_2CO_3 →
3	Balanced separated formulas	$(1\ Ca^{2+} + 2\ NO_3^-)$ + $(2\ Na^+ + 1\ CO_3^{2-})$ →
4	Solubility	

2. When the two solutions are mixed, two *new* combinations are possible. Each positive ion can attract either its original negative partner **or** the negative ion in the other reactant.

   In the underlined blanks for rows 1 and 3 of the chart below, write the new possible *products*. One way to do this is to "trade places:" in row 1, switch the places of the *negative* ion *names*, and in row 3, flip the *negative* ion *formulas*. In row 3, leave out the coefficients for now.

1	Names	(trade places) calcium nitrate + sodium carbonate → calcium _____ + sodium _____
2	Balanced solid formulas	$Ca(NO_3)_2$ + Na_2CO_3 →
3	Balanced separated formulas	$(1\ Ca^{2+} + 2\ NO_3^-)$ + $(2\ Na^+ + 1\ CO_3^{2-})$ → (Ca^{2+} +) + (Na^+ +)
4	Solubility	

Your paper should look like this:

1	Names	calcium nitrate + sodium carbonate → calcium **carbonate** + sodium **nitrate**
2	Balanced solid formulas	$Ca(NO_3)_2$ + Na_2CO_3 →
3	Balanced separated formulas	$(1\ Ca^{2+} + 2\ NO_3^-)$ + $(2\ Na^+ + 1\ CO_3^{2-})$ → (Ca^{2+} + **CO_3^{2-}**) + (Na^+ + **NO_3^-**)
4	Solubility	

3. In row 3, for the two *products, inside* the parentheses, add lowest whole number coefficients that balance the charges in the separated ion formulas.

4. Based on row 3, in row 2 write ionic solid formulas for the two *products,* then *balance* the equation in row 2.

Do those steps, then check below.

1	Names	calcium nitrate + sodium carbonate → calcium carbonate + sodium nitrate
2	Balanced solid formulas	$1\,Ca(NO_3)_2 + 1\,Na_2CO_3 \rightarrow 1\,CaCO_3 + 2\,NaNO_3$
3	Balanced separated formulas	$(1\,Ca^{2+} + 2\,NO_3^-) + \quad (2\,Na^+ + 1\,CO_3^{2-}) \rightarrow$ $(1\,Ca^{2+} + 1\,CO_3^{2-}) + \quad (1\,Na^+ + 1\,NO_3^-)$
4	Solubility	_____ + _____ → _____ + _____

5. In *front* of the parentheses in row 3, write the coefficients that you calculated for each formula in row 2.

6. In this type of experiment, each initial reactant solution always contains a *soluble* ionic compound. Some solutions may have a color due to the dissolved ions, but the *reactants* contain no solids. The rule is:

   In this type of precipitation experiment, the two *reactants* are always *soluble*.

   In row 4 above, in the first two blanks, write *soluble*.

7. Based on solubility rules, in row 4, label each *product* as *soluble* or *insoluble*.

8. Circle the combinations that are actually present in the solution using these steps.
   - Circle both *reactants* in row 3, because each reactant must consist of soluble, *separated* ions.
   - If a *product* is *insoluble*, circle it in row **2**, where it has a *solid* formula. If a possible product is insoluble, it will be a *solid* precipitate.
   - If a product is soluble, circle it in row **3**, because if the new combination is soluble, its ions will remain *separated* in water.

9. Will there be a precipitate in this case?   If so, what is its solid formula?

Do those steps, then check your work below.

1	Names	calcium nitrate + sodium carbonate → calcium carbonate + sodium nitrate
2	Balanced solid formulas	$1\,Ca(NO_3)_2 + 1\,Na_2CO_3 \rightarrow \boxed{1\,CaCO_3} + 2\,NaNO_3$
3	Balanced separated formulas	$\boxed{1\,(1\,Ca^{2+} + 2\,NO_3^-)} + \boxed{1\,(2\,Na^+ + 1\,CO_3^{2-})} \rightarrow$ $1\,(1\,Ca^{2+} + 1\,CO_3^{2-}) + \boxed{2\,(1\,Na^+ + 1\,NO_3^-)}$
4	Solubility	soluble + soluble → **insoluble** + **soluble**

Our solubility scheme predicts calcium carbonate ($CaCO_3$) is **insoluble** by the rule for carbonates, so it *will* precipitate.

Sodium nitrate ($NaNO_3$) is also a possible new combination, but it is soluble, and soluble combinations do not precipitate. The sodium and nitrate ions will remain dissolved in the solution.

For a case in which ions are mixed but both of the new products are soluble, no precipitate will form, and no precipitation reaction will occur.

This chart process is lengthy the first few times, but the rules and steps will become intuitive quickly—with practice.

To Predict Products in Precipitation Reactions

1. Write a four-row chart and add these labels to the rows: *Names*, Balanced *solid* formulas, Balanced *separated* formulas, and Solubility.

2. Fill in the *reactants* in rows 1–3, then change partners in rows 1 and 3 to write the new possible *products*.

3. Complete rows 1–3, *balancing* atoms and charges as you go.

4. In row 4, write the solubility of each compound, then circle the formulas above that are actually present in the reactants and products. The rules are:
   * The two *reactants* are always soluble, separated ions.
   * If a new product is possible that is **in**soluble, it *will* precipitate.
   * If a new product is possible that is soluble, its ions will remain separated.

Let's apply the rules to one more problem.

━━━━━━━━━ ▶ **TRY IT**

Q. When roughly equal volumes of 0.2 M lead(II) nitrate, and 0.2 M potassium iodide solutions are mixed, will a precipitate form? If so, what is its formula?

Answer by completing the chart below.

1	Names	+	→			
2	Balanced solid formulas	+	→			
3	Balanced separated formulas	(	+	) + (	+	) →
4	Solubility					

Answer:
Part way, your chart should look like this:

1	Names	lead(II) nitrate + potassium iodide →
		lead(II) iodide + potassium nitrate
2	Balanced solid formulas	$1\,Pb(NO_3)_2 + 2\,KI \rightarrow 1\,PbI_2 + 2\,KNO_3$
3	Balanced separated formulas	$(1\,Pb^{2+} + 2\,NO_3^-) + \quad (1\,K^+ + 1\,I^-) \rightarrow$ $(1\,Pb^{2+} + 2\,I^-) + \quad (1\,K^+ + 1\,NO_3^-)$
4	Solubility	soluble + soluble → _____ + _____

If needed, adjust your work and finish.

1	Names	lead(II) nitrate + potassium iodide →
		lead(II) iodide + potassium nitrate
2	Balanced solid formulas	$1\,Pb(NO_3)_2 + 2\,KI \rightarrow \boxed{1\,PbI_2} + 2\,KNO_3$
3	Balanced separated formulas	$\boxed{1\,(1\,Pb^{2+} + 2\,NO_3^-)} + \boxed{2\,(1\,K^+ + 1\,I^-)} \rightarrow$ $1\,(1\,Pb^{2+} + 2\,I^-) + \boxed{2\,(1\,K^+ + 1\,NO_3^-)}$
4	Solubility	soluble + soluble → **insoluble + soluble**

The solid precipitate is PbI_2.

PRACTICE   **A**

1. Label these as soluble or insoluble.

   a. Silver carbonate   b. Ammonium chloride   c. $MgCO_3$ d. $PbBr_2$

   e. Na_2S f. $Pb(NO_3)_2$ g. $AgCl$

2. When 0.2 M $AgNO_3$ is mixed with 0.2 M Na_2CrO_4, a brick-red precipitate forms. Complete a four-row chart for this reaction.

Total and Net Ionic Equations

For any reactions in which the solubility of the products can be predicted, the four-row chart will answer our first two questions about precipitation:

1. If 0.2 M solutions of two soluble ionic compounds are mixed, will a precipitate form?

2. If a precipitate forms, what is its formula?

The chart can also answer our two remaining questions:

3. What is the *total* ionic equation for the possible reaction between the two solutions?

4. If a precipitate forms, what is the *net* ionic equation for the reaction?

TRY IT

Let's start with the answer to the first question in the section above, then add two rows to our table.

1	Names	calcium nitrate + sodium carbonate → calcium carbonate + sodium nitrate
2	Balanced solid formulas	$1\, Ca(NO_3)_2 + 1\, Na_2CO_3 \rightarrow$ ⓵ $CaCO_3$ $+ 2\, NaNO_3$
3	Balanced separated formulas	⟮**1** $(1\, Ca^{2+} + 2\, NO_3^-)$⟯ $+$ ⟮**1** $(2\, Na^+ + 1\, CO_3^{2-})$⟯ $\rightarrow$ **1** $(1\, Ca^{2+} + 1\, CO_3^{2-}) +$ ⟮**2** $(1\, Na^+ + 1\, NO_3^-)$⟯
4	Solubility	soluble + soluble → **insoluble** + **soluble**
5	Total ionic equation	
6	Net ionic equation	

Q. In the table above, complete the following steps.

1. In row 5, for the *total* ionic equation, write all of the ions involved in the reaction: as a separated ion formula for soluble compounds, but as a solid formula for precipitates. To do so, use the four *circled* compounds in rows 2 and 3. Start by writing in row 5 the *separated* ions for the two circled *reactants*, but take *out* the parentheses. The state of separated ions is (*aq*), but in these equations you may leave out the (*aq*) as understood.

   In row 5, for the products, place an (*s*) for solid after a precipitate formula (when a solid is formed by precipitation, you may also see it labeled (*ppt.*) for precipitate.) For the soluble products, write separated ions.

2. In row 6, for the *net* ionic equation, rewrite row 5, including only the particles that changed: the ions on the left that react, and the solid precipitate on the right, all with their coefficients.

3. Check your work: The equations in both row 5 and row 6 must be balanced for atoms and charge. If they are not, check your coefficient calculations in the rows above.

STOP   **Answers:**

5	Total ionic equation	$1\, Ca^{2+} + 2\, NO_3^- + 2\, Na^+ + 1\, CO_3^{2-} \rightarrow$ $1\, CaCO_3(s) + 2\, Na^+ + 2\, NO_3^-$
6	Net ionic equation	$1\, Ca^{2+}(aq) + 1\, CO_3^{2-}(aq) \rightarrow 1\, CaCO_3(s)$

Try one more.

Q. In the chart below showing the answer to the second question in the section above, fill in rows 5 and 6.

1	Names	lead(II) nitrate + potassium iodide → lead(II) iodide + potassium nitrate
2	Balanced solid formulas	$1\, Pb(NO_3)_2 + 2\, KI \rightarrow$ ⓵ PbI_2 $+ 2\, KNO_3$

3	Balanced separated formulas	$\boxed{\mathbf{1}\,(1\,Pb^{2+}\,+\,2\,NO_3^-)} + \boxed{\mathbf{2}\,(1\,K^+\,+\,1\,I^-)} \rightarrow$ $\mathbf{1}\,(1\,Pb^{2+}\,+\,2\,I^-) + \boxed{\mathbf{2}\,(1\,K^+\,+\,1\,NO_3^-)}$
4	Solubility	soluble + soluble → **insoluble** + **soluble**
5	Total ionic equation	
6	Net ionic equation	

Answers:

5	Total ionic equation	$\mathbf{1}\,Pb^{2+} + \mathbf{2}\,NO_3^- + \mathbf{2}\,K^+ + \mathbf{2}\,I^- \rightarrow \mathbf{1}\,PbI_2(s) + \mathbf{2}\,K^+ + \mathbf{2}\,NO_3^-$
6	Net ionic equation	$\mathbf{1}\,Pb^{2+}(aq) + \mathbf{2}\,I^-(aq) \rightarrow \mathbf{1}\,PbI_2(s)$

The chart method provides a complete view of the precipitation process, and these methodical steps can help to build your understanding of reactions between ions. With practice, you will be able to answer questions about precipitation automatically, or by writing only parts of the chart. That's the goal.

Flashcards

Add these to your collection. Practice until you can answer the questions perfectly for three days.

You must also commit to memory a solubility scheme and a method to predict the products when two solutions of ionic compounds are mixed.

One-Way Cards (with Notch)	Back Side—Answers
Chemical equations must be balanced for	Atoms and charge
Net ionic equations leave out	Spectator ions
When will mixing two soluble ionic compounds produce a precipitate?	If a new possible combination is insoluble
When will mixing two soluble ionic compounds not result in a precipitate?	If both new possible combinations are soluble

Two-Way Cards (without Notch)	
Ionic compounds with these *positive* ions are nearly always soluble	(alkali metals)$^+$, NH_4^+
Ionic compounds with these *negative* ions are nearly always soluble	Nitrate, acetate, chlorate, perchlorate

PRACTICE B

1. When potassium hydroxide and cobalt(II) nitrate solutions are mixed, an intense blue precipitate of cobalt(II) hydroxide forms. Knowing the precipitate, complete a six-row chart, including the total and net ionic equations.

2. When 0.2 M solutions of nickel(II) bromide and potassium phosphate are mixed,

   a. What are the ionic solid formulas for the two new combinations that are possible?

   b. Will a precipitate form? If so, what is its name and formula?

   c. Write the total and net ionic equations for the reaction.

3. Write the predicted products and the total and net ionic equations for this reaction:

$$K_2S(aq) + BaCl_2(aq) \rightarrow$$

4. Combining solutions of magnesium chloride and sodium nitrate,

   a. Name the two new combinations that are possible.

   b. Which of the new combinations will be soluble in aqueous solutions?

   c. Which of the new combinations will precipitate?

SUMMARY

Solubility and Precipitation

1. Most ionic solids can be characterized as soluble or insoluble in water. Rules for solubility must be memorized, and exceptions occur. The most frequently used rules are these:
   - Compounds containing alkali metal atoms, plus NH_4^+, NO_3^-, ClO_3^-, ClO_4^-, and CH_3COO^- ions, are nearly always *soluble* in water.
   - *Except* in the above cases, compounds containing Pb^{2+}, Hg_2^{2+}, Ag^+, CO_3^{2-}, PO_4^{3-}, S^{2-}, or CrO_4^{2-} ions will generally be *in*soluble in water.

2. When solutions of two soluble ionic compounds are *mixed*,
   - If a new combination is possible that is *in*soluble, it *will* precipitate
   - If a new combination is soluble, its ions will remain separated and dissolved in the solution

3. When two solutions containing soluble ionic compounds are mixed, to represent the precipitation reactions that may occur,
   - *Total* ionic equations include all of the solids and separated ions present in the reactants and products
   - *Net* ionic equations omit the separated spectator ions that do not change their state or formula in the reaction

4. In general, to understand the reactions of ionic compounds dissolved in water,
   - Rewrite the reactants in their separated-ions format
   - Look for reactions among the possible new ion combinations

ANSWERS

Lesson 13.1

Practice 1a. $K^+ + S^{2-}$: **Soluble**. By rule 2, sulfides are insoluble, but by rule 1 alkali metal ions are soluble, and rule 1 has precedence.

1b. $Sr^{2+} + Cl^-$: **Soluble** by rule 3 for chloride.

1c. $Ca^{2+} + CO_3^{2-}$: **Insoluble** by rule 2 for carbonates (if only one ion is in table, use it).

1d. $Ag^+ + CrO_4^{2-}$: **Insoluble** by rule 2 for chromate ion and/or silver ion.

2b. Barium carbonate $\rightarrow Ba^{2+} + CO_3^{2-}$; **insoluble** by rule 2 for carbonates.

2c. Sodium hydroxide $\rightarrow Na^+ + OH^-$; **soluble** by rule 1 for alkali metal ions.

2d. $SrBr_2 \rightarrow Sr^{2+} + 2\,Br^-$; **soluble** by rule 3 for bromides.

2e. Silver nitrate $\rightarrow Ag^+ + NO_3^-$; **soluble** by rule 1 for nitrates. Rule 1 has precedence.

2f. Ammonium hydroxide $\rightarrow NH_4^+ + OH^-$; **soluble** by rule 1 for ammonium ion.

2g. $Fe_3(PO_4)_2 \rightarrow 3\,Fe^{2+} + 2\,PO_4^{3-}$; **insoluble** by rule 2 for phosphate ions.

2h. $Pb(CH_3COO)_2 \rightarrow Pb^{2+} + 2\,CH_3COO^-$; **soluble** by rule 1 for acetates.

2i. $Ni(ClO_3)_2 \rightarrow Ni^{2+} + 2\,ClO_3^-$; **soluble** by rule 1 for chlorates.

2j. $RbBr \rightarrow Rb^+ + Br^-$; **soluble** by rule 1 for alkali metal ions.

2k. $Fe_2S_3 \rightarrow 2\,Fe^{3+} + 3\,S^{2-}$; **insoluble** by rule 2 for sulfide ions.

3. Which ions in combination with Pb^{2+} form soluble compounds? **Nitrate, chlorate, perchlorate, and acetate**.

4. Are all nitrates soluble? **YES** (*almost* always). This rule is used frequently.

5. Are all phosphates insoluble? **NO**. When the phosphate anion is combined with either alkali metal ions or ammonium ions, the phosphate is soluble. Rule 1 has precedence.

Lesson 13.2

For reactions run in aqueous solution, assume (*aq*) if no state is shown. Coefficients of 1 may be omitted as understood.

Practice A 1. $Pb^{2+}(aq) + \cancel{2\,NO_3^-(aq)} + \cancel{Cu^{2+}(aq)} + 2\,Cl^-(aq) \rightarrow \cancel{Cu^{2+}(aq)} + \cancel{2\,NO_3^-(aq)} + \boxed{PbCl_2(s)}$

Net ionic equation: $Pb^{2+}(aq) + 2\,Cl^-(aq) \rightarrow PbCl_2(s)$

2. $\cancel{6\,Na^+} + 2\,PO_4^{3-} + 3\,Mg^{2+} + \cancel{3\,SO_4^{2-}} \rightarrow \boxed{Mg_3(PO_4)_2(s)} + \cancel{6\,Na^+} + \cancel{3\,SO_4^{2-}}$

Net ionic equation: $3\,Mg^{2+}(aq) + 2\,PO_4^{3-}(aq) \rightarrow Mg_3(PO_4)_2(s)$

Practice B 1. $1\,[2\,K^+ + 1\,SO_4^{2-}] + 1\,[1\,Sr^{2+} + 2\,NO_3^-] \rightarrow 1\,SrSO_4(s) + 2\,[1\,K^+ + 1\,NO_3^-]$

Total ionic equation: $2\,K^+ + 1\,SO_4^{2-} + 1\,Sr^{2+} + 2\,NO_3^- \rightarrow 1\,SrSO_4(s) + 2\,K^+ + 2\,NO_3^-$

2. $3\,[1\,Fe^{2+} + 2\,Br^-] + 2\,[3\,Na^+ + 1\,PO_4^{3-}] \rightarrow 6\,[1\,Na^+ + 1\,Br^-] + 1\,Fe_3(PO_4)_2(s)$

Total ionic equation: $3\,Fe^{2+} + 6\,Br^- + 6\,Na^+ + 2\,PO_4^{3-} \rightarrow 6\,Na^+ + 6\,Br^- + 1\,Fe_3(PO_4)_2(s)$

3. $2\,[1\,Ag^+ + 1\,NO_3^-] + 1\,[2\,K^+ + 1\,CrO_4^{2-}] \rightarrow 2\,[1\,K^+ + 1\,NO_3^-] + 1\,Ag_2CrO_4(s)$

Total ionic equation: $2\,Ag^+ + 2\,NO_3^- + 2\,K^+ + 1\,CrO_4^{2-} \rightarrow 2\,K^+ + 2\,NO_3^- + 1\,Ag_2CrO_4(s)$

Practice C 1a. $Fe(NO_3)_3(aq) + 3\,NaOH(aq) \rightarrow 1\,Fe(OH)_3(s) + 3\,NaNO_3(aq)$

1b. $3\,Ca(NO_3)_2 + 2\,K_3PO_4 \rightarrow 6\,KNO_3 + 1\,Ca_3(PO_4)_2(s)$

2a. Total ionic equation: $1\,Fe^{3+} + 3\,NO_3^- + 3\,Na^+ + 3\,OH^- \rightarrow 1\,Fe(OH)_3(s) + 3\,Na^+ + 3\,NO_3^-$

2b. Total ionic equation: $3\,Ca^{2+} + 6\,NO_3^- + 6\,K^+ + 2\,PO_4^{3-} \rightarrow 6\,K^+ + 6\,NO_3^- + 1\,Ca_3(PO_4)_2(s)$

3a. Net ionic equation: $1\,Fe^{3+}(aq) + 3\,OH^-(aq) \rightarrow 1\,Fe(OH)_3(s)$

3b. Net ionic equation: $3\,Ca^{2+} + 2\,PO_4^{3-} \rightarrow 1\,Ca_3(PO_4)_2(s)$

Practice D   Coefficients of 1 may be omitted.

1. $1\,Co(NO_3)_2 + 2\,OH^- \rightarrow 1\,Co(OH)_2 + 2\,NO_3^-$
2. $2\,AgNO_3 + 1\,S^{2-} \rightarrow 1\,Ag_2S + 2\,NO_3^-$
3. $1\,CO_3^{2-} + 2\,HCl \rightarrow 1\,H_2CO_3 + 2\,Cl^-$
4. $1\,Al^{3+} + 1\,NO + 2\,H_2O \rightarrow 1\,Al + 1\,NO_3^- + 4\,H^+$

Lesson 13.3

Practice A 1a. Silver carbonate: **Insoluble**

1b. Ammonium chloride: Ammonium compounds are always predicted to be **soluble**.

1c. $MgCO_3$: **Insoluble** by the carbonate ion rule.

1d. $PbBr_2$: **Insoluble** by the lead(II) ion rule.

1e. Na_2S: Compounds containing alkali metal ions are **soluble**.

1f. $Pb(NO_3)_2$: Nitrates are **soluble**.

1g. AgCl: **Insoluble** due to silver ion.

2. Assume the phase is (*aq*) if not shown.

1	Names	silver nitrate + sodium chromate → silver chromate + sodium nitrate
2	Balanced solid formulas	$2\,AgNO_3 + 1\,Na_2CrO_4 \rightarrow \boxed{1\,Ag_2CrO_4} + 2\,NaNO_3$
3	Balanced separated formulas	$\boxed{2\,(1\,Ag^+ + 1\,NO_3^-)} + 1\,(\boxed{2\,Na^+ + 1\,CrO_4^{2-}}) \rightarrow$ $\qquad\qquad 1\,(2\,Ag^+ + CrO_4^{2-}) + \boxed{2\,(1\,Na^+ + 1\,NO_3^-)}$
4	Solubility	soluble + soluble → **insoluble** + **soluble**

Practice B 1. The problem identifies the precipitate (ppt.) as cobalt(II) hydroxide.

1	Names	potassium hydroxide + cobalt(II) nitrate → $\qquad\qquad$ potassium nitrate + cobalt(II) hydroxide (*ppt.*)
2	Balanced solid formulas	$2\,KOH + 1\,Co(NO_3)_2 \rightarrow 2\,KNO_3 + \boxed{1\,Co(OH)_2}$
3	Balanced separated formulas	$\boxed{2\,(1\,K^+ + 1\,OH^-)} + \boxed{1\,(1\,Co^{2+} + 2\,NO_3^-)} \rightarrow$ $\qquad\qquad \boxed{2\,(1\,K^+ + 1\,NO_3^-)} + 1\,(1\,Co^{2+} + 2\,OH^-)$
4	Solubility	soluble + soluble → **soluble** + **insoluble**
5	Total ionic equation	$2\,K^+ + 2\,OH^- + 1\,Co^{2+} + 2\,NO_3^- \rightarrow 2\,K^+ + 2\,NO_3^- + 1\,Co(OH)_2(s)$
6	Net ionic equation	$1\,Co^{2+}(aq) + 2\,OH^-(aq) \rightarrow 1\,Co(OH)_2(s)$

2.

1	Names	**nickel(II) bromide + potassium phosphate →** **potassium bromide + nickel(II) phosphate**
2	Balanced solid formulas	$3\,NiBr_2 + 2\,K_3PO_4 \rightarrow 6\,KBr + \boxed{1\,Ni_3(PO_4)_2}$
3	Balanced separated formulas	$\boxed{3\,(1\,Ni^{2+} + 2\,Br^-)} + \boxed{2\,(3\,K^+ + 1\,PO_4^{3-})} \rightarrow$ $\boxed{6\,(1\,K^+ + 1\,Br^-)} + 1\,(3\,Ni^{2+} + 2\,PO_4^{3-})$
4	Solubility	soluble + soluble → **soluble + insoluble**
5	Total ionic equation	$3\,Ni^{2+} + 6\,Br^- + 6\,K^+ + 2\,PO_4^{3-} \rightarrow 6\,K^+ + 6\,Br^- + 1\,Ni_3(PO_4)_2(s)$
6	Net ionic equation	$3\,Ni^{2+}(aq) + 2\,PO_4^{3-}(aq) \rightarrow 1\,Ni_3(PO_4)_2(s)$

The precipitate is **nickel(II) phosphate: $Ni_3(PO_4)_2$.**

3.

1	Names	**potassium sulfide + barium chloride → potassium chloride + barium sulfide**
2	Balanced solid formulas	$1\,K_2S + 1\,BaCl_2 \rightarrow 2\,KCl + \boxed{1\,BaS}$
3	Balanced separated formulas	$\boxed{1\,(2\,K^+ + 1\,S^{2-})} + \boxed{1\,(1\,Ba^{2+} + 2\,Cl^-)} \rightarrow$ $\boxed{2\,(1\,K^+ + 1\,Cl^-)} + 1\,(1\,Ba^{2+} + 1\,S^{2-})$
4	Solubility	soluble + soluble → **soluble + insoluble**
5	Total ionic equation	$2\,K^+ + 1\,S^{2-} + 1\,Ba^{2+} + 2\,Cl^- \rightarrow 2\,K^+ + 2\,Cl^- + 1\,BaS(s)$
6	Net ionic equation	$1\,Ba^{2+}(aq) + 1\,S^{2-}(aq) \rightarrow 1\,BaS(s)$

4a. Names: magnesium chloride + sodium nitrate → **magnesium nitrate + sodium chloride**

For the names of the possible products, each positive ion is combined with a new negative partner.

4b. **Both** new combinations will be soluble.

4c. **Neither** will precipitate.

14

Acid–Base Neutralization

Lesson 14.1 # Ions in Acid–Base Neutralization

Terminology

Many substances can be classified as acids or bases (and some can act as both). A variety of definitions exist for acids and bases, with each definition helpful in certain types of reactions and calculations.

Because acids and bases can react with many substances, they are often termed **corrosive**: they may damage a surface, metal, or fabric. When an acid or base is **neutralized**, at least some of its reactive ions are used up, weakening its corrosive power.

In an acid–base **neutralization** reaction, an acid and a base are *reactants* that are mixed together, react, and are both used up to some extent. In the process, both the acid and the base are said to be neutralized.

For the limited purpose of studying neutralization, we will use the following definitions:

- An **acid** is a substance that forms H^+ ions when dissolved in water.
- A **base** is a substance that can react with (use up) H^+ ions.

Understanding Neutralization

As substances, acids and bases may be ionic or covalent compounds. However, when dissolved in water, acids and bases form *ions*. A key to understanding neutralization in aqueous solutions is to write the acid and base reactants as *separated ions*.

Acids

When they are dissolved in water, acids form **H^+ ions** plus other ions. The ions formed are always present in ratios that guarantee electrical neutrality.

A **strong acid** is one that separates into ions essentially 100% when it dissolves in water.

Example:

Nitric acid: $1\ HNO_3 \xrightarrow{\text{H}_2\text{O}} 1\ \textbf{H}^+ + 1\ NO_3^-$ (The reaction goes ~ 100%.)

Other frequently encountered strong acids include these:

- HCl, *hydrochloric* acid, which ionizes to form an H^+ ion and a chloride ion (Cl^-).
- H_2SO_4, *sulfuric* acid, which is used in lead-acid car batteries. It is termed a **diprotic acid** because each neutral H_2SO_4 molecule can ionize to form two H^+ ions.

Strong acids can neutralize both strong and weak bases.

Weak acids dissociate only slightly in water. An example is acetic acid, the active ingredient in vinegar. The formula for acetic acid is written as CH_3COOH by organic (carbon chemistry) chemists to convey structural information, or as $HC_2H_3O_2$ by inorganic (non-carbon) chemists. Acetic acid ionizes slightly in water to form one H^+ ion and one acetate anion (written as CH_3COO^- or $C_2H_3O_2^-$). Weak acids are neutralized by strong bases, but not necessarily by weak bases.

For these four acids that are encountered frequently, given the name, be able to write the formula, and given the formula, be able to write the name:

- Hydrochloric acid (HCl)
- Nitric acid (HNO_3)
- Sulfuric acid (H_2SO_4)
- Acetic acid (CH_3COOH *or* $HC_2H_3O_2$)

Bases

Strong bases can neutralize both strong and weak acids. Examples of strong bases include NaOH (sodium hydroxide) and KOH (potassium hydroxide).

Many other particles can act as bases, such as ammonia (NH_3), carbonate ions (CO_3^{2-}), and bicarbonate ions (HCO_3^-), but bases containing OH^- ions are the most frequently encountered in acid–base neutralization.

Conductivity

Because acids and bases form ions when they dissolve in water, they are **electrolytes**: their solutions can conduct electricity. Strong acids and strong bases are termed **strong electrolytes** because in water they separate into ions ~100%.

The Structure of H^+: A Proton

Neutral hydrogen atoms contain one proton and one electron. (A small percentage of naturally occurring H atoms also contain one or two neutrons, but the neutrons have no impact on the types of reactions in which hydrogen atoms participate.)

An **H^+ ion** is a hydrogen atom without an electron, so most H^+ ions in terms of structure are single **protons**. The terms *H^+ ion* and *proton* are often used interchangeably to describe the active particle in an acid.

The ion formed by acids = H^+ = proton.

In aqueous solutions, the proton released by an acid is nearly always found attached to a water molecule, forming a **hydronium ion** (H_3O^+). This reversible reaction can be represented as

$$1\,H^+ + 1\,H_2O \rightleftharpoons 1\,H_3O^+$$ (Reversible, but favors the right.)

However, though a proton released by an acid attaches to water, it is loosely bound and *behaves* in most respects as if it were a free H^+. This means that the symbols H^+ and H_3O^+ in most cases can be considered to be equivalent. For now, to simplify our work, we will use H^+ as the symbol for the ion contributed to solutions by acids.

Identifying Acidic Hydrogens

Hydrogen atoms in compounds can be divided into two types:

- **Acidic** hydrogen atoms are generally defined as those that react with hydroxide ions.
- **Non-acidic** hydrogen atoms are those that do not.

Compounds often contain both acidic and non-acidic hydrogen atoms. Example: In CH_3COOH (acetic acid), the H atom at the end of the formula reacts with NaOH, but the other three H atoms do not. The H at the *end* is termed an *acidic* hydrogen and the other H atoms are *not* acidic hydrogens.

In an acid formula, how can you predict which hydrogen atoms will be acidic and which will not? In most cases, the rules are:

- If one or more H atoms is written at the *front* of a formula, while other H atoms are not, the H atoms at the *front* will be acidic and the others will not.
    Example: Acetic acid is often written as $HC_2H_3O_2$. Only the H in front is acidic.
- The H at the *end* of a $-COOH$ group (also written $-CO_2H$) is acidic.
    Examples: In C_6H_5COOH, only the H at the end is acidic.
    In $C_3H_7CO_2H$, only the H at the end is acidic.
- If H is the *second* atom in the formula, written after a metal atom but before other atoms, it is acidic.
    Examples: In $KHC_8H_4O_4$, the *first* H (and only the first) is acidic.
    In NaH_2PO_4, the two Hs after the metal atom are acidic.
- If a substance with only one H reacts with hydroxide ion, the H is acidic.

PRACTICE   A

1. Draw an arrow pointing toward the acidic hydrogens in each compound; then, after each formula, write the number of acidic hydrogens in the compound.

    a. NaH_2PO_4 b. $C_{12}H_{25}COOH$ c. $H_2C_4H_4O_6$

    d. H_3AsO_4 e. $KHC_8H_4O_4$ f. $NaHSO_4$

2. Fill in the blanks to show the number of protons formed when these compounds ionize in water.

    a. $C_3H_7CO_2H \rightarrow$ ____ H^+ b. $HC_2H_3O_2 \rightarrow$ ____ H^+

    c. $NaH_2(C_3H_5O(COO)_3) \rightarrow$ ____ H^+

Identifying Acids and Bases

From a chemical formula, how can you tell whether a substance is an acid or a base? It is not always easy to determine. A general rule is this:

To Identify Whether a Compound Is an Acid or a Base

Write the balanced equation for the substance formula separating into *familiar* ions.

- Compounds that ionize to form H^+ ions are **acids**.
- Compounds that contain OH^- ions can act as **bases**.

In the reactions below, assume that each compound on the left is being added to water and separates in the water to form ions. Write balanced equations showing the ions formed, then label each initial reactant as either an *acid* or a *base*. For help, review Lesson 7.3.

1. $\text{LiOH} \xrightarrow{\text{H}_2\text{O}}$

2. $\text{HCN} \rightarrow$

3. $\text{Ca(OH)}_2 \rightarrow$

4. $\text{HC}_2\text{H}_3\text{O}_2 \rightarrow$

Lesson 14.2 Balancing Hydroxide Neutralization

Hydroxide Neutralization

Acids can be neutralized by many different types of bases, including carbonates, sulfites, and ammonia. However, the acid–base neutralization reactions that are encountered most frequently are those between acids and substances containing hydroxide ions.

In an aqueous solution, for the reaction of an acid and a hydroxide, the neutralization reaction forms liquid *water*. Water can be written as H—OH *or* HOH *or* H_2O.

$$\text{H}^+ + \text{OH}^- \rightarrow \text{H—OH}(\ell) \tag{14.1}$$

The reaction of an acid and a hydroxide ion is *driven to completion* by the formation of water: a low potential-energy molecule. Whenever a physical or chemical system can go to lower potential energy, there is a strong tendency to do so.

The reaction of acids with compounds containing hydroxide ions can be represented by this general equation:

$$\text{An } acid + \text{a } base \text{ containing OH}^- \rightarrow \text{H}_2\text{O} + \text{a salt} \tag{14.2}$$

Historically in chemistry, the term **salt** in equation 14.2 was a general term for any product, in addition to water, formed when an acid and base react. In modern usage, *salt* is often used as a synonym for *ionic compound*.

A typical acid–hydroxide neutralization is the reaction of hydrochloric acid with sodium hydroxide to form sodium chloride (also known as table salt).

$$\text{HCl}(aq) + \text{NaOH}(aq) \rightarrow \text{H—OH}(\ell) + \text{NaCl}(aq) \tag{14.3}$$

Most (but not all) acid–base reactions are carried out in water. In the acid–base solution reactions in these lessons, you may assume that the water is liquid and the other compounds and ions are *aqueous* (*aq*) unless otherwise noted. Writing water as H—OH helps to emphasize the reaction that occurs between acids and hydroxide ions. The water that forms has the behavior of a covalently bonded molecule: it separates to form ions only very slightly in the solution.

Equation 14.3 above is one way that this reaction is represented. However, both HCl and NaOH, when dissolved in water, separate completely into ions. The table salt in the solution after the reaction also exists as separate ions of Na^+ and Cl^-. Rewriting the equation to show the separated ions that actually exist in the solution, the reaction is

$$H^+ + Cl^- + Na^+ + OH^- \rightarrow H\!-\!OH(\ell) + Na^+ + Cl^- \qquad (14.4)$$

Note in equation 14.4 that the sodium and chloride ions are *spectators*: they remain unchanged at the end of the reaction.

Partial versus Complete Neutralization

In compounds that contain more than one acidic hydrogen or basic group, it is usually possible to mix the acid and base so that some of the acidic or basic groups are neutralized, but others are not. However, in these lessons, unless otherwise noted, you should assume that *neutralization* means *exact* and *complete* neutralization: the point in the reaction at which the acid and base are mixed in amounts such that *all* of the acidic hydrogens and basic groups are completely used up.

Predicting the Products of Hydroxide Neutralization

If formulas are supplied for all the reactants and products, neutralization equations can be balanced by trial and error using the methods in Lesson 9.2. However, in acid–base neutralization problems, often the product formulas are *not* supplied. In these cases, you can often *predict* the products and balance the reaction equation by using these steps:

- Rewrite the acid and base as separated ions.
- Predict one of the neutralization products.

In hydroxide neutralization, a key rule is:

> If an acid reacts with OH^-, one of the products is $H\!-\!OH$.

Knowing that water is one product, you can usually make a "best guess" for the other product formula(s).

▷ **TRY IT**

(See "How to Use These Lessons," point 1, p. xv.)
Complete the steps below in your notebook.
Q. Write a balanced equation (using ionic solid formulas) for the complete neutralization of aluminum hydroxide by nitric acid. Begin with

$$HNO_3 + Al(OH)_3 \rightarrow$$

Steps:

1. Write the acid and base reactants in their ionic *solid* (or *molecular*) formula formats (done above).

2. On the line below, rewrite each reactant inside parentheses ( ) in its *separated*-ion format.

3. After the reactants, on both lines, add "$\rightarrow$ H—OH+" to show water as one product.

Answer:

Solid: $HNO_3 +$ $Al(OH)_3 \rightarrow$ **H—OH +**

Separated: $(1\,H^+ + 1\,NO_3^-) +$ $(1\,Al^{3+} + 3\,OH^-) \rightarrow$
                                        H—OH +

4. In the separated formula, add the lowest whole-number coefficients in front of the parentheses ( ) that balance the H^+, OH^-, and water.

Separated: $\underline{3}\,(1\,H^+ + 1\,NO_3^-) + \underline{1}\,(1\,Al^{3+} + 3\,OH^-) \rightarrow \underline{3}\,H\text{—}OH +$
$\qquad\qquad\quad \wedge \qquad\qquad\qquad\qquad\qquad\quad \wedge \qquad\qquad\quad \wedge$
$\qquad\quad [(3\,H^+) \qquad\qquad\qquad\qquad\qquad (3\,OH^-) \rightarrow (3\,H\text{—}OH)]$

The total H^+ ions must equal the total OH^- ions and must equal the total H_2O.

5. On the products side, fill in the remaining ions. Because the focus is neutralization, assume at this point that the product ions will be spectators that are unchanged from the reactants (though in some cases those ions undergo additional reactions). Make sure the ion coefficients are added as well.

6. Move the left-side coefficients that are in *front* of the ( ), plus the coefficient of the water, up to the *solid* line.

7. Finish adding coefficients to the equation using *solid* formulas, balancing by trial and error. To write a formula for the remaining product, use the rules for writing ionic *solid* formulas in Lesson 7.3.

Solid: $\underline{3}\,HNO_3 + \underline{1}\,Al(OH)_3 \rightarrow \underline{3}\,H_2O + 1\,Al(NO_3)_3$

Separated:

$3(1\,H^+ + 1\,NO_3^-) + 1\,(Al^{3+} + 3\,OH^-) \rightarrow 3\,H_2O + 1\,Al^{3+} + 3\,NO_3^-$

In this case, the products that are not H^+ and OH^- are aluminum and nitrate ions: a soluble combination of "spectator ions."

8. Check that the final equation is balanced.

Assuming that the ions that are *not* H^+ and OH^- will be non-reacting spectators is a prediction that will often be accurate. In some cases, additional reactions occur that further change the final product formulas, but knowing one product must be water, we can balance the equation for the *reactant* ratios correctly. That's usually all that we need to solve most acid–base neutralization calculations.

PRACTICE   A

In your notebook, balance by inspection *or* by the methods above. Assume that the acids and bases are completely neutralized. Do every other part, and more if you need more practice.

(continued)

1. Write the predicted product formulas in their ionic solid format and balance the equation.

   a. $\quad\quad$ HNO_3 + $\quad\quad$ KOH $\rightarrow$

   b. $\quad\quad$ KOH + $\quad\quad$ H_2SO_4 $\rightarrow$

   c. $\quad\quad$ H_2SO_4 + $\quad\quad$ $Al(OH)_3$ $\rightarrow$

2. Write equations with reactant formulas and the predicted product formulas. At the final step, write a balanced equation with all formulas in their ionic-solid (molecular) format.

   a. Cesium hydroxide + H_2SO_4 $\rightarrow$

   b. $Ca(OH)_2$ + nitric acid $\rightarrow$

   c. Barium hydroxide reacting with sulfuric acid

   d. Hydrochloric acid reacting with magnesium hydroxide

Balancing with an Unknown Formula

In some neutralization calculations, the formula for an acid is not supplied, but the *number* of protons in the acid is provided. In these cases, the acid formula can be represented by a formula such as **H_3R**, where the R stands for an *unknown* group and **H_3** represents three acidic hydrogens. Using these formulas, you can usually balance the acid–base reactant ratio, and that *partial balancing* will be all that you need in solving most neutralization stoichiometry.

 TRY IT

Q. A solid acid has an unknown formula but is known to contain three acidic hydrogens. What will be the ratio for the complete neutralization of this acid by NaOH?

STOP

STOP

Answer:

The two reactants can be written as H_3R + NaOH $\rightarrow$ $\quad$ Finish from here.

$$H_3R \ + \ NaOH \ \rightarrow \ \mathbf{1}\,(3\,H^+) \ + \ \mathbf{3}\,(OH^-) \ \rightarrow \ \mathbf{3\,H{-}OH} \ + \ ...$$

The *acid–base* ratio must be

$$\mathbf{1}\,H_3R \ + \ \mathbf{3}\,NaOH \ \rightarrow$$

A "best guess" formula could be predicted for the other products above, such as

$$1\,H_3R \ + \ 3\,NaOH \ \rightarrow \ 3\,H{-}OH \ + \ \mathbf{1\,Na_3R\,(?)}$$

but this would be speculation. The anion from the original acid could remain intact, or it could decay in some fashion when it is produced (as happens when hydrogen carbonates are reacted with acids). However, for most neutralization calculations, an exact formula for the non-water product is not required: the acid–base ratio on the left side of the equation is the *bridge* ratio. That bridge conversion is all that is needed to solve most acid–base stoichiometry.

SUMMARY

This is the synopsis of the neutralization rules so far.

1. **Acid–base *neutralization*** is an *ionic* reaction. To understand ionic reactions, write the *separated*-ion formulas.

2. **Ions:** Acids contain H^+.
   The reacting particle in acids $= H^+ =$ a proton
   Bases include compounds with hydroxide (OH^-) ions. Hydroxide bases will neutralize both strong and weak acids.

3. **Products:** For acids $+ OH^-$, one product is water: $H\!-\!OH$.

$$H^+ + OH^- \rightarrow H\!-\!OH$$

4. **Balancing:** To *predict* the products and *balance* the equations,
   • Write the *separated*-ion formulas in ( )
   • Write a known product
   • Finish by balancing atoms and charge

5. **To balance** hydroxide neutralization when the acid formula is **unknown**,
   • If a substance has X acidic hydrogens, write its formula as H_xR
   • Balance with the hydroxide ions to produce $H\!-\!OH$

PRACTICE   B

Learn the rules above, then do these problems. Assume that all reactants are completely neutralized.

1. Supply the ratios of reaction for the two reactants.

   a. HR + $KOH \rightarrow$

   b. H_2R + $Sr(OH)_2 \rightarrow$

   c. H_4R + $NaOH \rightarrow$

2. An unknown acid with two acidic hydrogens is totally neutralized by potassium hydroxide. Supply the coefficients for the two reactants.

Lesson 14.3   Neutralization and Titration Calculations

Solution Stoichiometry and Acid–Base Reactions

In any reaction that has two reactants, at the point where the *moles* of the reactants have been mixed in the exact ratio that matches the coefficients of the balanced equation, the reactants are *stoichiometrically equivalent* (see Lesson 9.6). Calculations involving equivalent amounts can be solved using the stoichiometry steps in Lessons 9.5 and 12.3.

For a reaction in which *acid* and *base* reactants have been mixed in stoichiometrically equivalent amounts, both the acid and the base are said to be *neutralized*. (In a calculation problem, unless otherwise noted, assume that *neutralization* means *exact* neutralization: the acid and base have been mixed in stoichiometrically equivalent amounts.)

Neutralization problems can be divided into two types: those that supply

- All of the reactant and product formulas, or
- Only the reactant formulas

If all of the reactant and product formulas are supplied, any equation can be balanced by trial and error, and calculations involving reaction amounts can then be solved by stoichiometry. You solved several neutralization calculations that supplied the reactants and products in Lesson 12.3.

We are now prepared to solve calculations in which the acid and hydroxide formulas are supplied but the product formulas are not. Using the balancing strategies in Lesson 14.2, the coefficients of the *acid* and *base* can be determined even when the formula of the non-water product may not be certain. In most neutralization calculations, the ratio of the two *reactants* is all that is needed to solve conversion stoichiometry.

▶ TRY IT

Apply the acid–hydroxide balancing and stoichiometry steps to this example.

Q. How many milliliters of a 0.200 M sodium hydroxide solution are needed to neutralize all of the acidic protons in 2.34 g of arsenic acid (H_3AsO_4)?

If you get stuck, read a *portion* of the answer until you are unstuck, then try again.

Reaction calculations for two substances that are at an equivalence point (including when an acid and base are both exactly neutralized) are solved by conversion stoichiometry (Lesson 9.5). Start with WDBB.

Answer:

Your paper should look like this, minus the parenthetical comments.

WANTED: ? mL NaOH soln.

DATA: 0.200 mol NaOH = 1 L NaOH   (M prompt.)

2.34 *g* H_3AsO_4 (Single unit *given*.)

141.9 *g* H_3AsO_4 = 1 *mol* H_3AsO_4 (g prompt.)

(Include formulas after all units to distinguish the two substances being measured.)

Balance: **1** H_3AsO_4 + **3** NaOH → **3** H—OH + Na_3AsO_4 (?)

(The product formula Na_3AsO_4 is a "best guess" for an unfamiliar ion; additional reactions might occur.)

Bridge: **1** mol H_3AsO_4 = **3** mol NaOH

SOLVE: (Want a single unit?)

$$? \text{ mL NaOH} = 2.34 \text{ g H}_3\text{AsO}_4 \cdot \frac{1 \text{ mol H}_3\text{AsO}_4}{141.9 \text{ g H}_3\text{AsO}_4} \cdot \frac{3 \text{ mol NaOH}}{1 \text{ mol NaOH}}$$

$$\cdot \frac{1 \text{ L NaOH}}{0.200 \text{ mol NaOH}} \cdot \frac{1 \text{ mL}}{10^{-3} \text{ L}} = \textbf{247 mL NaOH soln.}$$

(For detailed rules on solution stoichiometry, see Lesson 12.3.)

What's the logic of these steps? We want to know the mL of base that will neutralize a known amount of acid.

- From the grams of the acid and its g/mol, we can find the moles of acid in the sample.
- From the balanced equation, we can find the moles of base that will neutralize the acid.
- From the moles of base needed and its mol/L, we can find the liters and mL of base that are needed to exactly neutralize the acid.

The steps of stoichiometry allow us to precisely predict the amounts used up and formed in chemical reactions.

PRACTICE   A

If all of the acidic hydrogens in a 2.00 M H_3PO_4 solution are neutralized by 1.50 L of 0.500 M KOH, how many milliliters of H_3PO_4 solution were neutralized?

Titration Terminology

Titration is an experimental technique that can supply the data needed for stoichiometry calculations. In titration, calibrated **burets** are used to precisely measure the amounts of solution added as a chemical reaction takes place.

Indicators are dyes used in titration that change color at the instant the *moles* of two reacting particles are equal or have reacted in a simple whole-number ratio. When this **equivalence point** is reached, a change in indicator color signals the **endpoint** of the titration.

An **acid–base titration** is a precise neutralization in which the amounts of acid and base are carefully measured. If one or both of the reactants is a *strong* acid or base, the reaction goes to completion: it will proceed until both reactants are completely used up.

When titrating a *weak* acid or base, the opposite solution must be strong, and a careful selection of the indicator dye is required to show a sharp endpoint. However, for all acids and bases, if one or both is strong, stoichiometry calculations are done using the same steps.

Acid–Hydroxide Titration Calculations

In the titration of an acid and a hydroxide, because OH^- is a strong base, the reaction goes to completion. The acid and base ions react as soon as they are mixed, and as additional solution is added, the reaction continues to occur until the equivalence point is reached. At that point, the indicator changes color.

For an acid with one acidic hydrogen, at the endpoint of the titration, the moles of acid in the sample will equal the moles H^+ supplied to the reaction mixture, which will equal the moles of OH^- supplied by the base solution.

> At an acid–hydroxide endpoint, **moles H^+** from acid = **moles OH^-** from base.

For acids that contain more than one acidic hydrogen, neutralization produces a series of equivalence points. For many of these **polyprotic** acids, by carefully selecting an appropriate indicator, it may be possible to titrate to an equivalence point for different numbers of acidic hydrogens in the acid.

At the endpoint of the titration, you should assume that the moles of acid and base have been mixed in the exact ratio shown by the coefficients of the balanced equation for the neutralization, and that both are exactly used up (at the uncertainty indicated by the significant figures in the data). At that point, the acid and base are stoichiometrically equivalent: both reactants are limiting. Calculations using titrated amounts can then be solved by conversion stoichiometry.

> For calculations involving a *reaction* to an *equivalence point* or *titration* to an *endpoint*, use conversion stoichiometry. Start with WDBB.

PRACTICE B

1. A 25.0 mL sample of a 0.145 M hydrochloric acid solution is titrated by a 0.200 M potassium hydroxide solution. How many milliliters of the base must be added to reach the point where the indicator changes color?

2. Oxalic acid ($H_2C_2O_4$, 90.03 g · mol^{-1}) is a solid at room temperature. A sample of oxalic acid is titrated by 0.0500 molar sodium hydroxide, and an indicator is chosen that changes color at the point when both of the acidic hydrogens are neutralized. How many milliliters of NaOH are needed to titrate 0.100 g of oxalic acid crystals?

3. If you have not already done so, review Lessons 14.1–14.3 and prepare flashcards that cover the fundamentals.

SUMMARY

1. **Acid–base neutralization** is an *ionic* reaction. To understand ionic reactions, write the *separated*-ion formulas, then look for new ion combinations that react.

2. **Ions:** Acids contain H^+. The reacting particle in acids = H^+ = a proton.
   Bases include compounds that contain hydroxide OH^- ions.

3. **Products:** For acids + OH^-, one product is water: **H—OH**.

$$H^+ + OH^- \rightarrow H-OH$$

4. **Balancing:** To predict the products and balance the equations,
   - Write the separated-ion formulas in ( )
   - Write one product (for acid–hydroxide reactions, one product is always water)
   - Finish by balancing atoms, familiar ions, and charge

5. In neutralization, at the equivalence point where both the acid and base are exactly used up, both reactants are limiting, and either reactant amount can be used as a *given* to solve reaction calculations.

6. For calculations involving *reaction amounts* at an *equivalence point* or a *titration endpoint*, use conversion stoichiometry. Start with WDBB.

ANSWERS

Lesson 14.1

Practice A 1a. NaH_2PO_4 **Two**   1b. $C_{12}H_{25}COOH$ **One**   1c. $H_2C_4H_4O_6$ **Two**

1d. H_3AsO_4 **Three**   1e. $KHC_8H_4O_4$ **One**   1f. $NaHSO_4$ **One**

2a. $C_3H_7CO_2H \rightarrow \underline{1}\,H^+$ 2b. $HC_2H_3O_2 \rightarrow \underline{1}\,H^+$ 2c. $NaH_2(C_3H_5O(COO)_3) \rightarrow \underline{2}\,H^+$

Practice B 1. $LiOH \rightarrow Li^+ + OH^-$; **base**   (In compounds, alkali metals are 1^+ ions.)

2. $HCN \rightarrow H^+ + CN^-$; **acid**   (This anion is the cyanide ion.)

3. $Ca(OH)_2 \rightarrow Ca^{2+} + 2\,OH^-$; **base**   (In compounds, alkaline earth metals—in column 2 of the periodic table—are 2^+ ions.)

4. $HC_2H_3O_2 \rightarrow 1\,H^+ + C_2H_3O_2^-$; **acid**   (The H in front is acidic; this acid is acetic acid.)

Lesson 14.2

Practice A Coefficients of one may be omitted. Lowest whole-number coefficients are preferred.

1a. $\mathbf{1}\,HNO_3 + \mathbf{1}\,KOH \rightarrow \mathbf{1}\,H_2O + \mathbf{1}\,KNO_3$ 1b. $\mathbf{2}\,KOH + \mathbf{1}\,H_2SO_4 \rightarrow \mathbf{2}\,H_2O + \mathbf{1}\,K_2SO_4$

1c. $\mathbf{3}\,H_2SO_4 + \mathbf{2}\,Al(OH)_3 \rightarrow \mathbf{Al_2(SO_4)_3} + \mathbf{6}\,H_2O$

2a. $\mathbf{2}\,CsOH + \mathbf{1}\,H_2SO_4 \rightarrow \mathbf{2}\,H-OH + \mathbf{1}\,Cs_2SO_4$

2b. $\mathbf{1}\,Ca(OH)_2 + \mathbf{2}\,HNO_3 \rightarrow \mathbf{2}\,H-OH + \mathbf{1}\,Ca(NO_3)_2$

2c. $\mathbf{1}\,Ba(OH)_2 + \mathbf{1}\,H_2SO_4 \rightarrow \mathbf{2}\,H_2O + \mathbf{1}\,BaSO_4$

2d. $\mathbf{2}\,HCl + \mathbf{Mg(OH)_2} \rightarrow \mathbf{2}\,H_2O + \mathbf{1}\,MgCl_2$

Practice B 1a. $\mathbf{1}\,HR + \mathbf{1}\,KOH \rightarrow \mathbf{1}\,H-OH + \mathbf{1}\,KR$ (?)   1b. $\mathbf{1}\,H_2R + \mathbf{1}\,Sr(OH)_2 \rightarrow \mathbf{2}\,H-OH + \mathbf{1}\,SrR$ (?)

1c. $\mathbf{1}\,H_4R + \mathbf{4}\,NaOH \rightarrow \mathbf{4}\,H-OH + \mathbf{1}\,Na_4R$ (?)

2. $\mathbf{1}\,H_2R + \mathbf{2}\,KOH \rightarrow \mathbf{2}\,H_2O + \mathbf{1}\,K_2R$ (?)

Lesson 14.3

Practice A   WANTED: $?\ mL\ H_3PO_4$ (You want a single unit.)

DATA: $2.00\ mol\ H_3PO_4 = 1\ L\ H_3PO_4$ (M prompt.)

1.50 L KOH soln.   (The single unit *given*.)

$0.500\ mol\ KOH = 1\ L\ KOH$ (M prompt.)

Because all volumes in this problem are for aqueous solutions, you may label one volume as "soln." but omit the other "solution" labels after volumes as understood.

Balance: **1** H_3PO_4 + **3** KOH → **3** H_2O + **1** K_3PO_4

Bridge: **1 mol** H_3PO_4 = **3 mol** KOH

SOLVE: $? \, mL \, H_3PO_4 = 1.50 \, L \, KOH \cdot \dfrac{0.500 \, mol \, KOH}{1 \, L \, KOH} \cdot \dfrac{1 \, mol \, H_3PO_4}{3 \, mol \, KOH} \cdot \dfrac{1 \, L \, H_3PO_4}{2.00 \, mol \, H_3PO_4} \cdot \dfrac{1 \, mL}{10^{-3} \, L}$

$= \textbf{125 mL } \mathbf{H_3PO_4}$

Practice B   1. WANTED: **?** mL KOH solution

DATA: 25.0 mL HCl   (Single-unit *given.*)

0.145 mol HCl = 1 L HCl   (M prompt.)

0.200 mol KOH = 1 L KOH   (M prompt.)

Balance: 1 HCl + 1 KOH → 1 H_2O + 1 KCl

Bridge: 1 mol HCl = 1 mol KOH

SOLVE: $? \, mL \, KOH = 25.0 \, mL \, HCl \cdot \dfrac{10^{-3} \, L}{1 \, mL} \cdot \dfrac{0.145 \, mol \, HCl}{1 \, L \, HCl} \cdot \dfrac{1 \, mol \, KOH}{1 \, mol \, HCl} \cdot \dfrac{1 \, L \, KOH}{0.200 \, mol \, KOH} \cdot \dfrac{1 \, mL}{10^{-3} \, L}$

$= \textbf{18.1 mL KOH solution}$

2. WANTED: **?** mL NaOH

DATA: 0.0500 mol NaOH = 1 L NaOH

0.100 *g* $H_2C_2O_4$ (Single-unit *given.*)

90.03 *g* $H_2C_2O_4$ = 1 *mol* $H_2C_2O_4$

Balance: **1** $H_2C_2O_4$ + **2** NaOH → 2 H_2O + 1 $Na_2C_2O_4$ (?)   (Two acidic hydrogens.)

Bridge: 1 mol $H_2C_2O_4$ = 2 mol NaOH   (Relate the WANTED and *given* substances.)

SOLVE: $? \, mL \, NaOH \, soln. = 0.100 \, g \, H_2C_2O_4 \cdot \dfrac{1 \, mol \, H_2C_2O_4}{90.03 \, g \, H_2C_2O_4} \cdot \dfrac{2 \, mol \, NaOH}{1 \, mol \, H_2C_2O_4} \cdot \dfrac{1 \, L \, NaOH}{0.0500 \, mol \, NaOH}$

$\cdot \dfrac{1 \, mL}{10^{-3} \, L} = \textbf{44.4 mL NaOH soln.}$

3. Your flashcards might include the following.

One-Way Cards (with Notch)	Back Side—Answers
Names and formulas for three strong acids	Hydrochloric: HCl, sulfuric: H_2SO_4, nitric: HNO_3
Ion symbol for a proton	H^+
One product in acid–hydroxide neutralization	H_2O
Solve calculations to a titration endpoint with	Conversion stoichiometry

Two-Way Cards (without Notch)	
An acid	A substance that produces H^+ in water
A base	A substance that neutralizes H^+

15

Redox Reactions

<table>
<tr><td>**Lesson 15.1**</td><td></td></tr>
</table>

Oxidation Numbers

Electron Transfer Reactions

In Chapters 13 and 14, we have studied two types of reactions that can occur in aqueous solutions.

- In *precipitation* reactions, soluble cations react with soluble anions to form a compound that is insoluble in water.
- In *neutralization* reactions, a proton (H^+) leaves an acid and combines with a particle from a base.

In this chapter, we will consider **redox reactions**. As neutralization can be viewed as a proton transfer, redox is an electron transfer. As with acid–base reactions, redox reactions can occur both in aqueous solutions and in other circumstances, such as reactions between gases.

Redox reactions are a combination of *red*uction and *ox*idation.

- **Oxidation** is the *loss* of electrons.
- **Reduction** is the *gain* of electrons.

In a redox reaction, one reactant particle must be oxidized (lose its electrons) and another reactant must be reduced (gain the transferred electrons).

The term *oxidation* is derived from the fact that the most reactive major component of the air around us, elemental oxygen (O_2), tends to remove electrons in its reactions. The term *reduction* reflects that when metal ions found in certain rocks (termed minerals or ores) are heated with substances that tend to lose electrons, the volume of the ore is seemingly *reduced* as its positively charged metal ions react to form the neutral atoms of metals. Because metals can be shined, shaped, and sharpened, they have been of value throughout human history in uses from jewelry to weaponry.

Oxidation Numbers

Oxidation numbers are numeric values related to electrical charge that can be assigned to each individual atom in a chemical particle. Those numbers can then assist in tracking the transfer of electrons that occurs during redox reactions.

Some sets of rules for assigning oxidation numbers are more complex than others, and those rules will predict behavior for a larger number of substances. However, in all cases, oxidation numbers should be viewed as a simplified model: helpful in predicting redox behavior, but imprecise in describing the actual distribution of charges in particles. In these lessons, we will use a limited set of rules that will help us to predict redox behavior for *most* chemical particles.

Rules for Assigning Oxidation Numbers

1. In an *element*, each atom is assigned an oxidation number (O.N.) of *zero*.
   Elements are substances that contain only one kind of atom. Chemical formulas for elements have one kind of atom and no charge.
   Examples: Element formulas include C, Na, O_2, S_8, and Cl_2 (see Lesson 7.2 for additional information on elements).

2. In a *monatomic ion*, the oxidation number of the atom is the *charge* on the ion.

    Examples:

    For Na^+, the Na is assigned an oxidation number of +1.

    For Al^{3+}, the Al is assigned an oxidation number of +3.

    For S^{2-}, the S is assigned an oxidation number of −2.

    Note these conventions, as shown in the above examples:
    - If a particle has a *charge* of zero, the 0 charge is usually omitted.
    - If a particle is an ion, its charge is shown as a superscript with the *number* of charges *followed* by a + or −, with values of 1 omitted as understood.
    - For the *oxidation numbers* of atoms, a value of 0 is written.
    - For oxidation numbers that are not zero, a + or − sign must be written *first*, and numbers, including 1, are shown.

3. For particles that are *not* elements or monatomic ions, oxidation numbers are assigned to the atoms below as follows.

    a. Each oxygen (O) atom is assigned an O.N. of −2 (except in compounds named *peroxides*, where O is −1).

    b. Each hydrogen (H) atom is assigned a +1, except in metallic hydrides (compounds of a metal atom and hydrogen), where H is −1.

    c. Each alkali metal atom is assigned a +1 and each column 2 atom is assigned a +2.

        Example: In NaOH, by the rules above, these oxidation numbers are assigned to the individual atoms.

        Oxidation number of each atom: +1 −2 +1

        Formula for the particle: Na O H

4. Oxidation numbers not covered by the rules above are selected so that the *sum* of all of the oxidation numbers equals the *overall charge* on the molecule or ion.

    In these lessons, we will use a labeling system where the oxidation number of *each individual atom* in a particle is shown above the particle symbol, and the *sum* for the atoms of that kind is shown below the symbol.

    Example: What is the oxidation number for the Mn atom in MnO_4^-?

    Each O has a −2 oxidation number. Because there are 4 oxygen atoms in the particle, the sum of the O.N. for the oxygens is −8.

    Oxidation number of **each** atom: −2

    Formula for the particle: MnO_4^-

    Sum of oxidation numbers for those atoms: −8

    The sum of all of the oxidation numbers must add up to the −1 charge on the ion. The oxidation number on the Mn atom must therefore be **+7**.

    Each: +7 −2

    Formula: MnO_4^-

    Sum: +7 −8   (Must add up to −1.)

TRY IT

(See "How to Use These Lessons," point 1, p. xv.)

In your notebook, apply the steps in the example above to this problem.

Q. What is the oxidation number for the nitrogen atom in NO_2?

Answer:

For any particle that is not an element or monatomic ion, first write the *individual* oxidation numbers for **O, H,** and column **one** and **two** atoms above the atom symbol, then write the *sum* of the oxidation numbers for each of those atoms below the symbol.

Each oxygen atom in NO_2 has an oxidation number of -2; for the two oxygens, the sum of the O oxidation numbers is -4.

Each: -2
Formula: NO_2
Sum: -4 [(individual O.N.) $\times$ (subscript) = sum]

For the nitrogen atom, start with the rule that the *sum* of all of the numbers on the bottom must add up to the overall charge on the particle, which in this case is zero.

If needed, finish from here.

The *sum* of the oxidation numbers for all of the nitrogens must be **+4**. Because we have only one nitrogen, its *individual* oxidation number must be **+4**.

Each: **+4** -2
Formula: NO_2
Sum: $+4$ -4 (Must sum to equal charge on particle.)

5. The oxidation number of an individual *atom* in a chemical particle is the sum of the oxidation numbers for the atoms of that kind, *divided* by the number of atoms of that kind.

TRY IT

Q. What is the oxidation number for each of the nitrogen atoms in N_2O_4?

Answer:

Each: $+4$ -2
Formula: N_2O_4
Sum: $+8$ -8 (Must add up to 0.)

The sum of all of the oxidation numbers must add up to the zero charge on the molecule. The sum of the oxidation numbers of the two N atoms must be $+8$, so *each* N atom must be **+4**.

6. For ionic compounds, oxidation numbers may be assigned based on applying the rules above to the ionic solid formula or to the formulas for each separate ion in the compound. Both methods will result in the same oxidation number for the individual atoms.

Apply the rules to one more problem.

▸ **TRY IT**

Q. Find the oxidation number of the nitrogen atom in an ammonium ion, NH_4^+.

 Answer:

Each: -3 +1

Formula: NH_4^+

Sum: −3 +4   (Must add up to +1.)

In NH_4^+, each hydrogen atom has an oxidation number of +1; the sum of the oxidation numbers for the four hydrogens is **+4**.

Since the overall charge on the NH_4^+ particle is **+1**, the sum of the oxidation numbers for all the atoms (on the bottom) must add up to **+1**. The sum for all nitrogens must therefore be −3, and since there is only one nitrogen atom, its individual oxidation number must be **−3**.

SUMMARY

Assigning Oxidation Numbers

1. Each atom in an element is assigned an oxidation number of zero.

2. The oxidation number of an atom in a monatomic ion is the charge on the ion.

3. To assign oxidation numbers to other individual atoms,
   • Write the *individual* oxidation numbers above the atom symbols for *O, H,* and column *one* and *two* atoms, then write below each atom symbol the sum of the oxidation numbers for the atoms of that kind.
   • Write the remaining *sums* below each atom symbol so that the sum of sums is the charge on the particle.
   • Write the oxidation numbers for each *individual* remaining atom above the atom symbol.

PRACTICE

Find the oxidation number (O.N.) of each individual atom for the atoms specified in these compounds. Do at least every other problem.

1. CO_2 C = _____   2. Br_2 Br = _____   3. Mn^{2+} Mn = _____

(continued)

4. $CaBr_2$ Br = _____

5. NO_3^- N = _____

6. Na_2CrO_4 Cr = _____

7. $Cr_2O_7^{2-}$ Cr = _____

8. S_8 S = _____

9. H_2O_2 O = _____   (H_2O_2 is hydrogen peroxide.)

10. $C_2O_4^{2-}$ C = _____

11. H_3AsO_4 As = _____

12. $KMnO_4$ Mn = _____

13. ClO_3^- Cl = _____

14. Mn_2O_7 Mn = _____

15. Hg_2^{2+} Hg = _____

Lesson 15.2   Balancing Charge

Half-Reactions

In balanced redox equations, the number of electrons lost by one reactant particle must equal the number gained by another reactant particle. However, this key balanced transfer of electrons is hidden when a redox equation is written. To understand redox reactions, it is often necessary to separate the reaction into two parts in order to see and balance the number of electrons lost and gained.

A redox **half-reaction** is a balanced equation that includes free electrons. The free electrons can be in the reactants (on the left of the arrow) or the products.

Every redox reaction is a result of *two* half-reactions: one that loses electrons, and the other that gains the same number of electrons. Half-reactions break a redox reaction into its two components: the oxidation and the reduction.

Example: The oxidation of lead by silver ion can be represented as

$$Pb + 2\,Ag^+ \rightarrow 2\,Ag + Pb^{2+} \tag{15.1}$$

This reaction is a result of two *half*-reactions. Each reactant lead atom loses two electrons, and each reactant silver ion gains one electron.

$$Pb \rightarrow Pb^{2+} + \mathbf{2\,e^-} \quad \text{(The lead metal is oxidized, losing two electrons.)} \tag{15.2}$$

$$\mathbf{2\,e^-} + 2\,Ag^+ \rightarrow 2\,Ag \quad \text{(For each Pb oxidized, two } Ag^+ \text{ are reduced.)} \tag{15.3}$$

In both half-reactions and complete reactions, the number and kinds of *atoms*, and the sum of the *charges*, must be the same on each side. Check each of the three reactions above to see that both atoms and total charge are conserved.

Balancing Charge in Half-Reactions

To balance redox reactions, it is often necessary to balance the half-reactions first.

> To balance half-reactions, use these steps:
>
> 1. First *write coefficients* that balance the atoms.
>
> 2. Then *add electrons* (e^-) to balance charge.

 TRY IT

Q. Apply those steps to balance this half-reaction: $Al \rightarrow$ Al^{3+}

Answer:

Each side has one Al atom, so the atoms are already balanced. The left side is neutral, and the right has a charge of 3+. For charge to be the same on each side, *you* must add three negative electrons to the right.

$$Al \rightarrow Al^{3+} + 3\,e^-$$ 1 Al atom and a zero total charge on *each side*.

Balanced.

The total *charge* must be the same on each side of an equation. In a half-reaction, each electron serves as a -1 charge that is used to balance charge. The rule is: both atoms and charge must balance.

 TRY IT

Q. Balance this half-reaction: $Br_2 \rightarrow$ Br^-

Answer:

First, to balance *atoms*, you must add *coefficients* (you cannot change the particle formulas). A coefficient of 2 must be added on the right.

$$Br_2 \rightarrow 2\,Br^-$$

That balances atoms, but the charge is not balanced.

The right side has a total charge of 2−, and the left is neutral. To balance charge, two electrons must be added to the left.

$$2\,e^- + Br_2 \rightarrow 2\,Br^-$$

The half-reaction is now balanced for atoms and charge.

PRACTICE

Check your answers as you go.

1. Add electrons to balance these half-reactions.

   a. $Co \rightarrow \quad Co^{2+}$

   b. $S \rightarrow \quad S^{2-}$

   c. $Cu^{2+} \rightarrow \quad Cu^{+}$

2. Balance these half-reactions.

   a. $H_2 \rightarrow \quad H^{+}$

   b. $I_2 \rightarrow \quad I^{-}$

   c. $Hg_2^{2+} \rightarrow \quad Hg^{2+}$

   d. $Sn^{4+} \rightarrow \quad Sn^{2+}$

Lesson 15.3   Oxidizing and Reducing Agents

Redox Agent Terminology

Redox reactions are a combination of **red**uction and **ox**idation. One reactant must be oxidized (*lose* its electrons), and another reactant must be reduced (*gain* the transferred electrons).

In a redox reaction, the reactant particle that is **oxidized** is the one that contains an atom that *loses* electrons. This particle is termed the **reducing agent (RA)**, because, by giving its electrons to another particle, it acts as an agent causing the other particle to be reduced.

The reactant particle that is **reduced** is the one that contains an atom or ion that *gains* electrons. This particle is termed the **oxidizing agent (OA)**, because, by taking electrons from another particle, it is the agent that causes the other particle to be oxidized. On each side of a redox reaction, there must be one oxidizing agent and one reducing agent.

The following definitions and behaviors must be committed to memory.

Redox Definitions

- **Oxidation** is the loss of electrons.
- **Reduction** is the gain of electrons.
- An **oxidizing agent** is a particle that *accepts* electrons in a reaction; it removes electrons from another atom.
- A **reducing agent** is a particle that *loses* electrons in a reaction; it donates electrons to another atom.

Redox Behavior

- In a **redox reaction**, one reactant must be an OA and one must be an RA.
- In a redox reaction, the **reducing agent** is **oxidized** as it loses electrons, and the **oxidizing agent** is **reduced** as it gains electrons.

TRY IT

Write a summary of the six bulleted points above. Practice writing your summary until you can list all six points from memory. Then write answers to the following questions.

For the reaction

$$Cu + 2\,Ag^+ \rightarrow 2\,Ag + Cu^{2+}$$

Q. Which symbol represents silver metal?

Answer:

The symbol for silver metal is **Ag**. By definition, a pure metal is an element: it is composed of electrically neutral atoms. Atom symbols written without a charge are understood to have a charge of zero.

The symbol Ag^+ represents silver *ion*. Compounds that contain metal ions do not polish, shape, or conduct electricity in the way that metals do.

Q. Is the copper metal being oxidized or reduced?

Answer:

In this reaction, the neutral Cu metal atom becomes a Cu^{2+} ion; it must lose two electrons to do so. As it loses electrons, the neutral Cu is being oxidized.

Q. Is the Ag^+ being oxidized or reduced?

Answer:

Since the neutral Cu reactant is being oxidized, the Ag^+ reactant must be being reduced.

The Ag^+ becomes Ag in the reaction. To do so, it must gain an electron. Particles that gain electrons in a reaction are being reduced.

Q. Is the Ag^+ acting as an oxidizing agent or a reducing agent?

Answer:

Since Ag^+ is being reduced, it must be an oxidizing agent. By accepting the electron from neutral copper, Ag^+ is the agent that causes Cu to be oxidized.

Add a label below each *reactant* identifying it as an oxidizing agent (OA) or reducing agent (RA), then circle the reactant particle being *oxidized*.

1. $Sn^{4+} + Ni \rightarrow Ni^{2+} + Sn^{2+}$

2. $Mg + 2H^+ \rightarrow H_2 + Mg^{2+}$

Identifying Agents Using Oxidation Numbers

In complex reactions, it is often helpful to assign oxidation numbers to atoms to identify which particles are being oxidized and reduced.

> **TRY IT**

Q. For the following redox reaction, do the following:

1. Assign oxidation numbers to each atom.

2. Write the formula for the reactant particle that is the oxidizing agent.

3. Write the formula for the reactant being oxidized.

$$2 MnO_4^- + Cl^- + H_2O \rightarrow 2 MnO_2 + ClO_3^- + 2 OH^-$$

Answer:

1. Assigning oxidation numbers:

   Each: $\underset{+7\ -2}{2\,MnO_4^-} + \underset{-1}{Cl^-} + \underset{+1-2}{H_2O} \rightarrow \underset{+4\ -2}{2\,MnO_2} + \underset{+5-2}{ClO_3^-} + \underset{-2\ +1}{2\,OH^-}$

   Sum: $\quad\ +7\ -8 \qquad -1 \qquad +2-2 \qquad +4\ -4 \qquad +5-6 \qquad -2\ +1$

2. The Mn atom is $+7$ on the left and $+4$ on the right, so it gains three electrons. MnO_4^- contains the atom that is being reduced, so **MnO_4^-** ion is an **oxidizing agent**.

3. The **Cl** is -1 in the reactants and $+5$ in the products. The **Cl^-** ion is therefore losing six electrons in the reaction and is the reactant being **oxidized**.

In general, strong oxidizing agents include O_2, CrO_4^{2-}, $Cr_2O_7^{2-}$, and substances or ions that include *per-* in their names, such as permanganate. Elements that are metals are often used as reducing agents. However, rather than memorize a list, to determine the role of particles in a reaction it is best to assign oxidation numbers and then analyze the gain and loss of electrons.

Flashcards

Make needed cards from the list below. Add cards of your own design that are helpful. Run the new cards three days in a row, and again before each quiz or test on this material.

One-Way Cards (with Notch)	Back Side—Answers
Oxidation number for atoms in an element	Zero
Usual oxidation number for oxygen atoms	−2
Oxidation number for O atoms in peroxides	−1
Normal oxidation number for hydrogen atoms	+1
Oxidation number for H atoms in hydrides	−1
Oxidation number for alkali metals in compounds	+1
Oxidation number for column 2 atoms in compounds	+2
The oxidation number of a monatomic ion is	Its charge
Sum of oxidation numbers must add up to	The charge on the particle
In reactions, reducing agents are	Oxidized
In reactions, oxidizing agents are	Reduced
In redox reactions, the reactants must include	One OA and one RA
The fundamental law of redox balancing	Electrons lost must equal electrons gained

Two-Way Cards (without Notch)

Reduction	The gain of electrons
Oxidation	The loss of electrons
Oxidizing agent	Particle that accepts electrons
Reducing agent	Particle that gives away electrons

PRACTICE B

Practice your redox flashcards until you can answer each card correctly. Then, for the problems below, assign oxidation numbers to each atom, label one reactant as an oxidizing agent (OA) and one reactant as a reducing agent (RA), and circle the reactant being *reduced*.

1. $2 \, Al + 3 \, NiCl_2 \rightarrow 2 \, AlCl_3 + 3 \, Ni$

2. $4 \, As + 3 \, HClO_3 + 6 \, H_2O \rightarrow 4 \, H_3AsO_3 + 3 \, HClO$

Lesson 15.4   Balancing Redox Using Oxidation Numbers

Trial and error can balance all chemical equations, but for many redox reactions, trial and error can be a slow process. Oxidation numbers can speed balancing by suggesting coefficients. A key principle in oxidation–reduction reactions is

The number of electrons lost by the reducing agent must equal the number gained by the oxidizing agent.

By assigning oxidation numbers, the gain and loss of electrons among atoms can be seen. Coefficients can then be assigned that balance the electron gain and loss, and balancing a redox equation can be completed by trial and error.

The oxidation number method of balancing is best learned by example.

► TRY IT

Q. In your notebook, write this redox equation, then complete the following steps.

$$NaClO_3 + HI \rightarrow I_2 + H_2O + NaCl$$

Using Oxidation Numbers To Balance Redox Reactions

1. Above the atom symbols, write the oxidation numbers for *each atom* in the equation. Write the sums below if needed.

   Each: $\quad$ +1 +5 −2 $\quad$ +1 −*1* $\quad$ 0 $\quad$ +1 −2 $\quad$ +1 −1
   $$NaClO_3 + HI \rightarrow I_2 + H_2O + NaCl$$

2. Identify the *two* atoms on *each side* that *change* their oxidation number in going from reactants to products.

   The two atoms that change oxidation numbers are **I** and **Cl**.

3. For each of those two atoms, draw arrows connecting the atom on one side to the same atom on the other side. Draw one arrow above and the other below the equation. Label each arrow with the *electron change* that must take place in going from one oxidation number to the other.

   To go from + 5 to −1, each chlorine must gain six negative electrons. To go from −1 to zero, each iodine atom must lose one electron.

   gained 6 e⁻

   Each: $\quad$ +1 +5 −2 $\quad$ +1 −*1* $\quad$ 0 $\quad$ +1 −2 $\quad$ +1 − 1
   $$NaClO_3 + HI \rightarrow I_2 + H_2O + NaCl \text{ (One arrow must lose}$$
   electrons, and the other
   lost 1 e⁻ $\qquad\qquad$ must gain electrons.)

4. Calculate the multiplier, based on the lowest common multiplier (LCM) by which each loss or gain can be multiplied so that the *total* electrons lost and gained are *equal*.

   Multiply the **I** arrow by 6 and the **Cl** arrow by 1.

   gained 6 e⁻ (*1x*)

   Each: $\quad$ +1 +5 −2 $\quad$ +1 −*1* $\quad$ 0 $\quad$ +1 −2 $\quad$ +1 −1
   $$NaClO_3 + HI \rightarrow I_2 + H_2O + NaCl$$
   lost 1 e⁻ (*6x*)

5. Rewrite the original equation, using the two *arrow multipliers* as trial coefficients for the two particles connected by the arrow.

   $$\textbf{1 } NaClO_3 + \textbf{6 } HI \rightarrow \textbf{6 } I_2 + \underline{\quad\quad} H_2O + \textbf{1 } NaCl$$

6. The **subscript tweak**. For each *atom* that *changed* oxidation number, if its subscript is *not* 1, divide the trial coefficient in front of its *particle* by the subscript of that atom.

STOP

    In this problem, we have a subscript tweak in one case out of four. One of the I atoms has a subscript of 2, so divide the trial coefficient of its particle by 2.

$$1\,NaClO_3 + 6\,HI \rightarrow {}^{3}\!\!\!\diagup\!\!\!_{6}\,I_2 + \underline{\quad}\,H_2O + 1\,NaCl$$

7. Write coefficients that finish the balancing by trial and error.

STOP

$$1\,NaClO_3 + 6\,HI \rightarrow 3\,I_2 + \underline{\mathbf{3}}\,H_2O + 1\,NaCl$$

8. **Check:** 1 Na, 1 Cl, 3 O, 6 H, and 6 I atoms on each side. Neutral charge on each side. The equation is balanced. Done!

Oxidation number balancing gives a few *trial* coefficients. These trial coefficients are good hints, but they are often not final answers.

In *all* methods of balancing, it is necessary to modify trial coefficients if needed, by trial and error, until the atoms and overall charge are the same on each side.

This method will become easier with practice.

PRACTICE

Write a "key word" outline or summary of the steps above. Learn it, then use it to balance these redox equations in your *notebook*. If you need help, peek at the answer and try again. Save one or two problems for your next practice session.

1. $Mn^{2+} + BiO_3^- + H^+ \rightarrow MnO_4^- + Bi^{3+} + H_2O$
2. $KI + O_2 + H_2O \rightarrow I_2 + KOH$
3. $Co + H^+ + NO_3^- \rightarrow Co^{2+} + N_2 + H_2O$
4. $Zn + H^+ + MnO_4^- \rightarrow Zn^{2+} + Mn^{2+} + H_2O$
5. $Al + HCl \rightarrow Al^{3+} + Cl^- + H_2$
6. Write the formula for the reactant that is the reducing agent in each of your balanced equations.
7. Write the formula for the reactant that is being oxidized in each of your balanced equations.

Lesson 15.5   Review Quiz for Chapters 13–15

You may use a calculator and a periodic table. Do the work needed to solve each problem on your own paper. Set a 25-minute limit, then check your answers at the end of the chapter.

1. Lesson 13.1: Label each of these compounds as soluble or insoluble in water.
   a. $(NH_4)_3PO_4$ b. $AgBr$ c. $Pb(NO_3)_2$
   d. $BaCO_3$ e. $CaCl_2$

2. Lesson 13.3: For the reaction $Pb(NO_3)_2 + KCl \rightarrow$

   a. Write a total ionic equation.   b. Write the net ionic equation.

3. Lesson 14.2: Supply coefficients for these reactants:

   $$HCl + \quad\quad Al(OH)_3 \rightarrow$$

4. Lesson 14.3: In the titration of a solid acid, an endpoint is reached after 22.0 mL of 0.120 M NaOH has been added. Assuming that each acid particle contains two acidic hydrogens (call it H_2R) and both are neutralized, how many moles of acid were in the acid sample?

5. Lesson 15.4: Balance:

   $$FeCl_2 + KMnO_4 + HCl \rightarrow MnCl_2 + FeCl_3 + H_2O + KCl$$

6. Lesson 15.3: In problem 5, which reactant is the reducing agent?

SUMMARY

1. Redox definitions:
   * **Oxidation** is the loss of electrons. **Reduction** is the gain of electrons.
   * An **oxidizing agent** is a particle that *accepts* electrons in a reaction; it removes electrons from another atom.
   * A **reducing agent** is a particle that loses electrons in a reaction by donating its electrons to another atom.
   * In redox reactions, **reducing agents** are **oxidized** as they lose electrons, and **oxidizing agents** are **reduced** as they gain electrons.

2. Rules for assigning oxidation numbers:
   * Each atom in an element is assigned an oxidation number of zero.
   * The oxidation number of an atom in a monatomic ion is the charge on the ion.
   * For particles that are not elements or monatomic ions, oxidation numbers are assigned as follows: Each oxygen (O) atom is assigned a -2 (except in peroxides, O is -1). Each hydrogen (H) atom is assigned a $+1$, except in metallic hydrides (compounds of a metal atom and hydrogen), where H is -1. Each alkali metal atom is assigned a $+1$. Each column 2 atom is assigned a $+2$. All other oxidation numbers are selected to make the sum of the oxidation numbers equal the overall charge on the molecule or ion.
   * The oxidation number of each individual atom in a chemical particle is the sum of the oxidation numbers for the atoms of that kind, divided by the number of atoms of that kind.

3. A key principle in oxidation–reduction reactions is this:

   The number of electrons lost by the reducing agent must equal the number gained by the oxidizing agent.

4. In balancing redox reactions, assigning oxidation numbers to assist in balancing provides *trial* coefficients. Always adjust the coefficients by trial and error at the last step in order to complete the balancing.

ANSWERS

Lesson 15.1

Practice   1. **C = +4**

Do individual, then sum of O.N., for O, H, and columns 1 and 2. Here, the sum for O is $2 \times (-2) = -4$
Do sum, then individual O.N., for remaining atoms. C must be +4, because CO_2 has a zero charge.

Oxidation number of *each* atom: $\boxed{+4}$ -2

Formula for the particle: CO_2

Sum of oxidation numbers for those atoms: $+4$ -4 (Must sum to zero.)

2. **Br = Zero**. Br_2 has one kind of atom and is neutral: it is an element. In elements, each atom has an oxidation number of zero.

3. **Mn = +2**. The oxidation number on a monatomic ion is equal to its charge.

4. **Br = −1**. If you recognize that $CaBr_2$ is ionic, write its monatomic ions. Each ion charge is its oxidation number. OR do individual, then sum of O.N. for O, H, and column 1 and 2. Ca is column 2: $1 \times (+2) = +2$
Do sum, then individual, for remaining atoms. Br must be −1 owing to the zero charge on $CaBr_2$.

Each: $+2$ -1

Formula: $CaBr_2$

Sum: $+2$ -2 (Must sum to zero.)

5. **N = +5**. Do O first. Each O is -2. $3 \times -2 = -6$. N must be **+5** to equal the **−1** overall charge.

Each: $+5$ -2

Formula: NO_3^-

Sum: $+5$ -6 (Must sum to −1.)

6. **Cr = +6**. Each Na is +1 (Na is an alkali metal). Each O is −2.

Each: $+1$ $+6$ -2

Formula: Na_2CrO_4

Sum: $+2$ $+6$ -8 (Must sum to zero.)

7. **Cr = +6**. Each O is −2: $\times 7 = -14$. The 2 Cr must sum +12, so each is +6.

Each: $+6$ -2

Formula: $Cr_2O_7^{2-}$

Sum: $+12$ -14 (Must sum to −2.)

8. **S = Zero**. S_8 is an element: one kind of atom, and neutral. Atoms in elements have an O.N. of zero.

9. **O = −1**. In *per*oxides, oxygen has an O.N. of −1.

10. **C = +3**. Each O is −2: $4 \times -2 = -8$. The two carbons must sum to +6, so **each** is +3

Each: $+3$ -2

Formula: $C_2O_4^{2-}$

Sum: $+6$ -8 (Must sum to −2.)

11. **As = +5**. Each H is +1. Each O is −2. (Individual O.N.) times (subscript) = O.N. sum

Each: $+1$ $+5$ -2

Formula: $H_3As O_4$

Sum: $+3$ $+5$ -8 (Must sum to zero.)

12. **Mn = +7**   13. **Cl = +5**   14. **Mn = +7**   15. **Hg = +1**

Lesson 15.2

Practice   1a. $Co \rightarrow Co^{2+} + 2e^-$ 1b. $S + 2e^- \rightarrow S^{2-}$ 1c. $1e^- + 1Cu^{2+} \rightarrow 1Cu^+$

2a. $H_2 \rightarrow 2H^+ + 2e^-$ 2b. $I_2 + 2e^- \rightarrow 2I^-$ 2c. $Hg_2^{2+} \rightarrow 2Hg^{2+} + 2e^-$ 2d. $2e^- + Sn^{4+} \rightarrow Sn^{2+}$

Lesson 15.3

Practice A   1. $Sn^{4+} + \boxed{Ni} \rightarrow Ni^{2+} + Sn^{2+}$
       OA   RA

2. $\boxed{Mg} + 2\,H^+ \rightarrow H_2 + Mg^{2+}$
   RA   OA

In problem 1, Sn^{4+} changes to Sn^{2+}, so it must gain two electrons and is being reduced. Because Sn^{4+} is accepting electrons, it is the oxidizing *agent*.

The Ni loses two electrons in reacting, so it is being *oxidized*. Because Ni gives electrons to the Sn^{4+}, it is a reducing agent. The particle being oxidized is the reducing agent.

In problem 2, the Mg on the left loses two electrons as it forms Mg^{2+}. It is therefore being *oxidized*. Because it is giving away its electrons, it is the reducing agent.

That means the H^+ must be the oxidizing agent. Two H^+ ions gain two electrons to form neutral hydrogen. In a redox reaction, the reactants must contain *both* an RA and an OA.

Practice B   1. Each: **0**   **+2**−1   **+3**−1   *0*

$2\,Al + 3\boxed{NiCl_2} \rightarrow 2\,AlCl_3 + 3\,Ni$
     RA   OA

Neutral Al and Ni atoms are elements. Atoms in elements have a zero oxidation number.

$NiCl_2$ and $AlCl_3$ are both *ionic* compounds because they combine metal and non-metal atoms. All of their atoms are monatomic ions. The oxidation numbers for the atoms are the charges of the ions.

Because the **Al** goes from neutral to 3+, it loses three electrons. A particle that loses electrons is being oxidized and is acting as a reducing agent.

The other reactant ($NiCl_2$) must therefore be the oxidizing agent and is being **reduced**. The Ni^{2+} ion gains two electrons in the reaction.

2. Each: **0**   +1**+5**−2   +1 −2   +1**+3**−2   +1**+1**−2

$4\,As + 3\boxed{HClO_3} + 6\,H_2O \rightarrow 4\,H_3AsO_3 + 3\,HClO$
     RA   OA

The **As** atom is an element with a zero oxidation number. After reaction, it has a +3 oxidation number. The As loses three electrons, is oxidized, and acts as a reducing agent.

The **Cl** atom goes from a +5 oxidation number in the reactants to +1 in the products. It gains four electrons. Particles that contain atoms that gain electrons in a reaction are being **reduced**. Because they accept electrons, those particles are oxidizing agents.

Lesson 15.4

Practice

Problem 1:

1. Calculate the oxidation numbers for each atom in the equation.

Each: **+2**   +5 −2   +1   **+7** −2   **+3**   +1−2
$$Mn^{2+} + BiO_3^- + H^+ \rightarrow MnO_4^- + Bi^{3+} + H_2O$$

2. Identify two atoms that change their oxidation numbers. **Mn and Bi**

3. Add arrows connecting each atom that changed O.N. Label with the *electron change*.

lost 5 e⁻

Each: **+2**   +5 −2   +1   **+7** −2   **+3**   +1 −2
$$Mn^{2+} + BiO_3^- + H^+ \rightarrow MnO_4^- + Bi^{3+} + H_2O$$
gained 2 e⁻

4. Label the lowest common multipliers on the arrows that result in the electron loss and gain being equal.

− 5 e⁻ (2x)

Each: **+2**   +5 −2   +1   **+7** −2   **+3**   +1 −2
$$Mn^{2+} + BiO_3^- + H^+ \rightarrow MnO_4^- + Bi^{3+} + H_2O$$
+ 2 e⁻ (5x)

5. Rewrite the original equation. Use each arrow multiplier value as a *coefficient* for the particles connected by the arrows.

$$2\,Mn^{2+} + 5\,BiO_3^- + ?\,H^+ \rightarrow 2\,MnO_4^- + 5\,Bi^{3+} + ?\,H_2O$$

6. The *subscript tweak*. Not needed—none of the Bi or Mn atoms have subscripts.

7. Complete balancing by trial and error.

   15 O on the left. Must be 7 H_2O on right to get 15 O on right. 7 H_2O means 14 H^+ on left.

$$2\,Mn^{2+} + 5\,BiO_3^- + \mathbf{14}\,H^+ \rightarrow 2\,MnO_4^- + 5\,Bi^{3+} + \mathbf{7}\,H_2O$$

8. Check: 2 Mn, 5 Bi, 15 O, and 14 H atoms on each side. **+13** charge on each side. Balanced.

Problem 2:

1. Calculate the O.N. for each atom.

   Each: $+1$ $\boxed{-1}$ $\boxed{0}$ $+1{-}2$ $\boxed{0}$ $+1\boxed{-2}+1$

$$KI + O_2 + H_2O \rightarrow I_2 + KOH$$

   O_2 and I_2 are *elements*: neutral particles containing only one kind of atom.

2. Identify two atoms on each side that change their oxidation number. **I** and **O**.

3. Write the electron change for each atom.

4. Calculate an arrow multiplier for each to get the same number of electrons in both.

5. In the original equation, add the multipliers as coefficients for connected particles.

$$\underline{2}\,KI + \underline{1}\,O_2 + \underline{?}\,H_2O \rightarrow \underline{2}\,I_2 + \underline{1}\,KOH$$

6. The *subscript tweak*. Because both the **O** on the left and the **I** on the right have a subscript of 2, divide the coefficient of their particles by 2. Fractions are OK at this point.

$$2\,KI + {}^{1/2}\underline{1}\,O_2 + \underline{?}\,H_2O \rightarrow {}^{1}\underline{2}\,I_2 + 1\,KOH$$

7. Finish by trial and error. With 2 K on the left, there must be 2 on the right, so adjust the trial KOH coefficient. With 2 H now on the right, one water is needed on the left.

$$2\,KI + 1/2\,O_2 + \underline{1}\,H_2O \rightarrow 1\,I_2 + \mathbf{2}\,KOH$$

   Coefficients that are fractions are sometimes required in balanced equations, and this *is* a legitimate balanced equation. However, if fractions are not required, lowest whole-number coefficients are preferred. Multiplying all coefficients by 2 to eliminate the fraction results in

$$4\,KI + 1\,O_2 + \mathbf{2}\,H_2O \rightarrow 2\,I_2 + \mathbf{4}\,KOH$$

   Both equations in step 7 are acceptable answers.

8. *Check*. Based on the second step 7 equation: 4 K, 4 I, 4 O, and 4 H atoms on each side. Zero charge on each side. Balanced!

   Note that the trial coefficients had to be adjusted in step 7 for the K atoms to balance. The coefficients found by oxidation numbers are good *hints*; but always adjust by trial and error at the end, if needed, until the equation is balanced.

Problem 3:

1. Calculate the O.N. for each atom.

2. Identify two atoms that change oxidation number. **Co** and **N**

3. Write the electron change for each atom.

$$\overset{\displaystyle \overset{-2\,e^-}{\overbrace{}}}{\underset{\displaystyle \underset{+5\,e^-}{\underbrace{}}}{\text{Each: } \boxed{0} \quad +1 \quad \boxed{+5}\,-2 \quad \boxed{+2} \quad \boxed{0} \quad +1\,-2}}$$

$$\text{Co} + \text{H}^+ + \text{NO}_3^- \rightarrow \text{Co}^{2+} + \text{N}_2 + \text{H}_2\text{O}$$

4. Add the lowest arrow multiplier that results in electron loss = electron gain.

$$\overset{\displaystyle \overset{(5\text{x}) \,-2\,e^-}{\overbrace{}}}{\text{}}$$

$$\boxed{0} \quad +1 \quad \boxed{+5}\,-2 \quad \boxed{+2} \quad \boxed{0} \quad +1\,-2$$

$$\text{Co} + \text{H}^+ + \text{NO}_3^- \rightarrow \text{Co}^{2+} + \text{N}_2 + \text{H}_2\text{O}$$

$$\underset{\displaystyle \underset{(2\text{x}) \,+5\,e^-}{\underbrace{}}}{\text{}}$$

5. To the original equation, add the multipliers as *coefficients* for both connected particles.

$$5\,\text{Co} + \underline{\qquad}\,\text{H}^+ + 2\,\text{NO}_3^- \rightarrow 5\,\text{Co}^{2+} + 2\,\text{N}_2 + \underline{\qquad}\,\text{H}_2\text{O}$$

6. The *subscript tweak*. Since N on the right has a subscript 2, divide its coefficient by 2.

$$5\,\text{Co} + \underline{\qquad}\,\text{H}^+ + 2\,\text{NO}_3^- \rightarrow 5\,\text{Co}^{2+} + 1\!\!\!\!/2\,\text{N}_2 + \underline{\qquad}\,\text{H}_2\text{O}$$

7. Complete by trial and error.

The 6 O atoms on the left require 6 H_2O on the right. That means 12 H^+ must be on the left.

$$5\,\text{Co} + \mathbf{12}\,\text{H}^+ + 2\,\text{NO}_3^- \rightarrow 5\,\text{Co}^{2+} + \mathbf{1}\,\text{N}_2 + \mathbf{6}\,\text{H}_2\text{O}$$

8. *Check*: 5 Co, 12 H, 2 N, and 6 O atoms on each side. +**10** charge on each side. Balanced!

Problem 4: $5\,\text{Zn} + \mathbf{16}\,\text{H}^+ + 2\,\text{MnO}_4^- \rightarrow 5\,\text{Zn}^{2+} + 2\,\text{Mn}^{2+} + 8\,\text{H}_2\text{O}$

Problem 5: $2\,\text{Al} + 6\,\text{HCl} \rightarrow 2\,\text{Al}^{3+} + 6\,\text{Cl}^- + 3\,\text{H}_2$

Problem 6:   The reducing agent is the reactant particle that is losing electrons and donating them to another particle.

1. **Mn^{2+}** is losing five electrons.   2. In **KI**, I^- is losing one electron.   3. **Co**   4. **Zn**   5. **Al**

Problem 7:   The particle being oxidized is the reducing agent. Answers are the same as in problem 6.

Lesson 15.5

1a. Soluble   1b. Insoluble   1c. Soluble   1d. Insoluble   1e. Soluble

2a. $\text{Pb}^{2+} + 2\,\text{NO}_3^- + 2\,\text{K}^+ + 2\,\text{Cl}^- \rightarrow \text{PbCl}_2(s) + 2\,\text{K}^+ + 2\,\text{NO}_3^-$ [Phase is (*aq*) if not specified.]

2b. $\text{Pb}^{2+} + 2\,\text{Cl}^- \rightarrow \text{PbCl}_2(s)$

3. $\mathbf{3}\,\text{HCl} + \mathbf{1}\,\text{Al(OH)}_3 \rightarrow$

4. **0.00132 mol** or $\mathbf{1.32 \times 10^{-3}}$ **mol** of the acid H_2R

5. $5\,\text{FeCl}_2 + 1\,\text{KMnO}_4 + 8\,\text{HCl} \rightarrow 1\,\text{MnCl}_2 + 5\,\text{FeCl}_3 + 4\,\text{H}_2\text{O} + 1\,\text{KCl}$

6. **FeCl_2**   (*Or* the **Fe^{2+}** ion.)

16

Ideal Gases

Lesson 16.1 — Gas Fundamentals

Gas Quantities and Their Units

Chemistry is most often concerned with matter in three states: gas, liquid, and solid. The gas state is in most respects the easiest to study, because by most measures, gases have similar and highly predictable behaviors. Gas quantities can be measured using **four** variables: pressure, volume, temperature, and moles of gas particles.

The symbols for these variables are P, V, T, and n.

Gas Pressure

In an experiment, a glass tube about 100 centimeters (1 meter) long is sealed at one end and then filled with mercury, an element that is a dense, silver-colored metal (symbol Hg) and is a *liquid* at room temperature. The open end of the filled tube is covered, the tube is turned over, and the covered end is placed under the surface of additional mercury in a partially filled beaker. The tube end that is under the mercury in the beaker is then uncovered.

What happens? In all experiments at standard atmospheric pressure, the result is the same. The top of the mercury in the tube quickly falls from the top of the tube until it is *about* 76 centimeters above the surface of the mercury in the beaker. There, the mercury descent stops. The result is a column of mercury inside the glass tube that is about 76 centimeters high. What is in the tube *above* the mercury column? A bit of mercury vapor, but no air. The space above the mercury in the tube is mostly empty: close to a **vacuum**.

What happens if the top of the tube is snapped off? The mercury inside the tube behaves the same as a straw full of liquid when you take your finger off the top. The liquid mercury in the tube falls quickly until it reaches the same level as the mercury in the beaker. However, as long as the tube is sealed and is longer than 76 centimeters, the top of the mercury in the tube in will remain about 76 centimeters above the top of the mercury in the beaker.

Why? The device created in this experiment is a mercury **barometer**. It measures the **pressure** of the air outside the tube.

A barometer is a kind of balance, like a playground teeter-totter or seesaw. The pressure of the dense column of mercury *in* the tube, pressing down on the top of the pool of mercury in the beaker, is *balanced* by the pressure of the 20-mile-high column of air, the atmospheric pressure, pressing down on the pool of Hg *outside* the tube.

We could construct our barometer using water as the liquid. However, for water to balance the air, since water is about 13.5 times less dense than mercury, our tube would need to be 13.5 times higher, about 34 feet high, roughly three to four stories on a typical building.

If the air pressure outside the mercury column increases, the mercury is pushed higher in the tube. If the surrounding atmospheric pressure is lowered, the mercury level in the tube falls.

When a weather forecast states that "the barometric pressure is 30.04 inches and falling," it is describing the height of the mercury column in a barometer (76 centimeters is about 30 inches). In meteorology, a high pressure system is usually associated with fair weather. Falling barometers and low pressure systems are often associated with clouds, storms, and precipitation.

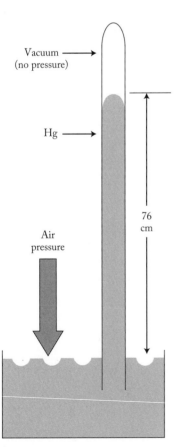

Vacuum
(no pressure)

Hg

Air
pressure

76
cm

Measuring Pressure

In the branches of science, pressure is measured in a variety of units.

Chemistry defines **standard pressure** as a gas pressure of exactly 760 millimeters (76 centimeters) of mercury as measured by a barometer. This is also known as exactly **one atmosphere (atm)** of pressure. Normal atmospheric pressure at sea level on a fair-weather day is *about* one atmosphere.

The SI unit for pressure is termed the **pascal** (abbreviated as Pa) in honor of the 17th-century French mathematician and scientist Blaise Pascal, whose experiments with gases led him to propose the concept of a vacuum. (Pascal's contemporary, the scientist and mathematician René Descartes, disagreed with the vacuum concept, writing that Pascal had "too much vacuum in his head.")

The following table of pressure units must be memorized. These equalities will be used to convert among pressure units.

Pressure Units
Standard pressure $\equiv$ 1 atm ($\equiv$ means "is *defined* as equal to")
$\equiv$ 760 mm Hg (mercury) $\equiv$ 760 torr
= 101 kilopascals (kPa) (not exact; not a definition)
= 1.01 bars (abbreviated *bar*)

Any two of those measures can be used in a *conversion factor* for pressure units. Examples: 1 atm = 760 mm Hg *or* 101 kPa = 760 torr

PRACTICE **A**

Practice until you can write the table of pressure unit definitions from memory, then use the definitions as conversions in these calculations.

1. The lowest atmospheric pressure at sea level was recorded in 1979 during Typhoon Tip; a pressure of 870. millibars. What is this pressure in kilopascals?

2. If 2.54 cm $\equiv$ 1 in. (exactly), and standard pressure is defined as exactly 760 mm Hg, what is standard pressure in inches of mercury?

3. Standard pressure in English units is 14.7 pounds per square inch (psi). If a bicycle tire has a pressure of 72 psi, how many atmospheres would this be?

4. How many kilopascals is 25.0 torr?

Gas Volumes

Gas volumes are measured in metric volume units such as liters (dm^3) and mL (cm^3).

Gas Temperature

Temperature is defined as the **average kinetic energy** of particles.

Kinetic energy is energy of **motion**, calculated by the equation

$$\text{Energy of motion} = \tfrac{1}{2}(\text{mass})(\text{velocity})^2, \text{ or } \mathbf{KE = 1/2\ mv^2}$$

Since the chemical particles of a substance have a constant mass, this equation means that when molecules move *twice* as fast, they must have "2 squared," or *four* times as much kinetic energy, and their **absolute temperature** must be four times higher.

One of the implications of this equation is that, though particles cannot have zero mass, they *can* have zero velocity: they can (in theory) stop moving.

If all of the molecules in a sample had zero velocity, the temperature of the sample would be $-273.15°C$, which is defined as **absolute zero**. Absolute zero is the *bottom* of the temperature scale. Nothing can be colder than absolute zero.

The **Kelvin** (or **absolute**) temperature scale simplifies the mathematics of calculations based on gas temperatures. The Celsius scale (abbreviated °C) defines **0** degrees as the melting and freezing temperature of water and 100 degrees as the boiling temperature of water at standard pressure. The Kelvin scale keeps the *same size degree* as Celsius, but defines **0** as absolute zero.

The equation relating the Kelvin and Celsius scales is $K = °C + 273$ (use 273.15 in calculations if additional precision is needed). This equation must be memorized. The SI unit that measures absolute temperature is termed the **kelvin**. A temperature in kelvins is always 273 degrees *higher* than the temperature in degrees Celsius.

Temperature measurements must be converted to kelvins when using any equation which specifies a capital *T*.

In measurements, the *abbreviation* for kelvins is a capital **K** without a degree symbol. In equations, the *symbol* for kelvins is a capital *T*. A lowercase *t* is generally used as a symbol for temperature in degrees *Celsius*.

When measuring gases, it is preferred to record gas volumes at **standard temperature** (for gas laws), which is the temperature of an ice–water mixture: zero degrees Celsius (273 K).

PRACTICE B

1. Commit to memory the equation relating kelvins to degrees Celsius, then complete the following chart (assume boiling and freezing points are at *standard* pressure):

	In Kelvins	In Degrees Celsius
Absolute Zero	_____	_____
Water Boils	_____	_____
Nitrogen Boils	_____	$-196°C$
Table Salt Melts	1074 K	_____
Water Freezes	_____	_____
Standard Temperature	_____	_____

2. In a problem involving gases, you calculate a temperature for the gas of $-310°C$. What do you notice about your answer?

Lesson 16.2 | Gases at STP

Standard Temperature and Pressure

In experiments with gases, it is preferred to measure gas *volumes* at **standard temperature and pressure**, abbreviated **STP**. The following values must be committed to memory.

> Standard temperature for gases $\equiv 0°C = 273.15$ K
>
> Standard pressure $\equiv 1$ atm $\equiv 760$ mm Hg $\equiv 760$ torr
>
> $= 101$ kPa $= 1.01$ bar

Molar Gas Volume at STP

Gases have remarkably consistent behavior. One important example involves gas volumes: if two samples of gas have the same number of gas molecules and they are at the same temperature and pressure, the two samples will (in most cases) occupy the same volume. This rule is true under most conditions even if the gases have different molar masses or different molecular formulas. It is also true for mixtures of different gases.

In addition, if we know the temperature, pressure, and number of moles of gas particles in a sample, the gas volume will be predictable.

Example: In *most* cases, for one mole of any gas or any mixture of gas molecules, the volume at STP will be 22.4 liters.

The volume of a mole of gas at STP (the molar volume) provides us with a conversion factor for gas calculations at STP. Our rule will be

> **The STP Prompt**
>
> In calculations, if a gas is stated to be *at STP or* is at a T and P that are equivalent to STP, write in the DATA:
>
> **1 mol gas = 22.4 L gas at STP**

For gases at STP, we will use the STP prompt to solve calculations by conversions.

Attaching P and T to V

Note that in the above equality, "at STP" is attached only to the gas *volume*. Because gas volumes vary with temperature and pressure, *all* gas volumes *must* have pressure and temperature conditions stated if the volume is to be a measure of the number of particles in the sample.

The rule is: Gas *volumes* must be labeled with a T and P if the T and P are known.

In calculations, when you write gas volumes (with units such as L or mL), attach a P and T.

Nonideal Behavior

When pressure is increased and/or temperature is decreased, at some point all gases condense (become liquids or solids). When gases approach a pressure and temperature at which they condense, the *ideal* gas assumptions of the STP prompt begin to lose their validity. At pressures above standard pressure, variations from *ideal* behavior can also become substantial. However, unless a problem indicates nonideal gas behavior, if the conditions are at STP, you should assume that the STP mole-to-volume relationship applies.

PRACTICE   A

1. Write values for standard pressure using five different units.

2. Write values for standard temperature using two different units.

Calculations for Gases at STP

The STP prompt (relating gas liters to moles) is often used in conjunction with

*   The grams prompt (write the grams-to-moles relationship)
*   The Avogadro prompt (write the particles-to-moles relationship)

The purpose of these prompts is to get into your DATA table the conversions that will allow you to convert between the units in the problem and *moles*. In chemistry, most fundamental relationships are defined as ratios that include moles. Having the relationships in the DATA table that relate the units to moles will allow you minimize interruptions while applying conversions to solve.

Together, these prompts allow you to solve most STP gas calculations using conversions. You will also need to recall that if a problem asks for

*   Molar mass, you WANT grams *per* 1 mole
*   Density, you WANT one mass unit (such as g or kg) over one volume unit (such as L, dm^3, mL)

 TRY IT

(See "How to Use These Lessons," point 1, p. xv.)

Complete the following problem in your notebook, then check your answer below.

Q. 2.0×10^{23} molecules of NO_2 gas would occupy how many liters at STP?

STOP   **Answer:**

WANTED: ? L NO_2 gas at STP =   (First, write the *unit* WANTED.)

DATA: 2.0×10^{23} molecules of NO_2 gas

6.02×10^{23} molecules = 1 mol   (Avogadro prompt.)

1 mol any gas = 22.4 L any gas at STP   (STP prompt.)

Strategy: Want a single unit? Start with a single unit, and find moles first.

SOLVE:

$$? \, \mathbf{L} \, NO_2(g) \, STP = 2.0 \times 10^{23} \text{ molecules } NO_2 \cdot \frac{1 \text{ mol}}{6.02 \times 10^{23} \text{ molecules}}$$

$$\cdot \frac{22.4 \, \mathbf{L} \text{ gas STP}}{1 \text{ mol gas}} = \mathbf{7.4 \, L} \, NO_2(g) \text{ at STP}$$

Note the difference between these calculations and stoichiometry. The above problem involved only *one* substance. If stoichiometry steps are needed, you will see DATA for *two* substances involved in a chemical *reaction* and you will need a mol–mol conversion when you solve.

Working the Examples

An effective technique in learning to solve problems in the physical sciences and math is to *work the examples* in the textbook. To do so: *cover* the answer, read the question, and try to solve. If you need help, peek at the answer, and then try the question again.

 TRY IT

Apply that method to this problem.

Q. Determine the density of O_2 gas at STP in grams per milliliter.

Hint: When solving for a *ratio*, if a calculation involves only *one* substance, the rule "want a ratio, start with a ratio" will usually solve faster than "solve for the top and bottom units separately."

Answer:

WANTED: $? \dfrac{g \, O_2 \text{ gas}}{mL \text{ at STP}}$ (It helps to write complete labels.)

DATA: $32.0 \, g \, O_2 = 1 \text{ mol } O_2$ (*Grams* prompt in WANTED unit.)

1 mol any gas = 22.4 L any gas at STP   (STP in WANTED unit.)

SOLVE: (To review solving for a ratio, see Lesson 10.2.)

$$? \, \frac{\mathbf{g \, O_2} \text{ gas}}{mL \text{ gas STP}} = \frac{32.0 \, \mathbf{g \, O_2}}{1 \text{ mol } O_2} \cdot \frac{1 \text{ mol gas}}{22.4 \text{ L gas STP}} \cdot \frac{10^{-3} \, L}{1 \, \mathbf{mL}} = \frac{0.00143 \, \mathbf{g \, O_2}}{mL \text{ at STP}}$$

Your answer should have those three conversions, right-side up, in any order.

Note that STP was attached to each gas *volume* unit (mL and L).
Note also that

- The conversions for *molar mass* and *particles per mole* are valid whether the substance is a gas, liquid, or solid.
- The *STP* prompt, however, only works for gases and only works at STP.

Make certain that you can write all three of the above prompts from memory. Then do all of problem 1 below. After that, do every other problem, and more if you need more practice.

1. Write one or more possible metric *units* WANTED when you are asked to find

    a. Molarity   e. Density

    b. Molar mass   f. Speed or velocity

    c. Volume   g. Gas pressure

    d. Mass   h. Temperature

2. Calculate the density of SO_2 gas in g/L at STP.

3. The density of a gas at STP is $0.00205 \text{ g} \cdot \text{mL}^{-1}$. What is its molar mass?

4. If 250. mL of a gas at STP weighs 0.313 g, what is the molar mass of the gas?

5. Calculate the number of molecules in 1.12 L of CO_2 gas at STP.

6. Calculate the volume in milliliters of 15.2 g of F_2 gas at 273 K and standard pressure.

7. If 0.0700 mol of a gas has a volume of 1,760 mL, what is the volume of one mole of the gas in liters under the same temperature and pressure conditions?

8. Calculate the density in $\text{kg} \cdot \text{L}^{-1}$ of radon gas at STP.

Lesson 16.3   Cancellation of Complex Units

Solving gas law calculations requires working with **complex units**: those with *reciprocal* units or *more* than one unit in the numerator or denominator. Let's review the mathematics of working with fractions.

Why Not to Write "*A/B/C*"

When solving complex fractions, the notation for the fractions must be handled with care.
    Why?

TRY IT

Q. On this page, write answers to the following.

    A. $\dfrac{8}{4}$ divided by 2 =

    B. 8 divided by $\dfrac{4}{2}$ =

    C. 8/4/2 =

STOP **Answer:**

A. 2 divided by 2 = **1**   B. 8 divided by 2 = **4**

C. Could be **1** or **4**, depending on which is the fraction: 8/4 or 4/2.

The format *A/B/C* for numbers or units is ambiguous unless you know, from the context or from prior steps, which is the fraction.

Let's try problem C again, this time with the fraction identified by parentheses.

 **TRY IT**

Use the rule: Perform the operation inside the parentheses *first*.

D. (8/4)/2 =   E. 8/(4/2) =

STOP **Answer:**

D. 2/2 = 1   E. 8/2 = 4

With the parentheses, the problem is easy. But in part C, for 8/4/2 without the parentheses, you cannot be sure of the right answer. For this reason, writing numbers or units in a format *A/B/C* should be avoided.

In these lessons, we will either use a *thick* underline or group fractions in parentheses to distinguish a numerator from a denominator.

Multiplying and Dividing Fractions

By definition:

$$1 \text{ divided by } (1/X) = \text{ the } \textbf{reciprocal} \text{ of } (1/X) = \frac{1}{\dfrac{1}{X}} = X$$

Recall that $1/X$ can be written as X^{-1}. Another way to state the relationships above is to apply the rule: When you take an exponential to a power, you multiply the exponents.

$$(X^{-1})^{-1} = (X^{+1}) = X$$

The rule $1/(1/X) = X$ is an example of this general rule:

Complex Fraction Rule 1

To simplify the *reciprocal* of a *fraction*, invert the fraction.

TRY IT

Apply rule 1 to the following problem.
   Q. Remove the parentheses and simplify: $1/(B/C) =$

STOP   **Answer:**

The reciprocal of a fraction simplifies by inverting the fraction.
   $1/(B/C)$ simplifies by inversion (flips the fraction over) to **C/B**.

In symbols: $\dfrac{1}{\dfrac{B}{C}} = \dfrac{C}{B}$

In exponents: $(B/C)^{-1} = (B \cdot C^{-1})^{-1} = B^{-1} \cdot C$

Calculations in chemistry may involve fractions in the numerator, denominator, or both. To handle these cases systematically, use the following rule.

Complex Fraction Rule 2

When a term has *two* or more fraction lines (either _____ **or** /), *separate* the terms that are *fractions*. To do so, apply these steps in this order:

   a. If a term has a fraction in the *denominator*, separate the terms into (1/fraction in the denominator) multiplied by the remaining terms.

   b. If there is a fraction in the *numerator*, separate and multiply that fraction by the other terms in the fraction.

   c. Then simplify: **invert** any reciprocal *fractions*, cancel units that cancel, and multiply the terms.

Let's learn these rules by completing some examples.

TRY IT

Solve the questions below in your notebook.
   Q. Remove the parentheses and simplify: $A/(B/C)$.
   Apply rule 2a, then 2c.

STOP   **Answer:**

$$A/(B/C) = \dfrac{A}{\dfrac{B}{C}} = A \cdot \dfrac{1}{\dfrac{B}{C}} = A \cdot \dfrac{C}{B} = \dfrac{A \cdot C}{B}$$

                    (rule 2a)   (rule 2c)

Since there is a fraction in the denominator, separate that fraction into a reciprocal times the other terms, then invert the reciprocal fraction and multiply.
   You can also solve using exponents: $A \cdot (B \cdot C^{-1})^{-1} = A \cdot B^{-1} \cdot C$

Q. Simplify $(A/B)/C$; apply rule 2b.

 Answer:

Note these three different but equivalent ways of representing this problem, and then how the answer differs from the previous example.

$$(A/B)/C = \frac{\dfrac{A}{B}}{C} = \frac{A}{B} \cdot \frac{1}{C} = \frac{A}{B \cdot C}$$

(rewrite for clarity)   (rule 2b)

Since there is a fraction in the numerator, separate that fraction from other terms. A way to summarize rule 2 is:

To simplify a term with more than one fraction, separate the fractions.

Q. Simplify $(A/B)/(C \cdot (D/E)) =$

 Answer:

$$(A/B)/(C \cdot (D/E)) = \frac{\dfrac{A}{B}}{C \cdot \dfrac{D}{E}} = \frac{A}{B} \cdot \frac{1}{\dfrac{D}{E} C} = \frac{A}{B} \cdot \frac{1}{C} \cdot \frac{1}{\dfrac{D}{E}} = \frac{A \cdot E}{B \cdot C \cdot D}$$

(rewrite for clarity)   (rule 2a)   (rule 2b)   (rule 2c)

In solving problems, when *you* are *writing* terms with two or more fractions, you will need to develop a systematic way to distinguish fractions in the numerator and denominator in cases where confusion may occur.

PRACTICE   **A**

Learn the rules, then simplify these.

1. $X/(Y/Z) =$

2. $(D/E)^{-1} =$

3. $\dfrac{\text{meters}}{\dfrac{\text{meters}}{\text{second}}} =$

4. $\dfrac{(\text{meters/second})}{\text{second}} =$

5. meters/second/second $=$

Dividing Complex Numbers and Units

For some derived quantities and constants, units are complex fractions. Examples that we will soon encounter include

$$\text{The gas constant} = \boldsymbol{R} = 0.0821 \; \frac{\text{atm} \cdot \text{L}}{\text{mol} \cdot \text{K}}$$

$$\text{The specific heat capacity of water} = c_{\text{water}} = 4.184 \; \text{joule/gram} \cdot \text{K}$$

Joule (J) is a unit that measures energy.

In those units, the *dot* between two units means that the two units are multiplied together in either the numerator or denominator.

Example: In 4.184 joule/**gram** · K, grams and kelvins are both in the *denominator*.

$$4.184 \; \text{J/g} \cdot \textbf{K} = 4.184 \; \text{J/(g} \cdot \textbf{K}) = 4.184 \; \text{J} \cdot \textbf{g}^{-1} \cdot \textbf{K}^{-1}$$

When multiplying and dividing terms with complex units, you must do the math for both the numbers and the units. When units are found in *fractions*, unit cancellation follows the rules of algebra reviewed in the section above.

✏️━━━━━▶ TRY IT

Q. Solve for the number and the unit of the answer.

$$\frac{360 \; \text{J}}{18.0 \; \text{K} \cdot 0.50 \; \dfrac{\text{J}}{\text{g} \cdot \text{K}}} =$$

Answer:

Note how the units are rearranged in each step.

$$\frac{360 \; \text{J}}{18.0 \; \text{K} \cdot 0.50 \; \dfrac{\text{J}}{\textbf{g} \cdot \textbf{K}}} = \frac{360 \; \text{J}}{18.0 \; \text{K}} \cdot \frac{1}{0.50 \; \dfrac{\text{J}}{\textbf{g} \cdot \textbf{K}}} = \frac{360 \; \text{J}}{18.0 \; \text{K}} \cdot \frac{\textbf{g} \cdot \textbf{K}}{0.50 \; \text{J}} = \textbf{40. g}$$

Solving uses rule 2a: if there is a unit that has a fraction in the denominator, *separate* the fraction into *1/fraction in the denominator* multiplied by the other terms. Then, invert the reciprocal, cancel units that cancel, and multiply.

Mark the unit cancellation in the final step above.

Cancellation Shortcuts

When canceling numbers and units in complex fractions, you can often simplify by first canceling separately *within* the numerator and denominator, and then canceling *between* the numerator and denominator. However, when using this shortcut, you must remember that canceling a number or unit does *not* get rid of it: it replaces it with a 1.

TRY IT

Apply that rule to this problem.

Q. First cancel units in the denominator, then between the numerator and denominator:

$$\frac{360\text{ J}}{12\text{ K} \cdot 0.20\ \dfrac{\text{J}}{\text{g} \cdot \text{K}}} =$$

STOP

Answer:

$$\frac{360\text{ J}}{12\textbf{ K} \cdot 0.20\ \dfrac{\text{J}}{\text{g} \cdot \textbf{K}}} = \frac{360\ \cancel{\text{J}}}{12\textbf{ K} \cdot 0.20\ \dfrac{\cancel{\text{J}}}{\text{g} \cdot \textbf{K}}} = \frac{360}{12 \cdot 0.20}\ \cdot \frac{1}{\dfrac{1}{\textbf{g}}} = \textbf{150 g}$$

In the third term, the **1s** are important. *1/g* and *1/1/g* are not the same.

When in doubt about how to apply this type of unit cancellation, skip the short-cuts and use systematic rules 1–2c.

PRACTICE **B**

Apply the rules from memory to simplify these fractions.

1. $\dfrac{\text{calories}}{\dfrac{\text{calorie} \cdot \text{g}}{\text{g} \cdot {}^\circ\text{C}}} =$

2. $\dfrac{\text{atm} \cdot \text{L}}{(\text{mol}) \cdot \dfrac{\text{atm} \cdot \text{L}}{\text{mol} \cdot \text{K}}} =$

3. $\dfrac{(\text{mol}) \cdot \dfrac{\text{atm} \cdot \text{L}}{\text{mol} \cdot \text{K}} \cdot (\text{K})}{\text{L}} =$

Lesson 16.4   The Ideal Gas Law and Solving Equations

Ideal Gases

In calculations for a gas at STP, conversions using the STP prompt will often be the fastest way to solve. However, if gas conditions are at temperatures or pressures that are *not* STP, you will need to use gas *equations* to solve.

The gas equation encountered most frequently is the **ideal gas law** (often remembered as *piv-nert*). For all "ideal" gases,

$$PV = nRT$$

where

- P and V are pressure and volume in any metric units
- T is temperature in *kelvins*
- n represents the number of *moles* of gas
- R is a constant ratio: a number with ratio units called the **gas constant**

In this equation, P, V, T, and n are **variables**. They can have any values, depending on the conditions in the problem.

When using R to solve problems, R must have the same units as those used that are used to measure P and V in the problem. The *number* and *units* used for R will change depending on which P and V units are used. However, just as 12 fluid ounces is the same as 355 milliliters (our soda can equality), the different numbers and units used for R do not change the constant ratio that R represents.

For any gas, if you know any three of the four variables in the ideal gas law, and you know a value and units for R (which you are usually supplied with a problem), you can use the ideal gas law to predict the fourth variable.

One of the interesting implications of the ideal gas law is that if any of the three variables in $PV = nRT$ are the same for samples of different gases, the fourth variable will have the same value even when the gases have very different molar masses and chemical formulas.

Nonideal Behavior

For most gases, the ideal gas law is a good *approximation* in predicting behavior, unless the gas is either at pressures substantially above one atmosphere *or* close to conditions of pressure and temperature where the gas condenses to form a liquid or solid. In reality, all real gases display "nonideal" behavior if the temperature is low enough and/or the pressure is high enough. However, for the gas calculations in this chapter, you should assume that *ideal* gas conditions apply unless nonideal behavior is specified.

Solving Problems Requiring Equations

Calculations in chemistry can generally be put into three categories: those that can be solved with conversions, those that require equations, and those requiring both.

We will use the gas laws to develop a *system* for solving calculations that require equations. This system will be especially helpful if you take additional science courses.

Let's start with an easy problem, one in which we know the correct equation to use. This example can be solved in several ways, but we will solve with a method that has the advantage of working with more-difficult problems. Please try the method used here.

> **TRY IT**
>
> For this problem, do the steps below in your notebook.
>
> **Q.** A table of R values is usually supplied when solving gas problems. However, R can also be easily calculated for any set of units, based on values you know for standard temperature and pressure and the volume of a gas at STP.
>
> For example, one mole of any gas occupies a volume of 22.4 L at standard pressure (one atmosphere) and standard temperature (273 K).
>
> Use this data and the ideal gas law equation to calculate a value for the gas constant (R), using the units specified in this problem.

Steps for Solving Equations

1. This problem says to use the ideal gas law. *Write* the fundamental, memorized equation. Put a box around it to say: this is important.

2. Next, make a data table that includes each *symbol* in the equation.

   DATA: $P =$

   $V =$

   $n =$

   $R =$

   $T =$

3. Put a **?** after the *symbol* WANTED in the problem.

4. Read the problem again, and write each number and its unit after a symbol. Use units to match the symbol to the data (one ***mole*** goes after ***n***, 273 **K** goes after ***T***, etc.).

Do those steps, then check below.

At this point, your paper should look like this:

$$\boxed{PV = nRT}$$

DATA: $P =$ 1 atm

$V =$ 22.4 L

$n =$ 1 mol

$\boldsymbol{R} =$ **?**

$T =$ 273 K

5. Write SOLVE, and, using algebra, solve the fundamental, memorized equation for the *symbol* WANTED.

   The rule is: *memorize* equations in *one* fundamental format, then use algebra to solve for the symbols WANTED.

   Example: Memorize **K** = **°C + 273**, but do *not* memorize **°C** = **K − 273**

   If Celsius is WANTED, knowing kelvins, write the memorized equation above, then solve the equation for Celsius.

Do not plug in numbers until *after* you have solved for the WANTED symbol using symbols.

   Symbols move more quickly (and with fewer mistakes) than numbers with their units.

6. Plug in the numbers and solve. Cancel units when appropriate, but write the units that do not cancel after the number that you calculate for the answer.

Apply those steps to the current problem, then check your work below.

On your paper, you should have added this:

SOLVE: $PV = nRT$

$$? = R = \frac{PV}{nT} = \frac{(1 \text{ atm})(22.4 \text{ L})}{(1 \text{ mol})(273 \text{ K})} = \mathbf{0.0821} \frac{\mathbf{atm} \cdot \mathbf{L}}{\mathbf{mol} \cdot \mathbf{K}}$$

S.F.: Assume 1 is exact.

Be certain that your answers include the *units* after the numbers.

SUMMARY

To solve an equation when you know which equation is needed, use these steps.

1. Write the memorized equation.

2. Make a data table. On each line, put one symbol from the equation.

3. Based on units, after each symbol, write the matching DATA in the problem.

4. Solve your memorized equation for the WANTED **symbol** *before* plugging in numbers and units.

5. Put both numbers and units into the equation when you solve. If units cancel give an answer unit appropriate for the WANTED symbol, it is a check that the algebra was done correctly. If the units do not cancel properly, check your work.

PRACTICE

Complete all of these problems.

1. After each of these gas measurements, write a corresponding *symbol* from $PV = nRT$.

    a. 0.50 mol   b. 202 kPa   c. 11.2 dm^3 d. 373 K   e. 38 torr

2. In your notebook, solve $PV = nRT$ in symbols for

    a. $n =$ b. $T =$ c. $V =$

3. If the pressure of one mole of gas at STP increases to 2.3 atm but the volume of the gas is held constant, what must the new temperature be?
    To solve, use your calculated value above for R ($R = 0.0821$ atm $\cdot$ L/mol $\cdot$ K), the ideal gas law, and the method used in this lesson.

Lesson 16.5   Choosing Consistent Units

Consistent Units

When solving *most* equations in science, numbers must have units attached, and the units must cancel to result in the unit WANTED. In order for units to cancel, the units must be **consistent**. This means

For each fundamental or derived quantity in an equation, you must choose *one* unit to measure the quantity, then convert all data for that quantity to that unit.

Example: As one part of solving calculations using $PV = nRT$, you must pick a *pressure* unit (such as atmospheres, pascals, or torr), and convert all DATA that involves pressure to the unit you choose.

In some cases, an equation will require certain units.

Example: Gas law equations using a capital T require temperature to be measured in an absolute temperature scale. In the metric system, this means *kelvins*. A DATA temperature that is not in kelvins must be converted to kelvins. If degrees Celsius is WANTED, kelvins must be found first.

When using $PV = nRT$, consistent units are also important:

- When you must pick an R value to use, and
- When you must convert DATA to match the units of R

Let's take these cases one at a time.

Choosing a Unit for *R*

The gas constant R is one quantity, but it can be expressed in different units. Some of the equivalent values for R are below.

$$R = 0.0821 \; \textbf{atm} \cdot \text{L/mol} \cdot \text{K}$$
$$= 8.31 \; \textbf{kPa} \cdot \text{L/mol} \cdot \text{K}$$
$$= 62.4 \; \textbf{torr} \cdot \text{L/mol} \cdot \text{K}$$

Note that the *number* in each of those R values depends on the units used to measure P and V. However, just as a speedometer shows that 55 miles/hour is the same as 88 kilometers/hour, the different values for R do not change the *ratio* of the dimensions that R measures.

In gas calculations you generally will not need to memorize values for R, but you will need to select which R to use from a list of R values. If you need to choose an R, the rule is:

Pick the R that has *units* that most closely match the units in the DATA.

When solving ideal gas calculations, R, P, V, and T must have **consistent units**: the units used to measure P, V, and T when substituted into $PV = nRT$ must match the units attached to the chosen R. In some problems, this will require converting P, V, and/or T in the DATA table to the units of the chosen R. However, if the R chosen from the table above has units that are close to the units in the DATA for the problem, the conversions needed for P, V, and T will be minimized.

Keeping those factors in mind, try the following problem in your notebook.

Q. A sample of an ideal gas at 293 K and 202 kPa has a volume of 301 mL. How many moles are in the sample? (Use one of the *R* values in the table above.)

STOP

Answer:

The WANTED unit has the symbol *n*, 293 K is a *T*, 202 kPa is a *P*, and 301 mL is a *V*. What equation relates *n*, *T*, *P*, and *V*?

STOP

$PV = nRT$ Complete the DATA table using those symbols.

DATA: $P = 202$ kPa

$V = 301$ mL

$n = ?$ mol = WANTED

$R = ?$

$T = 293$ K

Which *R* value should you choose?

STOP

$R = 8.31$ **kPa** · L/mol · K is not an exact match with the DATA units, but because it uses kilopascals it is the closest. Add that *R* value to the DATA table.

One more change is needed. The units for *P*, *T*, and *n* match the chosen *R* units, but our best match for *R* uses liters, while the supplied *V* is in milliliters. The volume unit must be consistent: we must pick mL or L, or the units will not cancel and the equation will not work. Since we don't see an *R* value that uses milliliters, what's the best option?

STOP

Convert the supplied DATA to *liters* to match the volume unit in the closest *R*. Do so in your DATA table, then complete the problem.

STOP

DATA: $P = 202$ kPa

$V = 301$ mL = **0.301 L**   (By inspection; see Lesson 12.1.)

$n = ?$ mol = WANTED

$R = $ **8.31 kPa · L/mol · K**

$T = 293$ K

Are all of the units now consistent?
Yes.
SOLVE: $PV = nRT$ for the WANTED symbol *then* plug in numbers and units.

STOP

$$? = n = \frac{PV}{RT} = \frac{(202 \text{ kPa})(0.301 \text{ L})}{(293 \text{ K}) \, 8.31 \frac{\text{kPa} \cdot \text{L}}{\text{mol} \cdot \text{K}}} = \frac{(202 \text{ kPa})(0.301 \text{ L})}{(293 \text{ K})} \cdot \frac{\text{mol} \cdot \text{K}}{8.31 \text{ kPa} \cdot \text{L}}$$

$$\boxed{= 0.0250 \text{ mol}}$$

Check that the units cancel properly. They must. If you have done the problem correctly, they will. This unit cancellation is an essential check on your work.

PRACTICE

Use one of the following values to solve the following problem.

R = 0.0821 atm · L/mol · K = 8.31 kPa · L/mol · K = 62.4 torr · L/mol · K

If 0.0500 mol of an ideal gas at 0°C has a volume of 560. mL, what will be its pressure in torr?

Lesson 16.6 | The Combined Equation

Many gas law calculations involve the special case where the gas is trapped (the *moles* of gas in a sample does *not* change) while *P*, *V*, and/or *T are* changed.

If the number of gas moles is held constant, we can rewrite $PV = nRT$ as

$$\frac{PV}{T} = nR = (constant\ \text{moles})(\text{gas } constant\ R) = (\text{a new } constant)$$

When you multiply two constants, the result is new constant. While *R* is always a constant, the new constant above will only be true for the number of moles in the problem.

The equation above means that if gas conditions are changed while the moles of gas are held constant, the ratio "*P* times *V* over *T*" will keep the same numeric value no matter how you change *P*, *V*, and *T*.

Another way to express this relationship: as long as the number of particles of gas does not change, if you have an *initial* set of conditions P_1, V_1, and T_1, and you change to a *new* set of conditions P_2, V_2, and T_2, the ratio *PV/T* must stay the same. Expressed in the elegant and efficient shorthand that is algebra,

$$nR = \frac{P_1V_1}{T_1} = \frac{P_2V_2}{T_2}$$

when the moles of gas is held constant.

The boxed equation means that if five of the six variables among P_1, V_1, T_1, P_2, V_2, and T_2 are known, the sixth variable may be found using algebra, *without* knowing *n or* needing *R*.

We will call this relationship the **combined equation** (because it combines three historic gas laws). You may also see it referred to as the **two-point equation** since it is based on *initial* and *final* conditions. For problems in which the moles of gas particles do not change, when one set of conditions is *changed* to *new* conditions, the quickest way to solve is usually to apply

$$\frac{P_1V_1}{T_1} = \frac{P_2V_2}{T_2}$$

where P_1, V_1, and T_1 are *starting* gas measurements, and P_2, V_2, and T_2 are the *final* gas measurements, and the moles of gas stay the same.

This equation is often memorized by repeated recitation of "*P* one *V* one over *T* one equals *P* two *V* two over *T* two."

It may help to remember this rule:

> When a problem says a sample of gas is *sealed* or *trapped* or has *constant moles*, and the sample has a change in conditions, see if the *combined equation* symbols fit the data.

To solve a calculation in which the equation needed is known, apply the system developed in the previous lesson to solve equation calculations.

- *Write* and box the fundamental equation.
- Make a *data* table that lists the *symbols* in the equation.

In a problem that requires the combined equation, start by writing

$$\frac{P_1 V_1}{T_1} = \frac{P_2 V_2}{T_2}$$

and this data table.

DATA: $P_1 =$ $P_2 =$
       $V_1 =$ $V_2 =$
       $T_1 =$ $T_2 =$

Then,

- Put the numbers and their units from the problem into the table.
- Label the symbol WANTED with a "?". Add the units WANTED if they are specified.
- SOLVE the fundamental equation in *symbols* for the WANTED symbol using algebra *before* you plug in numbers.
- *Then* plug in numbers *and* units, and solve. Apply unit cancellation as well as number math. Include the units that do *not* cancel with your answer. Make sure that the answer unit fits the WANTED symbol.

TRY IT

Apply the method to this problem. Use the combined equation to solve.

Q. If a sample of gas in a sealed but flexible balloon at 273 K and 1.0 atm pressure has a volume of 15.0 L, and the pressure is increased to 2.5 atm while the temperature is increased to 373 K, what will be the new volume of the balloon?

STOP

$$\boxed{\frac{P_1 V_1}{T_1} = \frac{P_2 V_2}{T_2}}$$

DATA: $P_1 = 1.0$ atm $P_2 = 2.5$ atm

       $V_1 = 15.0$ L $V_2 = ?$

       $T_1 = 273$ K $T_2 = 373$ K

SOLVE: $? = V_2 = \dfrac{P_1 V_1 T_2}{P_2 T_1} = \dfrac{(1.0 \text{ atm})(15.0 \text{ L})(373 \text{ K})}{(2.5 \text{ atm})(273 \text{ K})} = \textbf{8.2 L}$

If the units do not cancel to give the WANTED unit, check your work.

PRACTICE   A

Check your answers as you go.

1. In your notebook, solve the combined equation in symbols for

   a. T_2 b. P_1 c. V_2

2. A gas cylinder with a volume of 2.50 L is at room temperature (293 K). The pressure inside the tank is 100. atm. When the gas is released into a 50.0 L container, the gas pressure falls to 2.00 atm. What will be the new temperature of the gas in kelvins, and in degrees Celsius?

Consistent Units

In gas law equations, a capital T means *absolute* temperature. Because the combined equation is derived from $PV = nRT$, when using metric units, both of those equations *require* that

- Temperatures must be converted to kelvins before substituting into the equation
- If a temperature that is not kelvins is WANTED, kelvins must be solved first, and then converted to other temperature units

For P and V, each variable must be converted to a *consistent* unit to solve. In most *combined* law calculations, since there is no R value with complex units that we are required to match, it does not matter which V or P units you choose.

- **V**olume may be in mL or L.
- **P**ressure can be converted to kPa, atm, torr, bar, or other pressure units.

However, *you* must choose *one unit* for P and for V, and all DATA for that quantity must be converted to that unit. It usually simplifies calculations if you convert DATA to the WANTED unit if it is specified, but any units will work in the equation as long as units are consistent. If non-WANTED units are found using the equation, simply convert those units to the units WANTED.

TRY IT

Keeping that in mind, try this problem.

Q. A spray can contains 250. mL of gas under 4.5 atm pressure at 27°C. How many liters would the gas occupy at 50.5 kPa and standard (std.) T?

Answer:

When one set of gas conditions is changed to new conditions, but moles are held constant, try the combined equation. Write

$$\boxed{\dfrac{P_1V_1}{T_1} = \dfrac{P_2V_2}{T_2}}$$

(If you need a fundamental equation, write it first. The more often you write it, the longer it will be remembered.)

DATA: $P_1 = 4.5$ atm $P_2 = 50.5$ kPa = **0.500 atm**

$V_1 = 250.\,\text{mL} = 0.250$ L $V_2 = ?\,\text{L}$ = WANTED

$T_1 = ?\,\text{K} = 27°\text{C} + 273 = 300.\,\text{K}$ T_2 in K = 273 K = std. T

Unit conversions:
T_1 *must* use K. K = °C + 273 = 27°C + 273 = **300. K**
For P_1, choose one of the units in the DATA: kPa or atm. Either choice will result in the same answer. If you choose atm,

$$P_2 = ?\,\text{atm} = 50.5\,\text{kPa} \cdot \frac{1\,\text{atm}}{101\,\text{kPa}} = \textbf{0.500 atm}\ (or\ P_1 = 455\,\text{kPa})$$

Since liters is WANTED but the initial volume is in milliliters, pick a unit and convert the other to it. It's best to convert DATA to the *answer* unit if it is specified.

$V_1 = ?\,\text{L} = 250.\,\text{mL} = \textbf{0.250 L}$ (By inspection: See Lesson 12.1.)

Solve for the WANTED symbol, *then* substitute the DATA.

STOP

$$\text{SOLVE: } ? = V_2 = \frac{P_1 V_1 T_2}{P_2 T_1} = \frac{(4.5\,\text{atm})(0.250\,\text{L})(273\,\text{K})}{(0.500\,\text{atm})(300\,\text{K})} = \textbf{2.0 L}$$

PRACTICE   B

A sealed sample of hydrogen gas occupies 500. mL at 20.°C and 150. kPa. What would be the temperature in degrees Celsius if the volume of the container is increased to 2.00 L and the pressure is decreased to 0.550 atm?

Simplifying Conditions

In a calculation involving a *change* from initial to final conditions for a constant moles of gas (gas in a sealed container), the combined equation is used to solve. However, if temperature does not change, converting to kelvins is not necessary. Why?
If

$$\frac{P_1 V_1}{T_1} = \frac{P_2 V_2}{T_2}$$

and $T_1 = T_2$, then

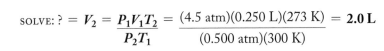

$$\frac{P_1 V_1}{\cancel{T_1}} = \frac{P_2 V_2}{\cancel{T_2}}$$

and the equation that is used is $P_1 V_1 = P_2 V_2$.
If the "before and after" temperatures are the same, T is not needed to solve.
When using the *combined* equation, if *any* two symbols have the same value in a problem, those symbols can be cancelled because they are the same on both sides.

TRY IT

Keep that rule in mind as you complete this question in your notebook.

Q. A sample of chlorine gas has a volume of 22.4 L at 27°C and standard pressure. What will be the pressure in torr if the temperature does not change but the volume is compressed to 16.8 L?

Answer:

$$\boxed{\dfrac{P_1V_1}{T_1} = \dfrac{P_2V_2}{T_2}}$$

(Write needed fundamentals *before* you get immersed in details.)

DATA: P_1 = Standard P—use 760 **torr** to match P_2 = **? in torr** = WANTED

$V_1 = 22.4\ \text{L}$ $V_2 = 16.8\ \text{L}$

$T_1 = 27°C$ = **T_2** T_2

SOLVE:

$$\dfrac{P_1V_1}{\cancel{T_1}} = \dfrac{P_2V_2}{\cancel{T_2}}; \quad \text{use } P_1V_1 = P_2V_2$$

$$? = P_2 \text{ in torr} = \dfrac{P_1V_1}{V_2} = \dfrac{(760 \text{ torr})(22.4 \text{ L})}{(16.8 \text{ L})} = \textbf{1,010 torr}$$

PRACTICE   C

A 0.500 L sample of neon gas in a sealed metal (rigid) container is at 30.°C and 380. mm Hg. What would be the pressure of the gas in kilopascals at standard temperature?

Lesson 16.7   Choosing the Right Equation

In physical science courses, there can be quite a list of equations that are needed to solve problems. How do you decide *which* equation to use *when*? The following system for choosing equations is easy to use and will work in both chemistry and other science courses.

A System for Finding the *Right* Equation

When you are not sure *which* equation to use, or whether to solve by equations or conversions, try these steps.

1. As you have been doing, first write the WANTED *unit* and/or *symbol*.

2. List the DATA with units, substances, and descriptive labels. Add any prompts.

3. Analyze whether the problem will require *conversions* or an *equation.*
   - Try conversions first. Conversions often work if *most* of the data are in *pairs* and *ratio units.*
   - If the data are mostly in *single* or in *complex* units, you will likely need an equation. Recent lessons in your course will likely indicate the equations that may be needed.
   - Watch for hints at the need for an equation. For example, the mention of *R* in a gas problem is likely a prompt that you will need a form of the ideal gas law to solve.

4. If a lecture or textbook is frequently using certain equations, for equations you are required to have in memory, learn them *before* you do the practice problems. Then write them at the top of your paper at the start of each assignment, quiz, or test.

5. If you recognize which equation is needed, write the equation, then make a DATA table listing each symbol in the equation. Fill in the table with the DATA from the problem.

6. If conversions don't work and you *suspect* you need an equation but are not sure *which* equation is needed, try this:

   a. *Label* each item of WANTED and DATA with a *symbol* based on its units. Use symbols that are used in the equations for the topic you are studying.
      Examples:
      - 25 **kPa** would be labeled with a *P* for pressure
      - 293 **K** is assigned a *T* for the Kelvin scale temperature
      - 20°**C** gets a lower case *t* for degrees Celsius
      - L or mL or cm^3 or dm^3 would be tagged with a *V* for volume

   b. Label the WANTED unit and each item of DATA with a symbol.

   c. *Compare* the symbols listed in the WANTED and DATA to your written, memorized list of equations for the topic. Find the equation that *uses* those symbols. Write that equation below the data.
      If no equations match exactly, see whether the problem's data can be converted to fit the symbols for a known equation (for example, degrees Celsius can be converted to kelvins, and grams can be converted to moles if you know a substance formula).

SUMMARY

Choosing an Equation

When you are not sure which equation to pick to solve a problem, label the data with symbols, then choose an equation that uses those symbols.

PRACTICE   A

Using the method above, try these in your notebook. If you get stuck, read a portion of the answer and try again.

1. Write symbols used in gas law equations to label each of these quantities.

   a. 122 K   b. 202 kPa   c. 50.5 bar   d. 30°C

2. If one mole of any gas occupies 22.4 L of volume at STP, calculate the value of R in units of kPa · L/mol · K.

3. Find the density of oxygen gas (O_2) in grams per liter at STP.

Gas Law Practice

From this point forward in these lessons, you will be expected to create appropriate flashcards on your own, a skill that will be helpful in this and perhaps other courses. To solve complex problems, fundamental facts and stepwise procedures (algorithms) must be in memory.

If you have not already done so, make a set of flashcards that cover the fundamental rules, equations, and relationships in this chapter. Include points in the chapter summary at the end of the chapter. Practice your flashcards and/or lists of relationships until they are firmly in memory.

Then try the problems below that cover the types of problems in this chapter. If you need additional practice, try problems in the earlier lessons that you may have saved for later review.

PRACTICE B

If you get stuck, read a part of the answer to get a hint, then try again ($R = 0.0821$ atm · L/mol · K = 8.31 kPa · L/mol · K = 62.4 torr · L/mol · K).

1. A gas in a sealed flexible balloon has a volume of 6.20 L at 30°C and standard pressure. What will be its volume at −10.°C and 740. torr?

2. How many gas molecules will there be, per milliliter, for all ideal gases at STP?

3. If gas in a sealed glass bottle has a pressure of 112 kPa at 25°C, and the temperature of the gas is increased to 100.°C, what will be the pressure?

4. If 70.4 g of UF_6 gas (352 g/mol) has a volume of 4.48 L and a pressure of 202.6 kPa, what is its temperature in kelvins?

SUMMARY

1. When measuring gases:

   Standard pressure ≡ 1 atm ≡ 760 mm Hg ≡ 760 torr
   = 101 kPa = 1.01 bar

   Standard temperature ≡ zero degrees Celsius (0°C) = 273 kelvins = 273 K

2. Temperatures must be converted to kelvins when using any equation that specifies a capital T. Use $\boxed{K = °C + 273}$

3. To be a measure of moles, gas **volumes** must be labeled with a P and a T.

4. **The STP prompt:**
   Gas measurements are often reported at standard temperature and pressure, abbreviated STP.

(continued)

> *If* a gas is at STP, write as DATA: 1 mol gas = 22.4 L gas at STP.

For gas calculations at STP, try the STP prompt and conversions first.

5. **The ideal gas law:** $\boxed{PV = nRT}$, where R = the gas constant

$$R = 0.0821 \text{ atm} \cdot \text{L/mol} \cdot \text{K} = 8.31 \text{ kPa} \cdot \text{L/mol} \cdot \text{K}$$
$$= 62.4 \text{ torr} \cdot \text{L/mol} \cdot \text{K}$$

In gas law calculations, values for R are usually supplied, but you will often need to choose which R value and units to use. Choose the R with units that most closely match the units in the DATA supplied in the problem.

6. **The combined equation.** *If* the gas conditions are changed but the *moles* of gas do not change,

$$\frac{P_1V_1}{T_1} = \frac{P_2V_2}{T_2}$$

7. To solve with equations, first convert to consistent units, solve in consistent units, then convert the consistent WANTED unit to other units if needed.

8. For solving *all kinds* of calculations, use a *system*:
   - List the WANTED and DATA
   - Try solving with conversions first

   If conversions do not work,
   - Add symbols to the WANTED and DATA, and then write a memorized fundamental *equation* that *uses* those symbols
   - Convert all DATA to *consistent* units
   - Solve for the WANTED *symbol* before plugging in numbers
   - Cancel units as a check of your work

9. For numbers, units, or symbols: $1/(1/X) = X$; $A/(B/C) = (A \cdot C)/B$

ANSWERS

Lesson 16.1

Practice A   1. WANTED: ? kPa

DATA: 870. millibars

SOLVE: $? \text{kPa} = 870. \text{ millibars} \cdot \dfrac{10^{-3} \text{ bar}}{1 \text{ millibar}} \cdot \dfrac{101 \text{ kPa}}{1.01 \text{ bars}} = \mathbf{87.0 \text{ kPa}}$

2. WANTED: ? in. Hg

DATA: 760 mm Hg (exact)
2.54 cm ≡ 1 in. (exact)

SOLVE: $? \text{ in. Hg} = 760 \text{ mm Hg} \cdot \dfrac{1 \text{ cm}}{10 \text{ mm}} \cdot \dfrac{1 \text{ in.}}{2.54 \text{ cm}} = \mathbf{29.92\dots \text{ in. Hg}}$ (All of the numbers are exact.)

3. WANTED: ? atm

    DATA: 72 psi

    1 atm = std. pressure = 14.7 psi

    SOLVE: ? atm = 72 psi · $\dfrac{1\ \text{atm}}{14.7\ \text{psi}}$ = **4.9 atm**

4. $? \text{kPa} = 25.0 \text{ torr} \cdot \dfrac{1\ \text{atm}}{760\ \text{torr}} \cdot \dfrac{101\ \text{kPa}}{1\ \text{atm}} = \textbf{3.32 kPa}$

Practice B   1.

	In Kelvins	In Degrees Celsius
Absolute Zero	**0 K**	**−273°C**
Water Boils	**373 K**	**100.°C**
Nitrogen Boils	**77 K**	**−196°C**
Table Salt Melts	**1,074 K**	**801°C**
Water Freezes	**273 K**	**0°C**
Standard Temperature	**273 K**	**0°C**

Note that when using $\boxed{\text{K} = \text{°C} + 273}$ you are adding or subtracting, and the significant figures are determined by the highest *place* with doubt. All of the numbers in the table have doubt in the one's place.

2. A temperature for a gas of −310°C is mistaken. −310°C is below absolute zero (−273°C). There cannot be a temperature colder than absolute zero.

Lesson 16.2

Practice A   1. Standard $P = 1\text{ atm} \equiv 760\text{ mm Hg} \equiv 760\text{ torr} = 101\text{ kPa} = 1.01\text{ bar}$

2. Standard $T = \textbf{0°C or 273 K}$

Practice B   1a. Molarity $\dfrac{\textbf{mol}}{\textbf{1 L solution}}$ 1b. Molar mass $\dfrac{\textbf{g}}{\textbf{1 mol}}$

1c. Volume **L, mL, dm³, cm³**   1d. Mass **kg, g, mg**

1e. Density $\dfrac{\text{any }\textbf{mass}\text{ unit (such as kg or g)}}{\text{any }\textbf{volume}\text{ unit (such as L, mL)}}$ 1f. Speed $\dfrac{\textbf{any distance}\text{ (such as cm, mi.)}}{\textbf{any time}\text{ (such as s, hr)}}$

1g. Gas pressure **atm, torr, mm Hg, kPa, bars**

1h. Temperature **°C** or **K**

2. WANTED: $? \dfrac{\text{g SO}_2\text{ gas}}{\text{L SO}_2\text{ gas at STP}}$ (Write *ratio* units WANTED as *fractions*.)

    DATA: 64.1 g SO₂ = 1 mol SO₂   (Grams prompt.)

    1 mol any gas = 22.4 L any gas at STP   (STP prompt.)

    SOLVE: (Want a ratio? Start with a ratio. Since grams is on top in the answer, you might start with grams on top as the given ratio.)

$$? \frac{\text{g SO}_2\text{ gas}}{\text{L SO}_2\text{ gas at STP}} = \frac{64.1\ \textbf{g SO}_2}{1\ \text{mol SO}_2} \cdot \frac{1\ \text{mol gas}}{22.4\ \textbf{L gas STP}} = \textbf{2.86}\ \frac{\textbf{g SO}_2(g)}{\textbf{L SO}_2\ \textbf{gas at STP}}$$

3. WANTED: ? $\frac{g}{mol}$ (Write the *unit* WANTED for molar mass: g/mol.)

   DATA: 0.00205 g gas = 1 mL gas at STP (Use $g \cdot mL^{-1}$ = g/mL; list *ratio* DATA as *equalities*.)

   1 mol any gas = 22.4 L any gas at STP (STP prompt.)

   (Note that the grams prompt only works if you know a substance *formula*.)

   SOLVE: (The conversions below may be in any order, so long as they are right-side up.)

$$? \frac{g}{mol} = \frac{0.00205 \text{ g gas}}{1 \text{ mL gas at STP}} \cdot \frac{1 \text{ mL}}{10^{-3} \text{ L}} \cdot \frac{22.4 \text{ L any gas at STP}}{1 \text{ mol gas}} = \mathbf{45.9} \frac{\mathbf{g}}{\mathbf{mol}}$$

4. WANTED: ? $\frac{g}{mol}$

   DATA: 0.313 g gas = 250. mL gas at STP (The period means 250. has three *s.f.*)

   1 mol any gas = 22.4 L any gas at STP (STP prompt.)

   SOLVE: (The conversions below may be in any order.)

$$? \frac{g}{mol} = \frac{0.313 \text{ g gas}}{250. \text{ mL gas at STP}} \cdot \frac{1 \text{ mL}}{10^{-3} \text{ L}} \cdot \frac{22.4 \text{ L any gas at STP}}{1 \text{ mol gas}} = \mathbf{28.0} \frac{\mathbf{g}}{\mathbf{mol}}$$

Reminders

- Attach temperature and pressure conditions, if known, to gas *volumes*.
- In the interest of readability, most unit cancellations in these answers are left for you to do. However, in your work, always ~~mark~~ your unit cancellations as a check on your conversions.

5. WANTED: ? molecules CO_2 gas

   DATA: 1.12 L CO_2 gas at STP

   6.02×10^{23} molecules anything = 1 mol anything (Avogadro prompt.)

   1 mol any gas = 22.4 L any gas at STP (STP prompt.)

   SOLVE:

$$? \text{ molecules } CO_2(g) = 1.12 \text{ L } CO_2(g) \text{ STP} \cdot \frac{1 \text{ mol gas}}{22.4 \text{ L gas STP}} \cdot \frac{6.02 \times 10^{23} \text{ molecules}}{1 \text{ mol}} = \mathbf{3.01 \times 10^{22} \text{ molecules } CO_2}$$

6. WANTED: ? mL F_2 gas at STP (These conditions are STP.)

   DATA: 15.2 g F_2 gas (A single unit *given*.)

   38.0 g F_2 gas = 1 mol F_2 gas

   1 mol any gas = 22.4 L any gas at STP

   SOLVE: (Want a single unit?)

$$? \text{ mL } F_2(g) \text{ STP} = 15.2 \text{ g } F_2(g) \cdot \frac{1 \text{ mol } F_2}{38.0 \text{ g } F_2} \cdot \frac{22.4 \text{ L gas STP}}{1 \text{ mol gas}} \cdot \frac{1 \text{ mL}}{10^{-3} \text{ L}} = \mathbf{8.96 \times 10^3 \text{ mL } F_2(g) \text{ STP}}$$

7. WANTED: ? $\underline{\text{L gas at given P and T}}$
   1 mol

   Strategy: If the problem is asking for a unit per one unit, it is asking for a ratio unit.
    This problem does not specify STP, but you can solve for the requested unit.
    Compare units: You WANT *liters* and *moles*. You are *given milliliters* and *moles*.

   DATA: 0.0700 mol gas $= 1{,}760$ mL gas at given T and P

   SOLVE: (Since answer unit *moles* is on bottom, and answer unit *liters* is not in the data, you might start with moles on the bottom in your *given* ratio.)

   $$\frac{?\text{ L gas at given }T\text{ and }P}{\textbf{1 mol}} = \frac{1{,}760 \text{ mL gas at given }T\text{ and }P}{0.0700\ \textbf{mol}} \cdot \frac{10^{-3}\text{ L}}{1\text{ mL}} = \frac{\textbf{25.1 L gas} \text{ at given }T\text{ and }P}{\textbf{mcl}}$$

8. Hint: The grams prompt applies to kilo*grams*, too. If needed, adjust your work and try again.

   WANTED: ? $\dfrac{\text{kg Rn gas}}{\text{L at STP}}$ (kg $\cdot$ L^{-1} = kg/L. Write ratio units WANTED as *fractions*.)

   DATA: 222 g Rn $= 1$ mol Rn (kg = g prompt.)

   1 mol any gas $= 22.4$ L any gas at STP (STP prompt.)

   SOLVE: (Start with a ratio. Try, in your *given*, to start with one unit where it belongs in the answer. . . , but your conversions in solving for a ratio may be in any order that cancels to give the WANTED unit.)

   $$\frac{?\ \ \textbf{kg}\text{ Rn gas}}{\textbf{L}\text{ Rn gas at STP}} = \frac{1\text{ mol gas}}{22.4\ \textbf{L}\text{ gas STP}} \cdot \frac{222\text{ g Rn}}{1\text{ mol Rn}} \cdot \frac{1\ \textbf{kg}}{10^3\text{ g}} = 9.91 \times 10^{-3}\ \frac{\textbf{kg}\text{ Rn}}{\textbf{L}\text{ Rn gas at STP}}$$

Lesson 16.3

Practice A 1. $X/(Y/Z) = X \cdot \dfrac{1}{\dfrac{Y}{Z}} = \dfrac{X \cdot Z}{Y}$
2. $(D/E)^{-1} = \dfrac{1}{\dfrac{D}{E}} = \dfrac{E}{D}$ or $(D^{-1} \cdot E)$

3. $\dfrac{\text{meters}}{\dfrac{\text{meters}}{\text{second}}} = \text{meters} \cdot \dfrac{1}{\dfrac{\text{meters}}{\text{second}}} = = \text{meters} \cdot \dfrac{\text{second}}{\text{meters}} = \textbf{seconds}$

4. $\dfrac{(\text{meters/second})}{\text{second}} = \dfrac{\text{meters}}{\text{second}} \cdot \dfrac{1}{\text{second}} = \dfrac{\textbf{meters}}{\textbf{second}^2}$

5. meters/second/second cannot be evaluated.

Practice B 1. $\dfrac{\text{calories}}{\dfrac{\text{calorie} \cdot \text{g}}{\text{g} \cdot {}^\circ\text{C}}} = \dfrac{1}{\dfrac{\text{calorie}}{\text{g} \cdot {}^\circ\text{C}}} \cdot \dfrac{\text{calories}}{\text{g}} = \dfrac{\text{g} \cdot {}^\circ\text{C}}{\text{calorie}} \cdot \dfrac{\text{calories}}{\text{g}} = {}^\circ\textbf{C}$

2. $\dfrac{\text{atm} \cdot \text{L}}{(\text{mol}) \cdot \dfrac{\text{atm} \cdot \text{L}}{\text{mol} \cdot \text{K}}} = \dfrac{\text{atm} \cdot \text{L} \cdot}{\text{mol}} \dfrac{1}{\dfrac{\text{atm} \cdot \text{L}}{\text{mol} \cdot \text{K}}} = \dfrac{\text{atm} \cdot \text{L}}{\text{mol}} \cdot \dfrac{\text{mol} \cdot \text{K}}{\text{atm} \cdot \text{L}} = \textbf{K}$

3. $\dfrac{(\text{mol}) \cdot \dfrac{\text{atm} \cdot \text{L}}{\text{mol} \cdot \text{K}} \cdot (\text{K})}{\text{L}} = (\text{mol}) \cdot \dfrac{(\text{atm} \cdot \text{L})}{\text{mol} \cdot \text{K}} \cdot (\text{K}) \cdot \dfrac{1}{\text{L}} = \textbf{atm}$

Lesson 16.4

Practice   1a. 0.50 mol n 1b. 202 kPa P 1c. 11.2 dm^3 V

1d. 373 K T 1e. 38 torr P

2. If $PV = nRT$:   a. $n = \dfrac{PV}{RT}$ b. $T = \dfrac{PV}{nR}$ c. $V = \dfrac{nRT}{P}$

> If you cannot do this algebra correctly *every* time, find a friend or tutor who can help you to review the algebra for this and the following lessons in this chapter. It will not take long to master. Gas laws are not difficult if you can do this algebra but are impossible if you cannot.

3. $\boxed{PV = nRT}$

DATA: $P = 2.3$ atm

$V = 22.4$ L

$n = 1$ mol

$R = 0.0821 \dfrac{\text{atm} \cdot \text{L}}{\text{mol} \cdot \text{K}}$

$T = \,?$

SOLVE: $PV = nRT$

$$? = T = \frac{PV}{nR} = \frac{(2.3 \text{ atm})(22.4 \text{ L})}{(1 \text{ mol})\; 0.0821\dfrac{(\text{atm} \cdot \text{L})}{\text{mol} \cdot \text{K}}} = \frac{(2.3 \text{ atm})(22.4 \text{ L})}{1 \text{ mol}} \cdot \frac{\textbf{mol} \cdot \textbf{K}}{0.0821 \text{ atm} \cdot \text{L}} = \textbf{630 K}$$

630 K is a large rise in temperature from 273 K. It is reasonable because for one mole of gas at 22.4 L volume to cause a pressure of 2.3 atm (rather than 1 atm, as at STP), the gas must be very hot, with molecules moving much faster on average than at standard temperature.

The unit cancellation: From Lesson 16.3, use rule 2a: Separate a fractional unit on the bottom into a reciprocal, and rule 2c: A *fraction* on the bottom flips over when you bring it to the top.

Mark the unit cancellation in the final step above.

When solving with equations, always include the unit cancellation. If the units cancel correctly, the numbers were probably put in the right place to get the right answer. This provides a check on your work.

Lesson 16.5

Practice $\boxed{PV = nRT}$

DATA: $P = \,?$ **torr** $=$ WANTED

$R = 62.4 \dfrac{\text{torr} \cdot \textbf{L}}{\text{mol} \cdot \textbf{K}}$ (Choose the R value that uses torr.)

$n = 0.0050$ mol

V must be in L to match $R = $ 560. mL $= \textbf{0.560 L}$ (By inspection, see Lesson 12.1.)

T must be in K: $\boxed{K = °C + 273} = 0°C + 273 = \textbf{273 K}$

SOLVE: $PV = nRT$ in symbols first.

$$? = P = \frac{nRT}{V} = nRT \cdot \frac{1}{V} = (0.0500 \text{ mol})(62.4 \frac{\text{torr} \cdot \text{L}}{\text{mol} \cdot \text{K}})(273 \text{ K}) \cdot \frac{1}{0.560 \text{ L}} = \textbf{1,520 torr}$$

Mark the unit cancellation in the last step.
To help with unit cancellation, a good rule is

> When solving an equation in *symbols* results in a *fraction* that includes any terms with *fractional* units (such as R above) on the top *or* bottom, rewrite the equation with the *symbols* in the denominator separated into a reciprocal, then plug in numbers into this separated format (see Lesson 16.3).

However, you may do the math for the numbers and units in any way you choose, provided you do *both* numbers *and* units.

Lesson 16.6

Practice A 1a. T_2: If $\dfrac{P_1V_1}{T_1} = \dfrac{P_2V_2}{T_2}$, then $T_2 = \dfrac{P_2V_2T_1}{P_1V_1}$

1b. $P_1 = \dfrac{P_2V_2T_1}{T_2V_1}$ 1c. $V_2 = \dfrac{P_1V_1T_2}{T_1P_2}$

As a check, note the *patterns* above.
The Ps and Vs, if multiplied together, have the same subscript, but if a T is multiplied with them, it will have the *opposite* subscript.
In the fractions, there is always one more term on the top than on the bottom.
If you substitute consistent units for all of the symbols, the units cancel correctly.

> If you have trouble solving for any of the six symbols, find a friend or tutor to help you learn the algebra.

2. If the moles of gas particles do not change, and one set of conditions is changed to new conditions, use

$$\boxed{\dfrac{P_1V_1}{T_1} = \dfrac{P_2V_2}{T_2}}$$

DATA: $P_1 = 100.$ atm $P_2 = 2.00$ atm

$V_1 = 2.50$ L $V_2 = 50.0$ L

$T_1 = 293$ K $T_2 = ?$ $t_2 = ?$

SOLVE: (In gas problems, to find Celsius, solve for kelvins first.)

$$? = T_2 = \dfrac{P_2V_2T_1}{P_1V_1} = \dfrac{(2.00 \text{ atm})(50.0 \text{ L})(293 \text{ K})}{(100. \text{ atm})(2.50 \text{ L})} = \textbf{117 K}$$

To find Celsius: $\boxed{K = °C + 273}$,°C $= K-273 = 117$ K$-273 = \textbf{−156°C}$

Practice B   For trapped gas moles changing from original to new conditions, use $\boxed{\dfrac{P_1V_1}{T_1} = \dfrac{P_2V_2}{T_2}}$

DATA: $P_1 = 150.$ kPa $P_2 = 0.550$ atm $= \textbf{55.6 kPa}$

$V_1 = 500.$ **mL** $V_2 = 2.00$ L $= \textbf{2,000 mL}$

$T_1 = 20.°C + 273 = 293$ K $T_2 = ?$ $t_2 = ?$

Needed unit conversions:

For T_1, K must be used. $\boxed{K = °C + 273}$ $= 20.°C + 273 = 293$ K

For P_1, kPa *or* atm could be used as units. If you choose kPa,

$$? \text{ kPa } = 0.550 \text{ atm} \cdot \frac{101 \text{ kPa}}{1 \text{ atm}} = 55.6 \text{ kPa}$$

SOLVE: (In gas problems, solve in kelvins first, then Celsius.)

$$? = T_2 = \frac{P_2 V_2 T_1}{P_1 V_1} = \frac{(55.6 \text{ kPa}) (2{,}000 \text{ mL}) (293 \text{ K})}{(150. \text{ kPa}) (500. \text{ mL})} = \mathbf{434\ K}$$

To find Celsius, *first* write the memorized equation, *then* solve:

$$\boxed{K = °C + 273}\ °C = K - 273 = 434\ K - 273 = \mathbf{161°C}$$

(Use of different consistent units may result in slightly different answers due to rounding.)

Practice C   For trapped gas changing to new conditions, use $\boxed{\dfrac{P_1 V_1}{T_1} = \dfrac{P_2 V_2}{T_2}}$

DATA: $P_1 = 380.$ mm Hg $= \mathbf{50.5\ kPa}$ $P_2 = ?\ \mathbf{kPa}$

$V_1 = 0.500$ L $= V_2$ (A sealed metal container will have a constant volume.)

$T_1 = 30.°C + 273 = \mathbf{303\ K}$ $T_2 = \mathbf{273\ K}$

Needed unit conversions:

For the pressures, you must change to consistent units: either kPa *or* mm Hg. If you choose kPa,

$$? \text{ kPa } = 380. \text{ mm Hg} \cdot \frac{101 \text{ kPa}}{760 \text{ mm Hg}} = \mathbf{50.5\ kPa}$$

SOLVE: $P_2 = \dfrac{P_1 V_1 T_2}{T_1 V_2} = \dfrac{(50.5 \text{ kPa}) (273 \text{ K})}{(303 \text{ K})} = \mathbf{45.5\ kPa}$

Lesson 16.7

Practice A 1a. 122 K T 1b. 202 kPa P 1c. 50.5 bar P 1d. 30°C t

2. WANTED: R in $\dfrac{\mathbf{kPa \cdot L}}{\mathbf{mol \cdot K}}$

DATA: std. $P = \mathbf{101\ kPa}$ in the units wanted for R P

22.4 L V

1 mol n

std. temperature $= 273$ K T

Strategy: Looking at all five symbols labeling WANTED *and* DATA, which equation works?

SOLVE: $\boxed{PV = nRT}$

$$? = R = \frac{PV}{nT} = \frac{(101 \text{ kPa})(22.4 \text{ L})}{(1 \text{ mol})(273 \text{ K})} = \mathbf{8.29} \ \frac{\mathbf{kPa \cdot L}}{\mathbf{mol \cdot K}}$$

Using less rounded values for P, T, and V, the accepted value is $\mathbf{8.31}$ kPa $\cdot$ L/mol $\cdot$ K

3. WANTED: ?   g O$_2$ gas
      ‾‾‾‾‾‾‾‾‾‾‾‾‾‾‾‾
      L O$_2$ gas at STP

   DATA: 32.0 g O$_2$ = 1 mol O$_2$ (Grams prompt.)

   1 mol any gas = 22.4 L any gas at STP   (STP prompt.)

Strategy: Analyze your units. Equalities lend themselves to conversions. You want grams over liters. You know grams to moles and moles to liters.

SOLVE: ?   g O$_2$ gas   = 32.0 **g O$_2$** · 1 mol gas   = **1.43** g O$_2$(g)
       ‾‾‾‾‾‾‾‾‾‾‾‾‾   ‾‾‾‾‾‾‾‾‾   ‾‾‾‾‾‾‾‾‾‾‾‾‾   ‾‾‾‾‾‾‾‾‾
       L O$_2$ gas at STP 1 mol O$_2$ 22.4 L gas at STP   **L O$_2$ at STP**

Practice B   1. WANTED: ? V at end V_2

   DATA: 6.20 L initial V_1

   30.°C + 273 = 303 K initial T_1

   std. P = 760 torr initial, using the P units in the problem P_1

   −10.°C + 273 = 263 K final T_2

   740 torr final P_2

Strategy: The WANTED and DATA symbols match $\boxed{\dfrac{P_1V_1}{T_1} = \dfrac{P_2V_2}{T_2}}$

SOLVE: ? = V_2 = $\dfrac{P_1V_1T_2}{P_2T_1}$ = $\dfrac{(760\text{ torr}) (6.20\text{ L}) (263\text{ K})}{(740.\text{ torr}) (303\text{ K})}$ = **5.53 L**

2. WANTED: ? molecules gas
      ‾‾‾‾‾‾‾‾‾‾‾‾‾
      mL gas at STP

   DATA: 6.02×10^{23} molecules = 1 mol   ("Molecules" calls the Avogadro prompt.)

   1 mol any gas = 22.4 L any gas at STP   (STP prompt.)

Strategy: Molecules and mL are wanted. The conversions use molecules, moles, and liters. The wanted is a ratio; all the data is in equalities. Try conversions.

SOLVE: ? molecules gas   = $\dfrac{6.02 \times 10^{23}\text{ molecules}}{1\text{ mol}}$ · $\dfrac{1\text{ mol gas}}{22.4\text{ L gas STP}}$ · $\dfrac{10^{-3}\text{ L}}{1\text{ mL}}$
       ‾‾‾‾‾‾‾‾‾‾‾‾‾
       mL gas at STP

   = **2.69 × 10^{19}** molecules
             ‾‾‾‾‾‾‾‾‾‾‾
             mL gas at STP

3. WANTED: ? P Problem uses kPa units and wants P at **end** = P_2

   DATA: P = 112 kPa = P at start = P_1

   t = 25°C initial;   K = 25°C + 273 = 298 K = T_1

   t = 100.°C final   K = 100.°C + 273 = 373 K = T_2

Strategy: There are two Ps and two ts, but no Vs. However, for a change in a sealed glass bottle, V_1 will equal V_2.

SOLVE: $\dfrac{P_1V_1}{T_1} = \dfrac{P_2V_2}{T_2}$; $\boxed{\dfrac{P_1}{T_1} = \dfrac{P_2}{T_2}}$? = P_2 = $\dfrac{P_1T_2}{T_1}$ = $\dfrac{112\text{ kPa} \cdot 373\text{ K}}{298\text{ K}}$ = **140. kPa**

4. WANTED: ? K T

   DATA: 70.4 g UF$_6$

   352 g UF$_6$ = 1 mol UF$_6$ (Grams prompt.)

   4.48 L gas V

202.6 **kPa**   *P*

$R = 8.31$ **kPa** $\cdot$ L / mol $\cdot$ K   (Since *P* is in kPa, choose *R* that include kPa.)

The symbols are T, V, P, R. You are missing n, but you can solve for moles from the data. Use which equation?

$$\boxed{PV = nRT}$$

$$n = ? \text{ mol UF}_6 = 70.4 \text{ g UF}_6 \cdot \frac{1 \text{ mol UF}_6}{352 \text{ g UF}_6} = 0.200 \text{ mol UF}_6$$

$$T = \frac{PV}{nR} = \frac{(202.6 \text{ kPa})(4.48 \text{ L})}{(0.200 \text{ mol})(8.31 \frac{\text{kPa} \cdot \text{L}}{\text{mol} \cdot \text{K}})} = \frac{(202.6 \text{ kPa})(4.48 \text{ L})}{(0.200 \text{ mol})} \cdot \frac{\text{mol} \cdot \text{K}}{8.31 \text{ kPa} \cdot \text{L}} = \mathbf{546 \text{ K}}$$

17

Phases and Energy

Lesson 17.1   Phases and Phase Changes

Three Phases

In first-year chemistry, we are concerned with *three* **phases** for pure substances: **solid**, **liquid**, and **gas** (the terms *states* and *phases*, in this context, have similar meaning).

In covalently bonded molecules, the forces that hold the atoms together within the molecule are relatively strong, but there are also relatively weak forces of attraction between molecules. These weak attractions mean that molecules are a bit "sticky:" they tend to attract each other somewhat like two weak attracting magnets (the weak molecular attractions are electrical, but the behavior is similar).

In the solid phase, molecules **vibrate**, but the weak attractions between molecules hold the molecules in a crystal structure where they are limited in the extent to which they can **rotate**, and they can **translate** (move from place to place) only very slightly.

In their liquid phase, the molecules gain some freedom: they can vibrate, rotate, and translate. However, the liquid phase molecules are still very close together: they have minimal space between them. This is why solids and liquids do not compress (or compress only very slightly) when pressure is applied.

In the gas phase, molecules are separated by a considerable distance. In a gas at room temperature, the distance between molecules is typically about 10 times the diameter of the molecule. This means that 99.9% of the gas is empty space. Gases can be compressed because the empty space between the molecules can be reduced.

In a gas, the molecules remain weakly attractive. If a gas is highly compressed, or if its temperature is lowered (which slows down the average speed at which the molecules move), the stickiness of the molecules becomes a larger factor. All gases condense into a liquid or a solid at some point as pressure is increased and/or temperature is decreased.

Three Phase Changes

In chemistry, a **chemical change** is defined as a process in which atoms rearrange to form new substances. A **physical change** is one in which characteristics of a substance change, but the substance does not change: it keeps the same chemical formula. A phase change is one example of a physical change.

There are six types of **phase changes** among the three phases, but most of these are familiar. The following terms are used to describe phase changes.

- **For solid/liquid changes:** Solids **melt** (or **fuse**) to become liquids; liquids **freeze** (or **solidify**) to become solids.
- **For liquid/gas changes:** Liquids **boil** or **evaporate** to form gases; gases **condense** to become liquids.
- **For solid/gas changes:** Solids **sublime** to become gases. In the reverse process, gases can undergo **deposition** to form solids.

Sublimation is a phase change that is less commonly encountered at room temperature and pressure, but you may be familiar with dry ice (solid carbon dioxide) or moth crystals (*para*-dichlorobenzene). At room temperature, these solids do not pass through a liquid phase as they convert from the solid to the gas phase. **Deposition** can be observed when water vapor forms ice crystals on a windshield that is below 0°Celsius.

Melting and Freezing

For a pure substance, the temperature at which it melts (its **melting point**) will *equal* the temperature at which it solidifies.

For pure substances: melting point ≡ freezing point

For pure substances, under atmospheric pressures that are at or near standard pressure, melting occurs at a *characteristic* temperature that can be used to identify the substance. However, even small amounts of impurity in a substance will weaken its crystal structure and cause it to melt and freeze at a lower temperature.

Vapor Pressure

Nearly every liquid or solid has a measurable tendency for its particles to become gas particles. At the surface of a liquid or solid, vibrating neutral particles can break free and become part of the vapor above the liquid or solid. The *evaporation* of liquids at temperatures far below their boiling points is one example of this tendency.

The gas molecules that leave a solid or liquid create a gas pressure called the vapor pressure. Vapor pressure can be *measured* by placing a liquid or solid in a vacuum, allowing the system to reach equilibrium at a given temperature, then measuring the gas pressure in the container. However, the vapor pressure of a liquid or solid is a characteristic property: a measure of the tendency of the particles of the substance to enter the gas phase. At a given temperature, the vapor pressure of a solid or liquid has the same value whether it is in a sealed container or not.

Vapor pressure is measured in pressure units, and the vapor pressure of a substance increases with increasing temperature. A liquid will **boil** at any temperature at which its **vapor pressure** equals the atmospheric pressure above it.

Boiling Temperature and Pressure

Boiling points are characteristic temperatures that can be used to identify a substance. A **normal** boiling point is recorded at *standard* pressure (one atmosphere).

Boiling points must be recorded at a known pressure because liquids boil at a temperature that depends on the surrounding atmospheric pressure. A liquid will boil at a temperature higher than its *normal* boiling point if the atmospheric pressure above it is higher than standard pressure. It will boil at a temperature lower than its normal boiling point if the pressure on the liquid is lower than standard pressure.

This means that there are three ways to boil a liquid: heat it until its vapor pressure rises to equal atmospheric pressure, or lower the lower pressure above the liquid, such as by using a bell jar and a vacuum pump, or do both.

Boiling Water

Atmospheric pressure is generally lower at a high altitude than at sea level. This means that at high elevations, when you heat a liquid, its vapor pressure will equal atmospheric pressure at a lower temperature. The liquid will therefore boil at a lower temperature at high altitude than at sea level.

Examples:

- Water boils at 100°C *if* the pressure above the water is 101 kPa = 760 torr = 1 atm = standard pressure, which is about the average atmospheric pressure on a fair weather day at sea level. However, water boils at about 95°C under the lower atmospheric pressure typically found in locations one mile above sea level (such as Denver, Colorado). At high altitude, it takes more time to "hard boil" an egg than at sea level because the boiling water around the egg is not as hot.

- At 20.°C, water has a vapor pressure of 17.5 torr. A *vacuum pump* can reduce the atmospheric pressure in a bell jar to below 17 torr, and water in this partial vacuum will boil at room temperature.

- In a *pressure cooker*, boiling water is at higher temperature than boiling water at room pressure, and the changes required to "cook" food occur more quickly.

Boiling temperatures are affected by relatively small changes in the surrounding air pressure, such as those caused by altitude changes. Melting points are substantially changed only by much larger changes in pressure.

Boiling versus Evaporating

Boiling is not the same as evaporating. Evaporation is a surface phenomenon. Measurable evaporation will occur from all liquids (and many solids) at any temperature. A liquid *boils* only when gas bubbles can form *below* the surface of the liquid and not just at its edges.

PRACTICE   **A**

Answer these questions before going on to the next section. Practice until you can answer the questions from memory.

1. Name three phases or states of matter. Name six different types of phase changes.

2. Which phases of matter can be significantly compressed in volume? Why?

3. Which has a higher temperature:

    a. The melting point or the freezing point of a pure substance?

    b. The melting point of a substance that is pure, or one that has impurities?

4. By definition, when does a liquid boil?

5. State three ways to boil a liquid.

6. At what temperature does water boil at an atmospheric pressure of 101 kPa?

7. At approximately what temperature does water boil in a city that is one mile above sea level? What explains the difference from the boiling temperature at sea level?

8. Why does it take longer to hard boil an egg at a high altitude?

Energy

Chemistry is primarily concerned with matter and energy. In chemistry, matter and energy can be considered to be separate entities. Matter has mass and can be described in terms of particles such as protons, neutrons, and electrons. Energy has no mass. Sunlight, heat, and radio waves are a few examples of the many forms of energy.

A fundamental principle of chemistry is the **law of conservation of energy**: energy can neither be created nor destroyed. However, during chemical or physical processes, energy can be *transferred* between substances and to and from the environment. Energy can also change its *form*.

There are many forms of energy, but they can generally be divided into two types:

- **Kinetic** energy (KE), defined as energy of *motion*
- **Potential** energy (PE), defined as *stored* energy

When a substance loses energy, the energy can do **work** as defined in physics (such as moving a piston against resistance), or energy can be can transferred as heat to the environment around the substance. A substance can gain energy when work adds energy to the substance, such as by compressing a gas, or when the environment supplies heat to the substance.

Kinetic Energy

Kinetic energy is energy of motion. The kinetic energy of an object is calculated by the equation:

$$KE = \tfrac{1}{2}(\text{mass})(\text{velocity})^2$$

This equation means that if particle B has twice the mass of particle A but is moving at the same speed, particle B has *twice* as much kinetic energy. If particle C has the same mass as particle A but is moving twice as fast as particle A, it has *four times* as much kinetic energy.

Temperature is a measure of the *average* kinetic energy of particles. When the temperature of particles goes up, their average kinetic energy has increased. For this to occur, since the particles of a substance cannot change their mass, they must, on average, *move faster*.

PRACTICE   B

Answer, and be able to answer from memory, these questions.

1. Define kinetic energy, in words, then using symbols in an equation.

2. Batter 2 hits a baseball with a bat three times as heavy as batter 1, swinging at the same speed. How much more energy will batter 2 impart to the ball than batter 1?

3. Batter 3 hits the ball with the same bat as batter 1, but swings three times as fast. How much more energy will batter 3 impart to the ball than batter 1?

4. Define temperature.

Potential Energy

Potential energy is stored energy. Lifting an object against gravity is one way to store energy in an object. If the object returns to its former lower position, it must release that added energy.

Example: To raise a hammer, you must add energy. The energy is *stored* in the raised hammer as energy of position. If the hammer falls down to its original position, it must release the energy used to raise it. It can do so by creating *heat* where it hits. The hammer can also do *work*, such as driving nails. Heat, work, and energy of position are simply different forms of energy.

In chemical processes, forms of energy that can be stored and released include heat energy (such as the energy stored in plants during photosynthesis) and electrical energy (as in rechargeable batteries). Compressing a gas can also store energy in a chemical system. When a compressed gas moves a piston against resistance, the potential energy stored in the gas is converted to work, another form of energy.

Energy and Particle Attractions

Molecules and ions are held together by the attractions (bonds) arising from the protons and electrons *within* those particles. Energy always must be *added* to break chemical bonds, or to change a solid substance to a liquid and then to a gas. This added energy is needed to "unstick" the attracting particles. If the "unstuck" particles return to the bonding condition which existed before they were separated, the same amount of energy that was added and stored during their separation must be *released*.

Energy, Reactions, and Phase Changes

Changes in the potential energy of a chemical system can be the result of *chemical reactions* or *phase changes*.

In a chemical reaction, bonds between atoms break and new bonds form. As a result of chemical reactions, there is nearly always a characteristic *net* change in the energy stored in the substances. Energy must be added to break a bond, but more or less energy will be released when a different bond forms. This means that in a chemical reaction, energy is nearly always stored or released to some extent.

In a phase change for a molecular substance, the bonds between the *atoms* in a molecule do not change, and the formulas for substances therefore do not change. However, as a molecular substance changes phase from *solid* to *liquid* to *gas*, the weak attractions between the *molecules* must be overcome during each phase change, so energy must be added. As the substance changes from *gas* to *liquid* to *solid*, that same amount of energy must be *removed* from the substance during each phase change.

Examples:

- When a molecular substance is melted, a characteristic amount of energy must be added per molecule to "unstick" the molecules so they can rotate and translate. The energy added to unstick the molecules is stored in the molecules that change from solid to liquid. If the liquid molecules change back to the solid state, the same amount of energy added to melt the solid must be *released* in order for the liquid molecules to solidify.

- When a substance boils, a characteristic amount of energy must be added per molecule. For that substance to condense from gas to liquid, that same amount of energy must be released.

PRACTICE **C**

Answer, and be able to answer from memory, these questions.

1. Define potential energy.

2. Name two types of chemical processes that can change the energy stored in molecules.

3. How does the heat of fusion of a substance (the heat/mole required to melt a substance) differ from the heat of freezing (the heat/mole released when the liquid changes to a solid)?

Energy and Phases

When energy is added to or removed from a pure substance, whether its *kinetic* or its *potential* energy changes depends on whether the substance is present in one phase or two.

Recall that by our definitions, a *substance* is composed of particles that all have the same chemical formula.

- When a substance is present in only *one* phase (all solid, all liquid, or all gas), adding or removing energy (such as by heating or cooling) changes the average *kinetic* energy of its particles (their temperature) but does *not* change the *potential* energy stored in the substance.

- During a phase *change* (such as melting or boiling), *two* phases must be present. If energy is added to or removed from a substance during a phase change, the *potential* energy stored in the substance *changes*, but the average *kinetic* energy of its particles does *not* change, and its temperature therefore stays constant.

During a phase change, if the two phases present are well mixed or in close contact, the temperature will be the same in both phases. If two phases are mixed and both phases are present after mixing, the temperature of the particles will have adjusted to become the same in both phases. A mixture of the solid and liquid phases of a substance will always adjust to the temperature that is its *melting point*.

Potential Energy and Phases

The *solid* phase of a substance will always have *less* stored (potential) energy than its *liquid* phase, which will always have less potential energy than its *gas* phase.

For a given substance: $PE_{solid} < PE_{liquid} < PE_{gas}$

Changes in potential energy may not be as apparent as changes in kinetic energy, which are evident as temperature changes. Let us therefore examine some examples of energy changes during phase changes.

Examples of Liquid–Gas Phase Changes

Boiling Water In a kitchen at standard pressure, consider a teakettle partially filled with cold water and placed on a lit gas stove. As long as the water in the kettle is *below* its *boiling* temperature, as it is heated by the flame its temperature rises. This increase in average kinetic energy is observable, and it means that the water molecules, on average, are moving faster. What is not observable, but is true, is that the *potential* energy stored in the water is *not* changing.

Boiling begins when the water temperature reaches 100°C. A thermometer will show that once the water begins a steady boil, *both* the liquid water *and* the steam above the boiling water have the same temperature (if precautions are taken to prevent "super-heating"). At standard pressure (101 kPa), for pure boiling water, that temperature will always be 100°C, by definition.

During any phase *change* for any substance, *both* phases will have the *same temperature* as long as they are in close contact. This means that the particles in both phases have the same average *kinetic* energy.

After five minutes of boiling, quite a bit of heat has been added by the flame to the water in the teakettle. However, a thermometer will show that both the water and steam remain at 100°C as long as any liquid water remains in the kettle.

Energy can neither be created nor destroyed. Where is all the energy being supplied by the flame going? The flame's energy is being *stored* in the *gas* particles (steam) that form during the phase change.

A characteristic amount of energy must be stored in molecules as they change from being as close together as possible in their liquid phase to being far apart in their gas phase.

If the flame remains lit beneath the kettle, the water will continue to boil until the last bit of liquid water is converted to steam. At that point, instead of two phases inside the teakettle, there is only *one* phase (steam). Adding energy with *one* phase present will increase the *temperature* of the steam (its average kinetic energy) instead of its potential energy. If *all* of the water is allowed to boil to steam, there is no longer a phase change to absorb the energy supplied by the flame, and the temperature of the steam in the kettle (and of the kettle itself) will increase *very* quickly.

However, as long as *some liquid* water remains in the kettle, the highest temperature possible for the water *or* the steam is 100°C, which is much cooler than the flame below.

Warming Leftovers On a practical note, this is why a little water should be added when heating leftovers in a loosely covered pan. As long as there is some liquid water between the heat and the food, the maximum temperature of the water and the food will be 100°C, enough to warm but not to burn most foods. If all of the water boils away, the food can burn quickly.

When water boils in the pan, it forms steam. When the steam reaches a cooler surface, it can condense to form water. When steam condenses on cool food in a pan, the same amount of potential energy which was stored in the steam as it formed from water must be lost from the steam. As the steam turns to water, energy is transferred to the food, and the "steamed food" heats quickly.

Thunderclouds—Water Condensing On a humid summer day, clear water vapor (a gas) in the atmosphere can condense to tiny drops of liquid water (clouds). As the water vapor condenses, the considerable amount of heat required to change liquid water to water vapor must be released, and the condensation of vapor to water heats the air around the water droplets. Since warmer air is less dense than colder air, it rises, creating an updraft that lifts both the moist air and the water droplets. Because the atmosphere generally cools with increasing altitude, more water vapor in the humid air forms more liquid water and more heat as it rises. As this cycle repeats, the cloud becomes a fast rising "thunderhead."

As the liquid water droplets become larger with increased condensation, the drops eventually become too heavy to be lifted by the updraft. The result is a thunderstorm. The falling raindrops create a powerful "downdraft" of air that strikes the ground and fans out ahead of and with the rain. The downdraft reverses the updraft feeding the thunderhead, eventually causing the thunderstorm to dissipate.

PRACTICE **D**

Answer these questions, then practice until you can answer the questions from memory before going on to the next section.

1. If substantial energy is added to a substance, and it remains the same substance but its temperature does not change, what does this tell you about the substance?

2. When does adding energy to a substance cause its temperature to rise?

3. For a given substance, which phase has the lowest amount of stored energy: solid, liquid, or gas?

4. In a kitchen where the atmospheric pressure is standard pressure, water is placed in a teakettle and heated on a gas stove. At the point where the water first starts to boil,

   a. What is the temperature of the liquid water in the kettle?

   b. What is the temperature of the steam above the water in the kettle?

5. After five minutes of heating, about half of the water in the kettle has boiled away.

   a. What is now the temperature of the liquid water in the kettle?

   b. What is the temperature of the steam above the water in the kettle?

6. During the five minutes of boiling, the gas stove adds considerable energy to the water in the teakettle.

   a. Has the kinetic energy of the water or steam changed?

   b. Has the potential energy of the molecules that are still liquid water changed?

   c. Has the potential energy of the molecules that were converted from water to steam changed?

   d. Where has the energy gone that was supplied by the stove in those five minutes? What kind of energy has it become?

Liquid–Solid Phase Changes

Mixing the Solid and Liquid Phases A stirred mixture of the solid and liquid phases of a substance will always adjust to the temperature that is the *melting point* of the substance.

Examples:

- H_2O melts and freezes at 0°C at pressures at or near typical atmospheric pressure. It is a characteristic of water molecules that a stirred *mixture* containing water and ice will always adjust to a temperature of 0°C: The melting point of H_2O.

A mixture of ice and water is a good *constant temperature bath* or *cold pack*. It will stay at 0°C for as long as both ice and liquid water are present.

- When warm water is added to an ice–water mixture, two phases are present, and the temperature of this new mixture must adjust toward the melting (= freezing) point of water. As ice melts, the warm water molecules become colder. The kinetic energy lost by the warm water is *equal* to the potential energy *stored* in the molecules of ice that become liquid. The warm water continues to cool, and ice continues to melt, until either the mixture reaches its melting point (0°C) or all of the ice melts.

Melting Ice   When ice melts, a solid becomes a liquid. To change a solid substance to its liquid phase, energy must be added. The liquid particles have a *characteristic* higher amount of stored energy, per particle, than the solid particles.

While ice is melting, its temperature does not change, but heat must be added from the environment. This is why a mound of packed snow can take quite a while to melt even when air temperatures are well above freezing. Considerable heat from the environment must be absorbed by the ice molecules that melt to become liquid water.

Freezing Water in an Ice Tray   The liquid phase of a substance has inherently more stored energy than its solid phase. To convert liquid molecules to solid molecules, stored energy must be removed.

To change water into ice, the *same* amount of energy, per molecule, must be taken *out* of the water that is put into the ice to melt it.

Example: When warm water in an ice tray is placed in a freezer, the temperature of the water drops rapidly as its heat transfers to the freezer environment. When the water's temperature reaches 0°C, it begins to freeze.

Unless potential energy leaves a liquid, the liquid cannot become solid. To freeze water, the air in a freezer must be colder than 0°C so that heat energy will flow out of the 0°C water molecules. The freezer compressor pumps heat out until the air is about −20.°C inside most household freezers. You can feel heat that is being pumped out if you place your hand in the space above the coils on the back or underside of a freezer while the compressor is running.

After freezing begins, the water/ice mixture in an ice tray will stay at 0°C until *all* of the water freezes. During this time, the water and ice mixture is the warmest spot in the freezer; warmer than the contents that are already frozen and at −20°C.

Once the ice tray water is completely frozen, one phase is present, and the temperature of the now solid ice cubes drops relatively quickly to the freezer's air temperature.

Ice cubes just removed from the freezer, at about −20°C, are cold enough to both cool and then freeze the moisture on your skin, which can cause the ice cubes to stick to your fingers. However, in a room temperature environment, −20°C ice warms quickly. When it reaches 0°C, the ice begins to melt. Ice at 0°C is not cold enough to freeze skin moisture: ice at 0°C will not stick to your skin.

PRACTICE **E**

Answer, and be able to answer from memory, these questions.

1. A mixture of crushed ice and water is added to an insulated container. After a minute of stirring, the temperature of the mixture no longer changes, and both ice and water remain.

   a. What is the temperature of the ice?   What is the temperature of the water?

   b. Which phase has the higher kinetic energy?

   c. Which phase has the higher potential energy?

2. Warm water is added to an ice–water mixture in an insulated cup. After stirring for one minute, the temperature is stable, and ice and water remain.

   a. What is the temperature of the water in the cup?

   b. What is the temperature of the ice?

3. During the one minute of stirring, the warm water in problem 2 lost some of its energy.

   a. What kind of energy did it lose?

   b. As the warm water lost its energy, what other change occurred?

   c. Where is the energy that was lost by the warm water, and what kind of energy is it?

SUMMARY

Phases, Phase Changes, and Energy

You may want to organize the following information into charts, numbered lists, and flashcards that will help with learning and retention in memory.

1. For substances, the three phases and six phase changes are these:
   - Solids *melt* (or *fuse*) to become liquids; liquids *freeze* (or *solidify*) to become solids.
   - Liquids *boil* or *evaporate* to form gases; gases *condense* to become liquids.
   - Solids *sublime* to become gases directly; gases that undergo *deposition* form solids.

2. *The law of conservation of energy* states that in chemical processes, energy can neither be created nor destroyed. However, energy can be transferred between substances and to and from the environment. Energy can also change its form during chemical or physical processes.

3. Two forms of energy are *potential* energy, defined as stored energy, and *kinetic* energy, defined as energy of motion.

$$\text{kinetic energy} = \tfrac{1}{2}(\text{mass})(\text{velocity})^2$$

4. Chemical substances can store energy in the attractions between atoms, molecules, and particles. During chemical reactions and phase changes, energy is stored or released.

(continued)

5. One way to store energy in a substance is to change its phase. The solid phase of a substance always has less stored (potential) energy than its liquid phase, which always has less potential energy than its gas phase.

Potential energy of a substance: **solid** $<$ **liquid** $<$ **gas**

6. When a substance is in *one* phase (all solid, liquid, or gas), adding or removing energy will change the average kinetic energy of its particles (its *temperature*), but *not* its potential energy.

7. For a substance during a phase change, adding or removing energy changes the potential energy, but not the average kinetic energy (temperature) of the particles.

8. During a phase change for a substance, temperature is constant, and the temperature is the same in both phases as long as they are in close contact.

9. The temperature at which a substance melts (its melting point) will equal the temperature at which it solidifies (melting point $\equiv$ freezing point).

10. In a stirred mixture of the solid and liquid phases of a substance, the temperature will adjust to the melting point of the substance.

11. The melting point is a characteristic of a substance. The melting point will be the same no matter how the substance is formed. The melting point can be used as evidence to identify a substance.

12. A liquid substance will boil at any temperature at which its vapor pressure equals the atmospheric pressure above it. If the atmospheric pressure on the liquid is lowered, the liquid will boil at a lower temperature. If the atmospheric pressure is raised, the liquid will boil at a higher temperature.

13. Boiling points are a characteristic which can be used to identify a substance, but only if the atmospheric pressure is known. Boiling points are far more sensitive to atmospheric pressure than melting points.

14. Evaporation is a surface phenomenon; measurable evaporation will occur from all liquids (and some solids) at any temperature. However, a liquid *boils* only when gas bubbles form below the liquid surface and not just at its edges.

PRACTICE **F**

At standard pressure, small cubes of ice are removed from a freezer and placed in a teakettle. A thermometer is inserted into the ice cubes and the kettle is placed on a lit gas stove. The kettle is heated until one minute after all of the water has boiled away.

The graph on the next page charts the changes in the temperature of the H_2O molecules as they change from ice to water to steam.

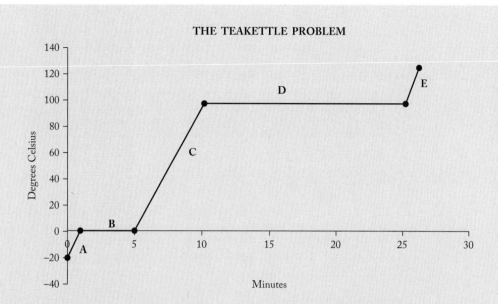

Use the graph and your knowledge in memory of phases and phase changes to answer these. If you write each question number and answer in your notebook, you can use this page again for later review.

1. How many phase changes occur during the above process?

2. How many phases will have been present by the time the above process is completed?

3. Which segment of the graph represents water boiling to steam?

4. How can a change in the kinetic energy of the H_2O be recognized during the process?

5. How can a change in the potential energy of the system be recognized?

6. In which lettered segments of the graph does potential energy remain constant?

7. In which segments of the graph does average kinetic energy remain constant?

8. In which portions of the graph do the H_2O molecules have the largest amount of stored energy?

9. Which portions of the graph show energy from the flame being converted into potential energy?

Lesson 17.2 — Specific Heat Capacity

If you have not already done so, before starting Lesson 17.2, complete Lesson 16.3 on cancellation of complex units.

Units That Measure Energy

In chemistry, energy is usually measured in joules or calories.

1. The **joule** (abbreviated **J**) is the SI unit measuring *energy*.

All forms of energy are equivalent and can be measured in any energy units. In chemistry, to measure the *heat energy* lost or gained in a chemical process, joules is the unit that is most often used.

2. The **calorie** (abbreviated **cal**) is a metric unit also used to measure energy.
    A chemical **calorie** is defined as the amount of heat needed to raise the temperature of 1 gram of liquid water by 1 degree (Celsius or kelvin).

3. Because all forms of energy are equivalent, all energy units can be related by equalities.
    The conversion between calories and joules is

$$1 \text{ calorie} = 4.184 \text{ joules}$$

4. In studies of nutrition, a **food Calorie** is the unit used to measure the heat released when food is burned completely and/or is metabolized in cells.

$$1 \textit{ food } \text{Calorie} = \textbf{1000 } \textit{chemical} \text{ calories} = 1 \textbf{ kilocalorie } (\text{kcal})$$
$$= \textbf{4.184 kilojoules } (\text{kJ})$$

The calories listed on nutritional labels are food Calories (chemical kilocalories).

In textbooks, *food* Calories are often written as **C**alories with a capital **C**, whereas *chemical* **c**alories are written with a lower case **c**.

Specific Heat Capacity

Heat (symbol q) can be defined as a form of energy that *transfers* due to a difference in temperature.

In these lessons, we will define the amount of heat required to raise one *gram* of a substance by one *degree* Celsius or kelvin as the **specific heat capacity** (symbol small c) of the substance. (Some textbooks use different terms and symbols for this quantity.)

In most calculations, specific heat capacity will be used to calculate the *total* heat energy change in a process. We will therefore learn the equation using specific heat capacity in this form:

For a substance in a single phase, $q = c \cdot m \cdot \Delta t$

This equation means: The *heat energy* (q) gained or lost by a substance = (specific heat capacity of the substance) × (its *mass*) × (its *change* in temperature).

When heat is added to a system of chemical particles, q is defined as positive. When chemical particles lose heat, q is defined as negative.

The units of c are joules *per* (gram · degree) *or* calories *per* (gram · degree).

Example: The specific heat capacity of liquid water is written $c_{\text{water}} =$ 4.184 J/g · K

Recall that the *dot* between gram and K means that the two units are multiplied together either in the numerator or the denominator. These three notations are equivalent:

$$4.184 \text{ joules/}\textbf{gram} \cdot \textbf{K} = 4.184 \frac{\text{J}}{\text{g} \cdot \text{K}} = 4.184 \, \text{J} \cdot \text{g}^{-1} \cdot \text{K}^{-1}$$

Specific heat capacity may also be measured in joules/*kilo*gram · degree. You will learn below how to convert data to the consistent units needed to solve equations. Note the difference indicated by the equation between *heat* and *temperature*.

- Temperature is an intensive property: When you measure a temperature, the value does not depend on the amount of matter being measured.
- Heat is an extensive property. When calculating the heat transferred in a process, what is being heated, the amount being heated, and how much heat is being transferred are all important. On a gas stove, to bring water at room temperature to the point that it begins to boil, for a large amount of water you must supply more heat than for a small amount.

Change in Temperature

In science, the symbol Δ (**delta**) means *change in*. The symbol Δt (read "delta *t*") means *change in temperature*. The definition is

$$\Delta t \equiv t_{\text{final}} - t_{\text{initial}}$$

This definition means that a change in temperature is labeled as positive when temperature increases and negative when it decreases.

A change in temperature will be the same *number* of degrees when measured in the Celsius *or* Kelvin scales. Why? The Kelvin and Celsius scales have the same *size* degree.

Example: If, in Celsius, $\Delta t = 25°C - 0°C = \textbf{25°C}$, the same measurements recorded in kelvins will result in the *same number* for the *change*: $\Delta t = 298 \text{ K} - 273 \text{ K} = \textbf{25 K}$.

This means that if the temperature units in a problem are based on Δt measurements, the word **degree** and the symbols **°C** and **K** are all *equivalent*.

Since specific heat capacity (*c*) is defined in the words and equation above as based on a *change* in temperature (Δt), a value for *c* of "4.184 joules/gram · **degree**" can also be written as "4.184 joules/gram · **K**."

When units are equivalent, they can cancel. In calculations using Δt values,

$$\frac{\text{joules} \cdot °C}{K} = \frac{\text{joules} \cdot °C}{K} = \textbf{joules} \quad and \quad \frac{\text{calories} \cdot K}{\text{degree}} = \frac{\text{calories} \cdot K}{\text{degree}} = \textbf{calories}$$

PRACTICE   A

Assume that the temperature units below are all measurements of Δt. Simplify these terms using unit cancellation, then write the final unit. For a review of the rules for unit cancellation, see Lesson 16.3.

1. $\dfrac{\text{J}}{\text{g} \cdot \text{K}} \cdot \text{g} \cdot °C =$

(continued)

2. $\dfrac{J}{\dfrac{J}{g \cdot K} \cdot g} =$

3. $\dfrac{cal}{\dfrac{cal}{g \cdot degree} \cdot {}^\circ C} =$

4. $\dfrac{J}{g \cdot {}^\circ C} =$

5. Name two metric units that can be used to measure q.

Values for Specific Heat Capacity

Chemical substances have a *characteristic* specific heat capacity (c) in each phase: a fixed amount of heat changes one gram of the substance by one degree. Some values for specific heat capacity are in the table below.

Substance	Specific Heat Capacity Values in J/g · K
H_2O **liquid**	4.184
H_2O **solid**	2.09
Cu solid	0.385
Fe solid	0.444

As in the case of H_2O in the table, for each substance, each *phase* has a different c value.

Heat capacity values apply only while a substance is in a single phase. A different measure for calculating heat changes will be needed when a substance is *changing* phase.

To solve calculations that include specific heat capacity, we will use the *equation* that includes specific heat capacity as follows.

The *c* Prompt

If you see the term "**specific heat capacity**" or its symbol c in a problem, write at the top of your DATA table this equation that uses c:

$$q = c \cdot m \cdot \Delta t$$

Solving Problems That Require Equations

In these lessons, we will refer to equalities that use symbols, such as $\boxed{q = c \cdot m \cdot \Delta t}$, as *equations* rather than *formulas* to distinguish them from chemical formulas, such as H_2O.

To solve calculations that include specific heat capacity, we will use the equation method that we employed for ideal gas equations. Let us start with an example.

TRY IT

(See "How to Use These Lessons," point 1, p. xv.)

Q. When 832 joules of heat is added to a sample of solid copper (Cu), the temperature rises from 15.0°C to 33.0°C. Based on the specific heat capacity in the table above, how many grams of copper were in the sample?

To solve, complete the following steps in your notebook.

1. As always, begin by writing "WANTED: ?" and the unit you are looking for.

2. This problem mentions "specific heat capacity." That's the *c prompt*. In the DATA section, write the memorized equation that includes *c*.

3. Below the equation, make a data table that lists *each symbol* in the equation.

Do those steps, then check below.

Answer:

For this problem, the data section should include:

DATA: $\boxed{q = c \cdot m \cdot \Delta t}$

$q =$

$c =$

$m =$

$\Delta t =$

4. After each symbol, based on *units*, write each number and unit supplied in the problem.

Try that step, then check below.

The symbol for heat energy is q, and energy is measured in *joules*, so in this problem $q = 832$ J From the table value for copper: $c = 0.385$ J/g · K

5. Put a **?** after the symbol in the table that you are *looking for*. Add the unit and label of the quantity you are looking for.

For this problem, fill in the data table completely, and then check below.

At this point, your paper should look like this:

WANTED: **?** g Cu

DATA: $\boxed{q = c \cdot m \cdot \Delta t}$

$q = 832$ J

$c = 0.385$ J/g · K

$m =$ **? g Cu**

$\Delta t = 33.0°C - 15.0°C = +18.0°C$

6. SOLVE the fundamental memorized equation, using algebra, for the *symbol* WANTED. Do not plug in numbers until you have solved for the WANTED symbol. (Symbols move more quickly than numbers and their units.)

STOP

Try that step, then check below.

SOLVE: Solving $\boxed{q = c \cdot m \cdot \Delta t}$ for the symbol m WANTED,

$$? = m = \frac{q}{c \cdot \Delta t}$$

7. *After* solving in symbols, plug in the numbers and solve. Cancel units that cancel, but leave the units that do not cancel, and include them after the calculated number.

Complete that step for this problem.

STOP

$$? = m = \frac{q}{c \cdot \Delta t} = \frac{832\ \text{J}}{0.385\ \dfrac{\text{J}}{\text{g} \cdot \text{K}} \cdot 18.0°\text{C}} = \frac{832\ \text{J}}{18.0°\text{C}} \cdot \frac{1}{0.385\ \dfrac{\text{J}}{\text{g} \cdot \text{K}}}$$

$$= \frac{832\ \text{J}}{18.0°\text{C}} \cdot \frac{\text{g} \cdot \text{K}}{0.385\ \text{J}} = 120.\ \text{g Cu}$$

Note that in solving, the term with a fraction in the denominator was separated and then simplified using the rules for reciprocals.

Double check the cancellation of the units for the last step above. Since *degrees* and $°C$ and K are all equivalent when they measure a Δt, as units they can cancel.

The above problem involved finding *grams* of Cu, but you did not need the molar mass to solve. This is because specific heat capacity is one of the rare quantities in chemistry that is defined based on grams rather than moles. Other heat problems *will* involve moles. Our rule will be

> In *heat* problems, if you see *both* grams *and* moles of a substance as units, write the molar mass in your data, because you will likely need the molar mass to solve the problem.

SUMMARY

Solving with Equations

If you need a mathematical *equation* to solve a problem, follow these steps.

1. Write the fundamental, memorized equation.

2. Make a data table with each of the *symbols* in the equation.

3. Based on *units*, write each item of data after a *symbol* in the data table.

4. Memorize equations in *one* format, then use algebra to solve for the symbol WANTED. This will minimize the need to memorize.

   Example: memorize the equation $\boxed{q = c \cdot m \cdot \Delta t}$

   Then if you WANT c or m or Δt, use algebra to solve the equation for that symbol.

   Don't memorize:

   $$m = \frac{q}{c \cdot \Delta t} \quad \text{and} \quad c = \frac{q}{m \cdot \Delta t} \quad \text{and} \quad \Delta t = \frac{q}{c \cdot m}$$

5. Solve the fundamental equation for the WANTED symbol *before* you plug in numbers.

6. Plug both numbers and units into equations. Rely on unit cancellation to check your work.

PRACTICE B

Make a set of flashcards covering the key definitions and equations in this lesson.

Note especially the boxed rules and equations. Practice until you can automatically answer each card.

Then, when doing the following problems, try to use your knowledge and your practice of the steps above to solve without looking back at the lesson.

1. Define what each of the symbols q, c, and Δt represent.

2. When 681 J of heat are added to 240. g of a pure solid, the temperature of the solid rises by 22.0 degrees. What is the specific heat capacity of the solid?

3. If 361 J are added to a 32.5 g sample of solid iron (Fe) at 20.0°C, use the value for c from the table above and solve for the final temperature of the sample.

Lesson 17.3 | Consistent Units in Heat Calculations

The substance most often used to supply or absorb heat in a chemical process is *liquid water*. Most courses require that the value for the specific heat capacity of water be memorized. In heat calculations that involve liquid water, we will use this rule:

The c Water Prompt

If a problem mentions *energy* or *heat* or *joules* or *calories*—**and liquid water**—write in your data table the equation using specific heat capacity, $\boxed{q = c \cdot m \cdot \Delta t}$. In the DATA table with those four symbols, write one of these equalities:

$$c = c_{water} = 4.184 \, J/g \cdot K \qquad or \qquad c_{water} = 1 \, \textbf{calorie}/g \cdot K$$

Using this prompt, problems involving heat and water can be solved in the same way as the specific heat problems in the previous lesson.

In calculations involving liquid water, assume that 1 **milliliter** *liquid* H_2O = 1.00 **gram** *liquid* H_2O unless a more precise density is noted.

Though the common name for H_2O is water, in problems dealing with energy it is important to distinguish between *ice*, *water*, and *steam*. These three phases for H_2O have different values for c. However, unless ice or steam is specified, in heat problems you should assume that *water* means *liquid* water.

Consistent Units

For an equation to work, the units must match the requirements of the equation.

Example: The equation for specific heat capacity requires *mass* (usually grams, occasionally kilograms). If the data is given in *moles*, you must convert moles to grams or kilograms before you solve.

In addition, when solving equations, units must be **consistent**.

Example: In an equation involving *mass*, grams *or* kilograms may be used, but not both. You must choose a mass unit, and then convert the other masses to that unit.

Which unit should you choose? In most cases, you may solve equations in any consistent units, but some ways of choosing consistent units will solve more quickly than others. To solve heat problems, we will use this rule:

> Convert DATA to the units in the most *complex* unit.

Example: If a heat calculation includes a unit of "joules/**kg** · K," convert the units in the DATA to joules and kilograms.

The best time to convert to consistent units is *early* in a problem. The best place to convert to consistent units is in the DATA table.

TRY IT

Keeping those points in mind, try this example in your notebook. If you get stuck, peek at a bit of the answer, then try again.

Q. 16.0 mol of water at 25.0°C is supplied with 28.0 kJ of heat from a bunsen burner. If all of the heat is absorbed by the water, what will be the water's final temperature?

Answer:

When you see *joules* and liquid water, that's the *c water prompt*. Write the equation that includes specific heat capacity, a data table to match its symbols, and fill in *c* for liquid water. Because the problem includes kilojoules, pick the *c* value for water that includes joules rather than calories.

DATA: $\boxed{q = c \cdot m \cdot \Delta t}$

$q =$

$c = \mathbf{4.184\,J/g \cdot K}$

$m =$

$\Delta t =$

Since the units supplied in the problem do *not* match the units of the complex unit, write a **?** and a *unit* beside each symbol that is consistent with the *complex* unit.

DATA: $q = \mathbf{?\,J} =$

$c = 4.184\,J/g \cdot K$

$m = \mathbf{?\,g} =$

$\Delta t = ?$ K or °C WANTED final temp. $= 25.0°C + \Delta t$

Note that a Δt value in degrees Celsius and kelvins has the same numeric value.
Add the remaining data from the problem to the table, converting the data to the consistent units as you go, then solve.

DATA: $q = ?\,\mathbf{J} = 28.0\,\mathbf{kJ} = 28.0 \times 10^3\,\mathbf{J}$ (Kilo means $\times\ 10^3$.)

$c = 4.184\,\text{J/g} \cdot \text{K}$

$m = ?\,\mathbf{g} = 16.0\ \text{mol}\ H_2O \cdot \dfrac{18.0\ \text{g}\ H_2O}{1\ \text{mol}\ H_2O} = 288\ \mathbf{g}\ H_2O$

$\Delta t = ?\,\mathbf{K}$ or $°\text{C}$ WANTED   final temp. $= 25.0°\text{C} + \Delta t$

SOLVE: $? = \Delta t = \dfrac{q}{c \cdot m} = \dfrac{28.0 \times 10^3\,\text{J}}{4.184\ \dfrac{\text{J}}{\text{g} \cdot \text{K}} \cdot 288\ \text{g}}$

$= 28,000\,\text{J} \cdot \dfrac{\text{g} \cdot \text{K}}{4.184\,\text{J}} \cdot \dfrac{1}{288\ \text{g}} = \mathbf{23.2\ K\ increase} = \Delta t$

Check the unit cancellation at the last step.

Since heat is being added to water, the temperature will increase. You may keep track of the direction of the temperature change using signs on Δt or with labels noting an increase or decrease in temperature.

The final temperature is $25.0°\text{C} + \mathbf{23.2}°\text{C}$ or K $= \boxed{\mathbf{48.2°C}}$

SUMMARY

Consistent Units

1. If an equation *requires* certain units, convert to those units in the DATA table.

2. Convert data to *consistent* units in the DATA table.

3. If an equation does not require certain units, but the units in the problem are not consistent,
   - Choose consistent units based on the most complex unit in the problem.
   - Write the consistent unit after each symbol in the DATA table.
   - In the DATA table, convert the supplied units to the consistent units.

4. If the WANTED unit is not consistent with the most complex unit, SOLVE using the chosen consistent unit, then convert to the WANTED unit.

PRACTICE

Add the new rules in this lesson to your previous flashcards for this chapter. Run the cards until you can answer each automatically. Then complete as many of the following problems as you need to feel confident. The more difficult problems are toward the end.

1. Convert these to the units WANTED. Try to do so by inspection (in your head).

   a. $?\,\text{J} = 0.25\,\text{kJ} = $ _____ = _____ = _____

   In: exponential notation   scientific notation   fixed notation

(continued)

b. ? g $H_2O(\ell)$ ≈ 75 mL $H_2O(\ell)$ = _____ in fixed notation.

c. In any notation: ? g $H_2O(\ell)$ = 8.9 L $H_2O(\ell)$

2. A 36.0 mL sample of water is raised in temperature by 15.0°C. How many joules are supplied?

3. A 15.0 mol sample of liquid water loses 6.70 kJ of heat. At the end of the process, the water temperature is 18.2°C. What was the original temperature of the water?

4. How much heat (in joules) would be required to raise 4.50 mol of ice from −20.0°C to the temperature at which it begins to melt (*c* for ice = 2.09 J/g · degree)?

SUMMARY

Energy Calculations

A summary for phases and phase changes is at the end of Lesson 17.1.

1. *The law of conservation of energy* means that energy can neither be created nor destroyed by chemical processes.

2. In chemistry, energy is usually measured in joules or calories.

   a. Joules (J) are the SI unit measuring energy.

   b. Calories (cal) are a metric unit also used to measure energy.
   A chemical calorie is defined as the amount of heat needed to raise the temperature of 1 gram of liquid water by 1 degree (Celsius or kelvin).

3. Because all forms of energy are equivalent, all energy units can be related by equalities. 1 cal = 4.184 J

4. 1 food Calorie = 1000 chemical calories = 1 kilocalorie (kcal) = 4.184 kJ
   Food calories are abbreviated with a capital C (Cal), chemical calories with a lowercase c (cal).

5. The *specific heat capacity* (symbol small *c*) of a substance in a given phase is defined as the amount of heat required to raise one gram of the substance by one degree (Celsius or kelvin).
   The units of *c* are joules *per* (gram · degree) *or* calories *per* (gram · degree).

6. The symbol Δ means *change in.* The symbol Δ*t* means change in temperature.

   $$\Delta t \equiv t_{final} = t_{initial}$$

   A Δ*t* value should be labeled as positive or negative, or as increasing or decreasing.
   A change in temperature is the same number of degrees whether measured in °C or K.

For Δt measurements, the terms *degree* and $°C$ and K are all equivalent. When terms are equivalent, they can cancel during unit cancellation.

7. **The c prompt** is used when you see the term *specific heat capacity* or its symbol c in a problem. Write at the top of your data this equation that includes c: $\boxed{q = c \cdot m \cdot \Delta t}$

    Heat energy (q) is defined as positive when energy is added to a chemical system, and negative when energy leaves.

8. **The c water prompt** is used when a problem mentions energy *or* heat *or* joules *or* calories—and liquid water. Write the c prompt equation, $\boxed{q = c \cdot m \cdot \Delta t}$, a data table listing those symbols, and enter $c = c_{water} = 4.184\,\text{J/g} \cdot \text{K}$ (*or* $= 1\,\text{cal/g} \cdot \text{K}$).

9. Units must be *consistent* when substituted into equations. When units are not consistent, or do not match what is needed in an equation,
    - Pick an appropriate unit for each DATA symbol (preferably those used in the most complex unit in the problem).
    - Write the consistent unit after each symbol in the DATA.
    - In the DATA table, convert the supplied units to the consistent unit, then solve.

ANSWERS

Lesson 17.1

Practice A 1. Phases: solid, liquid, gas. Changes: melting, freezing, boiling, condensing, sublimation, deposition.

2. Only the gas phase. There is substantial distance between particles only in the gas phase.

3a. Melting point $\equiv$ freezing point, by definition.

3b. A pure substance melts at a higher temperature than the same substance with impurities.

4. When its vapor pressure equals the atmospheric pressure above it.

5. Raise the liquid's vapor pressure by raising its temperature, or lower the atmospheric pressure above the liquid, such as by moving the liquid to higher altitude or into a partial vacuum or do both increasing T and lowering P.

6. 101 kPa is standard pressure: The pressure at which water boils at **100°C** by definition.

7. Approximately 95°C. At high altitude, atmospheric pressure is lower, and the water's vapor pressure will equal atmospheric pressure at a lower temperature.

8. The water boils at a lower temperature, and at a lower temperature the changes needed to "cook" food take longer to occur.

Practice B 1. Kinetic energy is energy of motion. KE $= \frac{1}{2}$(mass)(velocity)^2.

2. Batter 2 hits with three times as much energy.   3. Batter 3 hits with nine times more energy.

4. Temperature is a measure of the average kinetic energy of chemical particles.

Practice C 1. Stored energy   2. Chemical reactions and phase changes

3. Heat of melting $\equiv$ heat of freezing. The heat added in melting must be released when a liquid solidifies.

Practice D 1. Two phases are present, and the substance is undergoing a change to a phase with more stored energy.

2. When only one phase is present, which means the substance is not undergoing a phase change.

3. Solid

4a. and 4b.   Both the liquid water and the steam are at 100°C.

5a. and 5b.   Both the liquid water and the steam are still at 100°C.

6a. No   6b. No   6c. Yes

6d.   The energy is now potential energy stored in those molecules that changed phase from liquid to gas.

Practice E   1a. Both are at 0°C.   1b. Both have the same KE.   1c. The liquid water has higher PE.

2a. and 2b.   Water and ice are both at 0°C. If both solid and liquid are present, both must be at the melting point.

3a.   The warm water lost *kinetic* energy: its temperature fell.   3b.   The warm water melted some ice.

3c.   The kinetic energy lost by the water in cooling to 0°C is now *potential* energy that is stored in the molecules that were previously ice but were melted by the warm water.

Practice F   1. Two (melting and boiling)   2. Three (solid, liquid, and gas)   3. **D**

4.   The temperature **changes**. (This occurs when the line is not in a "plateau" region.)

5.   Where the graph has a "plateau" region, heat is being added from the stove for several minutes but the temperature (average kinetic energy) remains constant, so potential energy must be increasing.

6. **A, C, and E**—when the kinetic energy is changing.   7. **B and D**—the temperature stays constant.

8. **D and E**—the gas phase (the steam) that forms during D, and then heats during E, has the most potential energy.

9. **B and D**—during the two phase changes.

Lesson 17.2

Practice A   You may use other methods of unit cancellation (see Lesson 16.3) as long as you arrive at the same answers as these. *Degrees* and °*C* and *K* are all equivalent when they are used to measure a *change* in temperature (Δt). If one is on top and one is on the bottom when you multiply terms, they can cancel.

1. $\dfrac{J}{g \cdot K} \cdot g \cdot °C = \mathbf{J}$

2. $\dfrac{J}{\dfrac{J}{g \cdot K} \cdot g} = \dfrac{1}{\dfrac{1}{K}} = \mathbf{K}$

3. $\dfrac{cal}{\dfrac{cal}{g \cdot degree} \cdot °C} = \dfrac{1}{\dfrac{1}{g}} = \mathbf{g}$

4. $\dfrac{J}{g \cdot °C} = \dfrac{J}{\mathbf{g} \cdot °\mathbf{C}}$

(In problem 4, nothing cancels.)

5. Joules and calories.

Practice B   1. The symbol q stands for heat energy, c is specific heat capacity and Δt means change in temperature.

2. WANTED: $c = ?$

(Strategy: When "specific heat capacity" is mentioned, that calls the "c prompt." Write the equation that includes c in the DATA.)

DATA: $\boxed{q = c \cdot m \cdot \Delta t}$ (Make a data table to match all those symbols.)

$q = + 681 \, J$

$c = ? = $ WANTED

$m = 240. \, g$

$\Delta t = +22.0°C$

SOLVE: Since $\boxed{q = c \cdot m \cdot \Delta t}$

$$? = c = \frac{q}{m \cdot \Delta t} = \frac{681\,J}{240.\,g \cdot 22.0°C} = 0.129\ \frac{J}{g \cdot degree}$$

(Solve in symbols before plugging in numbers and units. Do the math for both numbers and units both. Make sure the answer unit matches what the unit should be for the symbol WANTED.)

3. WANTED: Final temperature. Since $\boxed{\Delta t \equiv t_{final} - t_{initial}}$; $t_{final} = t_{initial} + \Delta t$

Strategy: When c is mentioned, write the equation that includes c.

DATA: $\boxed{q = c \cdot m \cdot \Delta t}$ (Make a data table to match those symbols.)

$q = +\ 361\,J$

$c = 0.444\,J/g \cdot K$ for Fe

$m = 32.5\,g$ Fe

$\Delta t = ?$ WANTED **final** $t = 20.0°C + \Delta t$

SOLVE: $\boxed{q = c \cdot m \cdot \Delta t}$, and we want Δt.

$$? = \Delta t = \frac{q}{c \cdot m} = \frac{361\,J}{0.444\,\frac{J}{g \cdot k} \cdot 32.5\,g} = +\ 25.0\,K = \Delta t \quad \text{Are you done?}$$

WANTED $= t_{final} = t_{initial} + \Delta t$

Final $t = 20.0°C + \Delta t = 20.0°C + \textbf{25.0°C}$ or K $\boxed{= \textbf{45.0°C} \text{ final temperature}}$

(A Δt is the same *number* of degrees in the Celsius and kelvin temperature scales.)

Lesson 17.3

Practice 1a. $?\,J = 0.25\,kJ = \textbf{0.25} \times \textbf{10}^3\,\textbf{J} = \textbf{2.5} \times \textbf{10}^2\,\textbf{J} = \textbf{250 J}$

1b. $?\,g\,H_2O(\ell) \approx 75\,mL\,H_2O(\ell) = \textbf{75 g}\,H_2O(\ell)$ (1.00 g liquid water $\approx$ 1 mL at room temperature.)

1c. $?\,g\,H_2O(\ell) = 8.9\,L\,H_2O(\ell) = 8,900\,mL\,H_2O(\ell) = \textbf{8,900 g}\,H_2O(\ell) = \textbf{8.9} \times \textbf{10}^3\,\textbf{g}\,H_2O(\ell)$

2. WANTED: $?\,\textbf{J}$

$\boxed{q = c \cdot m \cdot \Delta t}$

DATA: $q = ?\,J$ WANTED

$c = 4.184\,J/\textbf{g} \cdot \textbf{K}$ for liquid water

$m = ?\,\textbf{g} = 36.0\,mL = 36.0\,g$ for liquid water   (1 mL liquid water $\approx$ 1.00 g water.)

$\Delta t = ?\,K$ or °C $= 15.0°C$

(The mL of water are converted above to the mass unit that matches the c unit.)

SOLVE: $? = q = c \cdot m \cdot \Delta t = 4.184\,\frac{J}{g \cdot K} \cdot 36.0\,g \cdot 15.0°C = \textbf{2,260 J}$

3. WANTED: *Initial* temp. As $\boxed{\Delta t \equiv t_{final} - t_{initial}}$ then $t_{initial} = t_{final} - \Delta t = $ WANTED.

Strategy: When you see joules and liquid water, write $\boxed{q = c \cdot m \cdot \Delta t}$ and c_{water} (The c water prompt.)

DATA: $q = ?\,J = -6.70\,kJ = -6.70 \times 10^3\,J = -6,700\,J$ (Convert to units of c.)

(In these problems, to track the increase or decrease in heat and temperature, you may use signs for q and Δt, *or* you may assign the signs using logic: if a substance loses heat, it must get cooler.)

$c = \textbf{4.184 J/g} \cdot \textbf{K}$ (The c water value that uses J.)

$$m = ?\,g = 15.0 \text{ mol } H_2O \cdot \frac{18.0 \text{ g } H_2O}{1 \text{ mol } H_2O} = \textbf{270. g } H_2O$$

(When grams are needed, but moles are given, do the conversion in the data table.)

$$\Delta t = \boxed{\Delta t \equiv t_{final} - t_{initial}} \text{ so } t_{initial} = t_{final} - \Delta t = \text{WANTED}$$

SOLVE: $? = \Delta t = \dfrac{q}{c \cdot m} = \dfrac{-6{,}700 \text{ J}}{4.184 \dfrac{J}{g \cdot K} \cdot 270. \text{ g}} = \textbf{-5.93 K} = \Delta t$

WANTED $= t_{initial} = t_{final} - \Delta t = 18.2°C - (-5.93°C \text{ or K}) = \textbf{24.1°C}$

(Since the water is losing heat, its initial temperature must be warmer than the final temperature, and it is.)

4. WANTED: q in joules

(Strategy: Note that the c value for ice is not the same as for liquid water, although both are H_2O.

Since the problem mentions c, that's our "c prompt." Write the equation that uses c.)

DATA: $\boxed{q = c \cdot m \cdot \Delta t}$

$q = ?\,J = $ WANTED

$c = \textbf{2.09 J/g} \cdot K$ for **ice**

[The equation requires converting to mass (m) in grams because the c unit includes grams. The relationship between grams and moles is the molar mass.]

$$m = ?\,g = 4.50 \text{ mol } H_2O \cdot \frac{18.0 \text{ g } H_2O}{1 \text{ mol } H_2O} = 81.0 \text{ g } H_2O(s)$$

(Ice begins to melt when it reaches 0.0°C.)

$$\Delta t = ?\,K \text{ or } °C = -20.0°C \text{ to } 0.0°C = +20.0°C$$

(The data have now been converted to the three units used by c. The equation will work because the units cancel properly.)

SOLVE: $? = q = c \cdot m \cdot \Delta t = 2.09 \dfrac{J}{g \cdot K} \cdot 81.0 \text{ g} \cdot 20.0°C = \textbf{3,390 J}$ of heat must be added.

18

Bonding

Prerequisites

For this chapter, you will need a set of molecular models. These can be purchased at college bookstores or online. As an alternative, patterns for cardboard models are provided in Lesson 18.2, but commercial models are recommended.

Introduction

Bonds are forces that hold atoms together. The nature of the chemical bond is a question at the heart of chemistry, but the answer is not completely understood. An explanation of bonding must take into account protons and electron pairs, electrical attraction and repulsion, and electrons that can behave as both particles and as waves. A theory that successfully unites all of those factors does not yet exist.

However, a variety of theoretical models predict typical bond behavior. We will begin with two of these models, Lewis structures and VSEPR, that allow us to predict the composition and shape of a significant percentage of the molecules within and around us.

Lesson 18.1 | Lewis Structures

Ionic versus Covalent Bonds

In *ionic* bonding, charged particles (ions) are held together by the electrical attraction of opposite charges.

Covalent bonding is often described as electron sharing. In covalent bonding, a pair of electrons between two atoms form the covalent bond that holds the two atoms together. Covalent bonds may be single bonds with one pairs, double bonds with two pairs, or triple bonds with three pairs of electrons.

If all of the atoms in electrically netural particle are bonded covalently, the particle is nearly always termed a molecule. The forces holding atoms together inside a molecule are strong compared to the forces between the molecules. Compared to the ions in ionic compounds, covalently bonded molecules are more easily pulled apart from each other when heated, so covalent molecules typically melt and boil at temperatures much below that of ionic compounds.

Lewis Structures

Lewis structures (also called **Lewis diagrams** or **electron dot diagrams**) can be drawn to represent covalent bonds in covalent molecules and polyatomic ions.

Example: The Lewis structure of H_2 is **H:H**

Lewis structures are useful for predicting the bonding, shape, and solubility of substances. To draw the Lewis structures for molecules, we begin by drawing the dot diagrams for atoms.

Rules for Drawing Lewis Structures for Neutral Atoms

1. Write the symbol for the atom.

2. Determine the number of *valence* electrons in the atom.

The number of valence electrons for an atom is equal to the number for the main group (the *tall* columns) in which the atom is located in the periodic table.

Examples:

- First column neutral atoms have *one* valence electron.
- Neutral atoms in the carbon family have *four* valence electrons.
- Noble gases have eight valence electrons (except *helium*, which has two).

3. Assume that each atom symbol has *four* sides. On each side can go at most two electrons. Using dots to represent the valence electrons, draw the valence electrons around the atom symbol. Put *one* electron on each of the four sides of the symbol *before* you start to *pair* electrons. The four sides are *equivalent*: you may place the paired and unpaired electrons on any side.

Examples: Boron, with three valence e⁻, is drawn as $\cdot \overset{\displaystyle \cdot}{B} \cdot$

Nitrogen (five valence e⁻) is $\cdot \overset{\displaystyle \cdot}{\underset{\displaystyle \cdot \cdot}{N}} \cdot$

The dot diagram for a neutral boron atom has three **unpaired** electrons. Nitrogen has three unpaired electrons and one **pair** of electrons.

The Octet Rule

To draw dot diagrams for molecules and ions, apply the **octet rule**: An atom has maximum stability when it is surrounded by *eight valence* electrons. (Hydrogen, however, is most stable with a **duet** of *two*.)

Combinations that result in all atoms being surrounded by eight valence electrons (two for H atoms) tend to be *stable*: those combinations are likely to be found in nature and formed in chemical reactions. A species that does not have a *satisfied octet* may exist, but it is likely to be unstable: it will tend to be a very *reactive* species.

Using Lewis Diagrams to Predict Bonding

To predict the bonding in stable molecules, depending on the type of problem, there are two methods for drawing dot diagrams.

Method 1 This method *predicts* the molecular formula that is likely if the formula is *not* supplied, but the substance has one or two kinds of atoms and all *single* bonds (*one pair* of electrons per bond). Let's learn method 1 with an example.

▶ **TRY IT**

(See "How to Use These Lessons," point 1, p. xv.)

Q. Draw the Lewis structure for a stable molecule that contains only chlorine atoms with single bonds.

Apply these steps for method 1:

1. Draw the dot diagram for each neutral atom.

2. *Combine* the diagrams of the atoms so that the *unpaired* electrons *pair* and are *shared* between two atoms. Combine the atom diagrams until each symbol is surrounded by *eight* valence electrons (except H, which needs two).

Answer:

A neutral chlorine atom has seven valence electrons. Place one on each of the four sides of the symbol, then start to pair electrons. This results in three *pairs* of dots and one *unpaired* dot representing the valence electrons around chlorine. The chlorine atom has seven valence electrons and it needs *eight* to be stable.

$$:\overset{\cdot\cdot}{\underset{\cdot\cdot}{Cl}}\cdot$$

To make a stable molecule, slide two chlorines together so that their unpaired electrons pair. *Each* chlorine is now surrounded by *eight* valence electrons. The *octet rule* is satisfied for both atoms. The two shared electrons are a **bonding pair**: they form a *bond* between the two chlorines. Each chlorine atom has three **lone pairs**.

$$:\overset{\cdot\cdot}{\underset{\cdot\cdot}{Cl}}\cdot \ + \ \cdot\overset{\cdot\cdot}{\underset{\cdot\cdot}{Cl}}: \ \longrightarrow \ :\overset{\cdot\cdot}{\underset{\cdot\cdot}{Cl}}:\overset{\cdot\cdot}{\underset{\cdot\cdot}{Cl}} \ = \ Cl-Cl \ = \ Cl_2$$

|(two *atom* dot diagrams)|(*molecule* dot diagram)|(structural and molecular formulas)|

In a particle with more than one atom, the *lone* pairs may also be termed **unshared pairs** or **non-bonding pairs**.

Method 1 is quite simplified, but it does predict the bonding in many cases when the molecular formula (and therefore the total number of valence electrons) is not known.

PRACTICE   A

Use a periodic table. If needed, check your answers after each part. The atoms in covalent compounds will most often be the nonmetals found toward the top right of the periodic table, plus hydrogen.

1. How many valence electrons are in these neutral atoms?

    a. Silicon   b. Phosphorus   c. Bromine   d. Sulfur

2. Draw the Lewis diagram for each atom in problem 1.

    a.   Si   b.   P

    c.   Br   d.   S

3. Using method 1, draw a Lewis diagram and then a structural formula for the predicted stable molecules formed by combinations of

    a. Fluorine atoms   b. Hydrogen and chlorine atoms

4. For each of the molecules in problem 3, list the number of covalent *bonds* and the total number of *lone pairs* of electrons.

    Bonds: 3a. _____   3b. _____

    Lone pairs: 3a. _____   3b. _____

Method 2 If the formula for a molecule or polyatomic ion is *supplied*, a model that better predicts the nature of bonds is to combine the valence electrons without regard to which atom contributes the electrons.

TRY IT

Let's learn method 2 with an example.

    Q. Draw the Lewis diagram for a water molecule, H_2O.

When you know the molecular formula, apply these steps for method 2:

1. *Count* the *total* number of valence electrons in the neutral atoms of the molecule.

2. Arrange the valence electrons around the central atom to satisfy the octet/duet rule: For maximum stability, each symbol needs to be surrounded by *eight* valence electrons (H needs two). Do not worry about which atom contributes which electrons.

    In simple molecules with more than two atoms, the **central atom** is the atom with the *most unpaired electrons* in its atom dot diagram. This means that the central atom in a formula is usually the atom closest to group 4A (the carbon family) in the periodic table. Carbon family atoms have four unpaired electrons.

Answer:

1. Each neutral hydrogen has one valence electron (and needs two for stability). The one neutral oxygen has six valence electrons (and needs eight). The total for the molecule is eight valence electrons. The central atom is oxygen because its atom dot diagram has two unpaired electrons, while hydrogen has only one.

2. The Lewis structure for water can be written in a variety of equivalent ways.

$$H\!:\!\overset{..}{\underset{..}{O}}\!:\!H \quad or \quad H\!:\!\overset{..}{\underset{..}{O}}\!: \quad = \quad H-\overset{..}{\underset{..}{O}}-H \quad or \quad H-\overset{..}{\underset{|}{O}}\!: \quad = \quad H-O$$
$$\qquad\qquad\qquad\quad H \qquad\qquad\qquad\qquad\qquad\qquad H \qquad\qquad\quad H$$

  (two equivalent Lewis diagrams)   (two equivalent *bond-dot* Lewis diagrams)   (a structural formula)

    The first two structures above are the electron dot form of the Lewis diagram showing all of the valence electrons. Because all four sides of the oxygen are equivalent, the 90° and 180° drawings of the two *Lewis* diagrams are *equivalent*: both represent the same molecule.

    Between adjacent atoms, the shared electrons are *bonding pairs*. In addition, the oxygen has two *lone pairs* of electrons.

    The third and fourth structures show an alternate way of writing a Lewis structure, with the bonding pairs (two shared electrons) written as a line to represent a bond, but the lone pairs (the *unshared* or *non-bonding* pairs) represented by dots.

    The last formula is a structural formula. A structural formula provides some information about the location of the atoms in a molecule, but often does not include the location of the lone pairs.

All of the diagrams show that in water, there are two bonds, and the oxygen atom is in the middle. The Lewis diagrams also show the lone pairs that will be needed to explain the shape of molecules.

The Lewis structure predicts that in water, *two* H and *one* O are a stable, favored combination because by sharing electrons, all of the atoms can be surrounded by the number of valence electrons needed for stability: two for H, eight for other atoms.

Many of the frequently encountered covalent molecules in first-year chemistry, as well as organic chemistry, consist of hydrogen plus the second-row *non*metals. In general, neutral atoms in the *second* row of the periodic table have the characteristics described below when they bond covalently. These patterns apply, with many additions and exceptions, for atoms below the second row. Learn this table so that given the terms in the first column you can fill in the blanks.

Lewis Structure Predictions for Neutral Single-Bonded Atoms

Second Row Symbol	Li	Be	B	C	N	O	F	Ne
Main Group Number	1	2	3	4	5	6	7	8
Valence Electrons	1	2	3	4	5	6	7	8 or 0
Bonds	(Primarily ionic bonds)		3	4	3	2	1	0
Lone Pairs			0	0	1	2	3	4

In predicting typical formulas for covalent molecules, it is helpful to remember: "Carbon bonds four times, nitrogen three times, oxygen twice, and hydrogen and halogens once."

Limitations on Lewis Structures

The rules for Lewis structures above assume that atoms are neutral and bond covalently by a relatively equal sharing of electrons. While all bonds have some covalent character that is evident in some circumstances, for bonds that have a high degree of ionic character, such as bonds to the metals in columns one and two of the periodic table, rules that apply to how ions arrange will better predict bonding behavior.

PRACTICE   B

Use a periodic table to answer these questions.

1. Using method 2, draw a Lewis diagram and then a structural formula for these.

   a. CH_4 b. PCl_3

2. For each of the molecules in problem 1, list the number of covalent *bonds* and the total number of *lone pairs* of electrons.

   Bonds: 1a. _____   1b. _____

   Lone pairs: 1a. _____   1b. _____

3. Predict how many bonds will typically be found around these neutral atoms.

   a. Sulfur   b. Iodine   c. Silicon   d. Nitrogen

Lesson 18.2   Molecular Shapes and Bond Angles

VSEPR

A key factor in the behavior of molecules is their shape: how the atoms and electrons are arranged in three-dimensional space.

The shapes and bond angles of most molecules can be predicted with reasonable accuracy based on Lewis diagrams. This technique is called **valence shell electron-pair repulsion** theory (VSEPR). The term simply means that all electron pairs, whether they are lone pairs or bonds, repel each other, and they will separate by the maximum possible angle around the nucleus of an atom.

Below, we will learn to predict how atoms and electrons are arranged based on the columns of the periodic table. A chart at the end of Lesson 18.2 will summarize these rules.

Predicting the Shape of a Covalent Molecule

To begin, draw the Lewis diagram for the molecule.

The general *shape* of a molecule is *named* based on the position of the electron pairs around the central atom. To determine the shape and bond angles, all of the electron pairs must be considered, but the shape is *named* based only on the positions of the atoms. The lone pairs are a factor in *determining* the shape, but they are ignored in *naming* the shape of a covalent molecule.

1. **One pair:** If a bonded atom is surrounded by only **one** electron pair, it has one bond to an adjacent atom. The shape around this atom is said to be **linear**. Because it takes three points to determine an angle, and two atoms are two points, an atom with only one bond has **no** bond angles.
   Example:

$$H{:}H = H—H = H_2$$

   Each H has one bond. The molecule has a *linear* shape with *no* bond angles.

2. **Two pairs:** If an atom in a molecule is surrounded by **two** electron pairs, both will be bonds. The two electron pairs will separate as much as possible by assuming a **linear** shape around the central atom. This shape results in three atoms in a line. With three points to determine an angle, the **bond angle** around the central atom is **180°**.
   Example:

$$:\!\overset{..}{Cl}\!:Be:\!\overset{..}{Cl}\!: = Cl—Be—Cl = BeCl_2$$

   The **Be** in $BeCl_2$ is surrounded by two electron pairs, and both are bonds. The arrangement of the bonds around the central atom, and the shape of the molecule, is **linear** with **180°** bond angles.

Note that $BeCl_2$ is *electron deficient*: it violates the octet rule. $BeCl_2$ does form, but as an electron-deficient molecule it has some unusual properties.

3. **Three pairs:** If a central atom is surrounded by electron pairs in **three** directions, the shape that allows the electron pairs to get as far apart as possible is termed **trigonal planar**. The three bonds are in a plane (flat) with **120°** bond angles.

Example:

The shape of the BBr_3 molecule is trigonal planar. All bond angles are 120°.

Like $BeCl_2$, a BBr_3 Lewis diagram can be drawn using single bonds, but it violates the octet rule. BBr_3 does form, but as you might predict with its electron-deficient structure, it has some unusual properties.

4. **Four pairs:** Due to the octet rule, *most* stable atoms are surrounded by *four* electron pairs. The three-dimensional **tetrahedral** shape allows those four pairs to get as far apart as possible. In a **tetrahedron**, all of the angles are **109.47°**, which we will express as *about* **109°** (**~109°**).

You will need a tetrahedral molecular model for the sections below. If you have not purchased models, build the cardboard model in the next section.

Building a Cutout Tetrahedral Model

If you do not have access to a commercial molecular model kit, a tetrahedral model can be constructed from the patterns below.

Steps

1. Obtain a sheet of foamboard or thick or corrugated cardboard at least one-half the size of this sheet of paper.

2. Either copy this page, or cover this page with thin paper and trace onto the paper, the three shapes below. Cut out the three paper patterns.

3. Using the patterns and blunt scissors, carefully cut the foamboard or cardboard to make **two** circles, **four** rectangles, and **three** ovals.

4. Cut slots in the nine pieces at the thick lines. The slots should be to the *depth* shown by the thick lines. Cut the slots to a *width* that matches the thickness of the cardboard, so that the pieces slide together in the slots tightly, but with minimal binding.

5. On the four *bonds*, round off the corners of the *atom* ends just a bit.

6. Push together the two circles using the deep slot in each. Arrange them so that they are at right angles, simulating a spherical shape.

7. Add two bonds to each circle to give four bonds total. Push the slots on the bonds into the shallow slots on each circle. Try to get the bonds to be perpendicular to the circle to which they are attached.

With four *bonds*, the model represents single-bonded central atoms in the *carbon* family. The four *electron pairs* around the central atom are in a tetrahedral shape. Since all of the electron pairs are bonds to atoms, the *atoms* around the central atom are in a tetrahedral shape. Since the position of the atoms decides the shape of the molecule, the molecular shape is termed tetrahedral, and the angle between any two bonds is ~109°.

Models for other families will be made by substituting lone pairs for bonds.

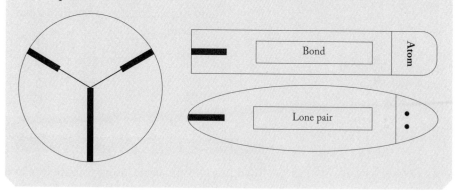

a. For a single-bonded central atom in the *carbon* family (group 4A), all four electron pairs around a central atom are bonds, and the arrangement of the *bonds* is said to be *tetrahedral*, with ~109° angles between all of the bonds.

Example:

$$\text{H}\!:\!\overset{\displaystyle\cdot\cdot}{\underset{\displaystyle\cdot\cdot}{\text{C}}}\!:\!\text{H} = \quad\overset{\text{H}}{\underset{\text{H}}{\text{H}-\!-\!\text{C}}}\!\overset{\sim\mathbf{109°}}{\diagdown}\text{H} = \text{CH}_4 = \text{tetrahedral, }\sim\mathbf{109°}\text{ angles}$$

A three-dimensional tetrahedron is difficult to represent on two-dimensional paper. In the diagram above, the dotted line represents a bond going behind the plane of the paper, and the wedge represents a bond coming out of the paper. A 3-D model will assist in working with this important shape.

━━━━━━━━━━▶ **TRY IT**

Build a CH_4 molecule from your molecular model pieces. Place the assembled model on a flat surface, then flip it so that it rests on three different points. Flip it again. Note the high symmetry of a three-dimensional tetrahedron: the shape of the molecule should be the same no matter which three atoms the model sits upon.

b. A single-bonded central atom in the *nitrogen* family (group 5A) is most often surrounded by *three bonds* and *one lone pair*.

There are four electron pairs around the central atom, and the pairs assume a tetrahedral shape to get as far apart as possible. Because this electronic geometry is tetrahedral, the angles between all of the electron pairs, bonds, and the atoms are tetrahedral (~109°).

However, the lone pairs, though they count in determining the shape around a central atom, are not considered when *naming* the shape. The shape is named based on the position of only bonds and atoms.

For this case of one lone pair and three bonds around a central atom, the four atoms are in the shape of a low pyramid. The central atom is above the plane of the three atoms to which it bonds. Since the pyramid rests on three points, the shape of the atoms is called a **trigonal pyramid**, and the *molecular* geometry is termed **trigonal pyramidal**.

Example:

$$H:\overset{\displaystyle ..}{\underset{\displaystyle H}{N}}:H \; = \; \underset{H}{\overset{}{}}\; \overset{\displaystyle ..}{N}\Big)^{\sim 109°}\; = \; NH_3$$

▶ **TRY IT**

Build this NH_3 molecule using your molecular models. Starting from CH_4, replace one bond with a lone pair. The four electron pairs are still in a tetrahedral shape. Then take off the lone pair to look at just the shape and angles of the *bonds* and *atoms*. With the central nitrogen atom on top, check that the atoms form a low *pyramid* with tetrahedral (~109°) angles.

---

c. A single-bonded neutral atom in the *oxygen* family (group 6A) is most often surrounded by *two bonds* and *two lone pairs*. These four electron pairs repel to assume a tetrahedral shape with ~109° angles around the central atom.

As always, the lone pairs count in deciding the shape, but do not count when naming the shape of the bonds around the central atom or the molecule. The two bonds and the three atoms are said to be in a **bent** shape, with ~109° angles.

Example:

$$H:\overset{\displaystyle ..}{\underset{\displaystyle H}{O}}: \; = \; \overset{\displaystyle ..}{O}\Big)^{\sim 109°}\; = \; H_2O$$

▶ **TRY IT**

Build this water molecule. Place two bonds and two lone pairs around the central atom. This puts the four electron pairs into a tetrahedral shape.

Switch the position of one bond and one lone pair. Does this create a different molecule?

STOP

No. Due to the symmetry of a tetrahedron, all four electron pairs around the central atom are in equivalent positions. The *same molecule* results no matter where the two bonds and two lone pairs are attached.

In Lewis diagrams, we treat four sides around an atom symbol as equivalent because four electron pairs repel into a tetrahedral shape, and the four sides of a tetrahedron are equivalent.

Now remove the two lone pairs. The geometric shape of the bonds and atoms, and of the H_2O molecule, is *bent*. Its one bond angle is ~109°.

d. If four electron pairs surround a central atom, but only one is a bond, the electron pair geometry is tetrahedral. However, since there are only two atoms, and the atoms determine the name of the shape, the shape of the molecule is *linear*. Since there is only one bond, there is no bond angle.

▶ TRY IT

Build an HCl molecule.

Place one bond and three lone pairs around the chlorine atom. Switch the position of the bond to hydrogen. Note that this does not create a different molecule.

Example:

$$H \!:\! \ddot{\underset{..}{Cl}} \!:\; = \; \underset{H}{\overset{..}{Cl}} \Big)^{\!\!\sim\mathbf{109}°} \; = \; H\!-\!Cl \; = \; HCl$$

The shape of the electron pairs around the chlorine is tetrahedral, but the two points of the H and Cl atoms form a line. The shape of the HCl molecule is therefore said to be **linear** with *no bond* angles (as three points are needed to determine an angle).

Lone Pair Repulsion

We can increase the accuracy of VSEPR bond-angle predictions with the following rule:

Lone pairs repel slightly more than bonds do. The lone pairs need more room.
If one or two lone pairs are present in a tetrahedral shape, the angles around the lone pairs will be slightly larger than 109°. This will push the angles between bonds *to be slightly* less than 109°.

Lone pairs tend to occupy slightly more space than bonds, because bonding pairs are more localized along the axis between the atomic nuclei. This means that a lone pair repels other pairs slightly more than bonds repel. The angle between bonds is therefore slightly smaller than other angles.

Example: The general model predicts that in a water molecule, the shape is bent with bond angles of ~**109°**. However, because water has two lone pairs, they repel each other and the bonds slightly more than the bonds repel each other. This "lone-pair scrunch" forces the bonds into an angle slightly *smaller* than 109°.

In water, the shape is bent as predicted, but the actual measured bond angle is **104.5°**, slightly less than the tetrahedral angle that the general rules predict. The angle in water is a typical value for a central atom surrounded by two lone pairs and two bonds.

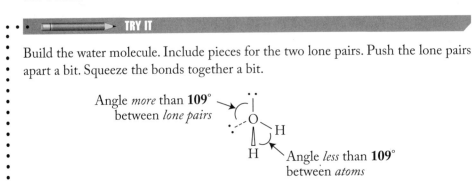

Build the water molecule. Include pieces for the two lone pairs. Push the lone pairs apart a bit. Squeeze the bonds together a bit.

Around single-bonded central atoms, when will bond angles be *less* than 109°? Only if the molecule has *one* lone pair and three bonds, or *two* lone pairs and two bonds: those molecules that have central atoms in the *nitrogen* or *oxygen* family (those in the fifth or sixth tall column of the periodic table).

Around single-bonded *carbon* family atoms, the bond angles are *~109°* rather than *less* than 109°. When there are no lone pairs around the central atom, there is no lone-pair effect on the angles.

We will call this the "lone-pair scrunch" rule:

> In neutral molecules containing only single bonds, around central atoms in the nitrogen family (tall column 5) or oxygen family (tall column 6), the bond angles are predicted to be 103°–107°, slightly less than 109°.

Summary

For the second row of the periodic table, and with frequent exceptions for rows below the second row, central atoms in neutral compounds will generally have the characteristics listed in the table below. Learn this table so that given the terms in the first column you can fill in the blanks, based on the rules for the behavior of electron pairs.

VSEPR Predictions for Neutral Single-Bonded Central Atoms

Second Row Symbol	Li	Be	B	C	N	O	F	Ne
Main Group Number	1	2	3	4	5	6	7	8
Valence Electrons	1	2	3	4	5	6	7	8
Bonds	(Primarily ionic bonds)		3	4	3	2	1	0
Lone Pairs			0	0	1	2	3	4
Shape			Trigonal planar	Tetra-hedral	Trigonal pyramidal	Bent	Linear	No bonds
Bond Angles			120°	109°	< 109°	< 109°	None	

VSEPR Limitations

As with the Lewis diagrams on which they are based, VSEPR is a model for predicting particle geometry that assumes covalent bonds: a relatively equal sharing of electrons. In the case of bonds that have a high degree of ionic character, rules that apply to how ions arrange will better predict the geometry of the particles in substances.

PRACTICE

Use a periodic table, plus models if needed. Check answers after each part.

1. For a molecule in which the central atom is surrounded by two bonds and two lone pairs,

    a. What is the shape of the electron pairs?

    b. What is the shape of the molecule?   c. What is the bond angle?

2. In your notebook, make a copy of the table below. Practice until, from memory, for the row 2 atoms in main groups 3 to 8 only, you can fill in the table with the VSEPR model predictions for the geometry of covalent compounds.

Central Atom			B	C	N	O	F	Ne
Main Group			3	4	5	6	7	8
Name of Shape of Molecule								
Bond Angles								

3. Complete this table based on VSEPR predictions for neutral, single-bonded atoms.

Molecule	NF_3	SiH_4	$AlCl_3$	SI_2
Lewis Structure				
Name of Shape of Electron Pairs				
Name of Shape of Molecule				
Bond Angles				

4. Which molecule in problem 3 would likely be the least stable and most reactive? Why?

Lesson 18.3   Electronegativity

The Electronegativity Scale

What decides if a particular bond will be ionic or covalent? Our previous "rule of thumb" has been that if a bond is between a metal and a nonmetal atom, it will likely be ionic, but if it is between two nonmetals, it will be covalent. A more precise view is that most bonds have a mixture of ionic and covalent character. This latter model for bonding will allow more accurate predictions of the properties and behavior of bonds and substances.

Atoms have differing attractions for electrons. **Electronegativity** (a model developed by the American chemist Linus Pauling) predicts how strongly each atom attracts the electrons in a bond.

The **electronegativity scale** assigns each atom an **electronegativity value** (EN) between 0.7 and 4.0. Fluorine (EN = 4.0) is the strongest electron attractor of all the atoms. Cesium and francium (EN = 0.7) are the weakest electron attractors.

The following table lists the electronegativity values of the atoms in those positions in the periodic table. (These numbers are termed the *Pauling values*. Other models may use slightly different EN values.)

To speed your work, the electronegativity values for the *second* row atoms should be memorized. This is easy, since the second row numbers start at 1.0 and increase by **0.5** for each atom to the right. The values for hydrogen (2.1) and chlorine (3.0) are also encountered frequently and should be committed to memory.

Electronegativity Values

Row 1	2.1								
Row 2	1.0	1.5		2.0	2.5	3.0	3.5	4.0	
Row 3	0.9	1.2		1.5	1.8	2.1	2.5	3.0	
Row 4	0.8	1.0	1.3–1.9	1.6	1.8	2.0	2.4	2.8	
Row 5	0.8	1.0	1.2–2.2	1.7	1.8	1.9	2.1	2.5	
Row 6	0.7	0.9	1.0–2.4	1.8	1.9	1.9	2.0	2.2	
Row 7	0.7	0.9							

In the table, note:

- Hydrogen's value of 2.1 is in the middle range of values
- Only four atoms have EN values of 3.0 and above: N, O, F, and Cl
- Values generally (but not always) increase toward the top right corner of the periodic table: to the right across a row and up a column

To predict bond behavior, the electronegativity model divides bonds into three types: ionic, polar covalent, and nonpolar covalent.

Ionic Bonds

In these lessons, our rule will be: if the *difference* between the electronegativities of two bonded atoms

- Is *greater* than 1.7, the bond will generally have *ionic* character
- Is 1.7 or *less*, the bond is likely to have *covalent* character

(Different textbooks may use different cutoff values, but 1.7 is a typical choice.)

An ionic bond can be thought of as a bond in which the difference in electron attraction is so strong that the more electronegative atom removes valence electrons from the other atom to form two ions.

Polar versus Nonpolar Covalent Bonds

Covalent bonds are divided into two types: **polar** and **nonpolar**.

For a covalent bond between two atoms that have the same or very similar electronegativity values, the electrons on average will be found at an equal distance between the two nuclei.

Examples: F—F   C≡C   N—Cl
EN: 4.0 4.0   2.5 2.5   3.0 3.0

A covalent bond in which the electrons are *equally shared* is said to be *nonpolar*.

Whether bonds are single, double, or triple bonds (as in C≡C above) does not affect the electronegativity difference.

As the difference in the electronegativity of two bonded atoms increases, the bond becomes more *polar*. The electrons are still shared, but they tend to be found closer to the more electron-attracting atom.

This uneven electron sharing creates a **dipole**: an uneven distribution of electric charge. The more electronegative atom can be described as having a *partial* negative charge, while the weaker electron attractor has a *partial* positive charge.

In chemistry, dipoles are generally represented using two types of notation. In math and science, a **δ** (a lowercase Greek *delta*) is often used as a symbol meaning *partial*. In this notation, a polar bond is labeled with two deltas: The atom that is the stronger electron attractor, with the higher electronegativity value, has its partial negative charge labeled **δ−** (pronounced *delta minus*), and the weaker electron attractor is labeled **δ+** (*delta plus*).

Examples: δ+ **C**—**O** δ−   δ+ **N**—**F** δ−   δ− **O**—**H** δ+

An alternate way to represent the dipole in a bond is to use an arrow in place of the bond. The arrow points toward the end of the bond with the stronger electron attractor: toward the side where on average the electrons are more likely to be found.

Examples: **C**→**O**   **N**→**F**   **O**←**H**

In these lessons, if the difference in electronegativity between two atoms is

- from **0 to 0.4**, we will consider the bond to be covalent and *nonpolar*
- from **0.5 to 1.7**, the bond is considered to be covalent but *polar*
- **above 1.7**, the bond will considered to be *ionic*

The choice of the breakpoints at 0.5 and 1.7 is arbitrary. Some textbooks use values such as 0.3 or 0.7 and 2.0.

A more accurate view of electronegativity is that as the difference in electronegativity in bonds rises, from zero to as high as 3.3, the character of the bond *gradually* shifts from nonpolar to polar to ionic. However, putting bonds into these three simplified categories will help us to make quick and general predictions about the behavior of substances that will be useful in *most* cases.

PRACTICE

Memorize the electronegativity values for H and Cl. Note the pattern for the values for the second row atoms. Then use a periodic table that does not include electronegativity values to solve the problems below. Check your answers after each part.

For each of the bonds in a–d,

1. Write the electronegativity value above each atom from memory.

2. Below the bond, label each atom as δ+, or δ−, or no δ.

(continued)

3. On the next line down, calculate the electronegativity difference.

4. On the next line down, label the bond as nonpolar, polar, or ionic.

5. On the next line down, rewrite the bond using an arrow in place of the bond to show the direction of the dipole.

   a. **C—H**   b. **N—Cl**   c. **C—F**   d. **O—B**

Lesson 18.4   Predicting Polarity

Polar versus Nonpolar Substances

- What determines whether a substance will be a solid, liquid, or gas at room temperature?
- Why do table salt and sugar dissolve in water, but most substances do not?
- Can we predict formulas for new pharmaceuticals that will relieve pain and cure disease?

The answers to these practical and important questions are often found by investigating the shapes and polarities of substances.

In the previous lesson, electronegativity was used to classify *bonds* as ionic, polar, and nonpolar. Substances (*molecules* and *ionic compounds*) can also be classified as having *ionic*, *polar*, and *nonpolar* character.

To predict the polarity of substances, we will use the following general rules.

1. A substance with just *one* ionic bond will usually behave as an ionic compound even if it also has many nonpolar bonds.

2. A substance with all covalent bonds will be a *polar* molecule if
   - It has polar bonds *and*
   - The dipoles do *not* cancel due to molecular symmetry

3. A substance will be a *nonpolar* molecule if
   - It has *all* nonpolar bonds, *or*
   - It has polar bonds, but the dipoles cancel due to symmetry

Predicting Substance Polarity

Knowing the chemical formula for a substance, we can often make general predictions about whether its molecules will have ionic, polar, or nonpolar behavior. The rules below comprise a simplified model, but they provide reasonably good predictions for the polarity of *most* substances.

To predict whether a substance will have ionic, nonpolar, or polar behavior, apply the steps of the following *flowchart* in order.

1. Draw the Lewis diagram for the substance.

2. Draw each *type* of *bond* in the substance and label each atom with its electronegativity (EN) value.

3. Based on the EN *differences*, label the *bonds* as ionic, nonpolar, or polar.

4. If the substance has *one* or more ionic bonds, predict that it will have *ionic* behavior.

5. If *all* of the bonds are *nonpolar*, predict the *substance* is *nonpolar*.

6. If the steps above do not apply, one or more of the bonds must be polar. To determine the substance polarity:

    a. Sketch the *shape* of the particle.

    b. On the sketch, replace the bonds with arrows representing the dipoles. Use geometry and symmetry to see whether the dipoles cancel. If needed, make a 3-D model.

    c. If the dipoles *cancel*, the molecule is *nonpolar*. If the dipoles do *not* cancel, there is a net dipole, and the molecule is *polar*.

Dipole Cancellation

You may have had practice *adding vectors* in math or physics classes. Dipoles are one of the types of quantities that add in two *or* three dimensions using vector addition. Even if you have not practiced vector addition, dipole addition can often by simplified by this rule:

> Equal but opposite dipoles *cancel.*

Let's learn the method by example.

 TRY IT

Based on the flowchart rules, apply the steps above to the following cases. After each part, check your answer below.

Q. Label these substances as *ionic, polar*, or *nonpolar.* Use a periodic table without EN values (all of these atoms have values you should know). Because these particles are *two*-dimensional, you should not need models to evaluate symmetry.

    a. Cl_2 b. LiCl   c. O=C=O (linear)   d. HCl

STOP **Answers:**

    a. In Cl_2, the shape must be **Cl—Cl**. Both have the same electronegativity value, so the difference in EN values between the two atoms is *zero*. When the EN difference is 0 to 0.4, the bond is nonpolar, and the *molecule* is **nonpolar**.

       When the electronegativity difference between two atoms is zero, there is *equal* sharing of the electrons in the bond between the two atoms. On average, the two electrons in the bond will be found halfway between each atom, so there is no bond dipole.

b. Li has a 1.0 EN and Cl has a 3.0 EN. The difference is 2.0, which is above 1.7, so the bond is likely to have *ionic* behavior. If one bond (or more) in a compound is ionic, the compound is **ionic**. The more electronegative atom will have the negative charge. In this case, the result is an Li^+ ion and a Cl^- ion.

c. CO_2 is a linear molecule with two double bonds. In calculating an EN difference, it does not matter whether the bond is single, double, or triple. The carbon EN is 2.5, oxygen's EN is 3.5, the EN difference is 1.0, so both *bonds* are *polar*.

  When bonds are polar, the symmetry test must be applied to see whether the dipoles cancel. If we add the dipole arrows to the molecular shape, the result is $O \leftarrow C \rightarrow O$. When dipoles are *equal* but in *opposite directions*, they *cancel* due to symmetry. The $O \leftarrow C \rightarrow O$ has polar bonds but is a **nonpolar** molecule because the dipoles cancel.

d. H has a 2.1 EN and Cl a 3.0 EN. The difference is 0.9, which is in the range of 0.5 to 1.7, so the bond is polar. The shape for this molecule must be **H — Cl**. Because Cl is more electronegative than H, the dipole points toward Cl: $H \rightarrow Cl$. Since this bond is polar and there are no other bonds to cancel its dipole, the *molecule* has a dipole and is **polar**.

PRACTICE   A

Use a periodic table that does not include electronegativity values (you should know these from their table position). If needed, check answers after each part.

  Based on VSEPR and electronegativity, predict whether these compounds will be *ionic*, *polar*, or *nonpolar*.

  1. BeH_2 2. LiF   3. $\underset{H}{\overset{H}{\diagdown}} C = O$ (flat shape, 120° angles)   4. BCl_3

Polarity in 3-D Molecules

In the section above, we considered two-dimensional molecules. For compounds that are three-dimensional, it helps to make a model to evaluate the dipoles and symmetry. For three-dimensional molecules with tetrahedral pairs, the following are general rules.

1. If a central atom in the *carbon* family has single bonds to four atoms that are the *same* kind of atom, even if the bonds are polar, the dipoles will cancel due to symmetry. The *molecule* will be *nonpolar*.
    Examples: CH_4 and SiF_4 are nonpolar molecules.

2. For a single-bonded central atom that obeys the octet rule in the nitrogen or oxygen family, the molecular shape is *trigonal pyramidal* or *bent*. If *any* of the three bonds are polar, *and if all* of the dipoles point to, or

all point away from, the central atom, the *molecule* is *polar* because the dipoles cannot cancel.

Examples: NH_3 and OF_2 are polar molecules.

3. In complex cases, a model should be made and the dipoles analyzed.

Some examples will help with these rules.

 TRY IT

Complete the parts below one at a time, checking your answers after each part.

Q. Using the flowchart and symmetry rules, label these compounds as *ionic, polar,* or *nonpolar.* Use a periodic table without electronegativity values. Make molecular models if needed.

a. H_2O b. CCl_4 c. NH_3 d. CHF_3

STOP

Answers:

To evaluate molecular polarity, use the flowchart. First evaluate bond polarity. *If* the bonds are *polar,* evaluate symmetry to see whether the dipoles cancel.

a. **H_2O:** H has an EN of 2.1, O has an EN of 3.5. The difference of 1.4 makes each bond polar. If one or more bonds are polar, evaluate the symmetry.

Water: :O:H = <109° = = *added* =

H_2O has tetrahedral 3-D electron pairs but 2-D bonds and atoms. In this model for predicting polarity, assume that bonds, not lone pairs, contain the dipoles. The bonds in water have a bent shape with <109° angles. Both of the dipoles point toward oxygen, so they do *not* cancel. The bonds are polar *and* the water *molecule* is **polar.** The dipole in water can be represented by an arrow (above) or using δ+ notation below.

$$\delta+ \begin{array}{c} H \\ \\ H \end{array}\!\!\!\!\Big\rangle O \ \delta-$$

The *net* dipole points from the H side toward O.

The H side of water is δ+ and the O side is δ−.

Even with its polar bonds, if water were H–O–H linear in shape, it would not have a net dipole. However, the bonds in water are *bent* rather than linear. This means that

Water is *polar.*

The polarity of water is an important factor in many reactions in chemistry and biology.

b. **CCl_4:** C is EN 2.5 and Cl is EN 3.0, so the C—Cl bond is weakly polar, and the dipoles point toward Cl.

CCl$_4$ is *tetrahedral* with four *equal* bond dipoles. Turning the model so that two bonds are up and two down, the top and the bottom two dipoles cancel side to side. The resultants are two dipoles, one pointing up and the other down. These two resultant dipoles also cancel, because they are equal but in opposite directions.

> When a central atom is surrounded by four tetrahedral bonds to the same atom, the molecule is always *nonpolar* due to symmetry.

c. **NH$_3$:** First evaluate *bond* polarity. The EN of N is 3.0, and of H is 2.1. The difference of 0.9 means that the bonds are polar, with the dipoles pointing toward N.

Polar bonds can mean polar *or* nonpolar molecules, depending on whether the dipoles cancel. To check for dipole cancellation, draw the Lewis diagram: NH$_3$ has tetrahedral electron pairs with one lone pair and three bonds. Make the tetrahedral model, then take off the lone pair, to focus on the bonds that determine the polarity. If the model is placed so that the central N is up, all of the dipoles point upward from the H's toward N. The dipoles are equal but not *opposite*: they do not cancel. NH$_3$ is **polar**.

> In trigonal pyramidal molecules, three bonds with dipoles in the same direction, pointing either to or from a central atom, result in polar molecules.

d. **CHF$_3$:** The bonds between C (2.5) and F (4.0) have a strong dipole toward F, with an EN difference of 1.5. The C—H bond is only slightly polar. The molecule may be nonpolar if the C—F dipoles cancel. To check for dipole cancellation, draw the Lewis diagram. CHF$_3$ has tetrahedral electron pairs with four bonds. Then assemble the tetrahedral model. If the model is held so that the H atom is up, all of the dipoles point down. The dipoles are *not* equal and opposite: they do not cancel. By VSEPR and electronegativity rules, CHF$_3$ is predicted to be a **polar** molecule.

Exceptions

The above rules classify substances as nonpolar, polar, and ionic. In reality, polarity is not that simple.

- The polarity of substances can be measured numerically to obtain a **dipole moment** and those measurements show a *continuum* of values from totally nonpolar to highly ionic.
- There are factors that affect polarity in addition to the electronegativity and atom geometry considered in our model above.
- Our rules classify molecules as nonpolar that in reality are slightly polar.

That said, our simplified model is a *starting point* for the prediction of properties based on polarity, such as whether or not a substance will be soluble in water. Those predictions will be accurate in most cases for a wide variety of substances.

PRACTICE   **B**

On these problems, use a periodic table *and* a table of electronegativity values. Be prepared to build tetrahedral models. If needed, check your answers after each part.

Based on VSEPR and electronegativity, predict whether these compounds will be *ionic*, *polar*, or *nonpolar*.

   1. SF_2 2. PCl_3 3. SiH_4 4. SiH_3Cl

Lesson 18.5   Predicting Solubility

How much of a substance will dissolve in a given liquid is complex: it depends on the size, geometry, electronic properties, temperature, and relative amounts of the particles of the substance and the liquid. However, some general rules can predict solubility for a large number of substances and solvents.

If a liquid composed of *non*polar molecules (such as a salad oil) is shaken with a liquid composed of *polar* molecules (such as vinegar, which is primarily water), when the shaking stops, the two liquids will slowly separate into two layers.

When two liquids composed of polar molecules, such as water and most alcohols, are mixed, the liquids dissolve in each other. The result is one solution without layers.

Why the difference?

For solubility, the general rule is: *like dissolves like.*

- *Polar* liquids tend to dissolve *polar* or *ionic* particles.
- Nonpolar liquids tend to dissolve nonpolar molecules.
- Polar and nonpolar substances tend not to dissolve in each other.

The most common *polar* solvent is water.

An example of a *non*polar liquid is carbon tetrachloride, in which the four equal and opposite dipoles cancel to give a nonpolar molecule.

If H_2O and CCl_4 are shaken together, after the shaking stops, the two liquids separate into two layers, just as with oil and vinegar dressing. In oil and vinegar, the water-based vinegar layer will sink to the bottom because water is more dense than salad oil. If water is mixed with CCl_4, the denser CCl_4 will be the bottom layer and the water will rise to the top.

When mixed liquids separate into layers, they are said to be **immiscible** (pronounced em-MISS-ible). When liquids dissolve in each other, as in the case of water and the ethanol in alcoholic beverages, they are termed **miscible** (MISS-ible).

Choosing a Solvent

A solvent can be any liquid that dissolves other materials, but different liquids dissolve different substances. By analyzing the polarity of substances and solvents, we can generally predict whether a solvent will dissolve a substance.

Because water is polar, it tends to dissolve ionic solids, plus sugars and alcohols that are polar.

Carbon tetrachloride (CCl_4) at one time was used as a "dry cleaning fluid." It cleans by dissolving from clothing the nonpolar oils that coat the surface of skin, and it does so without the use of the water that could damage some fabrics. (Modern dry cleaning liquids are also nonpolar but are less hazardous than CCl_4.)

Compounds that are classified as oils do not dissolve well in water. For this reason, water is a poor solvent for the oils produced by skin that can soil clothing. **Soaps** and **detergents** are generally long-chain molecules that have a polar group on one end of a long nonpolar chain. The polar group allows soaps and detergents to dissolve in water, while the nonpolar segment of the molecules can attract the oils on clothing fabrics. This "both polar and non-polar" structure allows soaps to both dissolve in water and dissolve oils at the same time.

To choose a solvent to dissolve a substance, we begin by analyzing whether its particles are ionic, polar, or nonpolar, then apply the solubility rule: "like dissolves like."

TRY IT

Q. Below, complete the PH_3 column first, check your answer, and then complete the remaining columns.

Molecule	PH_3	H_2	HBr
Lewis Structure			
Shape of Electron Pairs			

Molecule	PH_3	H_2	HBr
Shape of Molecule			
Bond Angles			
Bond Polarity			
Molecule Polarity			
Dissolves in Oil or Water?			

Answers:

$$PH_3: = \; H\!:\!\overset{\cdot\cdot}{P}\!:\!H \; = $$

PH_3 is predicted to have tetrahedral electron pairs, a trigonal pyramidal shape, and bond angles of $<109°$. For the bond polarity: EN of P = 2.1, EN of H = 2.1. The P—H bond is nonpolar. Because all of the bonds are nonpolar, the molecule is predicted to be nonpolar. Nonpolar molecules dissolve in nonpolar solvents, such as *oils*, gasoline, or CCl_4. They tend not to dissolve in water.

$$H_2: \qquad H\!:\!H = H\!-\!H$$

Two atoms in a molecule always have a linear shape with no bond angles.

For bond polarity: the EN difference is zero, so the bond is *nonpolar*. If all of the bonds are nonpolar, the molecule is *nonpolar*.

Nonpolar substances tend to dissolve in nonpolar solvents such as oils to a higher extent than in water. Our rules predict that hydrogen gas is not very soluble in water, and this is consistent with the laboratory behavior of H_2 gas.

$$HBr: \qquad H\!:\!\overset{\cdot\cdot}{\underset{\cdot\cdot}{Br}}\!: = H\!-\!Br$$

HBr has tetrahedral electron pairs around bromine, a linear shape, and no bond angles.

For the bond polarity: The EN of H = 2.1, the EN of Br = 2.8. The difference of 0.7 is in the range from 0.5 to 1.7, so the bond is *polar*.

Adding the dipole to the linear shape gives H → Br. The molecule is *polar*.

Polar compounds are predicted to be soluble in polar solvents such as *water* better than in nonpolar oils, and HBr, a gas at room temperature, does dissolve readily in water. For solubility, like dissolves like.

The Reliability of VSEPR and Solubility Predictions

In general, the models we have developed in this chapter are simplified, and there are many exceptions.

Example: In PH_3 above, VSEPR would predict bond angles of about 107° as in NH_3, but actual bond angles are about 94°.

Exceptions to the general rules make for interesting chemistry, and they often lead to the development of more sophisticated models to predict experimental results.

Solubility is another case that is more complex than indicated by "like dissolves like." Our simplified rules in these lessons for VSEPR, electronegativity, and "like dissolves like" should be considered as initial steps in predicting shape, polarity, and solubility. Those predictions will be reasonably accurate for many (but not all) substances.

PRACTICE

You may use a periodic table and a table of electronegativity values. Be prepared to build models, if needed.

Fill in the following chart for the substances shown. Based your predictions on the general rules for VSEPR, electronegativity, and solubility.

Molecule	SeF_2	SeI_2	BI_3
Lewis Structure			
Shape of Electron Pairs around Central Atom			
Name of Shape and Sketch of Shape			
Bond Angles			
Bond Polarity			
Molecule Polarity			
Dissolves in Oil or Water?			

SUMMARY

1. **The octet rule:** In stable particles, atoms tend to be surrounded by eight valence electrons (H and He tend to be surrounded by two). The electrons can be shared, as in covalent bonds, or gained or lost from neutral atoms, as in ionic bonds.

2. In covalent molecules, carbon tends to bond four times, nitrogen three times, and oxygen twice, and halogens and hydrogen bond one time.

3. In drawing Lewis structures, if the bonds around an atom are all single bonds, place the valence electrons on four equivalent sides around the atom.

4. In the *valence shell electron pair repulsion* (VSEPR) model for predicting shapes, electron pairs tend to get as far apart as possible around an atom. Lone pairs and double bonds repel other pairs slightly more than single bonds.

5. When there are four electron pairs around an atom, the pairs arrange in a tetrahedral shape. The shape of the molecule, however, is named based on the position of the atoms.

6. Atoms that have higher electronegativity (EN) values have more attraction for electrons. Electronegativity tends to increase for atoms toward the top right corner of the periodic table. Across row 2, the EN values increase by 0.5 per atom, from Li (EN 1.0) to F (EN 4.0).

7. A molecule will tend to be polar *if* its bonds are polar and *if* the bond dipoles do not cancel due to symmetry. If the bonds are nonpolar or if the dipoles cancel, the molecule is predicted to be nonpolar.

8. **The solubility rule:** like dissolves like. Polar solvents such as water tend to dissolve polar and ionic substances. Nonpolar molecules tend to be more soluble in nonpolar solvents.

ANSWERS

Lesson 18.1

Practice A   1a. Silicon: **4**   1b. Phosphorus: **5**   1c. Bromine: **7**   1d. Sulfur: **6**

2a. $\cdot \overset{\cdot\cdot}{\underset{\cdot}{Si}} \cdot$ 2b. $\cdot \overset{\cdot\cdot}{\underset{\cdot}{P}} \cdot$ 2c. $:\overset{\cdot\cdot}{\underset{\cdot\cdot}{Br}}:$ 2d. $:\overset{\cdot\cdot}{\underset{\cdot\cdot}{S}}:$

Which of the four sides have the paired or unpaired electrons does not matter: the sides are equivalent.

3a. $:\overset{\cdot\cdot}{\underset{\cdot\cdot}{F}}:\overset{\cdot\cdot}{\underset{\cdot\cdot}{F}}: = F-F$ 3b. $H:\overset{\cdot\cdot}{\underset{\cdot\cdot}{Cl}}: = H-Cl$

4. Bonds: 3a. **1**   3b. **1**   Lone pairs: 3a. **6**   3b. **3**

Practice B   1a. $H:\overset{H}{\underset{H}{C}}:H = H-\overset{\overset{H}{|}}{\underset{\underset{H}{|}}{C}}-H$ 1b. $:\overset{\cdot\cdot}{\underset{\cdot\cdot}{Cl}}:\overset{\cdot\cdot}{\underset{:\underset{\cdot\cdot}{Cl}:}{P}}:\overset{\cdot\cdot}{\underset{\cdot\cdot}{Cl}}: = Cl-\overset{}{\underset{\underset{Cl}{|}}{P}}-Cl$

2. Bonds: 1a. **4**   1b. **3**   Lone pairs: 1a. **0**   1b. **10**

3a. Sulfur: **2**   3b. Iodine: **1**   3c. Silicon: **4**   3d. Nitrogen: **3**

Lesson 18.2

Practice   1. The four electron pairs repel into a tetrahedral shape. The three atoms are bent, with a bond angle of slightly *less* than 109°.

2.

Central Atom			B	C	N	O	F	Ne
Main Group			3	4	5	6	7	8
Shape of Molecule			Trigonal planar	Tetrahedral	Trigonal pyramidal	Bent	Linear	No bonds
Bond Angles			120°	~109°	<109°	<109°	None	

3.

Molecule	NF_3	SiH_4	$AlCl_3$	SI_2
Lewis Structure	:F:N:F: :F:	H H:Si:H H	:Cl:Al:Cl: :Cl:	:I:S:I: (also could be drawn at 90°)
Name of Shape of Electron Pairs	Tetrahedral	Tetrahedral	Trigonal planar	Tetrahedral
Name of Shape of Molecule	Trigonal pyramidal	Tetrahedral	Trigonal planar	Bent
Bond Angles	$< 109°$	$\sim 109°$	$120°$	$< 109°$

4. **$AlCl_3$** would be predicted to be the least stable and most reactive because **Al** has an unsatisfied octet.

Lesson 18.3

Practice

	a.	b.	c.	d.
1.	**2.5 2.1**	**3.0 3.0**	**2.5 4.0**	**3.5 2.0**
	C—H	**N—Cl**	**C—F**	**O—B**
2.	$\delta-$ $\delta+$	no δ	$\delta+$ $\delta-$	$\delta-$ $\delta+$
3.	**0.4**	**0**	**1.5**	**1.5**
4.	**nonpolar or slightly polar**	**nonpolar**	**polar**	**polar**
5.	**C←H**	**N—Cl (no dipole)**	**C→F**	**O←B**

Lesson 18.4

Practice A   All of the molecules in Practice A are *two* dimensional: their shapes can be drawn on paper.

1. **BeH_2:** First assign EN values to categorize the bonds. 2.1 H−1.5 Be = **0.6** > 0.4 = **polar bonds**.
   If bonds are polar, draw the Lewis diagram and shape to see whether the dipoles cancel.
   The central atom **Be** is predicted by VSEPR to have a **linear** shape and **180°** bond angles.
   Add the dipoles to the diagram, then add the dipoles by vector addition. Because they are equal and in opposite directions, the dipoles cancel. The VSEPR prediction is that the molecule is **nonpolar**.

$$\text{EN: 2.1   1.5   2.1}$$
$$\textbf{BeH}_2 = \textbf{H:Be:H} = \text{ H—Be—H} = \text{H} \leftarrow \text{Be} \rightarrow \text{H} = \textbf{nonpolar}$$

2. **LiF:** First assign EN values to categorize the bonds. 4.0 F−1.0 Li = **3.0** > 1.7 = an **ionic bond**.
   If one bond is ionic, the substance is **ionic**.

3. **$H_2C{=}O$:** Assign EN values to categorize the bonds.
   2.1 H−2.5 C = **0.4** difference, which means a slight dipole toward C.
   3.5 O−2.5 C = **1.0** difference, which means a stronger dipole toward O.
   Electronegativity differences apply in the same way to single and double bonds.
   If one or more bonds are polar, draw the Lewis diagram, sketch the shape, add the dipoles and see whether the dipoles cancel. Because this molecule is flat, it can be analyzed on paper.

Because these dipoles are in the same direction, they do not cancel.

4. **BCl₃**: Assign EN values to categorize the bonds. 3.0 Cl−2.0 B = **1.0** = **polar bonds**.
   If bonds are polar, draw the Lewis diagram and shape to see whether the dipoles cancel.

$$BCl_3 = \text{Lewis diagram} = \text{trigonal planar shape, } 120° = \text{dipole vectors}$$

   The central atom **B** is predicted by VSEPR to have a **trigonal planar** shape for its bonds and **120°** bond angles. Adding the dipoles by vector addition, they are equal and in opposite directions, so they cancel. VSEPR predicts that the molecule is **nonpolar**.

Practice B   1. **SF₂**: Assign EN values to categorize the bonds. 4.0 F−2.5 S = **1.5** = a **polar bond**.
   If bonds are polar, draw the Lewis diagram and shape, add the vectors and see whether they cancel.
   Because fluorine, a halogen, bonds once, it must be at the ends of the molecule and not the central atom. The dot diagram predicts a molecule with sulfur in the middle, with a bent molecular shape and slightly less than 109° bond angles. *Bent* molecules are two dimensional, so their polarity can be evaluated on paper.
   Since the two dipoles are close to 109° apart rather than 180°, they are equal but not opposite. Adding the two dipoles by vector addition gives a net resultant dipole. The molecule is predicted to be polar.

   shape   dipoles   net resultant
                                    dipole

   Bent molecules with two or *trigonal pyramidal* molecules with three of the same polar bonds are predicted to be polar molecules.

2. **PCl₃**: First label the bonds using their electronegativities. 3.0 Cl−2.1 P = **0.9** = **polar bonds**.
   If any of the bonds are polar, draw the Lewis diagram and shape, then add the dipole vectors to see whether the dipoles cancel.
   Because each chlorine bonds only once, phosphorus must be the central atom. Single-bonded central atoms in the nitrogen family form neutral covalent molecules that are *trigonal pyramids* with slightly less than **109°** bond angles.

$$PCl_3 = \text{Lewis diagram} = \text{trigonal pyramid, } <109° = \text{dipole vectors}$$

   Lone pairs are not bonds and by our model do not have dipoles. The three bond dipoles are equal but *not* in opposite directions, so they do not cancel. The molecule is **polar**.

3. **SiH₄**: For a central atom in the carbon family, whenever four of the same atoms are attached, any dipoles will cancel due to tetrahedral symmetry. SiH₄ is a **nonpolar** molecule.

4. **SiH₃Cl**: Because silicon is in the carbon family, assume it bonds four times. Chlorine and hydrogen bond once, so silicon is the central atom. In the carbon family, molecules are *tetrahedral* with **109°** bond angles. Assign EN values to categorize the bonds.
   2.1 H−1.8 Si = **0.3** = relatively **nonpolar Si—H** bonds.
   3.0 Cl−1.8 Si = **1.2** = one **polar Si—Cl** bond with a relatively strong dipole toward Cl.
   If a molecule has only one strong dipole, the molecule will be **polar**, but let's look at the three-dimensional model to be sure.

$$SiH_3Cl = \text{Lewis diagram} = \text{tetrahedral, } {\sim}109° = \text{dipole vector}$$

   SiH₃Cl has one highly polar bond. The molecule is predicted to be **polar**.

Lesson 18.5

Molecule	SeF$_2$	SeI$_2$	BI$_3$
Lewis Structure	:F:Se:F: (also could be drawn at 90°)	:I:Se:I: (also could be drawn at 90°)	:I:B:I: :I:
Shape of Electron Pairs around Central Atom	Tetrahedral	Tetrahedral	Trigonal planar
Name of Shape and Sketch of Shape	Bent F—Se⁄F	Bent I—Se⁄I	Trigonal planar I—B(I)(I)
Bond Angles	< 109°	< 109°	120°
Bond Polarity	4.0 − 2.4 = 1.6 = **polar**	2.5 − 2.4 = 0.1 = **nonpolar**	2.5 − 2.0 = 0.5 = **polar**
Molecule Polarity	**Polar** (bent with polar bonds)	**Nonpolar** (bent but nonpolar bonds)	**Nonpolar** (the 3 dipoles cancel)
Dissolves in Oil or Water?	**Water**	**Oil**	**Oil**

19

Introduction to Equilibrium

Introduction to Equilibrium

Reaction Rates

In most chemical reactions, for a reaction to take place, two reactant particles must collide.

As the *temperature* at which a reaction is conducted increases, the reaction nearly always occurs at a faster rate. Why? Reactant particles at higher temperatures are traveling faster on average, and they have a higher average kinetic energy. This means that at higher temperatures, they collide at higher energy and are more likely to change in some way (react).

Reactions also proceed faster when the reactants are at higher concentrations. When the particles are more concentrated, they collide more often.

Reversible Reactions

Chemical reactions can be divided into three types.

1. **Reactions that go nearly 100% to completion.** Burning paper is one such reaction. Once it begins, the reaction continues until one of the reactants (the paper or oxygen) is essentially used up.

2. **Reactions that don't go.** Trying to convert carbon dioxide and water into paper is very difficult to do in a chemist's laboratory (though plants are able to accomplish most steps in this reaction by the remarkable process of photosynthesis).

3. **Reactions that are *reversible* and go *partially* to completion.** As a reversible reaction proceeds, the reactants are gradually used up. As a result, the forward reaction slows down. As product concentrations increase, they more frequently collide and react to reform the reactants. Finally, both the forward and reverse reactions are going at the same rate. As long as no substances or energy are added to or removed from the system, the two rates will remain equal and no further reaction seems to take place. The system is said to be at **equilibrium**.

For equilibrium to exist, both of these conditions must be met:

- All reactants and products must be present in at least small quantities.
- The reaction must be in a *closed* system: no particles or energy can be entering or leaving the reaction vessel.

At equilibrium, no reaction seems to be occurring, but this appearance is deceiving. Equilibrium is **dynamic**: the forward and reverse reactions *continue*. However, because the rate of the forward reaction is equal to the rate of the reverse reaction, there is no *net* change.

In theory, all reactions that occur are reversible, and *all* reactions go to equilibrium. In practice, many equilibria favor the products so much that the reaction is considered to go "to completion," and if the limiting reactant is known, the amounts of reactants used up and products formed can be calculated by conversion stoichiometry.

When reactions go only partially to completion, there is no limiting reactant that decides how much of the products form. Reactions that go to equilibrium require a careful accounting system to track the particles used up and formed.

1. State two ways to increase the rate of a reaction.

2. State two conditions that must be true if a chemical reaction is at equilibrium.

Lesson 19.2 | Le Châtelier's Principle

Shifts in Equilibrium

Equilibrium is important because many reactions, including those in biological systems, are in practice reversible. For reversible reactions, we want to be able to do both of the following:

- Predict what happens when a system at equilibrium is *disrupted*.
- *Shift* an equilibrium to make as much of a wanted substance as possible.

Shifts in equilibrium can be predicted by

> **Le Châtelier's Principle**
>
> If a system at equilibrium is subjected to a change, processes occur that tend to counteract that change.

Le Châtelier's principle predicts the direction that a reversible reaction will shift when a reaction mixture at equilibrium is subjected to change, including changes in concentration and temperature.

Changes in Concentration

To predict shifts in equilibrium owing to changes in *concentration*, it is helpful to restate Le Châtelier's principle in the following manner.

> **Predicting Shifts Due to Concentration Changes**
>
> For reversible reactions at equilibrium, write the balanced reaction equation using a *two-way* ($\leftrightarrows$) arrow, then apply these rules.
>
> a. *Increasing* a [substance] that appears on one side of an equilibrium equation shifts an equilibrium to the *other* side. The other substance concentrations on the *same* side as the [increased substance] are *de*creased, and the substance concentrations on the *other* side are *in*creased.
>
> b. *Decreasing* a [substance] that appears on one side of an equilibrium equation shifts the equilibrium *toward* that side. The other [substances] on the *same* side are *in*creased, and the [substances] on the *other* side are *de*creased.

TRY IT

(See "How to Use These Lessons," point 1, p. xv.)

Commit to memory the rules in the two boxes above, then apply the rules to the following problem.

Q. Chromate ions react with acids to form dichromate ions in this reversible reaction.

$$2\,CrO_4^{2-}(aq) + 2\,H^+(aq) \leftrightarrows Cr_2O_7^{2-}(aq) + H_2O(\ell)$$

(chromate ion—   (acid)   (dichromate ion—
yellow in solution)   orange in solution)

For the above reaction, if acid (H^+) is added to a yellow chromate ion solution at equilibrium,

     a. In which direction will the equilibrium shift (left or right)? _____.

     b. The [CrO_4^{2-}] will (increase or decrease?) _____.

     c. The [dichromate ion] will (increase or decrease?) _____.

     d. What color change will tend to occur? _____.

STOP

Answers:

     a. The equilibrium will shift to the **right**. Increasing the concentration of the H^+ found on the left side shifts the equilibrium toward the right side.

     b. The [CrO_4^{2-}] will **decrease**. Increasing the concentration of a substance that appears on one side decreases the concentration of the other substances on that side.

     c. The [dichromate ion] will **increase**. Increasing the [H^+] that appears on the left increases the concentration of the substances on the right.

     d. Because adding acid decreases the [chromate] and increases the [dichromate], the solution color shifts from yellow toward **orange**.

To explain these shifts, we examine what is happening in the reaction at the molecular level.

- When [H^+] increases, there will be more collisions between the H^+ and the chromate ions. Though the *percentage* of collisions that result in a reaction stays the same if the temperature remains constant, more collisions means more forward reaction. Increasing the [H^+] means that the rates of the forward and reverse reactions, equal at equilibrium, are thrown out of balance. The increased forward reaction uses up chromate and forms more dichromate.

- As more dichromate forms, its collisions with water increase, and the speed of the reverse reaction increases. A new balance is reached, but only after some of the yellow chromate has been used up and more orange dichromate has formed.

TRY IT

Q. For the same chromate–dichromate reaction, if a strong *base* is added to the orange solution that results after adding acid,

  a. The [acid] in the solution will (increase or decrease?) _____.
      (If you are unsure of your part a answer, check below before continuing.)

  b. In which direction will the equilibrium shift (left or right)? _____.

  c. The $[CrO_4^{2-}]$ will (increase or decrease?) _____.

  d. The $[Cr_2O_7^{2-}]$ will (increase or decrease?) _____.

  e. What color change will occur? _____.

STOP **Answers:**

  a. Bases neutralize acid (see Lesson 14.1). This **decreases** the $[H^+]$.

  b. Decreasing the concentration of H^+, which is in the equation on the left side, will shift the equilibrium *toward* that side.

  c. Shifting to the left means that the $[CrO_4^{2-}]$ will **increase**.

  d. Decreasing the concentration of a term on the left shifts the equilibrium to the left and decreases the concentration of the terms on the right. The $[Cr_2O_7^{2-}]$ will **decrease**.

  e. Adding base decreases the [acid]. The equilibrium shifts toward the side with the acid term, increasing the [yellow chromate] and decreasing the [orange dichromate]. The color shifts from orange toward **yellow**.

In terms of what is happening at the molecular level, these shifts are logical. When $[H^+]$ decreases, the balanced rates at equilibrium are upset. There will be fewer collisions between the acid and the chromate ions, so that the reverse reaction is now faster than the slowed forward reaction. The reverse reaction uses up orange dichromate and forms yellow chromate until a new balanced equilibrium is reached.

Driving a Reversible Reaction

Given a reversible reaction at equilibrium, if the system is opened and a reactant or product is allowed to escape (such as letting a gas product escape from a solution reaction), the reversible reaction is no longer in a closed system at equilibrium. In this case, the reaction and the mixture composition will shift toward the side that contains the escaping particle. This type of shift can be used to drive a reversible reaction toward a side with products that are wanted.

PRACTICE **A**

First learn the rules, then do the problems applying the rules from memory. Check your answers after question 1.

  1. For the Haber process reaction:

$$N_2(g) + 3\,H_2(g) \rightleftarrows 2\,NH_3(g)$$

(continued)

If the $[H_2]$ is increased,

   a. Equilibrium will shift to the (left or right?) —————.

   b. $[N_2]$ will (increase or decrease?) —————————.

   c. $[NH_3]$ will (increase or decrease?) —————————.

2. For the same reaction, if the $[N_2]$ is decreased,

   a. Equilibrium will shift to the (left or right?) —————.

   b. $[H_2]$ will (increase or decrease?) —————————.

   c. $[NH_3]$ will —————————.

Energy Changes

For the reaction of a system at equilibrium that includes an energy term, changes in energy produce changes that can be predicted by Le Châtelier's principle in a manner similar to changes in concentration.

- *Adding* energy shifts the equilibrium *away* from the side with the energy term.
- *Removing* energy shifts the equilibrium *toward* the side with the energy term.

One way to add energy to a system is to heat it. Energy can be removed by cooling a system.

Example: The melting of ice and freezing of water can be described by this reaction:

$$H_2O(s) + 6.03 \text{ kJ/mol} \leftrightarrows H_2O(\ell)$$

Adding heat to a mixture of ice and water drives the equilibrium to the right: some of the ice melts. Cooling the system (such as by refrigeration) removes energy and shifts the equilibrium to the left: liquid water changes to ice.

Le Châtelier's principle predicts the above changes only while both ice and liquid water remain in the system. To be at equilibrium, all of the reactants and products must be present.

TRY IT

Apply the rules for energy terms to the following problem.

   Q. For a system at equilibrium described by

$$N_2(g) + O_2(g) + 90.0 \text{ kJ} \leftrightarrows 2 \text{ NO}(g)$$

If the reaction vessel is heated,

   1. The equilibrium will shift to the (left or right?) —————.

   2. The $[O_2]$ will (increase or decrease?) —————————.

   3. The $[NO]$ will —————————.

Answers:

1. The equilibrium shifts to the **right**. Energy is a term on the left, so if energy is added, the system uses up energy by shifting the equilibrium to the right.

2. The $[O_2]$ **decreases**. If heat is needed to cause a reaction on one side, adding heat will cause the rate of that reaction to increase, using up other substances on that side.

3. The [NO] **increases**. Increasing energy (a term on the left) causes more of substances on the right to form.

Q. For the system in the question above, if the $[O_2]$ is decreased,

1. The equilibrium will shift to the ―――――――――.

2. The $[N_2]$ will ――――――――.

3. The temperature will ――――――――.

 Answers:

1. The equilibrium shifts to the **left**.   2. The $[N_2]$ **increases**.

3. Shifting the equilibrium to the left makes more of the terms shown on the left, including more energy. In this system, the increased heat will be manifested as an increase in average kinetic energy: the temperature **increases**.

PRACTICE   B

First learn the rules, then do these problems.

1. In the equilibrium for this synthesis of methyl alcohol,

$$CO(g) + 2\,H_2(g) \leftrightarrows CH_3OH(g) + \text{energy}$$

a. If the temperature of the system is decreased,

  i. Equilibrium will shift to the (left or right?) ――――――――.

  ii. $[CH_3OH]$ will (increase or decrease?) ――――――――.

b. If the $[H_2]$ is increased,

  i. Equilibrium will shift to the ――――――.   ii. $[CH_3OH]$ will ――――――――.

  iii. The temperature in the vessel will ――――――――.

2. In a closed system, for the reaction,

$$PCl_3(g) + Cl_2(g) \leftrightarrows PCl_5(g) + \text{energy}$$

a. If the temperature is increased,

  i. Equilibrium will shift to the ――――――.   ii. $[PCl_5]$ will ――――――.

(continued)

b. If the $[Cl_2]$ is increased,

   i. Equilibrium will shift to the ————.   ii. $[PCl_3]$ will ————.

   iii. The temperature in the vessel will (increase or decrease?) ————.

Concentration Changes: Special Cases

There are two special rules for concentration changes.

1. **For a reaction at equilibrium, adding or removing a *solid* or a *liquid* does not shift the equilibrium.**

   Solids have a constant concentration determined by their density. Adding or removing solid does not change the concentration of the solid, and therefore does not shift the equilibrium.

   Example: If solid table salt (NaCl) is added to a glass of water, initially all of the salt dissolves. However, if enough salt is added and stirred, the solution becomes **saturated**: at a given temperature it reaches the limit of dissolved salt it can hold.

   This result is in agreement with Le Châtelier's principle. The equilibrium for this reaction, $NaCl(s) \leftrightarrows NaCl(aq)$, cannot not be reached until some solid remains after stirring, since equilibrium requires that *all* reactants and products be present.

   However, once equilibrium is reached, adding more salt crystals increases the *amount* of solid salt on the bottom but does not increase the *concentration* of the solid salt or the *dissolved* salt. Adding more solid does not shift the equilibrium concentrations.

   In a solid substance, the particles are tightly packed, that is, about as concentrated as they can be. If you hit a solid substance with a hammer, its particles may separate into pieces, but they will not permanently compress. The concentration of a solid substance is constant.

   Pure liquids also have a constant concentration at a given temperature. Adding or removing a pure liquid from a system at equilibrium will not change the concentration of the liquid, and therefore will not shift the equilibrium.

2. **If the *solvent* for the reaction is a term in the equation, adding more solvent, or using up or forming solvent in the reaction, does not substantially change the solvent concentration.**

   By definition, a solvent is a substance present in very high concentration compared to the other substances present in a solution. This means that in reasonably dilute solutions, the solvent concentration remains very close to constant even if it is used up or formed by the reaction occurring in the solvent.

Catalysts: No Shift

A **catalyst** is a substance that is not used up in a reaction, but when added to a reaction mixture, increases the rate of the reaction.

Adding a catalyst to a mixture at equilibrium will not shift the equilibrium. However, adding a catalyst will cause a reaction to reach equilibrium more quickly.

PRACTICE   C

Commit the rules for Le Châtelier's principle to memory, then work these problems. Check your answers after each lettered section. Save the last problem as a review for your next study session.

1. For this reaction at equilibrium at high temperature,

$$C(s) + 2\,Cl_2(g) \leftrightarrows CCl_4(g) + energy$$

   a. If the $[Cl_2]$ is decreased,

      i. Equilibrium will shift to the (left or right?) _____.

      ii. $[CCl_4]$ will (increase or decrease?) _____.

      iii. $[C]$ will _____.

      iv. The amount of carbon will _____.

      v. The temperature in the reaction vessel will _____.

   b. If a catalyst is added, equilibrium will shift to the _____.

2. Consider this reaction at equilibrium:

$$CH_3COO^-(aq) + H_2O(\ell) \leftrightarrows CH_3COOH(aq) + OH^-(aq)$$

   a. If the $[OH^-]$ is increased,

      i. Equilibrium will shift to the _____.

      ii. $[CH_3COOH]$ will _____.   iii. $[H_2O]$ will _____.

   b. If H^+ is added,

      i. Equilibrium will shift to the _____.   ii. $[CH_3COO^-]$ will _____.

3. For the reaction at equilibrium,

$$2\,H_2O(g) + 2\,Cl_2(g) + energy \leftrightarrows 4\,HCl(g) + O_2(g)$$

   a. If the $[O_2]$ is increased,

      i. Equilibrium will shift to the _____.   ii. $[HCl]$ will _____.

(continued)

  b. If the [HCl] is decreased,

   i. Equilibrium will shift to the _____.

   ii. $[O_2]$ will _____.   iii. $[Cl_2]$ will _____.

   iv. The temperature in the reaction vessel will _____.

  c. If the temperature in the reaction vessel is decreased,

   i. Equilibrium will shift to the _____.

   ii. $[O_2]$ will _____.   iii. $[H_2O]$ will _____.

  d. If a catalyst is added to the reaction mixture,

   i. Equilibrium will shift to the _____.

   ii. $[O_2]$ will _____.

4. For the reaction at equilibrium,

$$4\,PCl_5(g) + \text{energy} \rightleftharpoons P_4(s) + 10\,Cl_2(g)$$

  If the temperature in the reaction vessel is decreased,

  a. Equilibrium will shift to the _____.   b. $[Cl_2]$ will _____.

  c. $[PCl_5]$ will _____.   d. $[P_4]$ will _____.

Lesson 19.3   Powers and Roots of Exponential Notation

> **PRETEST**   If you can solve these problems correctly, skip to Lesson 19.4. Express your final answers in scientific notation. Answers are in the answer section for this lesson at the end of the chapter.
>
> 1. Without using a calculator, write the answer to $(3.0 \times 10^3)^3 =$
>
> 2. Use a calculator. $(4.5 \times 10^{-4})^2 =$
>
> 3. Do *not* use a calculator. The cube root of $8.0 \times 10^{-24} =$
>
> 4. Use a calculator for all or part of this problem. $(2.5 \times 10^5)^{1/3} =$

Taking Numbers to a Power

Many calculators have an $\boxed{x^2}$ key. To calculate both squares and higher powers, most calculators also have power functions labeled $\boxed{x^y}$ or $\boxed{y^x}$ or $\boxed{(y)\!\wedge\!x}$ or $\boxed{\wedge}$.

To learn how to use the power keys, you may check your calculator manual, experiment using simple examples for which you know the answer, or (preferably) do both.

TRY IT

Q. Do this calculation in your head and write the answer: $2^3 =$

Answer:

$2^3 = 2 \times 2 \times 2 = 8$

Now, by entering **2** and using the keys noted above, find a *key sequence* that will give the same answer on your calculator.

Write your key sequence:

Now let's test your key sequence.

(In this lesson, assume numbers without a decimal are exact numbers for purposes of significant figures.)

 TRY IT

Q. First solve in your head, then apply your sequence to the following problems.

In your head,

a. $2^4 =$ b. $3^4 =$ c. $12^2 =$

On the calculator,

a. $2^4 =$ b. $3^4 =$

c. Do *two* ways on keys: $12^2 =$

Answers:

a. 16   b. 81   c. $12 \times 12 = 144$

Did you get the same answers in your head and on the calculator?

 TRY IT

Q. For each problem below, use your power-key sequence to do the calculation. Write the answer. Then use the calculator to multiply each number by the number of times indicated by the power (as in $2^3 = 2 \times 2 \times 2$) and compare your answers.

Using power keys,

a. $16^3 =$ b. $2.50^4 =$ c. $0.50^2 =$

Multiplying,

a. $16^3 =$ b. $2.50^4 =$ c. $0.50^2 =$

Answers:

a. 4,096   b. 39.1   c. 0.25

Note that each of the calculations above can be done in at least two ways. Complex calculator operations should always be done in two different ways as a check on your work.

Taking Exponential Notation to a Power without a Calculator

1. Recall the rule: To take *exponential* terms to a power, *multiply* the exponents. (As used here, *power* and *exponent* have the same meaning.)
   Examples:

$$(10^3)^2 = 10^6 \qquad (10^5)^{-2} = 10^{-10}$$

$$(10^{-3})^{-4} = 10^{+12} \qquad \text{(Recite: "A minus times a minus is a plus.")}$$

 TRY IT

Q. Without using a calculator, write answers to these.

   a. $(10^6)^2 =$ b. $(10^5)^{-5} =$

   c. $(10^{-12})^{-3} =$

STOP **Answers:**

   a. 10^{12} b. 10^{-25} c. 10^{36}

2. When taking exponential notation to a power, the fundamental rules apply:
   Do numbers by number rules and exponents by exponential rules.
   Example: $(2.0 \times 10^4)^3 = 8.0 \times 10^{12}$
   Treat numbers as numbers. 2 cubed is 8.
   Treat exponents as exponents. $(10^4)^3$ is 10^{12}.

 TRY IT

Q. Without using a calculator, write answers to these.

   a. $(3 \times 10^3)^2 =$

   b. $(2 \times 10^{-5})^3 =$

STOP **Answers:**

   a. 9×10^6 b. 8×10^{-15}

Q. Without using a calculator, write answers, then rewrite the answers converted to scientific notation.

   a. $(5 \times 10^4)^2 =$

   b. $(2 \times 10^{-3})^4 =$

STOP **Answers:**

   a. $25 \times 10^8 = \mathbf{2.5 \times 10^9}$ b. $16 \times 10^{-12} = \mathbf{1.6 \times 10^{-11}}$

Taking Exponential Notation to a Power with a Calculator

To take exponential notation to a power, most calculators use the same keys used to take fixed decimal numbers to a power. However, you should also know how to take exponential notation to a power without entering the powers of 10, and by estimating the answer without using a calculator.

Let's learn with an example.

TRY IT

Q1. Using a calculator as needed and any method you prefer, write an answer in scientific notation to

$(3.5 \times 10^{-4})^3 =$

Now let's check your answer.

Q2. Write your answers to parts a and b below with*out* converting to scientific notation at the end and with*out* a calculator.

a. $(3 \times 10^{-4})^3 =$ b. $(4 \times 10^{-4})^3 =$

Answers:

For both: $(10^{-4})^3 = 10^{-12}$

a. $3 \times 3 \times 3 = \mathbf{27} \times 10^{-12}$ b. $4 \times 4 \times 4 = \mathbf{64} \times 10^{-12}$

Q3. Based on your Q2 answers, write an *estimate* of what $(3.5 \times 10^{-4})^3$ should equal:

Answer:

3.5 is halfway between 3 and 4, so you might estimate that halfway between $3^3 = \mathbf{27}$ and $4^3 = \mathbf{64}\ldots$ is *about* $\mathbf{45} \times 10^{-12}$.

Q4. Without rounding, write an answer to $(3.5)^3$

a. Using a $\boxed{x^y}$ -type calculator function: _____.

b. Using the calculator with*out* the $\boxed{y^x}$ or $\boxed{\wedge}$ -type keys: _____.

For part b, multiply $3.5 \times 3.5 \times 3.5 =$ _____. Do the two answers to Q4 match?

Answer:

42.875

Q5. Based on your answer to Q4 but without using a calculator, write a precise answer to $(3.5 \times 10^{-4})^3$.

a. First without rounding or adjusting to scientific notation:_____.

b. Then apply the rules for significant figures: _____.

c. Then convert to scientific notation: _____.

Answers:

a. 42.875×10^{-12} combining $(3.5)^3$ on calculator with $(10^{-4})^3$ in your head.

b. 43×10^{-12} applying significant figures.

c. $\boxed{4.3 \times 10^{-11}}$ in scientific notation.

Q6. Now try $(3.5 \times 10^{-4})^3$ by plugging everything into the calculator. You will probably need keys labeled $\boxed{x^y}$ or $\boxed{y^x}$ or $\boxed{\wedge}$ or $\boxed{(y)^\wedge x}$.

- A standard TI-type calculator *may* use 3.5 $\boxed{\text{E } or \text{ EE}}$ 4 $\boxed{+/-}$ $\boxed{y^x}$ 3 $\boxed{=}$
- A graphing calculator *might* use 3.5 $\boxed{\text{EE}}$ $\boxed{(-)}$ 4 $\boxed{\text{enter}}$ $\boxed{\wedge}$ 3 $\boxed{\text{enter}}$
- On an RPN calculator, try 3.5 $\boxed{\text{E } or \text{ EE } or \text{ EXP}}$ 4 $\boxed{+/-}$ $\boxed{\text{enter}}$ 3 $\boxed{y^x}$

An online search with your calculator name and model number and "exponential notation" may offer a better approach. Try to work through the *logic* for key sequences. Then practice the calculation until you can repeat it without looking at hints or directions.

Write the calculator answer, rounded to proper *s.f.*: ―――――――――.

Answers:

Compare answers in Q5 and Q6. They should agree. They should also be close to the value of your estimate in Q3. Do they match your initial answer in Q1?

Which method is easier: Numbers on the calculator but exponents in your head, *or* all on the calculator? Which method is easier to remember?

You may do calculations using any method you choose, but doing numbers on the calculator and exponents by mental arithmetic can speed and simplify your work.

In addition, every calculation should be done two *different ways* as a check on your calculator use. Estimating the math "in your head" is a good way to check a calculator answer.

PRACTICE   A

1. Solve these. Do *not* use a calculator.

    a. $(10^{-3})^2 =$ b. $(10^{-5})^{-2} =$

2. Write answers without using a calculator, then convert the final answers to scientific notation.

    a. $(2.0 \times 10^4)^4 =$

    b. $(3.0 \times 10^{-1})^3 =$

3. For the problems below, first write an *estimated* answer, then rewrite the estimate in scientific notation. Then use the calculator for whatever parts you wish and write a *final* answer in scientific notation.

Try any two parts. Need more practice? Do more. Check answers as you go.

a. $(2.1 \times 10^6)^2$ Estimate =

  In scientific notation =   All or part on calculator =

b. $(3.9 \times 10^{-2})^3$ Estimate =

  In scientific notation =   All or part on calculator =

c. $(7.7 \times 10^4)^4$ Estimate =

  In scientific notation =   All or part on calculator =

d. $(5.5 \times 10^{-2})^3$ Estimate =

  In scientific notation =   All or part on calculator =

Roots

To calculate a *square* root, some calculators have a square root button: $\boxed{\sqrt{x}}$ or $\boxed{x^{1/2}}$.
Other calculators use this two-key sequence: $\boxed{\text{2nd or INV}}\ \boxed{x^2}$.

However, to take a *cube* root or higher, you will need to use a different key sequence. You will also need to know multiple ways to take roots in order to check your answers.

Roots as Exponents

"Taking the root" of a quantity is the same is assigning the quantity the **fractional exponent** (or **reciprocal exponent**) of the root.

  a. The *cube* root of x can be written as $x^{1/3}$.

  b. $16^{1/4}$ means the fourth *root* of 16.

  c. In general, $\sqrt[y]{x} \equiv x^{1/y}$.

Taking Higher Roots

On many calculators, both square and higher *roots* may be calculated using the $\boxed{\text{2nd or INV}}$ key followed by the keys used to calculate *powers*, such as $\boxed{y^x}$ or $\boxed{(y)^\wedge x}$ or $\boxed{\wedge}$.

It may also be possible to calculate roots by entering *reciprocal* exponents as powers using the reciprocal key $\boxed{1/x \text{ or } x^{-1}}$ or the division operation ($1/x = 1\ \boxed{\div}\ x$).

Knowing at least two ways to calculate a root is necessary in order to check your calculator answers. An online search with the name and model number of your calculator, plus the word *root*, may help.

Once you determine two key sequences that work, it is important to practice and test those sequences on sample calculations that are easy to check.

► TRY IT

Q. Do the following calculation in your head: The cube root of $8 = 8^{1/3} = $ _____.

Answer:
$8 = 2 \times 2 \times 2$, so $8^{1/3} = $ **2**

Now, by entering **8** and using the inverse or reciprocal or division and/or power keys, see what key sequences give the same *answer* for the root on your calculator.

One or more of these sequences *may* work. Others may work as well.

- On a standard TI calculator, try **8** [2nd *or* INV] [y^x] **3** [=]
  and/or try **8** [y^x] [( ] **1** [÷] **3** [ )] [=]
  and/or try **8** [y^x] **0.33333333** [=]
- On a graphing calculator (*if* allowed), try **8** [^] [( ] **1** [÷] **3** [ )] [enter]
  and/or try **8** [^] **3** [1/x or x^{-1}] [enter]
- On an RPN calculator, try **8** [enter] **3** [1/x] [y^x]
  and/or try **8** [enter] **0.33333333** [y^x]

Write one and if possible *two* key sequences that work and make sense to you.

To take a root: _____

_____

On the problems below, *check* your key sequence.

► TRY IT

Q. First try the problem "in your head" and write your answer. Then solve two different ways using the calculator.

a. $16^{1/4}$ in your head =   On the calculator =

b. $125^{1/3}$ in your head =   On the calculator =

c. $(0.001)^{1/3}$ in your head =   On the calculator =

d. $(0.008)^{1/3}$ in your head =   On the calculator =

Answers:

a. 2   b. 5   c. 0.1   d. 0.2

One way to *check* a root calculation is to reverse the process: Take the *answer* to the *power* of the root. This should result in the original number.

► TRY IT

Q. Apply that rule:

a. If the cube root of $125 = 125^{1/3} = $ **5**, then 5^3 should equal _____.

b. For problems c and d in the previous question, start from the answer, reverse the process using your power key sequences, and see whether the result is the original number.

PRACTICE **B**

Take these roots by entering the numbers into a calculator. Try each two ways.

1. The square root of 9,025 =

2. $(0.004096)^{1/3} =$

Roots of Divisible Powers of 10

When estimating roots for numbers in exponential notation, it is important to be able to take a root of evenly divisible powers of 10 without a calculator.

- Roots of exponential terms can be taken without a calculator if the power of 10, when multiplied by the fractional exponent, results in an *integer*.
- Another way to say this: the root of 10^x can be found without the calculator if x is *evenly divisible* by **2** to find a *square* root, and by **3** to find a *cube* root, etc.

To calculate the root of an evenly divisible exponential term, use these steps.

1. Write the *root* as a *fractional* exponent.
   Example:
   Write the *square* root of 10^x as $(10^x)^{1/2}$, and the *fourth* root of 10^x as $(10^x)^{1/4}$.

2. Apply the rule: To take an exponential term to a power, multiply the exponents.
   Examples:

$$\text{The square root of } 10^4 = (10^4)^{1/2} = \mathbf{10^2}$$

$$\text{The cube root of } 10^{-9} = (10^{-9})^{1/3} = \mathbf{10^{-3}}$$

PRACTICE **C**

1. Do not use a calculator. Write answers to these as 10 to a power.

   a. The square root of $10^{12} =$ b. $(10^6)^{1/2} =$

   c. The cube root of $10^{-6} =$ d. $(10^{-12})^{1/4} =$

2. Which root is equivalent to $x^{0.125}$?

3. Without a calculator, $81^{0.25} =$

Roots of Exponential Notation

To find a root of numbers written in exponential notation, apply the fundamental rule: treat numbers as numbers, and exponents as exponents.

TRY IT

Q. With*out* a calculator, solve $(8.0 \times 10^{15})^{1/3} =$

Answer:

Treat numbers as numbers. The cube root of 8 is 2.

Treat exponentials as exponentials. $(10^{15})^{1/3} = 10^5$

$$(8.0 \times 10^{15})^{1/3} = \mathbf{2.0 \times 10^5}$$

For roots that cannot be solved by inspection, use these steps.

1. If the exponential term is *not* evenly divisible by the root, make the exponent *smaller* until the exponent times the fractional power results in a positive or negative whole number.

TRY IT

Q. Apply only step 1 to $(8.04 \times 10^{-5})^{1/3} =$

Answer:

$$(8.04 \times 10^{-5})^{1/3} = (\mathbf{80.4 \times 10^{-6}})^{1/3}$$

To make the exponent divisible by 3, it is *lowered* from 10^{-5} to 10^{-6}. When you make the exponent smaller, make the significand larger.

2. Without a calculator, write a rough *estimate* of the root of the significand in front of the evenly divisible exponent. Find the exact root of the exponential term. Combine these two parts and write the estimate for the root.

TRY IT

Q. Apply step 2 to the step 1 answer above:

$(80.4 \times 10^{-6})^{1/3} \approx$

Answer:

$$(\mathbf{80.4 \times 10^{-6}})^{1/3} = (\mathbf{80.4})^{1/3} \times (10^{-6})^{1/3}$$

To estimate a cube root of 80, since $4 \times 4 \times 4 = 64$, and 80 is a *little* higher than 64, you might guess ≈ 4.2. (The symbol $\approx$ means *approximately equals*.)

Handle exponents as exponents. $(10^{-6})^{1/3} = \mathbf{10^{-2}}$

Combine the two parts. $\boxed{\text{Estimate} \approx 4.2 \times 10^{-2}}$

3. To get a precise answer for the root of a number in exponential notation,

   a. Write the value that has the *evenly divisible* exponent

   b. Find the *precise* root of the significand on the calculator

   c. Take the root of the exponential term with*out* the calculator

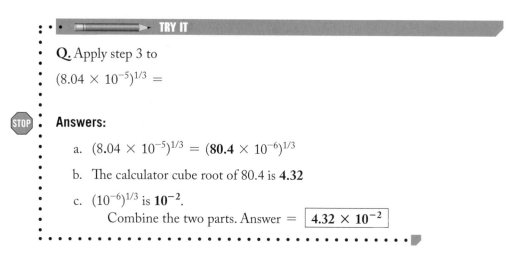

TRY IT

Q. Apply step 3 to

$(8.04 \times 10^{-5})^{1/3} =$

Answers:

a. $(8.04 \times 10^{-5})^{1/3} = (\mathbf{80.4} \times 10^{-6})^{1/3}$

b. The calculator cube root of 80.4 is **4.32**

c. $(10^{-6})^{1/3}$ is $\mathbf{10^{-2}}$.
   Combine the two parts. Answer $= \boxed{\mathbf{4.32 \times 10^{-2}}}$

4. Compare the step 3 calculator answer to the step 2 estimate. They should be close.

5. Now take the root by entering the *original* number in the problem into the calculator. One or more of these key sequences (and others) *may* work.

   • On a standard TI-type calculator, try

   8.04 ⌊EE⌋ **5** ⌊+/−⌋ ⌊2nd or INV⌋ ⌊yˣ⌋ **3** ⌊=⌋

   and/or try **8.04** ⌊E *or* EE *or* EXP⌋ **5** ⌊+/−⌋ ⌊yˣ⌋ ⌊(⌋ **1** ⌊÷⌋ **3** ⌊)⌋ ⌊=⌋

   • On a graphing calculator, try

   8.04 ⌊EE⌋ ⌊(−)⌋ **5** ⌊enter⌋ ⌊^⌋ ⌊(⌋ **1** ⌊÷⌋ **3** ⌊)⌋ ⌊enter⌋

   • On an RPN calculator, try

   8.04 ⌊E *or* EE *or* EXP⌋ **5** ⌊+/−⌋ ⌊enter⌋ **3** ⌊1/x⌋ ⌊yˣ⌋

   Your calculator answer should match the step 3 answer above.
   Circle or write one or two key sequences that work:

   _____

   Whatever sequence you use, work through the logic of *why* it works. Without the *why*, it will be difficult to remember the correct sequence.
   Once you have debugged and *practiced* a key sequence to calculate roots, doing the entire calculation on the calculator may be faster than converting to an evenly divisible root. However, using that conversion to estimate the root is a good way to check the calculator result.

Roots of Non-Divisible Powers of 10

Changing an exponent to make it divisible by the root can put a *number* in front of the exponential term that was not there before.

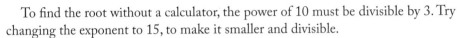

TRY IT

Q. First, estimate this answer with*out* a calculator.
The cube root of $10^{16} =$

To find the root without a calculator, the power of 10 must be divisible by 3. Try changing the exponent to 15, to make it smaller and divisible.
If the exponent is changed to 15, how should the value be written?

10×10^{15}. Estimate the cube root of this value.

Answer:

$$(10^{16})^{1/3} = (1.0 \times 10^{16})^{1/3} = (10 \times 10^{15})^{1/3} = (10^{1/3} \times 10^5) \approx 2.2 \times 10^5$$

Q. Now try taking the cube root of 10^{16} on the calculator.
You *may* need to enter $\underline{1} \times 10^{16}$ to take the root.

Answer:

2.15×10^5 Compare this to your estimate.

SUMMARY

Roots and Powers

1. If you are not certain that you are using calculator keys correctly, do a *simple* similar calculation, first on *paper* and then on the calculator.

2. For complex operations on a calculator, do each calculation a *second* time. Use estimates on paper with rounded numbers *or* use different steps or keys.

3. Using exponential notation, it is often easier to do the numbers on the calculator, but the exponents on paper.

4. In calculations using exponential notation, handle numbers and exponential terms separately. Do numbers by number rules and exponents by exponential rules.
   • When you multiply exponentials, you add the exponents.
   • When you divide exponentials, you subtract the exponents.
   • To take an exponent to a power, multiply the exponents.

5. To take roots of exponential notation without a calculator, follow these steps:
   • Convert roots to fractional exponents.
   • Adjust the exponential to be smaller and the significand to be larger to make the exponent evenly divisible by the root.

PRACTICE D

Fill in the blanks. Assume whole numbers are exact for significant figures.

1a. $4 \times 4 \times 4 =$ _____, so (_____)$^{1/3} = 4$.

1b. Using your calculator, find the cube root of 64: _____.

1c. $64 = 6.4 \times 10^1$, so $(6.4 \times 10^1)^{1/3}$ should equal _____.
   Find the root using the calculator: $(6.4 \times 10^1)^{1/3} =$ _____

1d. $0.4 \times 0.4 \times 0.4 = 0.064 = 6.4 \times 10^{-2}$, so $(6.4 \times 10^{-2})^{1/3}$ in your head = _____.
   See whether you get the same answer to $(6.4 \times 10^{-2})^{1/3}$ on the calculator.

2. Complete the problems below using the following steps. Do as many as you need to feel confident. Check your answer after each part.

A. First convert to an exponent with a divisible root.

B. Write an *estimated* answer for the root.

C. Starting from the divisible root, use the calculator for the root of the number, take the root of the exponential in your head, write the answer, then convert the answer to scientific notation. Round the significand to two digits.

D. Take the root of the original exponential notation on the calculator.

E. Compare your answers in steps B, C, and D.

2a. $(6.0 \times 10^{23})^{1/3}$ Divisible =   Root estimate =

   Estimate in scientific notation =   Calculator =

2b. $(10^{15})^{1/4}$ Divisible =   Root estimate =

   Estimate in scientific notation =   Calculator =

2c. The cube root of 1.25×10^{-7} Divisible =   Estimate =

   Estimate in scientific notation =   Calculator =

2d. $(1.6 \times 10^{-11})^{1/4}$ Divisible =   Root estimate =

   Estimate in scientific notation =   Calculator =

PRACTICE E

Use a calculator as needed.

1. $2^4 =$

2. $(0.25)^4 =$

3. $(4.5 \times 10^3)^5 =$

4. $(2.0 \times 10^5)^6 =$

5. $(3.3 \times 10^{-3})^8 =$

6. $(4.7 \times 10^{-4})^4 =$

7. $(81)^{1/4} =$

8. $\sqrt[5]{0.01024} =$

9. $(6.20 \times 10^4)^{1/8} =$

10. The sixth root of $9.5 \times 10^{15} =$

11. $(3.3 \times 10^{-3})^{1/9} =$

12. $(6.5 \times 10^{-3})^7 =$

Lesson 19.4   Equilibrium Constants

The Law of Mass Action

At a fixed temperature, for the general reversible reaction

$$aA + bB \leftrightharpoons cC + dD$$

the **Law of Mass Action** states that for a reaction at equilibrium, the ratio

$$\frac{[C]^c[D]^d}{[A]^a[B]^b} = \frac{\text{product of the [Products]}}{\text{product of the [Reactants]}}$$

is constant.

This ratio is called the **equilibrium constant**, which is given the symbol K. For the terms in an equilibrium constant,

- The *concentrations* of the particles in the *products* are *multiplied* on *top*
- The concentrations of particles on the left side (reactants) are multiplied on the bottom
- The *coefficient* for each substance becomes the *power* of its concentration

Example: For the reaction

$$S_2(g) + 2\,H_2(g) \leftrightharpoons 2\,H_2S(g), \qquad K = \frac{[H_2S]^2}{[S_2][H_2]^2}$$

An **equilibrium constant equation** has two parts:

- The equilibrium constant **expression** is the ratio that shows the *symbols* for the substance concentrations and their powers.
- The K **value** is a positive number that is the value of the ratio.

A K equation is the combination of the expression and value.
Example: For the reaction

$$S_2(g) + 2\,H_2(g) \leftrightharpoons 2\,H_2S(g)$$

the equilibrium constant *expression is*

$$K = \frac{[H_2S]^2}{[S_2][H_2]^2}$$

and the equilibrium constant *value* is 1.1×10^7 at 973 K.

The equilibrium constant *equation* at 973 K is

$$K = \frac{[H_2S]^2}{[S_2][H_2]^2} = 1.1 \times 10^7$$

As temperature changes, K values change, but the K expression stays the same. Note these conventions:

- We abbreviate kelvins as K and the equilibrium constant as an italicized K.
- If a concentration is written in a K expression, it must be a concentration measured at equilibrium: $[X]_{\text{at equilibrium}}$ or $[X]_{\text{at eq}}$. As a simplification, the label *at equilibrium* is often left off, but you should assume that all terms in a K expression are values *at equilibrium*.

K expressions can be written using units other than molar concentration. For this reason, a K equation based on molar *c*oncentrations may be given the symbol K_c. However, if no subscript after a K is written, you should assume the K value and expression is a K_c based on molarity. In this chapter we will restrict our attention to K_c equations based on concentrations.

Equilibrium Constant Expressions

To write a K expression for a reaction, all that is needed is a balanced equation.

 **TRY IT**

Q. For the Haber process reaction:

$$N_2(g) + 3\,H_2(g) \leftrightarrows 2\,NH_3(g)$$

write the equilibrium constant expression.

 Answer:

$$K = \frac{[NH_3]^2}{[N_2][H_2]^3}$$

with all concentrations measured at equilibrium.

The K equation means that at a given temperature, if a reaction is run until equilibrium is reached, the concentrations at equilibrium may vary, but the *ratio* calculated by this equilibrium constant expression will have the same numeric value.

PRACTICE **A**

Write the equilibrium constant expression for these reactions.

1. $CH_4(g) + 2\,O_2(g) \leftrightarrows CO_2(g) + 2\,H_2O(g)$
2. $2\,C_4H_{10}(g) + 13\,O_2(g) \leftrightarrows 8\,CO_2(g) + 10\,H_2O(g)$

Concentrations That Are Constant

The correlation between a balanced equation and its equilibrium constant expression is simple, but there is one important exception.

> Only concentrations that can *change* are included in K expressions.
> - *Solids, pure liquids,* and *solvents* have concentrations that do not change substantially during reactions.
> - In place of terms for the concentration of *solids, pure liquids,* and *solvents* (including *liquid* water), a **1** is substituted in the K expression.

By convention, if the concentration of a particle involved in a reaction is essentially constant, that constant value is included in the *value* of the equilibrium constant. Mathematically, this is equivalent to substituting a 1 for the term.

The concentrations that are assigned a value of **1** in a K expression are the same as those that do not shift an equilibrium when applying Le Châtelier's principle.

 TRY IT

Using the rules above, try the following example.

Q. Write the equilibrium constant expression for this reaction.

$$CaCl_2(s) \leftrightarrows Ca^{2+}(aq) + 2\,Cl^-(aq)$$

Answer:

$$K = [Ca^{2+}][Cl^-]^2$$

The reaction equation shows that the ions on the right are aqueous, meaning that the solvent for the reaction is water. When dissolved in a solvent, the concentration of ions can vary. Terms for concentrations that can vary are included in K expressions.

Because the reactant on the left is a solid, its concentration is assigned a value of 1.

Most K expressions will be fractions with a numerator and a denominator. For some reactions, however, including the one above, the numerator or denominator of K will be a **1**. The **1** must be written if it is *alone* in the numerator (on top), but it can otherwise be omitted when writing a K expression.

 TRY IT

Apply that rule to the following question.

Q. Write the equilibrium constant expression for this reaction.

$$BaO(s) + CO_2(g) \leftrightarrows BaCO_3(s)$$

Answer:

$$K = \frac{1}{[CO_2]}$$

Both sides of the equation have solids, and in the place of the terms for the concentration of a solid, we write a **1**. Because the concentrations of gases can vary, the term for the gas concentration must be included in the K expression. A **1** must be written if it is by itself in the numerator. In other cases, a 1 is omitted as understood when writing K expressions.

Substance States

The *phase* or *state* (solid, liquid, gas, or aqueous) of a substance must be known in the K expression. However if the balanced equation is written and includes states, the states can be omitted as understood in the accompanying K expression.

Water: The "constant concentration terms = 1" rule means that a [H_2O] term will be represented by a 1 in a *K* expression when the water is a solid (ice), a pure liquid, or a solvent. However, if the water is a reactant or product in its *gas* phase (as vapor or steam), since all gases are compressible, its concentration can *vary*, and the term for [$H_2O(g)$] must be included in the *K* expression.

In addition, when liquids are mixed with other liquids, they may dissolve in each other, as occurs with alcohols and water. If the concentrations are relatively close in value, the liquid concentrations *can* vary, and neither liquid is considered to be the solvent. Concentrations that can *vary* are *included* in *K* expressions.

The general rule is

> For *H_2O*, [$H_2O(g)$] is included in *K* expressions; but [$H_2O(s)$] and [$H_2O(\ell)$] if pure or the solvent] are replaced by a **1**.

Flashcards

In your notebook, make a list of the key rules covered so far in this chapter. As needed, design flashcards to help in moving the rules to long-term memory.

Here's a possible "one-way" card:

In a *K* expression, [products] are	Multiplied on top

After a few days of practice, you will be able to recall the rules quickly and automatically when needed to solve problems.

PRACTICE   B

Write the equilibrium constant expression for these reactions.

1. $2\ C_2H_6(g) + 7\ O_2(g) \leftrightarrows 4\ CO_2(g) + 6\ H_2O(\ell)$

2. $4\ Fe(s) + 3\ O_2(g) \leftrightarrows 2\ Fe_2O_3(s)$

3. $CH_3COOH(aq) + OH^-(aq) \leftrightarrows CH_3COO^-(aq) + H_2O(\ell)$

4. $Ca_3(PO_4)_2(s) \leftrightarrows 3\ Ca^{2+}(aq) + 2\ PO_4^{3-}(aq)$

Lesson 19.5   Equilibrium Constant Values

Why *K* Values Do Not Have Units

K values are *numbers without units*. This is in part because the *K* expression is a "shortcut" equation: a convenient simplification derived from more complex relationships in thermodynamics that are based on a unitless quantity called activity.

Concentration is related to activity but is easier to measure directly, so we use concentration as a practical way to solve calculations involving *K*. For "real" relationships in science, units must cancel properly, but in "shortcut" equations, units may not cancel properly.

For now, it is important to remember the following.

> When solving calculations based on a K equation,
> - K values are not assigned units.
> - When solving for a concentration based on K values, the unit M (mol/L) must be added to the answer.

K Values and the Favored Side

Equilibrium constant values are always positive numbers, and the values are most often written in scientific notation. At equilibrium,

- If the substances on the right side of an equation have higher concentrations than the substances on the left, the products are said to be **favored**, and the value of K will be a number *greater than one.*
- If the substances on the left have higher concentrations than those on the right, the reactants are favored, and the value of K will be a number between zero and one (in scientific notation, a positive number times 10 to a negative power).
- The more a reaction favors the products (goes to the right), the higher will be the value of its equilibrium constant.

  Examples: For these reactions at 25°C,

  A. $2\,N_2(g) + O_2(g) \leftrightarrows 2\,NO(g)$ $K = 1.0 \times 10^{-30}$

  B. $Ag^+(aq) + 2\,NH_3(aq) \leftrightarrows Ag(NH_3)_2{}^+(aq)$ $K = 1.7 \times 10^7$

  C. $HSO_4{}^-(aq) \leftrightarrows H^+(aq) + SO_4{}^{2-}(aq)$ $K = 0.013$

  Reaction A has a K value that is positive but is much less than one. At equilibrium, the substances on the left side (the reactants) will be favored.

  Reaction B has a K value that is much greater than one. At equilibrium, the substances on the right side (the products) will be favored.

  For K values much larger than one, the reactions go close to completion. For K values much smaller than one, the reaction goes only slightly.

  Reaction C has a K value that is smaller than one, favoring the left side, but compared to most K values, K is not far from one. At equilibrium in reaction C, you would expect to find a more balanced mixture of reactants and products than in reaction A or B.

K Values for Reversed Reactions

Equilibrium is the result of a reversible reaction. Reversible reactions can be written in either direction. For this reason, in every K problem, the reaction equation must be written in order to define the forward direction.

Example: The conversion of nitrogen dioxide to dinitrogen tetroxide is reversible. If we designate this as the forward reaction:

$$2\,NO_2(g) \leftrightarrows N_2O_4(g) \quad \text{then} \quad K_{\text{forward}} = \frac{[N_2O_4]}{[NO_2]^2}$$

By that designation, this means that the reverse reaction is

$$N_2O_4(g) \leftrightarrows 2\,NO_2(g) \quad \text{and} \quad K_{reverse} = \frac{[NO_2]^2}{[N_2O_4]}$$

In the above example, the two equilibrium constant expressions are different, but related:

$$K_{forward} = \frac{[N_2O_4]}{[NO_2]^2} = \frac{1}{\frac{[NO_2]^2}{[N_2O_4]}} = \frac{1}{K_{reverse}} \quad \text{or} \quad K_f = 1/K_r \quad \text{and} \quad \boxed{K_r = 1/K_f}$$

The relationship $K_r = 1/K_f$ will be true for all reactions. Stated in words,

> If a reaction equation is reversed, the new K value is the reciprocal of the original K value.

K Values When Coefficients Are Multiplied

When writing a K value, *either* the balanced equation *or* the K expression must also be written to indicate the coefficients used to balance the equation. Coefficients are ratios, and only one set of ratios will balance an equation, but different coefficients can be used to balance the equation as long as the coefficients are in the same ratio.

The value of K will depend on the coefficients used to balance the equation, but if the K value is known for any one set of coefficients, the value for K for different coefficients can easily be determined.

Example: Consider

$$2\,NO_2(g) \leftrightarrows N_2O_4(g) \qquad K_1 = \frac{[N_2O_4]}{[NO_2]^2}$$

and

$$4\,NO_2(g) \leftrightarrows 2\,N_2O_4(g) \qquad K_2 = \frac{[N_2O_4]^2}{[NO_2]^4} = (K_1)^2$$

The coefficients of the second equation are double the first. The ratios are the same, and both equations are balanced. The K values will be different, but related. If the coefficients are *doubled*, the new K value for the K expression based on those doubled coefficients is the original K value *squared*.

This relationship will be true for all K values:

> If a value of K is known for a reaction with one set of coefficients, those coefficients can be multiplied by any positive number, and the new value of K will be the original K to the power of the multiplier.

In equation form, this rule can be written,

If K = a numeric value (#) for the reversible reaction: $aA + bB \leftrightarrows cC + dD$

then $K = (\#)^n$ for the equation: $naA + nbB \leftrightarrows ncC + ndD$

The factor **n** can be any positive number: an integer, decimal, or fraction.

This rule illustrates why, when you write a K value, the reaction coefficients on which it is based must also be shown by writing either the K expression or the balanced equation.

PRACTICE

For the following reactions at 25°C,

$$Cu(s) + 2\,Ag^+(aq) \leftrightarrows Cu^{2+}(aq) + 2\,Ag(s) \qquad K = 1.0 \times 10^{15} \qquad \text{(Reaction A)}$$

$$PbCl_2(s) \leftrightarrows Pb^{2+}(aq) + 2\,Cl^-(aq) \qquad K = 2.0 \times 10^{-5} \qquad \text{(Reaction B)}$$

$$AgI(s) \leftrightarrows Ag^+(aq) + I^-(aq) \qquad K = 1.5 \times 10^{-16} \qquad \text{(Reaction C)}$$

1. Write the K expression for each reaction.

2. Which equilibrium most favors the substances on the right side of the equation?

3. Which reaction will form the least amount of product?

4. Write the K expression and K value for the equation

$$Cu^{2+}(aq) + 2\,Ag(s) \leftrightarrows Cu(s) + 2\,Ag^+(aq)$$

5. Write the K expression and K value for the equation

$$1/2\,PbCl_2(s) \leftrightarrows 1/2\,Pb^{2+}(aq) + Cl^-(aq)$$

6. Write the K expression and K value for the equation

$$1/2\,Cu^{2+}(aq) + Ag(s) \leftrightarrows 1/2\,Cu(s) + Ag^+(aq)$$

Lesson 19.6   Equilibrium Constant Calculations

The *WRECK* Steps

To solve equilibrium calculations in a systematic fashion, we add a step or two to our standard equation-solving methods. Our rule will be

For calculations involving K and concentrations, write the *WRECK* steps.

The *WRECK* steps are

1. *W* (WANTED): Write the WANTED unit and/or symbol.

2. *R* (Reaction): Write a balanced equation for the Reaction.

3. *E* (Extent): After the reaction, add the Extent of the reaction, such as "(Goes partially.)" or "(Goes ~100%.)."

4. *C*onc@Eq: List or calculate the *C*oncentrations at equilibrium for each of the particles in the reaction equation.

5. *K*: If the reaction goes to equilibrium, write the *K* equation.

6. Substitute numbers into the *K* equation, then SOLVE the *K* equation for the WANTED symbol.

7. If a *concentration* is calculated using an equilibrium equation, the unit moles/liter (M) must be added to the answer.

TRY IT

Apply the above seven-step method to the following problem. If you get stuck, read a bit of the answer until you are unstuck, and then complete your work.

Q. For the reversible reaction

$$H_2(g) + I_2(g) \rightleftharpoons 2\,HI(g)$$

if the concentrations at equilibrium are $[H_2] = 0.020$ M and $[I_2] = 0.32$ M and $K = 25$, find $[HI]_{\text{at eq}}$.

 Answer:

For calculations involving *K* and concentrations, write the *WRECK* steps.

1. *W* (WANTED): ? = $[HI]_{eq}$ in mol/L

2. *R* (Balanced Reaction Equation):

$$H_2(g) + I_2(g) \rightleftharpoons 2\,HI(g) \qquad \text{(Goes partially.)}$$

3. *E* (Extent): If a reaction goes to equilibrium, assume unless otherwise noted that it *goes partially* to completion.

4. *C*: $[HI]_{eq}$ = ? (WANTED)
   $[H_2]_{eq} = 0.020$ M
   $[I_2]_{eq} = 0.32$ M

5. *K*: $K = \dfrac{[HI]^2}{[H_2][I_2]} = 25$ at equilibrium.

6. Substitute numbers into the *K* equation, then SOLVE the *K* equation for the WANTED symbol.

    Until this point, our rules have been:

   • Always include units when you solve.
   • Solve in symbols before substituting numbers, because when doing algebra, symbols move more quickly than numbers with their units.

    For the special case of *K* calculations,

   • Units are omitted from values substituted into *K* equations.
   • You *may* plug numbers into the original equation and move the numbers as you solve the algebra, because numbers without units will likely move as quickly and accurately as symbols.

    However, because unit cancellation will not catch mistakes, in *K* calculations you must carefully check your substitution and algebra.

Substitute the data numbers directly into the K equation above, then solve.

$$\frac{[HI]^2}{(0.020)\,(0.32)} = 25 \text{ at equilibrium}$$

$[HI]^2 = (25)\,(0.020)\,(0.32) = 0.160$ (Carrying an extra *s.f.* until the final step.)

$[HI]_{eq} = $ square root of $[HI]^2 = ([HI]^2)^{1/2} = (0.160)^{1/2} = \boxed{\textbf{0.40 mol/L}}$

7. When solving K for a concentration, **M** (or **mol/L**) must be added to the answer.

Significant figures in K calculations: All numbers in the original data had two *s.f.*, so the answer is rounded to two *s.f.* Coefficients are exact, and in terms such as $[HI]^2$ above, since the 2 is based on coefficients, the 2 is exact. All numbers based on coefficients have infinite *s.f.* and do not restrict the *s.f.* in an answer.

Flashcards

Update the list in your notebook of the key rules covered so far in this chapter. Make new flashcards as needed, and practice until you can answer them from memory. Then try the problems below.

PRACTICE

1. Given the reaction

$$4\,NH_3 + 5\,O_2 \rightleftharpoons 4\,NO + 6\,H_2O \text{ (all gases)}$$

   at a certain temperature, the equilibrium concentrations are

   $[NH_3] = 0.050 \text{ M}, [O_2] = 0.0020 \text{ M}, [NO] = 0.50 \text{ M}, [H_2O] = 0.20 \text{ M}$

   What is the value of K at this temperature?

2. Given the reaction

$$2A + B \rightleftharpoons 4\,C \text{ (all gases)}$$

   concentrations at equilibrium are

   $[A] = 0.050 \text{ M and } [B] = 0.125 \text{ M}$

   If $K = 0.020$, find $[C]$.

3. For this system in a 20. L sealed container,

$$CO_2(g) + H_2(g) \rightleftharpoons CO(g) + H_2O(\ell)$$

   at equilibrium is found 0.40 mol CO_2, 0.60 mol H_2O, and 0.90 mol H_2. If the value of the equilibrium constant is 4.8, how many moles of CO are in the mixture?

Lesson 19.7   *K* and RICE Tables

So far in our *K* calculations, the concentrations at *equilibrium* have been known. However, if concentrations are known for a mixture of *reactants* initially (with no products yet formed), *as well as* for any *one* reactant or product after the reaction has reached equilibrium, concentrations at equilibrium for all substances, and a value for *K*, can be calculated.

To solve these equilibrium calculations, we need a "chemistry accounting system." We will call our system a ***RICE moles*** table: RICE for the labels of the rows (**R**eaction, **I**nitial, **C**hange, **E**quilibrium) and **moles** for the *numbers* that go into the table.

Our rule will be

> In calculations for reactions that go *to equilibrium,* to find the moles at equilibrium, use a RICE table.

Let's illustrate this method with a problem.

> **TRY IT**

Q. The morning chemistry lab assistant is filling lab drawers. The initial inventory contains 95 burners, 220 racks, and 2,500 test tubes. Into each top drawer is placed one Bunsen burner, two test tube racks, and 20 test tubes. When the afternoon assistant arrives, 60 racks remain in the inventory. How many drawers were filled? How many burners and test tubes remain in the inventory at the end of the process?

First balance this "equation" for the process:

_____ burners + _____ racks + _____ test tubes → **1** drawer

Then, to solve the problem, add numbers to the following table.

Reaction/Process	_____ Burner	_____ Racks	_____ Test tubes	_____ Drawer
Initial Count				
Change (use + and −)				
At End/Equilibrium				

Answer:

The initial data:

Reaction/Process	1 Burner	2 Racks	20 Test tubes	1 Drawer
Initial Count	95	220	2,500	0
Change (use + and −)				
At End/Equilibrium		60		

Adjust your work if needed, and then fill in all of the boxes in the table.

Calculate the one change for which there is sufficient data. From that number, use the ratios of the process to complete the *Change* row. Include a − *sign* for components used up and a + sign for those formed. Then calculate the amount of each component present at the end.

Reaction/Process	1 Burner	2 Racks	20 Test tubes	1 Drawer
Initial Count	95	220	2,500	**0**
Change (use + and −)	−80	−160	−1,600	+80
At End/Equilibrium	**15**	60	**900**	80

The bottom row numbers answer the question. Note a key to the table: the *ratios* in row 1 (Reaction) must match the ratios in row 3 (Change). The coefficients determine the *Change* ratios.

This same method can be used to find values for moles for chemical substances at equilibrium.

 TRY IT

Construct a *RICE moles* table to solve this problem.

Q. A reaction that occurs at high temperatures (and can cause air pollution from car engines) is

$$N_2 + O_2 \leftrightarrows 2\,NO \text{ (all gases)}$$

If 1.00 mol N_2, 2.00 mol O_2, and no NO moles are initially mixed, and at equilibrium, 1.80 mol O_2 remains, how many moles of N_2 and NO are present at equilibrium?

Answer:

WANTED: Moles of N_2 and NO at equilibrium.

Strategy: To solve for values at equilibrium, write a *RICE moles* table.

Reaction	1 N_2	1 O_2	2 NO
Initial	**1.00 mol**	**2.00 mol**	**0 mol**
Change (use + and −)			
At Equilibrium		**1.80 mol**	

- Calculate the *change* that you can.
- Use the coefficients to complete the **C**hange row. Coefficients show the *ratios* in which the moles of reactants are used up and moles of products form.
- Calculate the **E**quilibrium row.

The one change that can be calculated is below. Finish from here.

Reaction	1 N_2	1 O_2	2 NO
Initial	1.00 mol	2.00 mol	0 mol
Change (use + and −)		**−0.20 mol**	
At Equilibrium		1.80 mol	

From the O_2 moles change and the ratios of reaction (the coefficients), the other *changes* can be calculated. Be sure to include the + and − signs. Then complete the **E**quilibrium row.

Reaction	1 N₂	1 O₂	2 NO
Initial	1.00 mol	2.00 mol	0 mol
Change (use + and −)	*−0.20 mol*	−0.20 mol	*+0.40 mol*
At Equilibrium		1.80 mol	

Using the row labels, the WANTED moles at equilibrium can be found.

Reaction	1 N₂	1 O₂	2 NO
Initial	1.00 mol	2.00 mol	0 mol
Change (use + and −)	−0.20 mol	−0.20 mol	+0.40 mol
At Equilibrium	**0.80 mol**	1.80 mol	**+0.40 mol**

PRACTICE   A

For the reaction

$$H_2(g) + CO_2(g) \leftrightharpoons H_2O(g) + CO(g)$$

The initial gas mixture is composed of 2.00 mol H_2 and 1.00 mol CO_2. At equilibrium, 0.30 mol CO gas are found. Calculate the moles of the other substances present at equilibrium.

RICE Tables Using Concentration

Which *units* can be used in a RICE table?

1. Moles can always be used. RICE calculations are based on coefficients, and coefficients can always be read as moles.

2. Concentrations (in mol/L) can also be used in RICE tables *if* all of the moles in a problem are measured in the *same volume.*

   Why? Coefficients are mole ratios. However, if all of the moles are contained in the same volume, dividing each of the moles by that same volume will not change the ratios. The mole and the mol/L *ratios* will be the same, and the RICE table can be used to calculate either the *moles* or the *mol/L* used up and formed.

3. If gases or solutions are *added together* as part of a problem, measurements of the *initial* moles per liter cannot be substituted directly into a RICE table, because the volumes in which the moles are found are not the same during the process.

   Example: If 10 mL of 0.50 M Reactant A is *mixed* with 20 mL of 0.50 M Reactant B to conduct a reaction, both A and B are *diluted* as the solutions are mixed. Using concentrations before mixing as concentrations that apply to a reaction that occurs after mixing would cause an error.

4. If substances are mixed but *all* initial amounts are converted to *moles*, a *RICE moles* table can be used. Dilution does not change the moles of diluted particles. From moles, final mol/L can then be calculated if the total volume at equilibrium is known.

The bottom line?

- RICE tables can always be solved in moles.
- RICE tables can use values for *molarity* IF the *volume* in which the reactants and products are measured is the same in *all parts* of the problem.
- In RICE tables, the units must be written and must all be the same.

PRACTICE B

Check your answers after each part.

1. For the reaction

$$2 SO_2(g) + O_2(g) \leftrightarrows 2 SO_3(g)$$

in a sealed glass vessel with temperature held constant, the initial gas mixture contains $[SO_2] = 0.0100$ M, $[O_2] = 0.0030$ M, and no SO_3. At equilibrium, $[SO_3] = 0.0040$ M.

a. Calculate the concentrations of all reactants and products at equilibrium.

b. Use your part a answer to calculate a value for the equilibrium constant under these conditions.

c. Using your part b answer, calculate K for the reverse reaction.

Using RICE Tables with K Calculations

The values calculated in a RICE table can be used in K calculations, as was done in part b of the Practice B problem above. However, care must be taken to write the *units* in a RICE table. Why?

- *RICE moles* tables can always be solved in *moles*, and moles are often supplied in problem data. Our K equations, however, are K_c equations: they require concentration units of *mol/L*. K equations can be written based on certain other units in some circumstances, but K equations *cannot* be solved in moles.
- If data in a K calculation is supplied in moles, as is often the case, the data must be converted to moles *per liter at equilibrium* before it is substituted into the K equation.

This means that in both the RICE table and the DATA table that is used with K calculations, it is important to distinguish measurements in moles (mol) from moles/liter (M).

When data is supplied in moles, it is often easiest to solve the RICE table in moles, *then* to convert the moles at equilibrium found in the table to moles *per liter*, then to substitute those values into K calculations.

TRY IT

Apply those hints to the following problem.

Q. For the reaction

$$H_2 + I_2 \leftrightarrows 2 HI \text{ (all gases)}$$

initial amounts are 0.100 mol H_2, 0.090 mol I_2, and no HI. At equilibrium, 0.020 mol H_2 are present.

A. Calculate the moles of all of the substances present at equilibrium.

B. If the reaction takes place in a 2.0 L vessel, calculate the value of *K*.

Answers:

A. WANTED: moles of H_2, I_2, and HI at equilibrium.
Strategy: To solve for values at equilibrium, use a RICE table.

Reaction	1 H_2	1 I_2	2 HI
Initial	0.100 mol	0.090 mol	0 mol
Change (use + and −)	−0.080 mol	−0.080 mol	+0.16 mol
At Equilibrium	0.020 mol	0.010 mol	0.16 mol

In RICE tables, the units must be stated and must all be the same. The bottom row answers part A.

If you have not already done so, complete part B.

B. Part B involves a *K* value and concentrations. A *K* equation relates those terms. The rule is this: For calculations using *K* equations, write the ***WRECK*** steps.

1. WANTED unit or symbol.

2. **R**eaction: $H_2 + I_2 \leftrightharpoons 2\,HI$ (all gases)

3. **E**xtent: The reaction goes to equilibrium. Use a *K* equation to solve.

4. **C**onc@Eq: The important rule is

> To find *C*oncentrations at equilibrium (step *C* in the *WRECK* steps), calculate the values in the bottom row of a RICE table. Then convert to moles/liter if needed.

In the *RICE moles* table for this problem, we know moles at equilibrium, but our *K* equation requires mol/L. If needed, adjust your work.

All of these moles are in 2.0 L. In the DATA table, convert moles to the unit that measures the symbol: mol/L at equilibrium.

$$[HI]_{eq} = 0.16 \text{ mol at eq}/2.0 \text{ L} = \textbf{0.080 M at eq}$$

$$[H_2]_{eq} = 0.020 \text{ mol}/2.0 \text{ L} = \textbf{0.010 M}$$

$$[I_2]_{eq} = 0.010 \text{ mol}/2.0 \text{ L} = \textbf{0.0050 M}$$

5. *K* (write the *K* equation): $K = \dfrac{[HI]^2}{[H_2][I_2]}$ All measured at equilibrium.

6. SOLVE: $K = \dfrac{(0.080)^2}{(0.010)(0.0050)} = \dfrac{64 \times 10^{-4}}{5.0 \times 10^{-5}} = \boxed{130}$

K values are not assigned units.

At the point when values are substituted into a *K* equation, the units are omitted, but until that step, the units must be included with all DATA.

PRACTICE   C

If you find you need to look back at the lesson when solving Practice problems, try to write and box in your notebook a summary of the rule you need. Try to design a flashcard to help with remembering the rule.

1. For the reaction

$$2\,NO + Cl_2 \leftrightarrows 2\,NOCl \text{ (all gases)}$$

    0.40 mol NO and 0.60 mol Cl_2 are originally mixed in a 4.0 L sealed glass vessel. At equilibrium, 0.20 mol NO gas remains.

    a. Calculate the moles of Cl_2 and NOCl present at equilibrium.

    b. Calculate the value for *K* under the above conditions.

    c. Calculate the *K* value for this reaction: $NO + 1/2\,Cl_2 \leftrightarrows NOCl$

2. For this reversible reaction of gases in a sealed glass vessel at a constant temperature,

$$2\,HI \leftrightarrows H_2 + I_2$$

    assume that initially the vessel contains [HI] = 0.040 M but no H_2 or I_2. When the system reaches equilibrium, $[H_2]$ = 0.010 M. Calculate *K*. (Try completing the math on this problem without a calculator.)

SUMMARY

If you have not already done so, you may want to organize this summary into charts, numbered lists, and flashcards.

1. Reactions can be divided into three types: those that go nearly 100% to completion, those that don't go, and reactions that are in practice reversible and go partially to completion. Reversible reactions continue until both the forward and reverse reactions are going at the same rate, and no further change seems to take place. The reaction is then said to be at **equilibrium**.

    For equilibrium to exist,

    *   All reactants and products must be present in at least small quantities
    *   The reaction must be in a *closed* system: no particles or energy can be entering or leaving the system

2. **Le Châtelier's principle:** If a system at equilibrium is subjected to a change, processes occur that tend to counteract that change.

    Le Châtelier's principle *predicts* shifts in variables, including concentration and temperature.

    To apply Le Châtelier's principle, write the reactants and products of the reversible reaction with "two-way arrows" in between. Then,

a. *Increasing* a [substance] that appears on one side of a equilibrium equation shifts an equilibrium to the *other* side. The other substance concentrations *on* the same side as the [increased] are *de*creased, and the substance concentrations on the *other* side are *in*creased.

b. *Decreasing* a [substance] that appears on one side of a equilibrium equation shifts the equilibrium *toward* that side. The other [substances] *on* the same side are *in*creased, and the [substances] on the *other* side are *de*creased.

c. Adding energy shifts the equilibrium away from the side with the energy term, and removing energy shifts the equilibrium toward the side with the energy term.

d. Energy can be added to a system by increasing its temperature. Energy can be removed by cooling. When energy is produced by a shift in equilibrium, the temperature of the system goes up. When energy is used up, the system's temperature goes down.

e. Adding or removing a solid, pure liquid, or solvent does not shift an equilibrium, because shifting an equilibrium does not change the *concentration* of a solid, pure liquid, or solvent (though it may change the *amounts* present).

3. **The equilibrium constant:**

a. For the general reaction

$$aA + bB \leftrightharpoons cC + dD$$

at a given temperature, at equilibrium, the ratio

$$\frac{[C]^c[D]^d}{[A]^a[B]^b} = K$$

will be constant. (Remember: "In *K*, products are on top.")

b. The ratio with powers and *symbols* for concentrations is called the equilibrium constant (*K*) **expression**. The *K* **value** is the number that is the ratio. If the temperature changes, the *K* value will change, but the *K* expression does not.

c. Generally, only concentrations that can *change* are included in *K* expressions. Terms for solids, pure liquids, and solvents (including *liquid* water) are written as **1** in *K* expressions. This moves the constant value of those terms into the value of *K*.

4. An equilibrium constant value is a positive number without units.

a. When *K* values are calculated, the units are omitted.

b. When a concentration is calculated based on a *K* equation based on concentrations, the unit mol/L (M) must be added to the answer.

5. At equilibrium,

a. If the substances on the right side of an equation have higher concentrations than those on the left, the value of *K* will be greater than one.

(continued)

b. If the substances on the right side of an equation have lower concentrations than those on the left, the value of K will be a number between zero and one (in scientific notation, a positive number with a negative power of 10).

c. The more a reaction goes to the right, the higher will be the value of K.

6. If a value of K is known for a reaction written in one direction, the value of K for the reverse reaction will be the reciprocal of the original K.

$$K_r = 1/K_f$$

7. If a value of K (#) is known for a reaction with one set of coefficients, those coefficients can be multiplied by any positive number (n), and the new value of K will equal the original K to the *power* of the multiplier.

$$K = (\#)^n$$

8. *RICE moles tables*: In reaction calculations, we can track the counts of particles before, during, and at the end of a reaction with a *RICE moles* table. RICE tables have four rows: balanced *R*eaction equation, *I*nitial, *C*hange, and End/*E*quilibrium. RICE tables can always be solved in moles, and can also be solved in concentration units if the volume is the same during all measurements.

9. **Solving K calculations—the WRECK steps:**
   WRE. Write the *WANTED* unit, balanced *R*eaction equation, and *E*xtent to which the reaction goes to completion.
   *C*oncentrations at equilibrium. If concentrations at equilibrium are given in a problem, use them. A RICE table is not needed. When concentrations at equilibrium are not known, make a RICE table. The bottom row values are the concentrations at equilibrium, or can be used to calculate those concentrations, which are needed in the K equation.
   K: **Write the K equation.** Substitute the *E*quilibrium row RICE table terms into the K equation. **Solve** for the *WANTED* equation symbol.

ANSWERS

Lesson 19.1

Practice
1. Increase the temperature or increase the concentration of one or more of the reactants.

2. All of the reactants and products must be present, and the system must be closed: no particles or energy may be entering or leaving the system.

Lesson 19.2

Practice A   For $N_2(g) + 3 H_2(g) \leftrightarrows 2 NH_3(g)$

1. If [H_2] is increased: a. Equilibrium shifts **right**.   b. [N_2] will **decrease**.   c. [NH_3] will **increase**.

2. If [N_2] is decreased: a. Equilibrium shifts **left**.   b. [H_2] will **increase**.   c. [NH_3] will **decrease**.

Answers 447

Practice B
1. For $CO(g) + 2 H_2(g) \leftrightarrows CH_3OH(g) + $ energy

1a. If the temperature is decreased: i. Equilibrium shifts **right**.   ii. $[CH_3OH]$ will **increase**.

1b. If $[H_2]$ increases: i. Equilibrium shifts **right**.   ii. $[CH_3OH]$ **increases**.
iii. Energy is on the right, so the temperature **increases**.

2a. If temperature is increased: i. Equilibrium shifts **left**.   ii. $[PCl_5]$ will **decrease**.

2b. If $[Cl_2]$ is increased: i. Equilibrium shifts **right**.   ii. $[PCl_3]$ will **decrease**.
iii. Temperature **increases**.

Practice C
1. For $C(s) + 2 Cl_2(g) \leftrightarrows CCl_4(g) + $ energy

1a. If the $[Cl_2]$ is decreased: i. Equilibrium shifts **left**.   ii. $[CCl_4]$ will **decrease**.
iii. $[C]$ will **not change**, because C is solid.   iv. Amount of carbon **increases**.   v. Temperature **decreases**.

1b. Catalysts do not shift the position of an equilibrium.

2. For $CH_3COO^-(aq) + H_2O(\ell) \leftrightarrows CH_3COOH(aq) + OH^-(aq)$

2a. If $[OH^-]$ is increased: i. Equilibrium shifts **left**.   ii. $[CH_3COOH]$ will **decrease**.
iii. [liquid H_2O] will **not change**.

2b. If H^+ is added: i. H^+ neutralizes OH^-, lowering $[OH^-]$, so equilibrium shifts **right**.
ii. $[CH_3COO^-]$ **decreases**.

3. For $2 H_2O(g) + 2 Cl_2(g) + $ energy $\leftrightarrows 4 HCl(g) + O_2(g)$

3a. If $[O_2]$ is increased: i. Equilibrium shifts **left**.   ii. $[HCl]$ will **decrease.**

3b. If $[HCl]$ is decreased: i. Equilibrium shifts **right**.   ii. $[O_2]$ will **increase**.   iii. $[Cl_2]$ will **decrease**.
iv. Energy is used up when the reaction goes left, so temperature will **decrease**.

3c. If temperature is decreased: i. Equilibrium shifts **left**.   ii. $[O_2]$ will **decrease.**
iii. $[H_2O]$ will **increase**. If water is in the gas state as steam, its concentration can change.

3d. If catalyst is added: i. Equilibrium **will not shift.**   ii. $[O_2]$ will **not change**.

4. For: $4 PCl_5(g) + $ energy $\leftrightarrows P_4(s) + 10 Cl_2(g)$
If temperature is decreased: a. Equilibrium shifts **left**.   b. $[Cl_2]$ will **decrease**.
c. $[PCl_5]$ gas **increases**.   d. $[P_4]$, because it is a solid, will **not change**.

Lesson 19.3

Pretest   1. 2.7×10^{10} 2. 2.0×10^{-7} 3. 2.0×10^{-8} 4. 6.3×10^1

Practice A   1a. $(10^{-3})^2 = \mathbf{10^{-6}}$ 1b. $(10^{-5})^{-2} = \mathbf{10^{+10}}$

2a. $(2.0 \times 10^4)^4 = 16 \times 10^{16} = \mathbf{1.6 \times 10^{17}}$ 2b. $(3.0 \times 10^{-1})^3 = 27 \times 10^{-3} = \mathbf{2.7 \times 10^{-2}}$

3a. $(2.1 \times 10^6)^2 = \mathbf{4.4 \times 10^{12}}$ 3b. $(3.9 \times 10^{-2})^3 = 59 \times 10^{-6} = \mathbf{5.9 \times 10^{-5}}$

3c. $(7.7 \times 10^4)^4 = 3{,}515 \times 10^{16} = \mathbf{3.5 \times 10^{19}}$ 3d. $(5.5 \times 10^{-2})^3 = 166 \times 10^{-6} = \mathbf{1.7 \times 10^{-4}}$

Practice B   1. 95   2. 0.16

Practice C   1a. 10^6 1b. 10^3 1c. 10^{-2} 1d. 10^{-3}

2. $x^{0.125} = x^{1/8} = $ the **eighth** root   3. $81^{0.25} = 81^{1/4} = (81^{1/2})^{1/2} = (9)^{1/2} = \mathbf{3.0}$

Practice D   1a. 64   1b. 4   1c. 4   1d. **0.40**   or $\mathbf{4.0 \times 10^{-1}}$

2a. $(6.0 \times 10^{23})^{1/3} = (600 \times 10^{21})^{1/3} = (600^{1/3} \times 10^7) = \mathbf{8.4 \times 10^7}$

2b. $(10^{15})^{1/4} = (10^3 \times 10^{12})^{1/4} = (1000 \times 10^{12})^{1/4} = (1000^{1/4} \times 10^3) = \mathbf{5.6 \times 10^3}$

2c. $(1.25 \times 10^{-7})^{1/3} = (125 \times 10^{-9})^{1/3} = (125^{1/3} \times 10^{-3}) = \mathbf{5.00 \times 10^{-3}}$

2d. $(1.6 \times 10^{-11})^{1/4} = (16 \times 10^{-12})^{1/4} = (16^{1/4}) \times (10^{-12})^{1/4} = \mathbf{2.0 \times 10^{-3}}$

Practice E 1. 16   2. 3.9×10^{-3} 3. 1.8×10^{18} 4. 6.4×10^{31} 5. 1.4×10^{-20} 6. 4.9×10^{-14}
  7. 3.0   8. 0.4000   9. 3.97   10. 4.6×10^2 11. 0.53   12. 4.9×10^{-16}

Lesson 19.4

Practice A 1. $K = \dfrac{[CO_2][H_2O]^2}{[CH_4][O_2]^2}$ 2. $K = \dfrac{[CO_2]^8[H_2O]^{10}}{[C_4H_{10}]^2[O_2]^{13}}$

Practice B 1. $K = \dfrac{[CO_2]^4}{[C_2H_6]^2[O_2]^7}$ 2. $K = \dfrac{1}{[O_2]^3}$ 3. $K = \dfrac{[CH_3COO^-]}{[CH_3COOH][OH^-]}$ 4. $K = [Ca^{2+}]^3[PO_4^{3-}]^2$

Lesson 19.5

Practice 1A. $K = \dfrac{[Cu^{2+}]}{[Ag^+]^2}$ 1B. $K = [Pb^{2+}][Cl^-]^2$ 1C. $K = [Ag^+][I^-]$

2. Reaction A, with the largest K value, most favors the right side (products).

3. Reaction C, with the K value much smaller than the others, will most favor the left side (reactants), and will form the smallest concentrations of products.

4. This equation is reaction A written backwards. Both the K expression and K value will be the reciprocals of reaction A.

$$K \text{ expression} = \frac{[Ag^+]^2}{[Cu^{2+}]} \qquad K_{\text{reverse}} \text{ value} = \frac{1}{K_{\text{forward}}} = \frac{1}{1.0 \times 10^{15}} = \mathbf{1.0 \times 10^{-15}}$$

5. This reaction is reaction B with all coefficients multiplied by 1/2. In this K expression, the new coefficients become the powers of the concentrations. The K value for part 5 will be the K value for reaction B taken to the 1/2 power: the square root of the K in reaction B.

$$K \text{ expression} = [Pb^{2+}]^{1/2}[Cl^-] \qquad K \text{ value} = (2.0 \times 10^{-5})^{1/2} = (20 \times 10^{-6})^{1/2} = \mathbf{4.5 \times 10^{-3}}$$

6. This is reaction A written backwards and multiplied by 1/2. When writing the reaction backwards, invert the K value. When multiplying coefficients by 1/2, take the K value to the 1/2 power.

$$K \text{ expression} = \frac{[Ag^+]}{[Cu^{2+}]^{1/2}}$$

$$K \text{ value} = \text{square root of } \frac{1}{K_f} = (1.0 \times 10^{15})^{-1})^{1/2} = (10 \times 10^{-16})^{1/2} = \mathbf{3.2 \times 10^{-8}}$$

Lesson 19.6

Practice 1. For calculations involving K and concentrations, write the *WRECK* steps.

WANTED: $? = K$

*R*xn: $4\,NH_3 + 5\,O_2 \leftrightarrows 4\,NO + 6\,H_2O$ (all gases)

*E*xtent: Goes to equilibrium, use K equation.

*C*onc@Eq: *(See list in problem.)*

K: $K = \dfrac{[NO]^4[H_2O]^6}{[NH_3]^4[O_2]^5}$ (Since this H_2O is a gas, it is included in the K equation.)

SOLVE: The equation as written solves for the WANTED symbol. Plugging in numbers:

$$? = K = \frac{[0.50]^4[0.20]^6}{[0.050]^4[0.0020]^5} = \frac{(5.0 \times 10^{-1})^4(2.0 \times 10^{-1})^6}{(5.0 \times 10^{-2})^4(2.0 \times 10^{-3})^5} = \frac{(625 \times 10^{-4})(2.0)^6 \times 10^{-6}}{(625 \times 10^{-8})(2.0)^5 \times 10^{-15}}$$

$$= \frac{(2.0)^6 \times 10^{-10}}{(2.0)^5 \times 10^{-23}} = \boxed{\mathbf{2.0 \times 10^{+13}}} \quad (K \text{ values are written without units.})$$

The arithmetic may be done in any way that results in a correct answer. If you need practice at exponential notation calculations, review Lesson 1.3.

2. For calculations involving K and concentrations, write the *WRECK* steps.

   WANTED: ? = **[C]**

   *R*xn: $2A + B \leftrightharpoons 4C$ (all gases)

   *E*xtent: Goes to equilibrium, use K.

   Conc@Eq: (See list in problem.)

   K: $K = \dfrac{[C]^4}{[A]^2\,[B]}$

   SOLVE: First solve K for the term that *includes* the WANTED symbol.
   $[C]^4 = K \cdot [A]^2 \cdot [B] = (0.020)(0.050)^2(0.125) = 6.25 \times 10^{-6}$

   Then solve for the WANTED symbol.
   $[C] = (6.25 \times 10^{-6})^{1/4} = (625 \times 10^{-8})^{1/4} = \mathbf{5.0 \times 10^{-2}\,M = 0.050\,M}$

3. For calculations involving K and concentrations, write the *WRECK* steps.

   WANTED: ? = **mol CO**

   R+E: $CO_2(g) + H_2(g) \leftrightharpoons CO(g) + H_2O(\ell)$ (Goes to equilibrium, use K.)

   Conc@Eq: Data is given in moles, but the K equation requires concentration (mol/L). All of the substances are contained in a sealed 20. L container. If you needed that hint, adjust your work and continue.

   To find mol/L, divide mol by L.
   $[CO_2] = $ 0.40 mol in 20. L $= \mathbf{0.020\,mol/L}$
   $[H_2] = $ 0.90 mol/20. L $= \mathbf{0.045\,M}$

   K: $K = \dfrac{[CO]}{[CO_2][H_2]} = \mathbf{4.8}$ ([Liquids] = 1 in K expressions.)

   SOLVE: $[CO] = K \cdot [CO_2] \cdot [H_2] = (4.8)(0.020)(0.045) = \mathbf{0.00432\,M\,CO}$ (Carrying extra *s.f.*)

   Done? Always check your WANTED *unit* (especially after a long calculation).

   $? = \textbf{mol CO} = 20.\,L \cdot \dfrac{0.00432\ \text{mol CO}}{L} = \boxed{\textbf{0.086 mol CO}}$

   A good habit at the end of a long calculation is to (a) $\boxed{\text{box}}$ your final answers, but (b) each time you make the $\boxed{\text{box}}$, look back at the WANTED unit or symbol at the start of your answer to make sure that you found the unit WANTED.

Lesson 19.7

Practice A   WANTED: moles of H_2, CO_2, and H_2O at equilibrium.

Strategy: To find values at equilibrium when some of the data is *not* at equilibrium, use a RICE table.

Initial data:

Reaction	1 H₂	1 CO₂	1 H₂O	1 CO
Initial	2.00 mol	1.00 mol	0	0
Change				+0.30 mol
At Equilibrium				0.30 mol

Calculate the change row based on coefficients, then find the moles at equilibrium WANTED.

Reaction	1 H_2	1 CO_2	1 H_2O	1 CO
Initial	2.00 mol	1.00 mol	0	0
Change	−0.30 mol	−0.30 mol	+0.30 mol	+0.30 mol
At Equilibrium	1.70 mol	0.70 mol	0.30 mol	0.30 mol

Practice B   1a.   WANTED: Concentrations at equilibrium for all three substances, in mol/L.

DATA: Measurements at equilibrium are WANTED, so a *RICE moles* table is needed.

Since this reaction involves gases in a container with a *fixed* volume, moles or mol/L can be used in the *RICE moles* table. Because the data are in mol/L, and you want mol/L, use mol/L as the units in the RICE table.

Reaction	2 SO_2	1 O_2	2 SO_3
Initial	0.0100 M	0.0030 M	0 M
Change (use + and −)	−0.0040 M	−0.0020 M	+0.0040 M
At Equilibrium	0.0060 M	0.0010 M	0.0040 M

The bottom row of the RICE table shows the reactant and product concentrations at equilibrium.

1b.   For *K* calculations, write the *WRECK* steps.

WANTED = *K*

Reaction and Extent: 2 $SO_2(g)$ + $O_2(g)$ ⇌ 2 $SO_3(g)$ (Goes to *equilibrium* mixture.)

Conc@Eq: Use the values in the *bottom row* of the RICE table.

$$K = \frac{[SO_3]^2}{[SO_2]^2\,[O_2]}$$

These values substituted to find *K* must be concentrations at equilibrium, the values in the *bottom* row of the RICE table.

SOLVE: $K = \dfrac{(4.0 \times 10^{-3})^2}{(6.0 \times 10^{-3})^2 (1.0 \times 10^{-3})} = \dfrac{16 \times 10^{-6}}{36 \times 10^{-9}} = 0.444 \times 10^3 = \boxed{440}$

1c. $K_{\text{reverse}} = \dfrac{1}{K_{\text{forward}}} = \dfrac{1}{440} = 2.3 \times 10^{-3}$

Practice C   1a.   WANTED: Moles of reactants and products at equilibrium.

DATA: To find measurements at equilibrium when some supplied data is not at equilibrium, use a RICE table.

Since the WANTED unit is moles, solve the RICE table in *moles*.

Reaction	2 NO	1 Cl_2	2 NOCl
Initial	0.40 mol	0.60 mol	0
Change (use + and −)	−0.20 mol	−0.10 mol	+0.20 mol
At Equilibrium	*0.20 mol*	0.50 mol	0.20 mol

1b.   For calculations involving *K*, write the *WRECK* steps.

WANTED: *K* = ?

Reaction and Extent: 2 NO + Cl_2 ⇌ 2 NOCl (all gases)   (Goes to an *equilibrium* mixture.)

*C*onc@Eq: To find concentrations at equilibrium, use the bottom row of the *RICE moles* table.
In the above RICE table are *moles*. A K_c equation requires *mol/L* at equilibrium. All of these moles in the table are in **4.0 L**. In the data table, convert moles to the unit of each symbol: mol/L, then solve the *K* equation.

$$[NOCl]_{eq} = 0.20 \text{ mol}/4.0 \text{ L} = \mathbf{0.0500 \text{ M at eq}} \quad \text{(Carrying extra } s.f.\text{)}$$

$$[NO]_{eq} = 0.20 \text{ mol}/4.0 \text{ L} = \mathbf{0.0500 \text{ M}}$$

$$[Cl_2]_{eq} = 0.50 \text{ mol}/4.0 \text{ L} = \mathbf{0.125 \text{ M}}$$

$$K = \frac{[NOCl]^2}{[NO]^2[Cl_2]} \text{ with all concentrations measured at equilibrium.}$$

SOLVE: $$K = \frac{(0.0500)^2}{(0.0500)^2(0.125)} = \frac{1}{0.125} = \mathbf{8.0}$$

1c. This is the original reaction with all coefficients multiplied by 1/2.
The new value of *K* is the value for original *K taken* to the power of the multiplier. $\boxed{K = (\#)^n}$
For this reaction, this is the *K* value for the original reaction taken to the 1/2 power: the square root of the original *K*.
K value with coefficients halved $= (8.0)^{1/2} = \mathbf{2.8}$

2. For calculations involving *K*, write the *WRECK* steps.

WANTED: $$K = \frac{[H_2][I_2]}{[HI]^2}$$

To calculate *K*, these concentrations must be at equilibrium, the values in the *bottom* row of the RICE table.

Reaction and Extent: $2 HI \leftrightarrows H_2 + I_2$ (all gases)   (Goes to an *equilibrium* mixture).

*C*onc@Eq: To find measurements at equilibrium when some supplied data is not, use a RICE table.
Since the data is in mol/L, solve the RICE table in mol/L. RICE tables may be solved in mol/L as long as all substances are measured in the same volume, as is the case for gases in a sealed glass vessel.

Reaction	2 HI	1 H_2	1 I_2
Initial	0.040 M	0 M	0 M
Change (use + and −)	**−0.020 M**	**+0.010 M**	**+0.010 M**
At Equilibrium	**0.020 M**	0.010 M	**0.010 M**

Units must be included in RICE tables.
To calculate *K*, use concentrations at equilibrium: the values in the *bottom* row of this RICE table.

$$K = \frac{[H_2][I_2]}{[HI]^2} = \frac{(1.0 \times 10^{-2})^2}{(2.0 \times 10^{-2})^2} = \frac{1.0 \times 10^{-4}}{4.0 \times 10^{-4}} = \mathbf{0.25}$$

20

Acid–Base Fundamentals

Lesson 20.1 Acid–Base Math Review

> **PRETEST** This lesson reviews math needed for acid–base calculations. If your exponential math skills are good, try the *last* lettered part of Practice questions 3–6. If you get those right, you may skip this lesson.
>
> If review is needed, study the rules below, then do more of the practice set. For an additional, detailed review, see Lessons 1.1 to 1.3.

Rules for Exponential Notation

1. To *multiply* exponential terms, *add* the exponents.

2. To *divide* exponentials, *subtract* the exponents.

3. Numbers expressed in exponential notation have three parts:

$$\text{sign} \downarrow$$
$$-5.25 \times 10^3$$

significand   exponential

4. Answers in exponential notation should be converted to scientific notation. This places the decimal in the significand after the first digit that is not a zero.
   - When moving the decimal point Y times to make the significand *larger*, make the exponent of 10 *smaller* by a *count* of Y.
   - When moving the decimal point Y times to make the significand *smaller*, make the exponent *larger* by a count of Y.
   - "If you make one larger, make the other smaller."

5. When changing fixed decimal numbers to exponential notation,
   - The number of times you move the decimal becomes the positive or negative number in the exponential.
   - Values larger than one will have positive powers of 10.
   - Values between zero and one (such as 0.85) have negative powers of 10.

6. In calculations using exponential notation, handle the two parts separately: do number math by number rules and exponential math by exponential rules.

7. If an exponential term does not have a number in front, add a "**1 ×**" in front of the exponential so that the number–number division is clear.

PRACTICE

Do the problems below without looking at the rules. If you find that you need to look back, write a summary of the rules above, recite your rules, *then* continue with the practice.

To speed your progress, do every second or third problem. If you cannot solve easily, do more parts of that problem.

Try problems 1–5 without a calculator.

1. If you need help, see rule 1.

    a. $(10^{-8})(10^{+2}) =$

    b. $(10^{-3})(10^{-12}) =$

    c. $(x)(10^{-12}) = 10^{-14}; x =$

    d. $(x)(10^{-9}) = 10^{-14}; x =$

    e. $(10^{-3})(x) = 10^{-14}; x =$

2. Need a hint on the following? See rule 2.

    a. $\dfrac{10^{-14}}{10^3} =$

    b. $\dfrac{10^{-14}}{10^{-5}} =$

    c. $\dfrac{10^{-14}}{10^{-11}} =$

3. Convert these to scientific notation. For help, see rule 4.

    a. $324 \times 10^{+12} =$

    b. $0.050 \times 10^{-11} =$

4. Try to solve these without a calculator. Convert your final answer to scientific notation. Need a hint? See rule 6. Use all the paper you need for careful work. Solve in your notebook if needed.

    a. $(2.0 \times 10^1)(3.0 \times 10^{-11}) =$

    b. $\dfrac{1.0 \times 10^{-14}}{2.0 \times 10^4} =$

    c. $\dfrac{1.0 \times 10^{-14}}{3.0 \times 10^{-4}} =$

    d. $(x)(2.0 \times 10^{-8}) = 10. \times 10^{-15}; x =$

    e. $(2.5 \times 10^{-2})(x) = 10. \times 10^{-15}; x =$

5. Try to solve these without a calculator. It may help to convert all values to scientific notation.

    a. $\dfrac{1.0 \times 10^{-14}}{0.040} =$

    b. $\dfrac{1.0 \times 10^{-14}}{0.0030} =$

    c. $(x)(0.20) = 1.0 \times 10^{-14}; x =$

    d. $(0.0125)(x) = 1.0 \times 10^{-14}; x =$

6. Use a calculator for the numbers but not the exponentials. Convert answers to scientific notation.

    a. $\dfrac{1.0 \times 10^{-14}}{3.25 \times 10^{-5}} =$

    b. $\dfrac{1.0 \times 10^{-14}}{8.8 \times 10^{-4}} =$

(continued)

c. $\dfrac{1.0 \times 10^{-14}}{2.4} =$

d. $\dfrac{1.0 \times 10^{-14}}{4.3 \times 10^{-4}} =$

e. $(x)(6.7 \times 10^{-12}) = 1.0 \times 10^{-14}; x =$

f. $(1.25 \times 10^{-7})(x) = 1.0 \times 10^{-14}; x =$

Lesson 20.2 K_w Calculations

Acid–Base Review

In Chapter 14, we considered acid–base neutralization calculations. In those problems, the question was this: If we have a known amount of an acid or base, can we find its stoichiometric equivalent—the amount of the opposite that is needed to exactly neutralize the acid or base? These "reaction amount" calculations for stoichiometric equivalents were solved using conversion stoichiometry.

Beginning in this chapter, we return to acids and bases to ask additional questions important in both chemistry and biology. What is the nature of acid and base solutions before they react? Which particles and ions are present? How are concentrated acid or base solutions different from dilute solutions? How do strong acids like hydrochloric acid differ from weak acids such as vinegar that we frequently consume as food?

In most cases, we are interested in the behavior of acids and bases when they are dissolved in an aqueous solution. Let's start with the molecule that has the highest concentration in aqueous solutions: H_2O.

The Ionization of Water

A water molecule has oxygen in the middle, two bonds, and a bent shape.

At temperatures above absolute zero, the bonds in water bend and stretch. At room temperature, the liquid molecules also move and collide at high average speeds. In part as a result of these motions, for about one in 500 million molecules at room temperature, one of the bonds in liquid water is broken. The result is the formation of two ions, H^+ and OH^-, that also move about in the water solution.

When an ionic solid separates into ions that can move about freely, the reaction is termed **dissociation**. When a bond that separates to form ions has covalent character, the process may also be termed **ionization** (forming ions). In practice, all bonds have some covalent character, and the terms dissociation and ionization are often used interchangeably.

This separation of water into ions is reversible and can be represented by the equation

$$1\ H_2O(\ell) \leftrightarrows 1\ H^+(aq) + 1\ OH^-(aq)$$

>99.999% **un**-ionized   <0.001% ionized

In pure (**distilled**) liquid water, the *number* of H^+ and OH^- ions must be equal, since they must be formed in a 1:1 ratio when water ionizes. The *concentration* of the H^+ and OH^- ions must also be equal, since the equal moles of the two ions are contained in the same volume of solution. At room temperature, these *ion* concentrations are very small: 1.0×10^{-7} moles per liter for each. The concentration of the un-ionized water molecules is very large in comparison: about 55 mol/L.

However, even at these low concentrations, small changes in the balance between H^+ and OH^- can have a large influence on reactions that are important in chemistry and biology.

Water's Ionization Constant: K_w

For the reversible ionization of water,

$$H_2O(\ell) \leftrightarrows H^+(aq) + OH^-(aq)$$

the equilibrium constant expression *could* be written as

$$K = \frac{[H^+][OH^-]}{[H_2O]}$$

However, in aqueous solutions (those where water is the solvent), the concentration of the *non*-ionized water molecules is high (about 55 M) for the relatively dilute solutions used in *most* lab experiments, and it does not change substantially during dilution or reactions.

In equilibrium constants, concentrations that remain close to constant during reactions are generally included as part of the numeric value for K and are represented by a 1 in the K expression. The result is a simplified equilibrium constant expression (labeled K_w) for the relationship between H^+ and OH^- in an aqueous solution:

$$K_w = [H^+][OH^-]$$

At room temperature (25°C), the value for K_w is $\mathbf{1.0 \times 10^{-14}}$. This small numeric value (0.000 000 000 000 0**10**) indicates that the reaction favors the reactants: very few water molecules separate into ions.

As temperature increases, the molecules of water collide with higher energy, the bonds bend and stretch more vigorously, and the bonds break more often. Of importance in biology, at *body temperature* in mammals (37°C), the $[H^+]$ in water is about $\mathbf{1.6 \times 10^{-7}}$ M as opposed to $\mathbf{1.0 \times 10^{-7}}$ M at 25°C.

However, for acid–base calculations in chemistry, you may assume that conditions are at 25°C unless otherwise noted. In acid–base calculations, we will use this K_w equation:

$$\mathbf{K_w = [H^+][OH^-] = 1.0 \times 10^{-14}} \text{ at 25°C}$$

This relationship between $[H^+]$ and $[OH^-]$ is an inverse proportion. If substances are added to water that make one ion concentration increase, the other must decrease in the same proportion: if one triples, the other ion must become 1/3 of its original concentration.

Acid–Base Terminology

In pure water, the concentration of H^+ and OH^- ions must be equal. However, if acids or bases are dissolved into water, this balance is upset.

By what are termed the **classical** definitions of acids and bases,

- An **acid** is a substance that ionizes in water and forms H^+ ions
- A **base** is a substance that separates in water to form OH^- ions

These are also called the **Arrhenius** definitions for acids and bases, after the Swedish chemist Svante Arrhenius who first proposed the existence of electrically charged particles (ions).

When acids or bases are added to water, the water continues to ionize slightly, and the K_w equation will continue to predict the relationship between the $[H^+]$ and $[OH^-]$. This means that in an acidic or basic solution, if either $[H^+]$ or $[OH^-]$ is known, the concentration of the other ion can be calculated using the K_w equation.

A useful rule is this: In calculations that include *both* $[H^+]$ and $[OH^-]$ in the WANTED and DATA, write the K_w equation. Let's abbreviate the rule as follows:

The K_w Prompt

See $[H^+]$ *and* $[OH^-]$? *Write* $K_w = [H^+][OH^-] = 1.0 \times 10^{-14}$

Some problems will ask for the approximate $[H^+]$ and $[OH^-]$ in acid or base solutions, and those calculations can be done by mental arithmetic.

TRY IT

(See "How to Use These Lessons," point 1, p. xv.)

Apply the rule to this problem.

Q. In a solution with a $[OH^-]$ of about 10^{-2} M, what is the $[H^+]$? (Answer with a 10 to a power.)

 Answer:

WANTED: $[H^+]$

DATA: $[OH^-] \approx 10^{-2}$ M

Prompt: See $[H^+]$ *and* $[OH^-]$? Write $K_w = [H^+][OH^-] = 1.0 \times 10^{-14}$

SOLVE: Substitute the known $[OH^-]$:

$[H^+](\sim 10^{-2}) = 10^{-14}$; by inspection, $[H^+] \approx 10^{-12}$ M

Note that as always when solving K equations, units are omitted during calculations, but if the WANTED unit is a concentration, the unit mol/L (M) must be added to the answer.

In other problems, you will need to calculate the ion concentrations more precisely.

TRY IT

Complete this calculation in your notebook.

Q. If the $[H^+]$ in an aqueous solution is 5.0×10^{-3} M, find the $[OH^-]$.

Answer:

WANTED: $[OH^-] = ?$

DATA: $[H^+] = 5.0 \times 10^{-3}$ M

$K_w = [H^+][OH^-] = 1.0 \times 10^{-14}$ (K_w prompt.)

SOLVE: Solve the equation for the wanted *symbol, then* plug in the DATA.

$$? = [OH^-] = \frac{1.0 \times 10^{-14}}{[H^+]} = \frac{1.0 \times 10^{-14}}{5.0 \times 10^{-3}} = 0.20 \times 10^{-11}$$

$$= 2.0 \times 10^{-12} \text{ M}$$

K_w Check

When using the K_w equation, do a *check* at the end: *estimate* $[H^+]$ times $[OH^-]$ (circled above). The result must $\approx 10 \times 10^{-15}$ *or* 1.0×10^{-14}. Try multiplying the circled values above in your head. Does the answer check?

PRACTICE

Do parts 1b, 1d, 2b, 2d, 3a, and 3b. Need more practice? Do more parts.

1. In these aqueous solutions, find the $[H^+]$ if the $[OH^-]$ is

    a. 10^{-11} mol/L   b. 0.010 molar

    c. 5.0×10^{-11} M   d. 0.036 M

2. In these aqueous solutions, find the $[OH^-]$ if the $[H^+]$ is

    a. 10^{-9} mol/L   b. 1.0 M

    c. 3.0×10^{-5} M   d. 1.25 M

3. If 20.0 millimoles of OH^- ions are dissolved in 400. mL of solution, find

    a. $[OH^-]$ b. $[H^+]$

Lesson 20.3   Strong Acid Solutions

Definitions

By the classical (Arrhenius) definitions in chemistry, compounds can be characterized as follows:

- **Strong acids** ionize essentially 100% in water and form H^+ ions.
- **Strong bases** dissociate essentially 100% in water and form OH^- ions.
- **Weak** acids or bases dissociate (ionize) only partially when dissolved in water.

An H^+ ion has one proton and no electrons, so the H^+ ion is often referred to as a *proton*. Acids can be classified as **mono**protic (containing *one* hydrogen atom that can ionize) or **poly**protic (containing more than one).

Acid–Base States

Molecules that are acids and bases can react in a variety of ways.

Example: One type of acid–base reaction is the mixing of hydrogen chloride gas and ammonia gas to form solid ammonium chloride.

$$HCl(g) \ + \ NH_3(g) \rightarrow NH_4Cl(s)$$

However, *most* acid–base reactions are conducted in water, and in these lessons, if no state for an acidic or basic particle is shown, assume that the state is aqueous (*aq*).

Strong Acids

In chemistry, we often have a need for strong acid solutions. The strong acids most frequently encountered are HCl, HNO_3, and H_2SO_4.

- **HCl** and **HNO$_3$** are strong *mono*protic acids. Both are highly soluble in water and ionize essentially 100% to release one H^+ ion.

  Other strong monoprotic acids include HBr, HI, $HClO_4$, and $HMnO_4$, but because these substances may undergo redox as well as acid–base reactions, they are used less often for reactions that require a strong acid.

- **H$_2$SO$_4$** is a strong *di*protic acid. When H_2SO_4 is dissolved in water, the first proton ionizes essentially 100%, but the second ionizes only partially.

The ionization of acids is more complex than the ionization of other ionic compounds in part because the bond of the H in acids has a balance of ionic and covalent character. Because of the mixed nature of bonds to hydrogen, to calculate ion concentrations in acidic solutions, we need three sets of rules:

- One for strong acids (such as HCl and HNO_3) that ionize completely
- One for weak acids, in which ionization goes to equilibrium
- One for polyprotic acids (such as H_2SO_4) in which some H atoms ionize more easily than others

Let us start with the rules for solutions of the strong acids HCl and HNO_3.

Ion Concentrations in Strong Acid Solutions

The concentration of an HCl or HNO_3 solution is usually expressed in terms of an *un*-ionized acid formula, such as [HCl] = 0.45 M. However, the un-ionized formula does not represent the particles that are actually present in the solution.

Example: The gas hydrogen chloride (HCl) readily dissolves in water to form a solution of *hydrochloric acid.* If 0.20 moles of HCl is dissolved per one liter of solution, the solution concentration is written as "[HCl] = 0.20 M" based on how much HCl is added when mixing the solution. This 0.20 M is termed the [HCl]$_{as\ mixed}$.

However, there are *no* particles of HCl present in an "HCl solution." Why? As HCl dissolves in water, it immediately separates to form ions:

<u>**1**</u> HCl is used up $\rightarrow$ <u>**1**</u> H^+ is formed + <u>**1**</u> Cl^- is formed   (Goes ~100%.)

This reaction is reversible, but the right side is so strongly favored at equilibrium that in water, essentially all of the HCl is converted to ions.

In a solution *labeled* **[HCl] = 0.20 M**, the actual concentrations of the particles are

[HCl] = **0** M; [H^+] = 0.20 M; [Cl^-] = 0.20 M; and [H_2O] $\approx$ 55 M

That said, if a problem asks for the concentration of a strong acid, such as [HCl] or [HNO_3], you should assume it is asking for not 0 M, but the [strong acid]$_{as\ mixed}$.

In addition, a very small amount of OH^- ion is present in the solution (more on that later).

Calculating Ion Concentrations in Strong Acid Solutions

The ionization of a strong acid parallels what happens to other ionic compounds that separate into ions essentially 100% when dissolved in water (see Lesson 12.2). The key rule for calculations involving such compounds is

---

For Substances Separating 100% into Ions, Write the REC Steps

In calculations involving [ions], if a substance ionizes ~100%,

* **R**: Write the balanced ionization **R**eaction equation. After the equation, write
* **E**: The **E**xtent of the reaction ("goes ~100%"). Below each particle, write
* **C**: The **C**oncentration of the particle, based on the *coefficient* ratios.

If a reaction goes to completion, the coefficients supply the *mole* ratios of reactants used up and products formed. For the 100% dissociation of strong acids, since all of the reaction particles are in the same constant volume of solution, the coefficients also supply the mole *per liter* reaction ratios.

Example: In 0.15 M HNO_3 , what are the [ions]? Write the REC steps.

Reaction and **E**xtent: <u>**1**</u> HNO_3 $\rightarrow$ <u>**1**</u> H^+ + <u>**1**</u> NO_3^- (Goes ~100%.)

$\qquad\qquad\qquad\qquad\qquad\qquad \wedge \qquad\quad \wedge \qquad\quad \wedge$

Concentrations: ~~0.15 M~~ 0 M   0.15 M   0.15 M

In your notebook, apply the REC steps to this problem.

 TRY IT

Q. In a 0.45 M HCl solution, write the

    a. $[HCl]_{as\ mixed}$ b. $[H^+]_{in\ soln.}$ c. $[Cl^-]_{in\ soln.}$

Answers:

Since HCl is a strong acid, in water it ionizes ~100%. To find ion concentrations for substances that ionize ~100%, write the REC steps.

Rxn. and Extent: $\underline{1}$ HCl used up $\rightarrow$ $\underline{1}$ H$^+$ formed $+$ $\underline{1}$ Cl$^-$ formed (Goes ~100%.)

$\qquad\qquad\qquad\qquad\qquad\quad$ $\wedge$ $\qquad\qquad$ $\wedge$ $\qquad\qquad$ $\wedge$

Conc.:$\qquad\qquad$ ~~0.45 M~~ 0 M$\qquad$ **0.45 M**$\qquad$ **0.45 M**

The bottom row shows the WANTED H$^+$ and Cl$^-$ concentrations. The [HCl] *as mixed* is **0.45 M** (but in solution is 0 M).

In some problems, to find [ions], the [HCl] or [HNO$_3$] as mixed will need to be solved first.

 TRY IT

Solve this calculation in your notebook.

    Q. If 0.030 mol of HNO$_3$ is mixed with water to form 150 mL of solution, find

    a. $[HNO_3]_{as\ mixed}$ b. $[H^+]_{in\ soln.}$ c. $[NO_3^-]_{in\ soln.}$

Answers:

    a. WANTED: $? = \dfrac{mol\ HNO_3}{L\ soln.}$

        DATA: 0.030 mol HNO$_3$ = 150 mL soln.   (Two measures; same soln.)

                    (When a ratio unit is WANTED, all DATA will be in equalities.)

        SOLVE: (To review molarity calculations, see Lesson 10.3.)

$$? = \frac{mol\ HNO_3}{L\ soln.} = \frac{0.030\ mol\ HNO_3}{150\ mL\ soln.} \cdot \frac{1\ mL}{10^{-3}\ L} = \boxed{\textbf{0.20 M HNO}_3}$$

  b, c. WANTED: $[H^+]_{in\ soln.}$ and $[NO_3^-]_{in\ soln.}$

        HNO$_3$ ionizes ~100%. To find [ions], write the REC steps.

        Rxn. and Extent: $\underline{1}$ HNO$_3$ $\rightarrow$ $\underline{1}$ H$^+$ $+$ $\underline{1}$ NO$_3^-$ (Goes ~100%.)

                              $\wedge$ $\qquad$ $\wedge$ $\qquad$ $\wedge$

        Conc.   ~~0.20 M~~ 0 M **0.20 M** **0.20 M**

1. In a 0.50 M solution of nitric acid (HNO_3), what will be the

    a. $[H^+]$?   b. $[NO_3^-]$?

2. 7.30 g of HCl is dissolved in water to make 250 mL of solution. Find the

    a. Moles of HCl dissolved in the solution, per liter.

    b. $[H^+]$ =   c. $[Cl^-]$ =

Why [Acid] Determines [H⁺]

The separation of a strong acid in water can be represented by this general equation:

$$\text{Strong acid} \rightarrow H^+ + \textbf{conjugate base}$$

Conjugate base is the term for the particle remaining after the proton leaves the acid. Compared to the acid, the conjugate base will have a chemical formula that has lost both one H atom and one positive charge.

Example: $HNO_3 \rightarrow H^+ + NO_3^-$

In a nitric acid solution, the nitrate ion is the conjugate base.

A solution of a strong monoprotic acid will contain four particles.

- The strong acid separates into two particles: an H^+ ion and the acid's **conjugate base**.
- Water is the largest component in aqueous solutions, present primarily as non-ionized H_2O. Water also contains small amounts of H^+ and OH^- that form due to its ionization.

When a strong acid is dissolved in water, H^+ is contributed by *both* the acid and the water. However, if the strong acid is mixed in any significant concentration, the share of H^+ ions contributed by the water can be ignored.

Why? Consider a 0.20 M HCl solution. The strong acid contributes 0.20 moles of H^+ ions per liter to the water. Before the acid was added, the water contained only 10^{-7} moles of H^+ ions per liter. Note the uncertainty in these two amounts:

0.20 M H^+ from the acid, with doubt in the hundredths place, compares to

0.0000001 M H^+ initially in the water, with doubt in a much lower place.

This is one indication that any initial H^+ contribution from the water's ions is too small to be significant. The acid ionization is the *dominant* reaction. For this reason, the rule is

In an acid solution, use acid ionization rules to find $[H^+]$.

The HCl and HNO_3 *Quick Rule*

REC steps show concentrations for all ions formed when substances ionize 100%. However, in many strong acid calculations, only $[H^+]$ is wanted, and you can use this

> *Quick rule*: $[HCl]_{mixed}$ *or* $[HNO_3]_{mixed} = [H^+]_{in\ solution}$

Example: In a solution labeled 0.35 M HCl, $[H^+]_{in\ soln.} = \textbf{0.35 M}$

When a strong monoprotic acid dissolves in water, the ratio of the [acid as mixed] and the $[H^+]$ that forms in solution is always **1** to **1**.

Equivalency of H^+ and H_3O^+ Ions

In aqueous solutions, the proton released by an acid is nearly always found attached to a water molecule, forming a **hydronium ion** (H_3O^+). This reaction can be represented as

$$1\,H^+ + 1\,H_2O \leftrightharpoons 1\,H_3O^+ \qquad \text{(Goes} \sim 100\%.)$$

Textbooks often show acids in water forming H^+ in some reactions and H_3O^+ in others. In calculations involving acids in aqueous solutions, the symbols H^+ and H_3O^+ in most cases are considered to be equivalent. When H_3O^+ is encountered in calculations, apply this rule:

> If you see $\mathbf{H_3O^+}$, write: $\mathbf{H_3O^+ = H^+}$

Let's list the rules learned so far for acids. Design flashcards as needed that will help you to apply these rules from memory.

SUMMARY

Acid Rules

1. See $[H^+]$ *and* $[OH^-]$? Write $K_w = [H^+][OH^-] = 1.0 \times 10^{-14}$

2. Strong monoprotic acids ionize $\sim 100\%$ to form H^+. To find [ions], write the REC steps.

3. *Quick rule*: $[\text{HCl } or \text{ HNO}_3]_{mixed} = [H^+]_{in\ soln.}$

4. See H_3O^+? Write $H_3O^+ = H^+$

5. In an acid solution, use acid ionization rules to find $[H^+]$.

PRACTICE **B**

Commit to memory the rules above, then do these problems.

1. In 0.25 M HNO_3,

    a. $[H^+] =$ b. $[H_3O^+] =$

2. In 2.0 M HCl,

   a. $[H_3O^+]$ =

   b. $[H^+]$ =

3. In an HCl solution, what is the formula for the particle that is the conjugate base?

Lesson 20.4   The [OH⁻] in Strong Acid Solutions

The Impact of a Strong Acid on the [OH⁻] in Water

The ionization of an acid in water affects the other reaction occurring in an acid solution: the reversible "auto-ionization" of water.

$$1\ H_2O(\ell) \rightleftharpoons 1\ H^+(aq) + 1\ OH^-(aq) \quad \text{(Goes slightly.)}$$

Adding acid to water shifts this equilibrium in accord with Le Châtelier's principle. As the $[H^+]$ in the solution increases, the equilibrium shifts toward the side of the reaction equation opposite the H^+, decreasing the [OH⁻]. The *value* of the [OH⁻] after the shift can be calculated by the equilibrium constant for water's ionization:

$$K_w = [H^+][OH^-] = 1.0 \times 10^{-14} \text{ at } 25°C$$

To solve for [OH⁻] in an *acid* solution, let us add to rule 5 as follows.

> 5. In an acid solution, use acid ionization rules to find $[H^+]$, *then K_w to find* [OH⁻].

 **TRY IT**

Apply the rule to this problem.
  Q. In a 0.40 M HCl solution, find

   a. $[H^+]$ b. $[Cl^-]$ c. [OH⁻]

 Answers:

a, b. You may be able to solve parts a and b by the logic of the quick steps, or you can solve methodically using this rule: To find [ions] in a strong acid solution, write the REC steps.

    Rxn. and Extent: $1\ HCl \rightarrow 1\ H^+ + 1\ Cl^-$ (Goes ~100%.)

    Conc.:   ~~0.40 M~~ 0 M   0.40 M   0.40 M

c. WANTED: [OH⁻]

    In an acid solution, first use acid rules to find $[H^+]$, then K_w to find [OH⁻].

Part a (in the **C** row) found $[H^+] = 0.40$ M

The equation that relates $[H^+]$ *and* $[OH^-]$ is

$$K_w = \boxed{[H^+][OH^-] = 1.0 \times 10^{-14}}$$

SOLVE the boxed equation for the symbol WANTED.

$$? = [OH^-] = \frac{1.0 \times 10^{-14}}{[H^+]} = \frac{1.0 \times 10^{-14}}{0.40} = \boxed{2.5 \times 10^{-14} \text{M}}$$

K_w *check*: $[H^+] \times [OH^-]$ (circled above) must $\approx 10 \times 10^{-15}$ *or* 1.0×10^{-14}. Check.

PRACTICE

1. From memory, write the five acid–base rules learned so far in this chapter.

2. In a solution labeled 10^{-3} M HNO_3, find the

   a. $[H^+]$ b. $[NO_3^-]$ c. $[H_3O^+]$ d. $[OH^-]$

3. Solve in your notebook. In a solution labeled 0.020 M HCl,

   a. Write the balanced equation for the ionization of this strong acid.

   b. Which side will be favored in this ionization: products or reactants?

   c. What is the *mole-to-mole* ratio between HCl used up and H^+ formed?

   d. What is the *mol/L* ratio between HCl used up and H^+ formed?

   e. $[H^+] = ?$ f. $[Cl^-] = ?$

   g. $[OH^-] = ?$ h. $[H_3O^+] = ?$

Lesson 20.5   Strong Base Solutions

Strong Hydroxide Bases

Strong bases, by the classical (Arrhenius) definitions, dissociate essentially 100% in water to form OH^- ions. The substances most frequently used to make strong base solutions are two water soluble ionic solids: Sodium hydroxide (NaOH) and potassium hydroxide (KOH).

When mixed with water,

$$1\, NaOH(s) \rightarrow 1\, Na^+(aq) + 1\, OH^-(aq) \quad \text{(Goes ~100\%.)}$$

$$1\, KOH(s) \rightarrow 1\, K^+(aq) + 1\, OH^-(aq) \quad \text{(Goes ~100\%.)}$$

There are other types of strong bases, but their reactions can be more complex. For now, we will limit our attention to the strong bases that are *alkali metal hydroxides*.

[Ions] in Strong Base Solutions

As with strong acids, if a solution of a strong base is *labeled* [NaOH] = 0.15 M, this represents how the solution is *mixed*, but not the particles present in the solution. Because NaOH dissociates ~100% in water, what is actually present in "0.15 M NaOH" is **0** M NaOH, 0.15 M Na^+, and 0.15 M OH^-.

However as with strong acids, if a problem asks for the concentration of a strong base (SB) such as [NaOH] or [KOH], assume it is asking not for 0 M but for the $[SB]_{as\ mixed}$.

The rules for strong bases are similar to those for strong acids.

NaOH and KOH dissociate ~100% and form OH^-.

- For [ions], write the REC steps.
- If only $[OH^-]$ is needed, use the quick rule:

$$[NaOH\ or\ KOH]_{as\ mixed} = [OH^-]_{in\ soln.}$$

- In a *base* solution, use base dissociation rules to find $[OH^-]$, *then* K_w to find $[H^+]$.

Let's add these new rules to our list that, when committed to long-term memory, simplifies acid–base problem solving.

Fundamentals of Acid–Base Calculations

1. See $[H^+]$ *and* $[OH^-]$? Write $K_w = [H^+][OH^-] = 1.0 \times 10^{-14}$

2. Strong monoprotic acids ionize ~100% to form H^+. **Alkali metal hydroxides dissociate 100% to form OH^-.** For [ions], write the REC steps.

3. *Quick rules*:

$$[HCl\ or\ HNO_3]_{as\ mixed} = [H^+]_{in\ soln.}$$

$$\mathbf{[NaOH\ or\ KOH]_{as\ mixed} = [OH^-]_{in\ soln.}}$$

4. See $[H_3O^+]$? Write $[H_3O^+] = [H^+]$

5. In acid solutions, use acid ionization rules to find $[H^+]$, then K_w to find $[OH^-]$.

6. **In *base* solutions, use base rules to find $[OH^-]$, then K_w to find $[H^+]$.**

TRY IT

Using the new rules for base solutions, try this problem.

Q. In a solution labeled 0.0030 M NaOH,

a. What three ions are present in the solution?

b. What is the concentration of each ion?

Answers:

a. NaOH dissociates to form Na^+ and OH^- ions. Water ionizes to form H^+ and OH^-.

The three ions are $\mathbf{Na^+}$, $\mathbf{OH^-}$, and $\mathbf{H^+}$.

b. For NaOH or KOH and [ions], write the REC steps for ~100% ionization.

Rxn. and **E**xtent: $1\ NaOH(aq) \rightarrow 1\ Na^+(aq) + 1\ OH^-(aq)$ (Goes ~100%.)

Conc.:   ~~0.0030 M~~   0 M   0.0030 M   0.0030 M

which means $[NaOH]_{mixed} = \mathbf{0.0030\ M} = [OH^-]_{in\ soln.} = [Na^+]_{in\ soln.}$
To find the $[H^+]$ in a base solution, first use the base rules to find $[OH^-]$ (done), then use K_w to find $[H^+]$.

$$K_w = \mathbf{[H^+][OH^-] = 1.0 \times 10^{-14}}$$

$$[H^+] = \frac{1.0 \times 10^{-14}}{[OH^-]} = \frac{1.0 \times 10^{-14}}{3.0 \times 10^{-3}} = 0.33 \times 10^{-11} = \mathbf{3.3 \times 10^{-12}\ M\ H^+}$$

K_w check:

$$[H^+] \times [OH^-] = 3.0 \times 3.3 \approx \mathbf{10};\ 10^{-3} \times 10^{-12} = 10^{-15},$$
$$\text{combined} \approx \mathbf{10 \times 10^{-15}}$$

PRACTICE

1. In a solution labeled 10^{-1} M NaOH,

    a. Write the balanced equation for the dissociation of this strong base.

    b. Which side is favored in the reaction: products or reactants?

    c. $[OH^-] = ?$ d. $[Na^+] = ?$

    e. $[H^+] = ?$ f. $[H_3O^+] = ?$

Solve the following in your notebook.

2. In a solution labeled 5.0×10^{-3} M KOH, find the

    a. $[OH^-]$ b. $[K^+]$ c. $[H^+]$ d. $[H_3O^+]$

3. If 5.00 mmol NaOH are dissolved in 0.250 L soln., find

    a. $[OH^-]$ b. $[H^+]$ c. $[H_3O^+]$

Lesson 20.6

Base 10 Logarithms

In science, many relationships between variables can be represented by mathematical equations that include logarithms.

Examples: An equation that we will use to in calculations involving radioactive decay uses *natural* logarithms: **ln**(*fraction remaining*) $= -kt$

To measure the acidity of solutions, we will use pH, which is based on *base 10* logs: $\text{pH} \equiv -\log[\text{H}^+]$

Before we begin pH problems, let's review the rules for base 10 logarithm calculations.

Numbers, Bases, and Exponents

Any positive value that is written in fixed-decimal notation (as a regular number) can be expressed as a number (a base) to a power.

 **TRY IT**

Q. Computer science often calculates in *base 2*.

Examples: $2^{10} = \mathbf{1,024}$ and (fill in the blank) $2^4 = \underline{\hspace{1cm}}$.

 Answer:

$2^4 = 2 \times 2 \times 2 \times 2 = \mathbf{16}$

However, values can be represented by any base to any power.

 **TRY IT**

Q. Using your calculator, convert this base and power to a fixed decimal number: $3.5^{2.7} =$

 Answer:

- A standard TI-type calculator might use 3.5 $\boxed{y^x}$ 2.7 $\boxed{=}$
- On a graphing calculator (*if* allowed), try 3.5 $\boxed{\wedge}$ 2.7 $\boxed{\text{enter}}$
- On a reverse Polish notation (RPN) calculator, try 3.5 $\boxed{\text{enter}}$ 2.7 $\boxed{y^x}$

Write or circle a key sequence that gives a result of **29.4431....**

Powers of 10

Numeric values can be expressed by a variety of methods: fixed decimal numbers, exponential notation, or a number to a power. In scientific calculations, we often express numeric values as 10 to a power, where the power can be either an integer or a number with decimals.

━━━━ ▶ **TRY IT**

Q. Answer these without using a calculator.

    a. 10^2 = the fixed decimal number 100 and 10^3 = the number _____.

    b. Without a calculator, *estimate* the value of $10^{2.5}$ = _____.

 Answers:

    a. 10^3 = **1000**

    b. $100 = 10^2 < \mathbf{10^{2.5}} < 10^3 = 1000$

        Halfway between 100 and 1000 is 550 . . . , but the answer is sure to be "between 100 and 1000."

Now, use a calculator to find an exact answer.

━━━━ ▶ **TRY IT**

Q. $10^{2.5}$ = what fixed decimal number? _____.

 Answer:

  • On a standard TI–type calculator, you might try 2.5 $\boxed{10^x}$ *and/or* 10 $\boxed{y^x}$ 2.5 $\boxed{=}$ *and/or* 2.5 $\boxed{2^{nd} \text{ or INV}}$ $\boxed{\log}$. Try all three.
  • On a graphing calculator, you might try 10 $\boxed{\wedge}$ 2.5 $\boxed{\text{enter}}$
  • On an RPN scientific calculator, try 2.5 $\boxed{\text{enter}}$ $\boxed{10^x}$

Circle or write down a sequence that makes sense to you and gives this result: **316**

On significant figures, in calculations involving 10 to a decimal power, the statistical justification for significant figures breaks down. We will add a systematic rule when we study acid–base pH. Until then, use these rules:

  • If a value is an integer power of 10, assume it is an exact number
  • If an exponent has a decimal, round answers based on that value to three significant figures

━━━━ ▶ **TRY IT**

Q. Use your calculator key sequence tested above to convert these powers of 10 to values in scientific notation.

    a. $10^{23.7798}$ =   b. $10^{-3.9}$ =

Answers:

    a. 6.02×10^{23} (Rounded to three *s. f.*)

        Note how your calculator *displayed* the exponent. You will need to translate the calculator display into scientific notation when writing answers.

b. 1.26×10^{-4}

For *part b*, to enter a *negative* number, usually a $\boxed{+/-}$ or $\boxed{(-)}$ key must be used.

- On a standard TI-type calculator, try **3.9** $\boxed{+/-}$ $\boxed{10^x}$
- On an RPN calculator, try **3.9** $\boxed{+/-}$ $\boxed{\text{enter}}$ $\boxed{10^x}$

Checking Powers of 10

One outcome of the estimation logic above is this rule that can be used to check values expressed as powers of 10:

> When a value expressed as 10 to a power is compared to the same value written in scientific notation, the *exponents* of each **10** must be within **±1** of each other.

TRY IT

Q. Using your calculator, check that rule on these problems.

a. $10^{6.7}$ = (in scientific notation): _____
(exponents ± 1?) _____

b. $10^{-9.7}$ = (in scientific notation):
(exponents ± 1?) _____

c. $10^{-13.2}$ = (in scientific notation):
(exponents ± 1?) _____

STOP

Answers:

a. $10^{6.7}$ = **5.01 × 10⁶** (exponents ± 1? Yes: 6.7 and 6. ✓)

b. $10^{-9.7}$ = **2.00 × 10⁻¹⁰** (exponents ± 1?) ✓

c. $10^{-13.2}$ = **6.31 × 10⁻¹⁴** (exponents ± 1?) ✓

Note how your calculator displays exponents. Many calculators answer in exponential notation but display the exponent *far* to the right—where you may miss it when looking for the significant digits. That's another reason to apply the estimation "quick check" to your answers.

PRACTICE   A

Convert the following values to scientific notation with three significant figures.

1. $10^{+16.5}$ =   (exponents ± 1?) _____

2. $10^{-16.5}$ =   (exponents ± 1?) _____

3. $10^{2.2}$ =   (exponents ± 1?) _____

4. $10^{-11.7}$ =   (exponents ± 1?) _____

5. $10^{-0.7}$ =   (exponents ± 1?) _____

Logarithm Defintions

In words, we will define a logarithm in two ways.

> A logarithm is simply an exponent.

> A logarithm answers this question: If a number is written as a base number to a power, what is the power?

A logarithm can be a power of any base.

Example: Because $2^4 = 16$, the *base 2 log* of 16 can be written $\log_2 16 = 4$

In science, *base 10* and *base e* are the bases for logarithms that are used most often. The symbol for a base **10** logarithm is simply **log**. If no base is specified, you should assume that *log* means a *base 10* log. The symbol for a *base e* log (a *natural* log) is **ln**.

Base 10 Logs

The **log** function on a calculator finds a *base 10* log.

> The *log* function answers this question: If a value is written as **10** to a power, what is the power?

 TRY IT

Using that rule, answer these with*out* using a calculator.

Q. Write the log of

a. 10^2 b. 1000   c. 0.001

STOP **Answers:**

a. The log of 10^2 is **2**   b. $\log 1000 = \log 10^3 = 3$

c. $\log 0.001 = \log 10^{-3} = -3$

The equation defining a base 10 log is **$\log 10^x = x$**. It must be memorized. It is also helpful to remember this example: **The *log* of 100 is 2**.

For fixed-decimal numbers *greater* than 1, log values will be positive numbers.

For positive numbers between *0 and 1*, log values will be negative, as in part c above.

One way to check that you are doing calculator operations properly is to do a *simple* calculation, first in your head or on paper, then using the calculator. The two answers should agree.

TRY IT

Let's try that method on some simple examples.

Q1. Without using a calculator, write the log of

a. 100   b. 10,000   c. 0.01

Q2. Using a calculator, find the log of

a. 100   b. 10,000   c. 0.01

STOP **Answers:**

Without a calculator,

a. The log of 100 = the log of 10^2 = **2**

b. The log of 10,000 = log 10^4 = **4**

c. The log of 0.01 = log 10^{-2} = **−2**

On a calculator, for part a:

- A standard TI-type calculator might use 100 [log]
- On an RPN calculator, try 100 [enter] [log]
- Some graphing calculators may not have a log button. You can learn a workaround (log x = ln x/2.303) *or* buy an inexpensive calculator with a **log** button.

Did the calculator answers agree with your mental arithmetic? They must.

Checking Calculated Logarithms

To check a conversion between a base 10 log and a number, apply this rule.

> The *log* of a number and the *exponent* of the number when it is written in *scientific* notation must agree within **±1**.

 TRY IT

Q. Use your calculator to answer these, then apply the rule above to check your answers. Round answers to three *s. f.*

a. Log(7.4 × 10^6) = _____
    (Are log and exponent ± 1?) _____

b. Log(7.4 × 10^{-6}) =
    (Are log and exponent ± 1?) _____

c. Log 2,000 =
    (Log and exponent ± 1?) _____

STOP **Answers:**

a. Log(7.4 × 10^6) = **6.87** (±1 ? ✓)

b. Log(7.4 × 10^{-6}) = **−5.13** (±1 ? ✓)

    Keys: 7.4 [E or EE] 6 [+/−] [log] *or*
    On RPN: 7.4 [E or EE] 6 [+/−] [enter] [log]

c. Log 2,000 = log(2 × 10^3) = **3.30** (±1 ? ✓)

Log Rules

1. A *logarithm* is simply an exponent: the power to which a base number is raised.

2. A logarithm answers this question: If a number is written as a base to a power, what is the power?

3. Calculator **log** buttons find the *power* of a value expressed as 10 to a power.

4. The equation defining a base 10 log is **log 10^x = x**; the *log* of 100 is 2.

5. *Checking log results*: when a number is written in scientific notation, its power of 10 must agree with its base 10 logarithm within ± 1.

Design flashcards or other memory devices and practice as needed so that you can apply these rules intuitively and fluently in calculations.

PRACTICE   B

Save a few problems for your next study session. Round answers to three *s. f.*

1. $10^{-5.4}$ = (in scientific notation): _____ (± 1?) _____

2. $10^{-11.5}$ = _____ (± 1?) _____

3. $10^{-0.5}$ = (fixed decimal number): _____

   = (scientific notation): _____ (± 1?) _____

4. Log(6.8 × 10^{12}) =

5. Log(6.8 × 10^{-12}) =

6. Log 4.6 =

7. Log 0.0020 =

Converting from Logs to Numbers

Knowing a log, we need to be able to write the fixed-decimal number. This is called **taking the antilog** or taking the **inverse log**, but it is easier to remember what this means (and what buttons to press) if you remember what a log is.

▬▬▬▶ **TRY IT**

Q. A log is _____.

Answer:
A log is an exponent.

Q. If the log of a number is 2, the number is _____.

Answer:
Since a log is an exponent of 10, if the log is 2, the number is 10^2 = 100

As an equation, the rule for converting a log value to its corresponding number is

$$10^{\log x} = x$$

Repeat to remember: "10 to the log x equals x."

 TRY IT

Without a calculator, apply the rules above to these.
   Q. If these are the logs of numbers, write the numbers in fixed decimal notation.

   a. 6   b. −2   c. 0

STOP **Answers:**

   a. If the log is 6, the number is $10^6 = $ **1,000,000**

   b. If the log is −2, the number is $10^{-2} = $ **0.01**

   c. If log $= 0$, number $= 10^0 = $ **1**   (Anything to the 0 power is 1.)

Knowing the log, to find the number, take the *antilog*: write the log as a power of 10, then convert that exponential term to a number (or to scientific notation).
   On a calculator,

   • Input the log, then take the antilog: press [INV] [LOG] or [2nd] [LOG]
   • Alternatively, input the log, then press [10^x]. A log is simply an exponent of 10
   • On some calculators, the steps are: input 10, [$x^\wedge y$], input the log value, [=]

   The first key sequence is logical. To go from number to a log, take the log; to go from log to number, go backward (take the antilog).
   The second and third sequences are logical. To find the number, make the log what it is, a power of 10.

 TRY IT

Circle above or write in the following blank a key sequence that works on your calculator for the following calculation.
   Q. If the log is 2, the number is 100 (or 1.0×10^2) _____.
   Now using the same key sequences, try these on your calculator. Write your answer in scientific notation.

   a. If log $x = 8.7$, $x = $ _____ (± 1 ? ____)

   b. Log A $= -10.7$, A $= $ _____ (± 1 ? ____)

   Note the same "is it reasonable?" quick check. The *log* and the *exponent* of the number in scientific notation must agree within ± 1.

STOP **Answers:**

   a. Log $x = 8.7$, $x = $ **5.01×10^8** (± 1 ? ✓)

   b. Log $x = -10.7$, $x = $ **2.00×10^{-11}** (± 1 ? ✓)

Add these to your memorized log rule list.

6. Knowing the log, to find the number, take the antilog. A log is a power of 10.

7. Knowing the log, apply this rule: $10^{\log x} = x$
   Recite and repeat to remember: "10 to the log x equals x."

8. On most calculators, to convert a log value to a number, use these steps:
   - Input the log value, then press (INV)(LOG) or (2nd)(LOG)
   - Alternatively, input the log, then press (10^x). *Or* input 10, (x^y), input the log, (=)

PRACTICE **C**

Round answers to three *s.f.*

1. Log x = 12.4, x =   (±1?   )

2. Log A = −5.9, A =   (±1?   )

3. Log D = −0.25, D =

4. Log x = 1.1, x =

5. $10^{-3.3}$ =

6. Log (2.0×10^{-9}) =

7. Log 0.50 =

Lesson 20.7   # The pH System

Fundamentals of pH

pH is a measure of the proton (H^+) concentration in an aqueous solution. The pH is a number between −2 and 16 that indicates the acidity or basicity (bay-SIS-ah-tee) of the solution.

THE pH SCALE

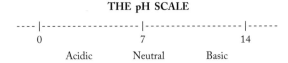

| 0 | 7 | 14 |
| Acidic | Neutral | Basic |

To simplify pH calculations, in these lessons we will use this simplified definition of pH:

> pH is the negative log of the hydrogen ion concentration: $pH \equiv -\log [H^+]$

If we rearrange this definition to solve for $[H^+]$, the result is $[H^+] \equiv 10^{-pH}$

It may help to remember pH as the "**p**ower of **H**": the number that is after the minus sign when $[H^+]$ is written as a negative power of 10.

To assist in calculations that involve pH, we will apply the following rule.

The pH Prompt

If you see **pH**, write $\mathbf{pH} \equiv -\mathbf{log}\,[\mathbf{H^+}]$ *and* $[\mathbf{H^+}] \equiv \mathbf{10^{-pH}}$

In addition, because our pH definitions are "shortcut" expressions of more complex equations, we will need this rule:

A pH value is not assigned a unit, but when a concentration is calculated based on a pH, the unit *mol/L* (M) must be added to the answer.

It is also important to know that at 25°C,

- Pure water has a pH of 7.00.
- In a *pH-neutral* solution at 25°C, $[H^+] = [OH^-] = 1.0 \times 10^{-7}\,M$; and pH = 7.00.
- In *acidic* solutions, the pH is *less* than **7**. In *basic* solutions, the pH is *greater* than **7**.
- *Lower* pH means a *higher acidity*. Higher pH means a higher basicity.
- The further from 7 is the pH, the higher is the acidity or basicity of the solution.
- Changing a fixed-decimal number by a factor of *10* changes its base 10 log by 1.
- Since pH is a *negative* log, *increasing* $[H^+]$ by a factor of 10 *lowers* the pH by one.

The first three bulleted rules above are especially important to be able to recall from memory.

Why Use the pH Scale?

The pH scale allows us to report acidity without using exponents. For non-chemists, "adjust the aquarium pH to 7.5" is easier to understand than "adjust the $[H^+]$ to 3.2×10^{-8} mol/L," though both statements have the same meaning. For chemists, pH is a quick way to convey solution acidity.

Integer pH

In aqueous solutions, if the $[H^+]$ *or* the $[OH^-]$ *or* the pH is known, the other two values can be determined.

When pH values are integers, pH problems do not require a calculator.

TRY IT

Apply the rules of the pH prompt to these.

Q. In an aqueous solution, if $[H^+] = 10^{-1}$ M, find the pH.

STOP **Answer:**

See pH? Write the pH prompt: $pH \equiv -\log [H^+]$ *and* $[H^+] \equiv 10^{-pH}$

$$[H^+] \equiv 10^{-pH} = 10^{-1} \text{ M, so } \mathbf{pH = 1}$$

Q. If $[H^+] = 0.001$ M, what is the pH?   Is the solution acidic or basic?

STOP **Answer:**

$$[H^+] \equiv 10^{-pH} = 0.001 \text{ M} = 10^{-3} \text{ M, so } \mathbf{pH = 3}$$

Because the pH is less than 7, the solution is *acidic*.

Q. If the pH = 9, what is the $[OH^-]$?   Is the solution acidic or basic?

STOP **Answer:**

Because $[H^+] = 10^{-pH} = 10^{-9}$ M, $[OH^-] = \mathbf{10^{-5}}$ **M** based on K_w. Note that the $[OH^-]$ is larger than the $[H^+]$, and that is the fundamental definition of a **basic** aqueous solution. The *quick* rule is this: Any pH greater than 7.00 is basic.

PRACTICE   **A**

Assume the following data apply to aqueous solutions of acids and bases. Write answers using mental or "pencil-and-paper" arithmetic rather than a calculator. Circle the correct answer in each part c.

1. If $[H^+] = 10^{-4}$ mol/L,

    a. $[OH^-] =$ b. pH =   c. Is the solution acidic or basic?

2. If pH = 8,

    a. $[H^+] =$ b. $[OH^-] =$ c. Is the solution acidic or basic?

3. If $[OH^-] = 10^{-3}$ molar,

    a. $[H^+] =$ b. pH =   c. Is the solution acidic or basic?

4. If $[OH^-] = 1.0 \times 10^{-14}$ M,

    a. $[H^+] =$ b. pH =   c. Is the solution acidic or basic?

5. If $[H^+] = 10$ M,

    a. $[OH^-] =$ b. pH =   c. Is the solution acidic or basic?

From [H⁺] to Decimal pH

If the pH *or* the exponent of the hydrogen ion concentration is *not* an integer, you can convert between $[H^+]$, $[OH^-]$, and pH using a calculator.

━━━━━▶ **TRY IT**

Q. For an aqueous solution, if $[H^+] = 5.0 \times 10^{-4}$ M, what is the pH?

Answer:

WANTED: pH   pH prompt: pH $\equiv -\log[H^+]$ *and* $[H^+] = 10^{-pH}$
Use the equation that solves *for* pH *from* **[H⁺]**:

$$pH \equiv -\log[H^+] = -\log(5.0 \times 10^{-4}) =$$

Solve using a calculator, rounding your answer to two *decimal* places.

$$pH \equiv -\log[H^+] = -\log(5.0 \times 10^{-4}) = 3.30$$

It will simplify calculator use to find the log without the minus sign in front, then change the sign manually. For any solution with a $[H^+]$ *less* than 1.0 M, the pH must be *positive*. If the $[H^+]$ is greater than 1.0 M, which can occur in relatively concentrated solutions of strong acids, the pH will be a negative number.

Checking pH Calculations

Does the answer for the last *Try It* above make sense?
For $[H^+]$:

$$1.0 \times 10^{-4} < 5.0 \times 10^{-4} < 10.0 \times 10^{-4} \equiv 1.0 \times 10^{-3}$$
$$pH = 4 \qquad pH = ? \qquad\qquad\qquad pH = 3$$

Based on the above, the pH should be between 4 and 3, which **3.30** is.
The estimation logic above can be used to *check* pH calculations by this rule.

> **The pH Check**
>
> The pH rounded *up* to the next whole number must equal the number after the minus sign of the exponential for the $[H^+]$ written in scientific notation.

For the above problem, pH = **3.30** rounds *up* to **4**. The $[H^+]$ exponential term in scientific notation should therefore be 10^{-4}, and it is: 5.0×10^{-4}.
The check rule can also be used to do a quick estimate of pH from $[H^+]$.

━━━━━▶ **TRY IT**

Q. If $[H^+] = 3.0 \times 10^{-10}$ M, estimate the pH.

Answer:
The pH must be a decimal number that rounds up to 10, which must be **9.xx**
The precise answer is pH $\equiv -\log[H^+] = -\log(3.0 \times 10^{-10}) =$ **9.52**
Check!

Rounding pH Calculations

Mathematically, the statistical basis for using significant figures to convey uncertainty does not directly apply to logarithmic functions. To convey approximate uncertainty, in these lessons we will apply the following conventions to pH calculations.

- When $[H^+]$ is written as simply a power of 10, write the pH as an integer.
     Example: $[H^+] = 10^{-4}\,M, pH = 4$
- When $[H^+]$ is written in scientific notation, round the pH so that the number of digits in the *significand* equals the number of digits *after the decimal* in the pH.
     Examples:

$$[H^+] = \mathbf{5} \times 10^{-6}\,M, pH = 5.\mathbf{3}$$

$$[H^+] = \mathbf{5.1} \times 10^{-6}\,M, pH = 5.\mathbf{29}$$

- The K_w value on which our pH rules are based assumes a solution temperature of 25°C, but small variations in temperature during experiments can cause significant changes in K values. Because of this inherent uncertainty, we will round each calculated pH to have at *most* two digits past the decimal.

PRACTICE   **B**

Complete the odd-numbered problems. Save the evens for your next practice session. Check your answers as you go.

1. Using the pH-check method above, *estimate* the pH for these questions below:

    a. Problem 2: pH ≈   b. Problem 3: pH ≈   c. Problem 5: pH ≈

Use your calculator and solve in your notebook for these solutions.

2. Find pH when $[H^+] = 2.1 \times 10^{-3}$ M.

3. If $[H_3O^+] = 8.2 \times 10^{-11}$ M, what is the pH?

4. If $[OH^-] = 2.0 \times 10^{-4}$ M, what is the pH?

5. In a 0.040 M HCl solution, solve for the pH.

6. If 0.012 g of NaOH is dissolved to make 100. mL of solution, what is the solution pH?

From Decimal pH to [H⁺]

Let's try a problem where a known pH value is a not an integer and the $[H^+]$ is WANTED.

Example: If the pH of a solution is 9.70, what is the $[H^+]$?
The pH prompt: $pH \equiv -\log[H^+]$ and $[H^+] = 10^{-pH}$
The equation that *finds* $[H^+]$ *from* pH is $[H^+] \equiv 10^{-\mathbf{pH}}$
When a concentration is calculated from a pH, the unit M must be added to the answer.

$[H^+]$ therefore equals $\mathbf{10^{-9.70}\,M}$. That's mathematically correct, but in chemistry it is preferred to express large or small values in scientific notation.

▶ **TRY IT**

Q. Convert $10^{-9.70}$ M to the $[H^+]$ in scientific notation using your calculator.

(STOP) **Answer:**

$$[H^+] = 10^{-9.70} \text{ M} = \mathbf{2.0 \times 10^{-10}} \textbf{ M}$$

Does this answer make sense? $[H^+] = \mathbf{10^{-9.70}}$ **M** is *close* to $[H^+] = \mathbf{10^{-10}}$ **M**. Since pH = 9.70 is a bit more acidic than pH = 10, the $[H^+]$ should be a bit higher than 10^{-10} M, and it is: $\mathbf{2.0 \times 10^{-10}}$ M.

The pH-check rule is a quick way to apply this logic.

▶ **TRY IT**

Q. Apply the *pH check* to the answer above.

(STOP) **Answer:**

The pH = 9.70 rounds up to the integer 10; the $[H^+]$ exponential term in scientific notation should be 10^{-10}. The answer is 2.0×10^{-10} M. Check!

PRACTICE C

Solve these in your problem notebook. Save a few for later review.

1. If pH = 5.5,

   a. $[H^+] =$ b. Does the pH check?

2. If pH = 8.20,

   a. $[H^+] =$ b. Does the pH check?   c. $[OH^-] =$

3. In an HCl solution of pH = 3.60, find

   a. $[H^+]$ b. $[OH^-]$ c. $[Cl^-]$ d. $[HCl]_{\text{as mixed}}$

4. If pH = 1.7, $[H^+] =$ 5. If pH = 7.22, $[H^+] =$

6. If pH = 12.5, $[H^+] =$ 7. If pH = −0.50, $[H^+] =$

8. Which solution in problems 3–7 is the most acidic?

SUMMARY

If you have not already done so, you may want to design flashcards for these rules.

1. See $[H^+]$ *and* $[OH^-]$? Write $K_w = [H^+][OH^-] = 1.0 \times 10^{-14}$

2. In water,
   - Strong monoprotic acids ionize ~100% and form H^+
   - Alkali metal hydroxides dissociate 100% and form OH^-
   - For [ions], write the REC steps or use the quick rules

(continued)

3. *Quick rules*:

$$[\text{HCl } or \text{ HNO}_3]_{\text{as mixed}} = [\text{H}^+]_{\text{in soln.}}$$
$$[\text{NaOH } or \text{ KOH}]_{\text{as mixed}} = [\text{OH}^-]_{\text{in soln.}}$$

4. See $[\text{H}_3\text{O}^+]$? Write $[\text{H}_3\text{O}^+] = [\text{H}^+]$

5. In *acid* solutions, use acid ionization rules to find $[\text{H}^+]$, then K_w to find $[\text{OH}^-]$.

6. In *base* solutions, use base rules to find $[\text{OH}^-]$, then K_w to find $[\text{H}^+]$.

7. The pH prompt: See pH? Write $\text{pH} \equiv -\log [\text{H}^+]$ *and* $[\text{H}^+] \equiv 10^{-\text{pH}}$

8. pH values are not assigned units. When a concentration is calculated based on pH, add *mol/L* (M) to the answer.

9. Using the pH scale, for aqueous solutions at 25°C,
   • Pure water has a pH of 7.00.
   • In a *neutral* solution, $[\text{H}^+] = [\text{OH}^-] = 1.0 \times 10^{-7}$ M and pH = 7.00.
   • In *acidic* solutions, pH is *less* than **7**. In *basic* solutions, pH is greater than 7.
   • A lower pH means a higher acidity. A higher pH means a higher basicity.
   • The further from 7 is pH, the higher is the acidity or basicity of the solution.

10. *pH check*: The pH value rounded up to the next whole number must equal the number after the minus sign of the exponent of the $[\text{H}^+]$ written in scientific notation.

ANSWERS

Lesson 20.1

Practice   1a. 10^{-6} 1b. 10^{-15} 1c. 10^{-2} 1d. 10^{-5} 1e. 10^{-11}

2a. 10^{-17} 2b. 10^{-9} 2c. 10^{-3} 3a. 3.24×10^{14} 3b. 5.0×10^{-13}

4a. 6.0×10^{-10} 4b. 5.0×10^{-19} 4c. 3.3×10^{-11} 4d. 5.0×10^{-7} 4e. 4.0×10^{-13}

5a. 2.5×10^{-13} 5b. 3.3×10^{-12} 5c. 5.0×10^{-14}

5d. $x = \dfrac{1.0 \times 10^{-14}}{1.25 \times 10^{-2}} = 0.80 \times 10^{-12} = \mathbf{8.0 \times 10^{-13}}$

6a. $\dfrac{1.0 \times 10^{-14}}{3.25 \times 10^{-5}} = 0.31 \times 10^{-9} = \mathbf{3.1 \times 10^{-10}}$

6b. $\dfrac{1.0 \times 10^{-14}}{8.8 \times 10^{-4}} = 0.11 \times 10^{-10} = \mathbf{1.1 \times 10^{-11}}$

6c. $\dfrac{1.0 \times 10^{-14}}{2.4} = \mathbf{4.2 \times 10^{-15}}$

6d. $\mathbf{2.3 \times 10^{-11}}$ 6e. $\mathbf{1.5 \times 10^{-3}}$

6f. $x = \dfrac{1.0 \times 10^{-14}}{1.25 \times 10^{-7}} = 0.80 \times 10^{-7} = \mathbf{8.0 \times 10^{-8}}$

Lesson 20.2

Practice 1a. $[H^+][OH^-] = 1.0 \times 10^{-14}$; $[H^+](10^{-11}) = 10^{-14}$; $\mathbf{[H^+] = 10^{-3}\,M}$

1b. $0.010\,M = [OH^-] = 1.0 \times 10^{-2}\,M$

$[H^+][OH^-] = 1.0 \times 10^{-14}$; $[H^+](1.0 \times 10^{-2}) = 1.0 \times 10^{-14}$; $\mathbf{[H^+] = 1.0 \times 10^{-12}\,M}$

1c. $5.0 \times 10^{-11}\,M = [OH^-]$; WANT $[H^+]$; the equation relating the two symbols is

$[H^+][OH^-] = 1.0 \times 10^{-14}$. Solve for the WANTED symbol.

$$\mathbf{[H^+]} = \frac{1.0 \times 10^{-14}}{[OH^-]} = \frac{1.0 \times 10^{-14}}{5.0 \times 10^{-11}} = 0.20 \times 10^{-3} = \mathbf{2.0 \times 10^{-4}\,M\,H^+}$$

1d. $0.036\,M = [OH^-]$; WANT $[H^+]$; the equation using those symbols is $[H^+][OH^-] = 1.0 \times 10^{-14}$

$$\mathbf{[H^+]} = \frac{1.0 \times 10^{-14}}{[OH^-]} = \frac{1.0 \times 10^{-14}}{3.6 \times 10^{-2}} = 0.28 \times 10^{-12} = \mathbf{2.8 \times 10^{-13}\,M\,H^+}$$

2a. $10^{-9}\,mol/L = [H^+]$, WANT $[OH^-]$. The equation relating those symbols is

$[H^+][OH^-] = 1.0 \times 10^{-14}$; $(10^{-9})[OH^-] = 10^{-14}$; $\mathbf{[OH^-] = 10^{-5}\,M}$

2b. $1.0\,M = [H^+]$, WANT $[OH^-]$, that calls the K_w prompt.

Write $[H^+][OH^-] = 1.0 \times 10^{-14}$; $(1.0)[OH^-] = 10^{-14}$; $\mathbf{[OH^-] = 1.0 \times 10^{-14}\,M}$

2c. $3.0 \times 10^{-5}\,M = [H^+]$, want $[OH^-]$, write $[H^+][OH^-] = 1.0 \times 10^{-14}$

$$\mathbf{[OH^-]} = \frac{1.0 \times 10^{-14}}{[H^+]} = \frac{1.0 \times 10^{-14}}{3.0 \times 10^{-5}} = 0.33 \times 10^{-9} = \mathbf{3.3 \times 10^{-10}\,M\,OH^-}$$

2d. $1.25\,M = [H^+]$; $\mathbf{[OH^-]} = \dfrac{1.0 \times 10^{-14}}{[H^+]} = \dfrac{1.0 \times 10^{-14}}{1.25} = 0.80 \times 10^{-14} = \mathbf{8.0 \times 10^{-15}\,M\,OH^-}$

3a. WANTED: $? = [OH^-] = \dfrac{mol\ OH^-}{L\ soln.}$

DATA: $20.0\,mmol\ KOH = 400.\,mL\ soln.$ (Two measures of same solution.)

(If you WANT a ratio, all of the DATA will be in equalities.)

SOLVE: $\dfrac{?\ mol\ OH^-}{L\ soln.} = \dfrac{20.0\ \cancel{mmol}\ OH^-}{400.\ \cancel{mL}\ soln.} = \mathbf{\dfrac{0.0500\ mol\ OH^-}{L\ soln.}}$

Because a prefix is an abbreviation for an exponential, if the same prefix is on the top and bottom, it can cancel, just as two equal exponentials can cancel.

3b. Knowing $[OH^-]$, $[H^+]$ can be solved with the K_w prompt.

$[H^+][OH^-] = 1.0 \times 10^{-14}$

$$\mathbf{[H^+]} = \frac{1.0 \times 10^{-14}}{[OH^-]} = \frac{1.0 \times 10^{-14}}{5.00 \times 10^{-2}} = 0.20 \times 10^{-12} = \mathbf{2.0 \times 10^{-13}\,M\,H^+}$$

Lesson 20.3

Practice A

1a, 1b. For problems involving HCl or HNO_3 and [ions], write the REC steps.

Rxn. and Extent: $\mathbf{1}\,HNO_3 \rightarrow \mathbf{1}\,H^+ + \mathbf{1}\,NO_3^-$ (Goes ~100%.)

$\qquad\qquad\qquad\qquad\quad \wedge \qquad\quad \wedge \qquad\quad \wedge$

Conc.: $\quad\cancel{0.15\,M}\,0\,M \quad 0.15\,M \quad 0.15\,M$

2a. WANTED: $?\ \dfrac{mol\ HCl}{L\ soln.} = [HCl]_{as\ mixed}$

DATA: $7.30\,g\ HCl = 250\,mL\ soln.$ (Equivalent: two measures; same soln.)

$\qquad\quad 36.5\,g\ HCl = 1\,mol\ HCl$ (Grams prompt.)

You want moles over liters. The data include grams, moles, and milliliters. Use conversions.

SOLVE: $? \text{ mol HCl} = \dfrac{7.30 \text{ g HCl}}{\text{L soln.}} \cdot \dfrac{1 \text{ mol HCl}}{250 \text{ mL soln.}} \cdot \dfrac{1 \text{ mL}}{36.5 \text{ g HCl}} \cdot \dfrac{1 \text{ mL}}{10^{-3}\text{L}} = \mathbf{0.80\,M\ HCl} = [\text{HCl}]_{\text{as mixed}}$

2b, 2c. For problems involving HCl or HNO_3 and [ions], write the REC steps.

Rxn. and Extent: $1 \text{ HCl} \rightarrow 1 \text{ H}^+ + 1 \text{ Cl}^-$ (Goes ~100%.)

$\wedge$ $\wedge$ $\wedge$

Conc.: ~~0.80 M~~ 0 M 0.80 M 0.80 M

Practice B 1. *Quick rule*: In HCl *or* HNO_3, for $[\text{H}^+]$, use $[\text{HCl } or \text{ HNO}_3]_{\text{mixed}} = [\text{H}^+]_{\text{in soln.}}$
   In 0.25 M HNO_3, a. $[\text{H}^+] = \mathbf{0.25\,M}$ b. $[\text{H}_3\text{O}^+] = [\text{H}^+] = \mathbf{0.25\,M}$

2. In 2.0 M HCl, a. $[\text{H}_3\text{O}^+] = [\text{H}^+] = \mathbf{2.0\,M}$ b. $[\text{H}^+] = \mathbf{2.0\,M}$

3. Cl^-. The conjugate base has one fewer H atoms and one fewer positive charges than the acid from which it is formed.

Lesson 20.4

Practice 1. See Lessons 20.3 and 20.4.

2a, 2b. For problems involving HCl or HNO_3 and [ions], solve by the REC steps.

Rxn. and Extent: $1 \text{ HNO}_3 \rightarrow 1 \text{ H}^+ + 1 \text{ NO}_3^-$ (Goes ~100%.)

$\wedge$ $\wedge$ $\wedge$

Conc.: ~~10^{-3} M~~ 0 M 10^{-3} M 10^{-3} M

2c. $[\text{H}_3\text{O}^+] = [\text{H}^+] = \mathbf{10^{-3}\,M}$

2d. To find $[\text{OH}^-]$ in an acid solution, use acid rules to find $[\text{H}^+]$, then K_w to find $[\text{OH}^-]$.
   Because $[\text{H}^+] = 10^{-3}$ M and $[\text{H}^+][\text{OH}^-] = 1.0 \times 10^{-14}$; $[\text{OH}^-] = \mathbf{10^{-11}\,M}$

3a, 3b. Equation: $1 \text{ HCl}(g) \rightarrow 1 \text{ H}^+(aq) + 1 \text{ Cl}^-(aq)$ (Goes 100%.) **Products favored**.

3c. 1 mol HCl used up = 1 mol H^+ formed.

3d. Because all moles are found in the same liters of solution, the mole and mol/L ratios are the same: **1 to 1**.
   1 mol/L HCl used up = 1 mol/L H^+ formed, which can be written $[\text{HCl}]_{\text{used up}} = [\text{H}^+]_{\text{formed}}$

3e, 3f. $[\text{HCl}]_{\text{as mixed}} = \mathbf{0.020\,M} = [\text{H}^+]_{\text{in soln.}} = [\text{Cl}^-]_{\text{in soln.}}$

3g. $[\text{OH}^-] = ?$ In an acid solution, use acid rules to find $[\text{H}^+]$ (done above), then K_w to find $[\text{OH}^-]$.
   Since $[\text{H}^+] = 0.020$ M $= 2.0 \times 10^{-2}$ M and $[\text{H}^+][\text{OH}^-] = 1.0 \times 10^{-14}$;

$$[\text{OH}^-] = \frac{1.0 \times 10^{-14}}{[\text{H}^+]} = \frac{1.0 \times 10^{-14}}{2.0 \times 10^{-2}} = 0.50 \times 10^{-12} = \mathbf{5.0 \times 10^{-13}\,M}$$

K_w check: $[\text{H}^+] \times [\text{OH}^-]$ must estimate to 10.0×10^{-15} or 1.0×10^{-14}

$[\text{H}^+] \times [\text{OH}^-] = 2 \times 5 = \mathbf{10}; 10^{-2} \times 10^{-13} = 10^{-15}$, combined $= \mathbf{10 \times 10^{-15}}$. Check.

3h. $[\text{H}_3\text{O}^+] = [\text{H}^+] = \mathbf{0.020\,M}\ or\ \mathbf{2.0 \times 10^{-2}\,M}$

Lesson 20.5

Practice 1a. $1 \text{ NaOH} \rightarrow 1 \text{ Na}^+ + 1 \text{ OH}^-$

1b. Strong bases dissociate 100%, this ionization favors the **products**.

1c, 1d. **R and E:** $1 \text{ NaOH} \rightarrow 1 \text{ Na}^+ + 1 \text{ OH}^-$ (Goes ~100%.)

$\wedge$ $\wedge$ $\wedge$

C: ~~10^{-1}~~ 0 M 10^{-1} M 10^{-1} M

The bottom REC row means that $[\text{NaOH}]_{\text{as mixed}} = \mathbf{10^{-1}\,M} = [\text{OH}^-]_{\text{in soln.}} = [\text{Na}^+]_{\text{in soln.}}$

1e, 1f. To find $[\text{H}^+]$ in a base solution, first find $[\text{OH}^-]$ (above), then $[\text{H}^+]$ using

$$K_w = [\text{H}^+][\text{OH}^-] = 1.0 \times 10^{-14}$$

$$[\text{H}^+] = \frac{1.0 \times 10^{-14}}{[\text{OH}^-]} = \frac{1.0 \times 10^{-14}}{10^{-1}} = \mathbf{10^{-13}\,M} = [\text{H}^+] = [\text{H}_3\text{O}^+]$$

2a, 2b. For NaOH or KOH, to find [ions], write the REC steps for ~100% ionization.

R and **E**: 1 KOH $\rightarrow$ 1 K$^+$ + 1 OH$^-$ (Goes ~100%.)

C: $\overset{\wedge}{\cancel{5.0 \times 10^{-3}}}$ 0 M $\overset{\wedge}{5.0 \times 10^{-3}}$ M $\overset{\wedge}{5.0 \times 10^{-3}}$ M

2c. [H$^+$] = ? [OH$^-$] is known. The relationship between those ions is [H$^+$][OH$^-$] = 1.0×10^{-14}

$$\mathbf{[H^+]} = \frac{1.0 \times 10^{-14}}{[OH^-]} = \frac{1.0 \times 10^{-14}}{5.0 \times 10^{-3}} = 0.20 \times 10^{-11} = \mathbf{2.0 \times 10^{-12}\,M\,H^+}$$

2d. [H$_3$O$^+$]? = [H$^+$] = **2.0 × 10^{-12} M**

3a. WANTED: ? [OH$^-$] = $\dfrac{mol\ OH^-}{L\ soln.}$ First find [NaOH], then apply [NaOH] = [OH$^-$].

DATA: 5.00 mmol NaOH = 0.250 L soln.   (Two measures of same solution.)

SOLVE: $\dfrac{?\ mol\ NaOH}{L\ soln.} = \dfrac{5.00\ mmol\ NaOH}{0.250\ L\ soln.} \cdot \dfrac{10^{-3}\ mol}{1\ mmol} = 2.00 \times 10^{-2}\dfrac{mol}{L} = [NaOH] = [OH^-]$

3b, 3c. $\mathbf{[H^+]} = \dfrac{1.0 \times 10^{-14}}{[OH^-]} = \dfrac{1.0 \times 10^{-14}}{2.00 \times 10^{-2}} = 0.50 \times 10^{-12} = \mathbf{5.0 \times 10^{-13}\,M} = [H^+] = [H_3O^+]$

Note that each of the *basic* solutions above contain both OH$^-$ and H$^+$, but in solutions of bases, [OH$^-$] is always higher than [H$^+$].

Lesson 20.6

Practice A 1. $10^{+16.5}$ = **3.16 × 10^{16}** (exponents agree ± ?) ✓ 2. $10^{-16.5}$ = **3.16 × 10^{-17}** (exponents ± 1?) ✓

3. $10^{2.2}$ = **1.58 × 10^2** (exponents ± 1?) ✓ 4. $10^{-11.7}$ = **2.00 × 10^{-12}**

5. $10^{-0.7}$ = **2.00 × 10^{-1}**

Practice B 1. $10^{-5.4}$ = **3.98 × 10^{-6}** (± 1?) ✓ 2. $10^{-11.5}$ = **3.16 × 10^{-12}** (± 1?) ✓

3. $10^{-0.5}$ = (number): **0.316** (scientific notation): **3.16 × 10^{-1}** (± 1?) ✓

4. Log (6.8 × 10^{12}) = **12.8** 5. Log (6.8 × 10^{-12}) = **−11.2**

6. Log 4.6 = **0.663** 7. Log 0.0020 = **−2.70**

Practice C 1. Log x = 12.4, x = **2.51 × 10^{12}** 2. Log x = −5.9, x = **1.26 × 10^{-6}**

3. Log x = −0.25, x = **0.562 = 5.62 × 10^{-1}** 4. Log x = 1.1, x = **12.6 = 1.26 × 10^1**

5. $10^{-3.3}$ = **5.01 × 10^{-4}** 6. Log (2.0 × 10^{-9}) = **−8.70**

7. Log 0.50 = **−0.301**

Lesson 20.7

Practice A 1. If [H$^+$] = 10^{-4} M,   1a. [OH$^-$] = **10^{-10} M**   1b. pH = **4**   1c. **Acidic**

2. If pH = 8,   2a. [H$^+$] = **10^{-8} M**   2b. [OH$^-$] = **10^{-6} M**   2c. **Basic**

3. If [OH$^-$] = 10^{-3} M,   3a. [H$^+$] = **10^{-11} M**   3b. pH = **11**   3c. **Basic**

4. If [OH$^-$] = 1.0 × 10^{-14} M,   4a. [H$^+$] = **1.0 M**

4b. pH = ?, [H$^+$] = 1.0 = 10^{-pH} = 10^0 = 10^{-0}; **pH = 0**

4c. **Acidic.** pH = 0, which is *less* than 7; [H$^+$] is higher than [OH$^-$].

5. If [H$^+$] = 10 M, 5a. [OH$^-$] = **10^{-15} M**   5b. [H$^+$] = 10^1 = 10^{-pH} = 10$^{-(-1)}$; **pH = −1**

pH will be lower than zero when [H$^+$] is greater than 1.0 M   5c. **Highly acidic**

Practice B 1. Using the quick-check method: 1a. Problem 2: **pH = 2.xx**   (Rounds up to 3.)

1b. Problem 3: **pH = 10.xx**   (Rounds up to 11.)

1c. Problem 5: 0.040 M HCl = 4 × 10^{-2} M H$^+$; **pH = 1.xx**

2. See pH? Write pH ≡ −log [H$^+$] *and* [H$^+$] ≡ 10^{-pH}

Since pH is WANTED, use pH ≡ −log [H$^+$]

pH = −log (2.1 × 10^{-3}) = **2.68**   Estimate was 2.xx in answer 1a. Check.

3. If $[H_3O^+] = 8.2 \times 10^{-11}$, what is the pH?

$[H_3O^+] = [H^+]$

See pH? Write $pH \equiv -\log [H^+]$ *and* $[H^+] \equiv 10^{-pH}$

If pH is WANTED, use $pH \equiv -\log [H^+] = -\log (8.2 \times 10^{-11}) = \textbf{10.09}$ Check.

4. See pH? Write $pH \equiv -\log [H^+]$ *and* $[H^+] \equiv 10^{-pH}$

WANT pH, need $[H^+]$ to find pH, but are given $[OH^-]$.

See H^+ and OH^-? Write $K_w = [H^+][OH^-] = 1.0 \times 10^{-14}$. Use K_w to find $[H^+]$:

$$[\textbf{H}^+] = \frac{1.0 \times 10^{-14}}{[OH^-]} = \frac{1.0 \times 10^{-14}}{2.0 \times 10^{-4}} = 0.50 \times 10^{-10} = \textbf{5.0} \times \textbf{10}^{-11}\,\textbf{M}\,\textbf{H}^+$$

Then $pH \equiv -\log [H^+] = -\log (5.0 \times 10^{-11}) = \textbf{10.30}$ (Rounds up to 11. Check.)

5. See pH? Write $pH \equiv -\log [H^+]$ *and* $[H^+] \equiv 10^{-pH}$

To find pH, need $[H^+]$ first. For quick $[H^+]$: $[HCl] = \textbf{0.040 M} = [H^+]_{in\,soln.}$

$pH \equiv -\log [H^+] = -\log (0.040) = -\log (4.0 \times 10^{-2}) = \textbf{1.40}$ (Rounded up = 2. Check.)

6. WANTED: pH. $pH \equiv -\log [H^+]$ *and* $[H^+] \equiv 10^{-pH}$

To find $[H^+]$, you first need to know $[NaOH]$.

WANTED: $[NaOH]_{as\,mixed} = ? \dfrac{mol\,NaOH}{L\,soln.}$

DATA: $0.012\,g\,NaOH = 100.\,mL\,soln.$ (Equivalent: two measures; same soln.)

$40.0\,g\,NaOH = 1\,mol\,NaOH$ (The grams prompt.)

SOLVE: $\dfrac{?\,mol\,NaOH}{L\,soln.} = \dfrac{0.012\,g\,NaOH}{100.\,mL\,soln.} \cdot \dfrac{1\,mol\,NaOH}{40.0\,g\,NaOH} \cdot \dfrac{1\,mL}{10^{-3}\,L} = \textbf{0.0030 M} = [\textbf{NaOH}]$

$\textbf{0.0030 M} = [\textbf{NaOH}] = [\textbf{OH}^-]$

To find $[H^+]$: $[H^+] = \dfrac{1.0 \times 10^{-14}}{[OH^-]} = \dfrac{1.0 \times 10^{-14}}{3.0 \times 10^{-3}} = 0.33 \times 10^{-11} = \textbf{3.3} \times \textbf{10}^{-12}\,\textbf{M}\,\textbf{H}^+$

$pH = -\log (3.3 \times 10^{-12}) = \textbf{11.48}$ (Base solution, basic pH; 11.48 rounded up = 12. Check.)

Practice C 1a. If pH = 5.5 See pH? Write $pH \equiv -\log [H^+]$ *and* $[H^+] \equiv 10^{-pH}$

The equation that finds $[H^+]$ from pH is: $[\textbf{H}^+] = 10^{-pH} = 10^{-5.5} = \textbf{3} \times \textbf{10}^{-6}\,\textbf{M}$

1b. pH = 5.5 rounded up is 6; $[H^+]$ estimated $= ? \times 10^{-6}$ M. Check.

2a. See pH? Write $pH \equiv -\log [H^+]$ *and* $[H^+] \equiv 10^{-pH}$

To find $[H^+]$ from pH, use $[\textbf{H}^+] = 10^{-pH} = 10^{-8.20} = \textbf{6.3} \times \textbf{10}^{-9}\,\textbf{M}\,\textbf{H}^+$

2b. pH = 8.20 rounds up to **9**, $[H^+]$ should $= ? \times 10^{-9}$ M. Check.

2c. WANT $[OH^-]$. Know $[H^+]$ and pH. K_w relates $[H^+]$ and $[OH^-]$: $[H^+][OH^-] = 1.0 \times 10^{-14}$

$? = [\textbf{OH}^-] = \dfrac{1.0 \times 10^{-14}}{[H^+]} = \dfrac{1.0 \times 10^{-14}}{6.3 \times 10^{-9}} = 0.16 \times 10^{-5} = \textbf{1.6} \times \textbf{10}^{-6}\,\textbf{M}\,\textbf{OH}^-$

3a. See pH? Write $pH \equiv -\log [H^+]$ *and* $[H^+] \equiv 10^{-pH}$

$[\textbf{H}^+] = ? = 10^{-pH} = 10^{-3.60} = \textbf{2.5} \times \textbf{10}^{-4}\,\textbf{M}$

3b. $[OH^-] = ?$ Knowing $[H^+]$, to find $[OH^-]$, use K_w.

$? = [\textbf{OH}^-] = \dfrac{1.0 \times 10^{-14}}{[H^+]} = \dfrac{1.0 \times 10^{-14}}{2.5 \times 10^{-4}} = 0.40 \times 10^{-10} = \textbf{4.0} \times \textbf{10}^{-11}\,\textbf{M}\,\textbf{OH}^-$

3c, 3d. $[HCl]_{mixed}$ from part a $= \textbf{2.5} \times \textbf{10}^{-4}\,\textbf{M} = [\textbf{Cl}^-] = [\textbf{HCl}]$

4. If pH = 1.7, $[\textbf{H}^+] \equiv 10^{-pH} = 10^{-1.7} = \textbf{2} \times \textbf{10}^{-2}\,\textbf{M}$

5. If pH = 7.22, $[\textbf{H}^+] \equiv 10^{-pH} = 10^{-7.22} = \textbf{6.0} \times \textbf{10}^{-8}\,\textbf{M}$

6. pH = 12.5, $[\textbf{H}^+] \equiv 10^{-pH} = 10^{-12.5} = \textbf{3} \times \textbf{10}^{-13}\,\textbf{M}$

7. pH = −0.50, $[\textbf{H}^+] \equiv 10^{-pH} = 10^{-(-0.50)} = 10^{0.50} = \textbf{3.2 M}$ or $\textbf{3.2} \times \textbf{10}^0\,\textbf{M}$

Solutions with a $[H^+]$ higher than 1.0 M have a negative pH.

8. The problem 7 solution has both the lowest pH and the highest $[H^+]$. By either measure it is the most acidic.

21

Weak Acids

Lesson 21.1 K_a Math and Approximation Equations

This lesson has two parts. The first part is a quick review of the math of powers and roots. The second part is new material. Even if you can do the Practice A problems quickly, devote extra care to the section after Practice A.

Squares and Square Roots

The calculations in this chapter require taking the square and square root of numbers in exponential notation. Rules for similar calculations were covered in Lesson 19.2. The following is a summary of rules that you need to be able to apply from memory.

Squares and Square Roots of Exponential Notation

1. To take an exponential term to a power, multiply the exponents.

2. "Taking the square root" has the same meaning as taking a quantity to the **1/2** power. When taking the square root of exponential notation, write $\sqrt{x}$ as $(x)^{1/2}$.

3. When taking exponential notation to a power, handle numbers by number rules and exponents by exponential rules.

4. A square root of a number in exponential notation can be calculated with*out* entering the exponent on the calculator. The steps are

   a. Write the square root as exponential notation taken to the 1/2 power.

   b. Make the *exponent even*. If the exponent is odd, make the *significand* 10 times *larger*, and the *exponent* 10 times (one number) *smaller*. Then,

   c. Take the square root of the *significand* on the calculator. Take the square root of the exponential term by multiplying the exponent by 1/2.
      Example:
      $$\sqrt{2.5 \times 10^{-7}} = (2.5 \times 10^{-7})^{1/2} = (25 \times 10^{-8})^{1/2} = 5.0 \times 10^{-4}$$

   d. In math, the square root of a number has two roots: $\sqrt{4} = \pm 2$. However, when solving science problems, usually only one root will make sense. The rule is: Choose the root that makes sense. In this chapter, assume the needed root is the *positive* root.

5. A square root can be estimated by making the exponent even, estimating the square root of the significand, and attaching the exact root of the exponential.

PRACTICE   A

Do the *last lettered part* of each numbered problem, then more *if* you need more practice. If additional review of the rules is needed, see Lesson 19.3.

1a. $(10^3)^2 =$ 1b. $(10^{-4})^{1/2} =$ 1c. $\sqrt{10^{-16}} =$

For problems 2–4, convert your final answers to scientific notation.

2. Solve without a calculator.

   a. $(7.0 \times 10^{-5})^2 =$

   b. $(36 \times 10^{-6})^{1/2} =$

   c. $\sqrt{0.0064} =$

3. Solve using the square and square root function on your calculator.

   a. $(7.0 \times 10^{-5})^2 =$

   b. $(36 \times 10^{-6})^{1/2} =$

   c. $\sqrt{0.0064} =$

Compare problems 2 and 3. Which was easier: with or without a calculator?

4. Use a calculator for the significand, but solve the exponential part by inspection.

   a. $(2.5 \times 10^{-5})^2 =$

   b. $(2.56 \times 10^{-4})^{1/2} =$

   c. $\sqrt{1.44 \times 10^{-6}} =$

5. *Estimate* the positive square root of

   a. 45   b. 95   c. 7

6. Use a calculator to take a square root, then compare your answers to problem 5.

   a. 45   b. 95   c. 7

For problems 7–8, convert final answers to scientific notation.

7. Estimate the square root. Need help? See rule 5 above.

   a. $(4.2 \times 10^{+5})^{1/2} =$

   b. $(8.1 \times 10^{-7})^{1/2} =$

   c. $\sqrt{7.2 \times 10^{-3}} =$

8. *Calculate* the square root. Need help? See rule 4. Compare answers to problem 7.

   a. $(4.2 \times 10^{+5})^{1/2} =$

   b. $(8.1 \times 10^{-7})^{1/2} =$

   c. $\sqrt{7.2 \times 10^{-3}} =$

Bottom Line on Calculator Use

With practice, you can quickly find squares and square roots of exponential notation on the calculator, but first write a quick estimate of the answer without the calculator. Check that the estimate and the calculator answer are *close*.

Approximation Equations

If mathematical expressions involve both large and small numbers, they can often be simplified to approximations that are easier to solve. The rule we will use is

> In *K* approximations, you *can ignore small* numbers *if* they are *added to* or *subtracted from* much larger numbers.

In the following questions, $\approx$ means *approximately equals*, *A* is any number, *B* is any different number, and *x* is any number that is much smaller than *A* and *B*.

TRY IT

(See "How to Use These Lessons," point 1, p. xv.)

Q. Write the simplified approximation for this expression by applying the rule above.

$$\frac{(A + x)\,(x)}{(B - x)} \approx$$

 Answer:

$$\frac{(A + \cancel{x})\,(x)}{(B - \cancel{x})} \approx \frac{(A)\,(x)}{B}$$

Let's test this rule using numbers.

TRY IT

Q1. Calculate an exact numeric answer (to three decimal places) for

$$\frac{(18.0 + 0.10)\,(0.10)}{(9.0 - 0.10)} =$$

 Answer:

$$\text{Exact: } \frac{(18.1)\,(0.10)}{(8.9)} = \mathbf{0.204}$$

Now let's simplify that equation by applying the approximation rules.

TRY IT

Q2. In the expression below, cross out the small terms added or subtracted from larger terms, calculate a numeric answer (to three decimal places), then compare to the answer above.

$$\frac{(18.0 + 0.10)(0.10)}{(9.0 - 0.10)} \approx$$

Answer:

$$\text{Approximation: } \frac{(18.0 + \cancel{0.10})(0.10)}{(9.0 - \cancel{0.10})} \approx \frac{(18.0)(0.10)}{(9.0)} = \mathbf{0.200}$$

Compare the answers to Q1 and Q2 above. Are they close?

If the small numbers added or subtracted from large numbers are removed, the difference between the two answers is 2%. In many science experiments and procedures, 2% would be more error than we would like to see. However, because K values often involve significant inherent uncertainty, in K calculations a change in an answer of up to 5% due to the use of an approximation is generally considered acceptable.

To solve K calculations involving weak acids quickly, our rule is: use approximation equations.

PRACTICE B

Apply the approximation assumptions above to simplify these expressions. Save a few problems for your next practice session.

1. $\dfrac{(A + x)(x)}{(A - x)} \approx$ 2. $\dfrac{(x)(x)}{(B - x)} \approx$

3. $\dfrac{(B + x)(x)}{(A - x)} \approx$ 4. $\dfrac{(A + x)}{(A - x)(x)} \approx$

5. Assuming x is much smaller than the other values in these equations, simplify these expressions.

a. $\dfrac{(0.050 + x)(x)}{0.020 - x} \approx$

b. $\dfrac{(2x)(x)}{0.10 - x} \approx$

c. $\dfrac{(x)(x)}{0.020 - x} \approx$

d. $\dfrac{(x + x)}{[WB] + x} \approx$

(continued)

6. Assuming x is very small compared to the other numbers, apply the approximation rules and then solve for x. Simplify the terms here, then finish in your notebook. Try to solve these without using a calculator.

a. $\dfrac{(x)\,(x)}{2.0 - x} = 8.0 \times 10^{-12}$

b. $\dfrac{(2x + x)}{0.50 + x} = 1.8 \times 10^{-5}$

c. $\dfrac{(0.60 + x)\,(x)}{0.20 - x} = 3.6 \times 10^{-9}$

d. $\dfrac{(2x)\,(x)}{4.0 - x} = 1.25 \times 10^{-11}$

Lesson 21.2 — Weak Acids and K_a Expressions

So far, our acid–base calculations have all involved either strong monoprotic acids (such as HCl and HNO_3) or strong bases (such as NaOH and KOH). Those compounds are widely used in laboratories and industrial processes when strong acids and bases are needed.

However, most acids and bases are *weak* rather than strong. An understanding of weak acids and bases is important in fields as interesting and diverse as the biological sciences and the culinary arts.

Weak Acid Solutions

Strong acids ionize 100% to form H^+ ions when dissolved in water. Weak acids can be defined as substances that form H^+ ions when dissolved in water, but do so only slightly.

The general equation describing the behavior of a weak acid dissolved in water is

1 weak acid ⇆ 1 proton + 1 conjugate base   (Left side favored.)

A specific example is the behavior of ammonium ion (a weak acid) in water:

$$NH_4^+(aq) \leftrightharpoons H^+(aq) + NH_3(aq) \qquad \text{(Goes slightly.)}$$

The loss of a proton by an acid is reversible: the proton can return to the conjugate base to re-form the acid. In a weak acid solution at equilibrium, the gain and loss of protons occurs continuously, but because both reactions proceed at the same rate, no *net* change occurs.

Because acid ionization is a reaction that in practice is nearly always reversible, we will represent the reaction *arrow* as ⇆ rather than →. For strong acid ionization, equilibrium favors the right-side products, but for weak acids, equilibrium favors the reactants.

A familiar weak acid solution is vinegar, which can be formed by the action of bacteria and oxygen on fruit juice. The oxidation of the sugar in fruit juice produces a mixture that is composed primarily of water and *acetic acid,* a weak acid that is soluble in water. Most vinegar solutions are about one volume of acetic acid per 10 to 20 volumes of water.

When dissolved in water, acetic acid ionizes slightly to produce H^+ ions. As the products form, the reverse reaction of the proton returning to the acetate ion also occurs, and the solution will quickly reach an *equilibrium* state where no further net change takes place.

The behavior of acetic acid in water can be represented as a separation into ions (termed *dissociation* or *ionization*):

$$1\ CH_3COOH(aq) \rightleftharpoons 1\ H^+(aq) + 1\ CH_3COO^-(aq) \qquad \text{(Goes ~1\%.)}$$

This reaction can also be described as a **hydrolysis** (a reaction with water):

$$1\ CH_3COOH(\ell) + 1\ H_2O(\ell) \rightleftharpoons 1\ H_3O^+(aq) + 1\ CH_3COO^-(aq)$$
$$\text{(Goes ~1\%.)}$$

Compare the two equations. Both represent a weak acid that is dissolved in water but ionizes only slightly. Recall that H^+ and H_3O^+ are equivalent ways of representing the proton released by acids. These two equations are *equivalent* ways of representing the same reaction.

In equations representing weak acid ionization, if no state is shown after a particle formula, assume the state is (*aq*).

[H⁺] in Weak versus Strong Acid Solutions

In water, strong acids ionize completely, but when *weak* acids dissolve in water, equilibrium favors the *un*-ionized species. A weak acid solution therefore has fewer H^+ ions and a *higher* pH (is less acidic) than a strong acid solution that has the same acid concentration.

Example: In **0.10 M** HCl as mixed, a solution of a *strong* acid, the approximate concentrations and pH include

$$[\text{HCl not ionized}] = \mathbf{0\,M};\ [\mathbf{H^+}] = \mathbf{0.10}\,M = 10^{-1}\,M;\ pH = \mathbf{1}$$

In a **0.10 M** *acetic* acid solution, roughly 1 acetic acid particle per 100 is ionized. This means that in this solution,

$$[\mathbf{CH_3COOH}] \approx 0.099\,M \approx \mathbf{0.10\,M};\ [\mathbf{H^+}] \approx \mathbf{0.001}\,M \approx 10^{-3}\,M;\ pH \approx \mathbf{3}$$

Compare the [H⁺] and pH in the two solutions. At the same mixed concentrations, the *hydrochloric* acid solution has about 100 times more protons than the *acetic* acid.

Vinegar can be safely used as a food ingredient because acetic acid solutions do not contain high concentrations of the reactive, corrosive particles in acids: H^+. However, even at low concentrations, the [H⁺] in aqueous solutions can have a major impact on reactions in chemistry and biology.

Conjugate Bases

An acid can be defined as any particle that can *lose a proton*. This means that acids can be positive or negative *ions* as well as electrically neutral particles.

Examples: HCl, NH_4^+, and HPO_4^{2-} can all act as an acid by losing a proton.

The ionization of an acid produces a proton and the acid's **conjugate base**. The formula for the conjugate base is the acid formula with *one less H* atom and *one less positive charge*.

Example: For the weak acid HF ionizing in water, the reaction can be represented in three ways.

As the general reaction: **1 weak acid ⇆ 1 proton + 1 conjugate base**

In the *ionization* format: $HF(aq) \leftrightarrows H^+(aq) + F^-(aq)$

In the *hydrolysis* format: $HF(aq) + H_2O(\ell) \leftrightarrows H_3O^+(aq) + F^-(aq)$

TRY IT

Use the definition of a conjugate base to answer these two questions.

Q. Assume the first particle in each reaction is acting as an *acid*. Complete the reaction by writing formulas for the products. Circle the conjugate base.

a. $HCN(aq) \leftrightarrows$

b. $NH_4^+ + H_2O(\ell) \leftrightarrows$

Answer:

a. $HCN(aq) \leftrightarrows H^+(aq) + \boxed{CN^-(aq)}$

An acid ionizes to produce a proton (H^+). The other particle is the conjugate base: the acid particle with one less H atom and one less positive charge.

b. $NH_4^+ + H_2O(\ell) \leftrightarrows H_3O^+ + \boxed{NH_3}$

When a weak acid ionization is written in the hydrolysis format (losing a proton by donating the proton to water), one of the product particles is always the hydronium ion (H_3O^+); the other is the conjugate base.

All reaction equations must be balanced for atoms *and* charge. Check the balancing in each of the answers above.

PRACTICE   A

In the reactions below, assume that the first particle is acting as a weak acid in an aqueous solution. Write the formulas for the products. Circle the conjugate base.

1. $HI \leftrightarrows$

2. $HCO_3^- \leftrightarrows$

3. $HS^- + H_2O \leftrightarrows$

4. $HPO_4^{2-} + H_2O \leftrightarrows$

The K_a Expression

For the reaction of an acid losing one proton, the equilibrium constant K is given a special name, the **acid-dissociation constant**, and a special symbol, K_a. For any reaction in which one acid particle ionizes to form one H^+ or reacts with water to form one H_3O^+, the symbol for the equilibrium constant is K_a.

By convention, a weak acid ionization or hydrolysis is written with one un-ionized acid particle on the left and one H^+ (or H_3O^+) on the right.

Example: The loss of a proton by acetic acid can be represented as either

$$CH_3COOH(aq) \leftrightharpoons H^+(aq) + CH_3COO^-(aq)$$

or

$$CH_3COOH(aq) + H_2O(\ell) \leftrightharpoons H_3O^+(aq) + CH_3COO^-(aq)$$

 TRY IT

Q. Write the K expression for the two reactions in the example above.

STOP

Answer:

Because each reaction consists of a weak acid releasing one proton, the K is a K_a.

> A K_a expression must include $[H^+]^1$ *or* $[H_3O^+]^1$ on *top* and $[\text{weak acid}]^1$ on the bottom.

For the above two reactions,

$$K_a = \frac{[H^+]_{eq}\,[CH_3COO^-]_{eq}}{[CH_3COOH]_{at\ eq}} \quad \text{or} \quad K_a = \frac{[H_3O^+][CH_3COO^-]}{[CH_3COOH]}$$

For all K_a expressions,

- As in standard K expressions: product concentrations are on top and reactant concentrations on the bottom.
- Because H^+ and H_3O^+ are considered to be equivalent ways to represent a proton in acid solutions, the two reaction formats and the two resulting K_a expressions are equivalent. One format may be substituted for the other.
- If the state of a particle is left out, it is assumed to be aqueous.
- Because a K is based on concentrations at equilibrium, if a label after a concentration term is omitted, the label $[\ ]_{\text{at equilibrium}}$ is understood.
- The powers of all of the concentrations are **1** and are omitted as understood.
- $[H_2O]$ is represented by a 1 in the K_a expression for the hydrolysis reaction.

$[H_2O]$ is represented by a 1 in K_a hydrolysis expressions for the same reason that it is represented by a 1 in the K_w expression. During acid hydrolysis, the concentration of liquid water remains high and close to constant, at about 55 M, except in solutions that have a very high concentration of the acid. By convention, if a particle concentration in a reaction equation cannot be significantly changed, its constant concentration is represented by a 1 in the K expression, and the 1 is omitted if it is in the expression denominator.

Pure acetic acid (CH_3COOH) is a *liquid* at room temperature. Because acetic acid is a polar molecule, it is soluble in water: these two liquids mix to form one layer, unlike the immiscible "oil and water." In the dilute acetic acid solutions used in most experiments, its concentration is low compared to the [water], and the *state* of the acetic acid is better described as aqueous (dissolved in a comparably large amount of water) rather than liquid. Depending on how much acetic acid is mixed with the water, the $[CH_3COOH(aq)]$ can *vary* substantially, so the $[CH_3COOH]$ term is *included* in the K expression.

In general,

> A [dissolved liquid] term is included in a K expression but the [solvent] is not.

Polyprotic Acids

A polyprotic acid can ionize to lose more than one proton, but in all polyprotic acid solutions, each successive hydrogen ionizes less than the one before. In K calculations, these successive ionizations are written separately, with the conjugate base of the first ionization becoming the weak acid in the second, etc. A K_a value always refers to the ionization of a weak acid to produce *one* H^+ or *one* H_3O^+.

PRACTICE **B**

In the reactions below, assume that the first particle is acting as a weak acid in an aqueous solution. Complete the reaction by writing the formulas for the products. Then write the K_a *expression* for each reaction.

1. $H_2CO_3 \leftrightarrows$

2. $HCO_3^- \leftrightarrows$

3. $H_2S + H_2O(\ell) \leftrightarrows$

4. $NH_4^+ + H_2O(\ell) \leftrightarrows$

K_a Values

Acids have *characteristic K_a values* at a given temperature.

Sample K_a Values at 25°C

Hydrochloric acid (strong)	$HCl \leftrightarrows H^+ + Cl^-$	K_a = very large
Hydrofluoric acid (weak)	$HF \leftrightarrows H^+ + F^-$	$K_a = 6.8 \times 10^{-4}$
Acetic acid (weak)	$CH_3COOH \leftrightarrows H^+ + CH_3COO^-$	$K_a = 1.8 \times 10^{-5}$
Hydrocyanic acid (weak)	$HCN \leftrightarrows H^+ + CN^-$	$K_a = 6.2 \times 10^{-10}$

For strong acids, ionization strongly favors the products. The K_a values for strong acids will be a number much greater than 1. Because strong acid ionization is considered to go to completion rather than equilibrium, the K value is not needed for calculations.

Weak acids have K_a values between 1.0 and 10^{-16}, written in scientific notation with a negative power of 10. The weaker the acid, the less it will ionize and the smaller will be its K_a.

TRY IT

Q. In the table above, which is the weakest acid?

(STOP) **Answer:**

HCN, because it has the lowest K_a.

As the temperature of a weak acid solution rises, more bonds to H break, more protons form, and K_a values increase. However, unless otherwise noted, you should assume that K_a calculations are based on reactions at 25°C (77°F).

For each H atom in a compound, a K_a value can be measured that represents the tendency of the H to ionize. However, if a K_a value is less than 10^{-16}, the H will not react as an acid except with bases that are *very* strong: stronger than hydroxides. Under most circumstances, an H with a K_a of less than 10^{-16} is considered to be a nonreactive (nonacidic) rather than an *acidic* hydrogen (see Lesson 14.1).

PRACTICE C

1. For the weak acid NH_4^+, the K_a value is 5.6×10^{-10}.

    a. Write the reaction for which this is the K value.

    b. Write the K_a expression for this reaction.

    c. What term does a K_a expression always have in its numerator?

2. The ion HSO_3^- has a K_a value of 6.2×10^{-8}.

    a. Write the reaction for which this is the K value.

    b. Write the K_a expression for this reaction.

3. Which ion above is the weaker acid: NH_4^+ or HSO_3^-?

Lesson 21.3 K_a Calculations

K_a Equations

The K_a *expression* and the K_a *value* together form a **K_a equation**.

For the ionization of acetic acid: $1\ CH_3COOH \leftrightharpoons 1\ H^+ + 1\ CH_3COO^-$ the K_a equation is

$$K_a = \frac{[H^+][CH_3COO^-]}{[CH_3COOH]} = \mathbf{1.8 \times 10^{-5}} \text{ at } 25°C$$

Calculations based on K_a follow the same rules as other K calculations:

- Units are not attached to K values.
- Units are omitted when substituting numbers into the K expression.
- When a concentration is WANTED, the *unit* moles/liter (M) must be added to the answer.

When to Use K_a

In solutions of strong acids, a K_a value is not needed to calculate particle concentrations at equilibrium. Since HCl and HNO_3 ionize completely in a 1 to 1 to 1 ratio, the original [HCl or HNO_3]$_{mixed}$ equals the [H^+] and [conjugate base] that form in the solution. These simple ratios make it possible to solve for ion concentrations in strong acid solutions by inspection.

Weak acids ionize to form one proton, but the reaction goes slightly: to establish equilibrium instead of to completion. The mixed original [weak acid] does not directly convey how much of the weak acid ionizes, nor how much of the products form. For reactions that go to equilibrium, a K equation is needed to calculate the particle concentrations at equilibrium.

We will call the following weak acid rule 1.

> 1. In calculations involving acids and [particles],
>   - For *100*% ionization (as in HCl or HNO_3 solutions), use the REC steps.
>   - For *slight* ionization (as in weak acid solutions), use WRECK steps.

The WRECK Steps for Weak Acid Ionization

For *all* weak acid (WA) ionizations, the general **R** and **E** (Reaction and Extent) can be written

$$1\,WA \leftrightharpoons 1\,H^+ + 1\text{ conjugate base (CB)} \qquad \text{(Goes slightly.)}$$

Since the reaction does not go to completion, the WRECK steps are needed to solve calculations.

 **TRY IT**

Q. Write the K expression for the general reaction above.

 Answer:

$$K_a \equiv \frac{[H^+]_{at\ eq} \cdot [CB]_{at\ eq}}{[WA]_{at\ equilibrium}} \qquad \text{(definition)}$$

Because this ionization forms H^+, the K is a K_a. The [ ]$_{at\ equilibrium}$ or "at eq" subscript is often omitted as understood in K expressions, but in this case it will be helpful in the discussion that follows.

In Lesson 19.7, to solve K calculations we used a RICE table. However, compared to general K calculations, K_a RICE tables are simplified because

- There is only one reactant and it is always a weak acid.
- One product is always [H^+] or [H_3O^+] and the other is the conjugate base.
- The initial quantities of the products are zero.
- The coefficients are always **1** weak acid particle used up *equals* **1** H^+ ion formed *and* **1** conjugate base formed.

This means

> For *all* weak acid ionization or hydrolysis, the general *RICE* table in moles or mol/L is the same.

We use RICE tables in K calculations to find particle moles or moles/liter at equilibrium: the numbers in the RICE table bottom row. However, because all weak acid ionization calculations are so similar, instead of writing a RICE table for each problem, we can write the bottom RICE table row by inspection as part of the WRECK steps.

Let's walk through the logic of applying the WRECK steps to weak acid ionization.

1. **Use x to represent small concentrations.** Because the coefficients in a weak acid ionization are all *one*,

$$1\,WA \rightleftharpoons 1\,H^+ + 1\text{ conjugate base}   \text{(Goes slightly.)}$$

the *ratio* of moles of weak acid used up to moles H^+ formed to moles of conjugate base (CB) formed is always **1** to **1** to **1**. Because all of the particles in the ionization reaction are dissolved in the same volume of solution, the moles *per liter* ratios are also **1** to **1** to **1**. This means that *all three* of the following concentrations can be represented by an x representing a relatively small concentration.

$$x = small = [WA]_{\text{that ionizes}} = [H^+]_{\text{formed}} = [CB]_{\text{formed}}$$

2. **The REC steps.** In a solution, the [WA] at equilibrium, *after* ionization, is usually difficult to measure directly. However, the mol/L of weak acid that we *originally added* to mix the solution is usually a quantity that we know.

   Substituting x from the equality above simplifies the math. By definition:

$$[WA]_{\text{at eq}} \equiv [WA]_{\text{mixed}} - small\ [WA]_{\text{that ionizes}} \equiv [WA]_{\text{mixed}} - x$$

If we can solve for x, we solve for four quantities:

$$x = [H^+]_{\text{formed}} = [CB]_{\text{formed}} = [WA]_{\text{that ionizes}} \text{ and } [WA]_{\text{at eq}} (\equiv [WA]_{\text{mixed}} - x)$$

TRY IT

Q. Substitute into the blanks in the **C** step below a symbol for each concentration. Write a term that *includes x* and uses the symbols for measurable concentrations in the equalities above.

Reaction and Extent: $1\,WA \rightleftharpoons 1\,H^+ + 1\,CB$ (Goes slightly.)
∧   ∧   ∧

Conc. at eq.: _____ ___ ___

Answer:

Rxn. and Extent: $1\,WA \rightleftharpoons 1\,H^+ + 1\,CB$ (Goes slightly.)
∧   ∧   ∧
Conc. at eq.: $[WA]_{\text{mixed}} - x$ x x

The **C** row will be the same for *all* weak acid ionization reactions.

3. **The K step.** We can substitute into the **K** definition, for each of the three *concentration* terms in the K_a below, a symbol for that term that includes *x* from the **C** row above.

TRY IT

Q. Substitute the terms from the **C** row of the REC steps to fill in the top and bottom of the *fraction* term below.

$$K_a \equiv \frac{[H^+]_{at\ eq} \cdot [CB]_{at\ eq}}{[WA]_{at\ eq}} \equiv \underline{\hspace{3cm}}$$

STOP **Answer:**

$$K_a \equiv \frac{[H^+]_{at\ eq}\,[CB]_{at\ eq}}{[WA]_{at\ eq}} \qquad \frac{x^2}{[WA]_{mixed} - x}$$

$$\wedge \qquad\qquad \wedge$$

Definition based on **R** row   *Exact* based on **C** row

Both of the K_a equations above are true at equilibrium, but the *exact* equation has an advantage. Our goal is usually to solve for *x*, and the exact equation can be solved for *x* if the K_a value and the [WA *as mixed*] are known, as they usually are.

However, because the exact equation has both x^2 and *x* terms, it is a quadratic equation. Quadratic equations can be converted to the general form

$$ax^2 + bx + c = 0$$

With **a, b,** and **c** known, you can solve for *x* using the quadratic formula.

$$x = \frac{-b \pm (b^2 - 4ac)^{1/2}}{2a}$$

The steps to solve a quadratic equation are not difficult, but for most weak acid calculations, the following approximation may be used that solves for *x* more *quickly.*

4. **The K approximation.** In a weak acid solution, the value of *x* is small compared to the value of the [WA] as mixed. This allows us to simplify the measurement of the [WA] at equilibrium.

TRY IT

Q. Using the approximation rule from the previous lesson, write the simplified, *approximate* term in the blank below.

$$[WA]_{at\ equilibrium} \equiv [WA]_{mixed} - x \approx \underline{\hspace{3cm}}$$

STOP **Answer:**

$$[WA]_{at\ equilibrium} \equiv [WA]_{mixed} - x \approx [WA]_{mixed}$$

When a small quantity is added to or subtracted from a larger quantity, the larger quantity remains approximately the same. In a weak acid solution, the weak acid concentration at equilibrium is not very different from the concentration as mixed.

Now fill in the box below.

 TRY IT

Q. Write an approximation starting from the exact equation.

$$K_a \equiv \frac{[H^+]_{eq}[CB]_{eq}}{[WA]_{at\ eq}} \equiv \frac{x \cdot x}{[WA]_{mixed} - x} \approx \boxed{}$$

$\wedge$ $\wedge$ $\wedge$

Definition   Exact   Approximation

Answer:

The K_a expression can be simplified to this approximation:

$$K_a \equiv \frac{[H^+]_{eq}[CB]_{eq}}{[WA]_{at\ eq}} \equiv \frac{x^2}{[WA]_{mixed} - x} \approx \boxed{\frac{x^2}{[WA]_{mixed}} \approx K_a}$$

$\wedge$ $\wedge$ $\wedge$

Definition   Exact   Approximation

The small $(-x)$ difference between the *exact* and the *approximation* equation means that solving the exact equation for x requires solving a quadratic equation, while solving the approximation requires only a square root. The approximation can therefore be solved more *quickly*. For K_a calculations, we will first solve the approximation equation, and in most cases the approximation will be an acceptable answer.

SUMMARY

Weak Acid Ionization Steps

For weak acid dissociation (ionization) reactions, these are always the *same*:

- The general reaction: $WA \leftrightharpoons H^+ + CB$
- The RICE table
- The WRECK steps
- The **C** step (which is the same as the bottom row of the RICE table)
- The **K** step: the definition, exact, and approximate expressions

In weak acid ionization (K_a) calculations, all that will vary are the specific formulas for WA and CB, and the values for $[WA]_{as\ mixed}$, x, and K_a. By representing each weak acid ionization with the *general* reaction equation, we can simplify problem solving.

PRACTICE   A

Check your answers as you go.

1. For the ionization of the weak acid (WA) to form a proton and its conjugate base (CB),

   a. Write the REC steps. In the **C** row, each term should include x.

   b. Write the K_a definition based on the **R** row reaction.

   c. Write the K_a exact equation using **C** row terms that include x.

   d. Write the K_a approximation equation.

(continued)

2. For the ionization of the weak acid NH_4^+,

  a. Write the **R**eaction and its **E**xtent, using the particle formulas in the reaction.

  b. Below the **R**eaction and **E**xtent, write the **C**oncentrations at equilibrium, defined so that each term includes x.

  c. Write the K_a definition equation, using the particle formulas.

  d. Write the K_a exact equation based on the terms defined in the **C** step.

  e. Write the K_a approximation equation, based on your answer to part d.

K_a Calculations

We can solve K_a calculations in the same way we solved K calculations: by writing the WRECK steps. To simplify K_a calculations, we will add a step: we will write *both* the reaction showing the particles for the *specific* weak acid and then the *general* reaction equation for *all* weak acids. These two **R** steps will define the terms in the K equation.

We can then solve all weak acid calculations using the following steps of weak acid rule 2:

The K_a Prompt

2. If a problem involves a weak acid or K_a and [ions], write the **WRECK** steps.
   • WANTED: Write the general *and* specific symbols WANTED.
   • Write the *specific* **R**eaction using the symbol for the particles in the problem, then write these general R, E, C, and K steps.

Rxn. and Extent: $1\,\text{WA} \;\leftrightharpoons\; 1\,\text{H}^+ \;+\; 1\,\text{CB}$ (Goes slightly.)

Conc. at eq.: $[\text{WA}]_{\text{mixed}} - x \qquad x \qquad\qquad x$

$$K_a \equiv \frac{[\text{H}^+]_{\text{eq}}\,[\text{CB}]_{\text{eq}}}{[\text{WA}]_{\text{at eq}}} \equiv \frac{x^2}{[\text{WA}]_{\text{mixed}} - x} \approx \frac{x^2}{[\text{WA}]_{\text{mixed}}} \approx K_a$$

Definition   Exact   Approximation

   • SOLVE the K_a approximation equation for the WANTED symbol.

In short: See weak acid or K_a and [ions]? Write the **WRECK** steps and solve the approximation.

TRY IT

Apply the rule above to this example.

Q. In a 2.0 M acetic acid (CH_3COOH) solution, calculate the $[H^+]$. Use ($K_a = 1.8 \times 10^{-5}$).

Answer:

WANTED: $[H^+] = x$

Specific **R:** $1\,CH_3COOH \leftrightharpoons 1\,H^+ + 1\,CH_3COO^-$

R and **E:** $1\,WA \leftrightharpoons 1\,H^+ + 1\,CB$ (Goes slightly.)

Conc. at eq.: $[WA]_{mixed} - x \quad x \qquad x$

$$K_a \equiv \frac{[H^+]_{eq}[CB]_{eq}}{[WA]_{at\ eq}} \equiv \frac{x^2}{[WA]_{mixed} - x} \approx \boxed{\frac{x^2}{[WA]_{mixed}}} \approx K_a$$

Definition   Exact   Approximation

Solve the boxed approximation equation for x. Start with a DATA table using the approximation symbols.

DATA: $x = [H^+] = ?$ = WANTED

$[WA]_{as\ mixed} = [CH_3COOH]_{as\ mixed} = 2.0$ M

$K_a = 1.8 \times 10^{-5}$

Use those values to solve the approximation for the WANTED symbol.

$$K_a \approx \frac{x^2}{[WA]_{mixed}} \text{; substituting: } 1.8 \times 10^{-5} = \frac{x^2}{2.0}$$

Solve for x^2, then x.

$$x^2 = (1.8 \times 10^{-5})(2.0) = 3.6 \times 10^{-5} = 36 \times 10^{-6}$$

$x = 6.0 \times 10^{-3}$ M $= [H^+]$ (When solving a K for [ ], add M as the unit.)

In the above problem, $x = \pm 6.0 \times 10^{-3}$, but concentration is always a positive value. In answering science problems, choose the root that makes sense.

PRACTICE   B

If in doubt, check your answers after each part.

1. Hydrogen cyanide (HCN) is both a weak acid and a notorious poison.

   a. Calculate the $[H^+]$ in a 0.50 M HCN solution. Use $K_a = 6.2 \times 10^{-10}$ for HCN.

   b. Calculate the pH of the HCN solution.

2. The weak acid hydrogen fluoride (HF) can be used to etch glass. If the $[F^-]$ in a 0.25 M HF solution is measured to be 0.012 molar,

   a. Find the K_a value for HF according to the data in this experiment.

   b. What is the pH of the solution?

   c. What is the $[OH^-]$ in this HF solution?

Lesson 21.4   Percent Dissociation

Moving Numbers Instead of Symbols

For past calculations based on equations, we have solved in symbols before plugging in numbers. We did so because when using algebra, symbols can be written more quickly than numbers with their units. In K calculations, however, you may choose to plug numbers into a K equation and *then* do the algebra. Why? The numbers *without* units used in K calculations often can be written more quickly than the symbols used in chemistry.

The downside is that in K_a as in other K calculations, you lose unit cancellation as a check on your algebra. This means that in K calculations, you must double check your math.

Solving the Approximation Directly

It is important to be able to solve K_a calculations by writing out the methodical WRECK steps. For problems that are more complex, we will need those methodical steps. However, once you master the WRECK steps, you may solve K_a calculations by writing these WA *quick* steps: WANTED, approximation, substitute, solve.

- Write the WANTED symbol.
- Write the K_a approximation equation:

$$K_a \approx \frac{x^2}{[WA]_{mixed}}$$

- Substitute the DATA and solve the approximation for the WANTED symbol.

> ▬▬▶ **TRY IT**

Apply the quick steps to this problem.

Q. Hypochlorous acid ionizes as follows.

$$HOCl \rightleftharpoons H^+ + OCl^- \qquad K_a = 3.5 \times 10^{-8}$$

Calculate the $[H^+]$ in a 0.40 M HOCl solution.

 Answer:

WANTED: $[H^+] = x$

  Since hypochlorous acid has a small K_a, it is a *weak acid*.

  Using the WA *quick* steps, write the K_a approximation, substitute, and solve.

$$K_a \approx \frac{x^2}{[WA]_{mixed}} \qquad \text{Substituting: } 3.5 \times 10^{-8} = \frac{x^2}{0.40}$$

SOLVE: To find x WANTED, first solve for x^2.

$$x^2 = (3.5 \times 10^{-8})(0.40) = 1.4 \times 10^{-8}$$

$$x = 1.2 \times 10^{-4} M = [H^+]$$

Using the quick steps, calculate the pH of a 0.50 M formic acid (HCOOH) solution ($K_a = 1.8 \times 10^{-4}$).

Measuring Dissociation

Percent dissociation, **percent ionization**, and **percent hydrolysis** are all terms that have the same meaning. The percent dissociation is simply the percentage of a weak acid concentration that ionizes. Since [WA that ionizes] = x, we can write the following as weak acid rule 3.

Percent Dissociation

$$3.\quad \% \text{ Dissociation} \equiv \frac{x}{[WA]_{mixed}} \cdot 100\% \equiv \frac{[WA]_{ionized}}{[WA]_{mixed}} \cdot 100\%$$

The dissociation terminology and the definition equations must be committed to memory.

When calculations involve percentages that must be solved using conversions, our rule has been to work in fractions rather than percentages. However, since problems with this equation generally involve straightforward substitution for symbols, we will solve calculations using the equation in the form above.

TRY IT

Apply the definition above to the following problem.

Q. Benzoic acid ionizes to form the benzoate ion.

$$C_6H_5COOH \rightleftharpoons H^+ + C_6H_5COO^- \qquad K_a = 6.3 \times 10^{-5}$$

a. Calculate the $[H^+]$ in a 0.040 M benzoic acid solution. Use the weak acid quick steps: solve the approximation directly.

b. Calculate the percent dissociation in the solution.

Answer:

a. WANTED: $[H^+] = x = ?$

By the quick steps,

$$K_a \approx \frac{x^2}{[WA]_{mixed}} \qquad \text{Substituting: } 6.3 \times 10^{-5} \approx \frac{x^2}{0.040 \text{ M}}$$

SOLVE: To find the x WANTED, first solve for x^2.

$$x^2 \approx (6.3 \times 10^{-5})(0.040) = 2.5 \times 10^{-6}$$

$$? = x \approx (\text{estimate } 1\text{--}2 \times 10^{-3}) \approx \mathbf{1.6 \times 10^{-3}\,M = [H^+]}$$

If needed, adjust your work and then try part b.

STOP

b. Watch for the answer from one part to be used as DATA for later parts. Since the dissociation equation is simple, you may solve without listing the symbols and values in a DATA table.

SOLVE:

$$\% \text{ Dissoc.} = \frac{x}{[WA]_{mixed}} \cdot 100\% = \frac{1.6 \times 10^{-3}\,M}{4.0 \times 10^{-2}\,M} \cdot 10^2\% = \mathbf{4.0\%}$$

In 0.040 M C_6H_5COOH at 25°C, 4.0% of the weak acid is ionized.

PRACTICE B

1. For a weak acid solution, $[H^+]$ is 2.0×10^{-3} M at $[WA] = 0.40$ M. Find the percentage of ionization of the weak acid.

2. A 2.0 molar solution of a monoprotic weak acid has a pH of 4.49.

   a. Calculate the $[H^+]$ in the solution.

   b. Calculate the percentage of ionization in this solution.

3. If a 0.50 M weak acid solution is 0.30% ionized, what is its K_a?

SUMMARY

You may want to organize this summary into charts and flashcards.

1. In acid and base solutions, to calculate [particles],
   • In *strong* acids such as HCl or HNO_3, or strong bases such as NaOH and KOH, either write the REC steps for 100% ionization, or use the quick steps to find [ions].
   • In *weak* acid calculations, write the WRECK steps. Use x to represent the small mol/L of weak acid that reacts (ionizes or hydrolyzes).

2. **K_a prompt:** If a problem has a K_a and [ions], to solve methodically, write the WRECK steps.
   • WANTED: Write the general and specific symbol.
   • Write the *specific* Reaction using the symbol for the particles in the problem, then write this *general* reaction and extent:

Rxn. and Extent: **WA** $\leftrightarrows$ **H$^+$** + **conjugate base**   (Goes slightly.)

Conc.: $[WA]_{mixed} - x$ x x

$$K_a \equiv \frac{[H^+]_{eq}[CB]_{eq}}{[WA]_{at\ eq}} \equiv \frac{x^2}{[WA]_{mixed} - x} \approx \frac{x^2}{[WA]_{mixed}} \approx K_a$$

Definition   Exact   Approximation

3. *Percent ionization*, *percent dissociation*, and *percent hydrolysis* of a weak acid all have the same meaning.

$$\% \text{ Dissociation} = \frac{x}{[WA]_{mixed}} \cdot 100\% \equiv \frac{[WA]_{ionized}}{[WA]_{mixed}} \cdot 100\%$$

4. **The K_a quick steps**: To solve K_a calculations quickly, solve the approximation directly.
   • Write the WANTED symbol.
   • Write the K_a approximation equation:

$$K_a \approx \frac{x^2}{[WA]_{mixed}}$$

   • Substitute the DATA, then solve for the WANTED symbol.

ANSWERS

Lesson 21.1

Practice A 1a. **10^6**   1b. **10^{-2}**   1c. **10^{-8}**

2, 3a. **4.9×10^{-9}**   2, 3b. **6.0×10^{-3}**   2, 3c. **$(64 \times 10^{-4})^{1/2} = 8.0 \times 10^{-2}$**

4a. **6.3×10^{-10}**   4b. **1.60×10^{-2}**   4c. **1.20×10^{-3}**

5a. $45: 6^2 = 36$ and $7^2 = 49$; estimate $\approx$ **6.7**

5b. $95: 9^2 = 81$ and $10^2 = 100$; estimate $\approx$ **9.7**

5c. $7: 2^2 = 4$ and $3^2 = 9$; estimate $\approx$ **2.7**

6a. **6.71**   6b. **9.74**   6c. **2.65** When estimating to *check* answers, close is good enough.

7, 8a. **6.5×10^2**   7, 8b. **9.0×10^{-4}**   7, 8c. **8.5×10^{-2}**

Practice B 1. $\dfrac{(A + \cancel{x})(x)}{(A - \cancel{x})} \approx \dfrac{(A)(x)}{(A)} \approx x$ 2. $\dfrac{(x)(x)}{(B - \cancel{x})} \approx \dfrac{x^2}{B}$

3. $\dfrac{(B + \cancel{x})(x)}{(A - \cancel{x})} \approx \dfrac{(B)(x)}{A}$ 4. $\dfrac{(A + \cancel{x})}{(A - \cancel{x})(x)} \approx x^{-1}$

5a. $\dfrac{(0.050 + \cancel{x})(x)}{0.020 - \cancel{x}} \approx \dfrac{(0.050)(x)}{0.020} \approx (2.5)\,x$ 5b. $\dfrac{(2x)(x)}{0.10 - \cancel{x}} \approx \dfrac{2x^2}{0.10} = (20.)\,x^2$

5c. $\dfrac{(x)(x)}{0.020 - \cancel{x}} \approx \dfrac{x^2}{0.020}$ *or* $(50.)\,x^2$ 5d. $\dfrac{(x + x)}{[WB] + \cancel{x}} \approx \dfrac{2x}{[WB]}$

On the following problems, you may do the steps differently, but you must get the same answers.

6a. $\dfrac{(x)\,(x)}{2.0 - \cancel{x}} = 8.0 \times 10^{-12}$

$\dfrac{x^2}{2.0} = 8.0 \times 10^{-12}$

$x^2 = 16 \times 10^{-12}$

$x = \mathbf{4.0 \times 10^{-6}}$

6b. $\dfrac{(2x + x)}{0.50 + \cancel{x}} = 1.8 \times 10^{-5}$

$\dfrac{3x}{0.50} = 1.8 \times 10^{-5}$

$3x = 9.0 \times 10^{-6}$

$x = \mathbf{3.0 \times 10^{-6}}$

6c. $\dfrac{(0.60 + \cancel{x})\,(x)}{0.20 - \cancel{x}} = 3.6 \times 10^{-9}$

$\dfrac{(0.60)\,(x)}{0.20} = 3.6 \times 10^{-9}$

$3x = 3.6 \times 10^{-9}$

$x = \mathbf{1.2 \times 10^{-9}}$

6d. $\dfrac{(2x)\,(x)}{4.0 - \cancel{x}} = 1.25 \times 10^{-11}$

$\dfrac{2x^2}{4.0} = 1.25 \times 10^{-11}$

$x^2 = 25.0 \times 10^{-12}$

$x = \mathbf{5.0 \times 10^{-6}}$

Lesson 21.2

Practice A 1. $HI \rightleftharpoons H^+ + \boxed{I^-}$

2. $HCO_3^- \rightleftharpoons H^+ + \boxed{CO_3^{2-}}$

3. $HS^- + H_2O \rightleftharpoons H_3O^+ + \boxed{S^{2-}}$

4. $HPO_4^{2-} + H_2O \rightleftharpoons H_3O^+ + \boxed{PO_4^{3-}}$

Practice B In the K expressions below, all concentrations are measured at equilibrium.

1. $H_2CO_3 \rightleftharpoons H^+ + HCO_3^-$

$K_a = \dfrac{[H^+][HCO_3^-]}{[H_2CO_3]}$

2. $HCO_3^- \rightleftharpoons H^+ + CO_3^{2-}$

$K_a = \dfrac{[H^+][CO_3^{2-}]}{[HCO_3^-]}$

3. $H_2S + H_2O(\ell) \rightleftharpoons H_3O^+ + HS^-$

$K_a = \dfrac{[H_3O^+]\,[HS^-]}{[H_2S]}$

4. $NH_4^+ + H_2O(\ell) \rightleftharpoons H_3O^+ + NH_3$

$K_a = \dfrac{[H_3O^+]\,[NH_3]}{[NH_4^+]}$

For problems 3 and 4, in aqueous solutions $[H_2O]$ is represented by a 1 in K expressions.

Practice C 1a. $NH_4^+ \rightleftharpoons H^+ + NH_3$

A K_a is always a K for the reaction where a weak acid loses an H^+ ion. In the equation for a K_a reaction, an H^+ or H_3O^+ ion will always be on the right side.

1b. $K_a = \dfrac{[H^+]\,[NH_3]}{[NH_4^+]}$

1c. A K_a expression always has $[H^+]$ or $[H_3O^+]$ in the numerator.

2a. $HSO_3^- \rightleftharpoons H^+ + SO_3^{2-}$

2b. $K_a = \dfrac{[H^+]\,[SO_3^{2-}]}{[HSO_3^-]}$

3. The weaker acid has the lower K_a value: NH_4^+.

Lesson 21.3

Practice A 1a. **R** and **E**: $\quad 1\,WA \rightleftharpoons 1\,H^+ + 1\,CB \quad$ (Goes slightly.)

$\qquad\qquad\qquad\quad \wedge \qquad\quad \wedge \qquad \wedge$

$\mathbf{C}_{\text{at eq}}$: $[WA]_{\text{mixed}} - x \quad x \qquad x$

$K_a \equiv \dfrac{[H^+]_{\text{eq}}\,[CB]_{\text{eq}}}{[WA]_{\text{eq}}} \equiv \dfrac{x^2}{[WA]_{\text{mixed}} - x} \approx \dfrac{x^2}{[WA]_{\text{mixed}}}$

1b, 1c, 1d. $\qquad\qquad \wedge \qquad\qquad\qquad \wedge \qquad\qquad\qquad \wedge$

$\qquad\qquad\qquad$ Definition $\qquad\qquad$ Exact $\qquad\qquad$ Approximation

2a. **R and E:** $1 NH_4^+ \leftrightarrows 1 H^+ + 1 NH_3$ (Goes slightly.)

2b. $C_{at\ eq}$: $[NH_4^+]_{mixed} - x$ x x

2c. $K_a \equiv \dfrac{[H^+]_{eq}\,[NH_3]_{eq}}{[NH_4^+]_{at\ eq}}$ 2d. $K_a \equiv \dfrac{x \cdot x}{[NH_4^+]_{mixed} - x}$ 2e. $K_a \approx \dfrac{x^2}{[NH_4^+]_{as\ mixed}}$

Practice B 1a. If K_a and [ions] are mentioned, write the WRECKs, solve the approximation.

WANTED: $[H^+] = ?$

Specific Reaction: $1 HCN \leftrightarrows 1 H^+ + 1 CN^-$ (Goes slightly.)

R and E: $1 WA \leftrightarrows 1 H^+ + 1 CB$ (Goes slightly.)

Conc. at eq.: $[WA]_{mixed} - x$ x x

$$K_a \equiv \underbrace{\frac{[H^+]_{eq}\,[CB]_{eq}}{[WA]_{at\ eq}}}_{\text{Definition}} \equiv \underbrace{\frac{x^2}{[WA]_{mxd} - x}}_{\text{Exact}} \approx \underbrace{\frac{x^2}{[WA]_{mixed}}}_{\text{Approximation}} \approx K_a$$

Solve the K_a approximation for the WANTED symbol.

$x^2 \approx (K_a)\,([WA]_{mixed}) = (6.2 \times 10^{-10})\,(0.50) = 3.1 \times 10^{-10}$

$[H^+] = x = (\text{estimate: } 1\text{-}2 \times 10^{-5}) = \mathbf{1.76 \times 10^{-5}\ M} = [H^+]$

(This answer carries an extra *s.f.* until the final step.)

1b. See pH? Write $pH \equiv -\log [H^+]$ and $[H^+] \equiv 10^{-pH}$

WANTED: **pH** $= -\log (1.76 \times 10^{-5}) = $ estimate **4.? = 4.75**   (4.75 rounded up = 5: Check.)

2a. If K_a *and* [ions] are mentioned, write the WRECKs.

WANTED: K_a

Specific Reaction: $1 HF \leftrightarrows 1 H^+ + 1 F^-$ (Goes slightly.)

R and E: $1 WA \leftrightarrows 1 H^+ + 1 CB$ (Goes slightly.)

Conc. at eq.: $[WA]_{mixed} - x$ x x

$$K_a \equiv \underbrace{\frac{[H^+]_{eq}\,[CB]_{eq}}{[WA]_{at\ eq}}}_{\text{Definition}} \equiv \underbrace{\frac{x^2}{[WA]_{mxd} - x}}_{\text{Exact}} \approx \underbrace{\frac{x^2}{[WA]_{mixed}}}_{\text{Approximation}} \approx K_a$$

If you solve the K_a approximation for the WANTED symbol,

$$K_a \approx \frac{x^2}{[WA]_{mixed}} \approx \frac{(1.2 \times 10^{-2})^2}{0.25} \approx \frac{1.44 \times 10^{-4}}{0.25} \approx \mathbf{5.8 \times 10^{-4}}$$

Or, since this problem supplies x, you can easily solve the exact equation,

$$K_a = \frac{x^2}{[WA]_{mixed} - x} = \frac{(1.2 \times 10^{-2})^2}{0.25 - 0.012} = \frac{1.44 \times 10^{-4}}{0.238} = \mathbf{6.1 \times 10^{-4}}$$

A K value is not assigned units. Note that the two answers are close. Since K values have high uncertainty, if the approximation is within 5% of the exact answer, solving with the approximation is considered "close enough."

2b. WANTED: pH $pH \equiv -\log [H^+]$ and $[H^+] \equiv 10^{-pH}$

$\mathbf{pH} = -\log [H^+] = -\log (1.2 \times 10^{-2}) = \mathbf{1.? = 1.92}$

2c. WANTED: $[OH^-]$

In an acid solution, find $[H^+]$ using acid rules, then $[OH^-]$ using K_w.

DATA: $[H^+] = x = 0.012\ M = 1.2 \times 10^{-2}\ M$ (From part a.)

SOLVE: $K_w = [H^+][OH^-] = 1.0 \times 10^{-14}$

$$[OH^-] = \frac{1.0 \times 10^{-14}}{[H^+]} = \frac{1.0 \times 10^{-14}}{1.2 \times 10^{-2}} = 0.83 \times 10^{-12} = \mathbf{8.3 \times 10^{-13}\ M}$$

(K_w check: $[H^+] \times [OH^-]$ must estimate to 10×10^{-15} or 1×10^{-14})

Lesson 21.4

Practice A   WANTED: pH $pH \equiv -\log[H^+]$ and $[H^+] \equiv 10^{-pH}$

First find $[H^+]$. Use the quick steps: Write the WANTED unit, K approximation, substitute, and solve.

$$K_a \approx \frac{x^2}{[WA]_{mixed}}\ ; \quad \frac{x^2}{0.50} = 1.8 \times 10^{-4}$$

$x^2 = (1.8 \times 10^{-4})(0.50) = 0.90 \times 10^{-4} = 90 \times 10^{-6}$ $x = [H^+] = 9.5 \times 10^{-3}\ M$

SOLVE: **pH** $= -\log[H^+] = -\log(9.5 \times 10^{-3}) = 2.? = \mathbf{2.02}$

Practice B   1. WANTED: Percent ionization = Percent dissociation
Write the known equation that includes the WANTED term.

$$\% \text{ Dissociation} = \frac{x}{[WA]_{mixed}} \cdot 100\%$$

Make a DATA table using the equation terms.

DATA: $[WA]_{mixed} = 0.40\ M$

$x = [H^+] = 2.0 \times 10^{-3}\ M$

SOLVE: **%Dissoication** $= \dfrac{x}{[WA]_{mixed}} \cdot \mathbf{100\%} = \dfrac{2.0 \times 10^{-3}\ M}{0.40\ M} \cdot 10^2\% = \mathbf{5.0 \times 10^{-1}\% = 0.05\%}$

2a. The pH Prompt: $pH \equiv -\log[H^+]$ and $[H^+] \equiv 10^{-pH}$

WANTED: $[H^+] = x$ (Rule for weak acids.)
For K_a and [ions], write the WRECKS, but K_a is not mentioned in this problem.
We want $[H^+]$ and know pH. The relationship that solves for $[H^+]$ is

$[H^+] = 10^{-pH} = 10^{-4.49} = \mathbf{3.2 \times 10^{-5}\ M}$ (Check: 4.49 rounded up = 5.)

2b. For weak acids, $x = [H^+]$ (Solved in part a.)

% Ionization $= \dfrac{x}{[WA]_{mixed}} \cdot 100\% = \dfrac{3.2 \times 10^{-5}\ M}{2.0\ M} \times 10^2\% = \mathbf{1.6 \times 10^{-3}\% = 0.0016\%}$

3. WANTED: K_a Write needed equations based on the terms used in the problem.

$$\% \text{ Dissociation} = \frac{x}{[WA]_{mixed}} \cdot 100\%$$

$$K_a \approx \frac{x^2}{[WA]_{mixed}} \quad \text{(Approximation)}$$

DATA: $[WA]_{mixed} = 0.50\ M$
% Dissociation $= 0.30\%$

In this problem, the percent dissociation equation above has one unknown value, and the K approximation has two. Begin by substituting into the top equation to find x.

$$0.30\% = \frac{x}{0.50\ M} \cdot 100\%; \ x = (0.30\%/100\%) \cdot 0.50\ M = \mathbf{0.0015\ M = 1.5 \times 10^{-3}\ M}$$

SOLVE for the WANTED symbol:

$$K_a \approx \frac{x^2}{[WA]_{mixed}} \approx \frac{(1.5 \times 10^{-3})^2}{0.50} \approx \mathbf{4.5 \times 10^{-6}}$$

22

Nuclear Chemistry

Lesson 22.1   The Nucleus—Review

Chemistry and Nuclear Reactions

The rules for nuclear reactions are quite different from those that govern chemistry. For example,

- During *chemical* reactions, the nucleus is not changed in any of the reacting atoms. Since the number of protons in the nucleus determines the identity of the atom, the number and kind of atoms cannot change during chemical reactions.

  In *nuclear* reactions, nuclei can combine or divide to form different atoms.
- Chemical reactions, such as the explosion of TNT (trinitrotoluene), release large amounts of energy. However, the energy added or released per atom in nuclear reactions is nearly always much higher than in chemical reactions. An atomic bomb is an example of a nuclear reaction.
- On Earth, chemical reactions occur with relative ease. In the human body, millions of chemical reactions occur each minute. Nuclear reactions do occur naturally on Earth, such as during radioactive decay, but compared to chemical reactions, nuclear reactions are relatively rare. (In a star, however, nuclear reactions are common.)

Because of this difference in rules, the detailed investigation of *most* nuclear reactions is generally assigned to physics. In chemistry, we limit our study to three nuclear reactions: radioactive decay, fission, and fusion. These three types of reactions explain how atoms are formed.

A Model for the Nucleus

Science has only a partial understanding of the nature of the atomic nucleus. When understandings in science are limited, *models* are developed that may be simplified, incomplete, or even speculative, but are useful in predicting how systems will behave.

To explain the nuclear reactions that are important in chemistry, the model for the nucleus that we use in chemistry is simplified compared to the models of physics, but this simplified model can nearly always predict the effect of the structure of the nucleus on processes of interest in chemistry and biology.

Our chemistry model for the atom and its nucleus was introduced in Lesson 6.2. Let's briefly review, with a focus on factors important in nuclear *reactions*.

1. **Nuclear structure:** Atoms are composed of three *subatomic particles*. In standard chemistry, our primary focus is on the behavior of the electrons in atoms. In nuclear chemistry, our focus is on the protons and neutrons in the nucleus of atoms.

2. **Protons**
   - Protons have a +1 electrical charge (1 unit of positive charge). Each proton has a mass of 1.007 amu (atomic mass units), which is equivalent to 1.007 grams per mole.
   - The number of protons is the *atomic number* of an atom. The number of protons determines the *name* (and thus the *symbol*) of a nucleus

or atom. The number of protons determines the *nuclear charge* of an atom's nucleus.
- The number of protons is a major factor in the atom's behavior.
- The number of protons in an atom is never changed by chemical reactions but can change during nuclear reactions.

3. **Neutrons**
   - Neutrons have an electrical charge of zero. A neutron has about the same mass as a proton: 1.009 amu.
   - Neutrons, like protons, are never gained or lost in chemical reactions, but the number of neutrons and protons in an atom can change in a nuclear reaction.
   - Unlike the number of protons, the number of neutrons in an atom has no influence on the types of chemical reactions that substances containing that atom will undergo. However, nuclei with the same number of protons but different numbers of neutrons will undergo different nuclear reactions.

4. **The nucleus:** All of the protons and neutrons in an atom are found at the center of the atom: in the nucleus. The diameter of a nucleus is roughly 100,000 times smaller than the effective diameter of most atoms. However, the nucleus contains all of an atom's positive charge and nearly all of its mass.

5. **Types of nuclei:** Only certain combinations of protons and neutrons form a nucleus that is stable. In a nuclear reaction, if a combination of protons and neutrons is formed that is unstable, the nucleus will decay. In terms of stability, nuclei can be divided into three types.
   - **Stable** nuclei are combinations of protons and neutrons that do not change in a planetary environment such as Earth over many billions of years.
   - **Radioactive** nuclei are *somewhat* stable. Some radioactive nuclei exist for only a few seconds, and others exist on average for several billion years, but they fall apart (**decay**) at a constant and characteristic rate.
   - **Unstable** nuclei, if formed in nuclear reactions, decay within a few seconds.
      Nuclei that exist on Earth include all of the stable nuclei plus some radioactive nuclei. All atoms in the Earth's crust with between 1 and 82 protons [except technetium (Tc) and promethium (Pm)] have at least one combination of protons and neutrons that is stable. Atoms with 83 to 92 protons exist in the Earth's crust but are always radioactive. Atoms with 93 or more protons exist on Earth only when they are created in nuclear reactors or nuclear weapons.
      Radioactive atoms comprise a very small percentage of the matter on Earth. More than 99.99% of Earth's atoms have stable nuclei that have not changed since atoms came together to form our planet billions of years ago.

6. **Terminology:** Protons and neutrons are termed the **nucleons**. The combination of a certain number of protons and neutrons is called a **nuclide**. The set of nuclides that have the same number of protons (so they are the same atom) but differing numbers of neutrons are called the **isotopes** of an atom.

Isotopes

Some atoms have only one stable nuclide; others have as many as 10 stable isotopes.

Examples: All atoms with one proton are called hydrogen. Two kinds of hydrogen nuclei are stable: those with

- One proton
- One proton and one neutron, an isotope that is referred to as **hydrogen-2**, or **deuterium**, or **heavy hydrogen**

Most hydrogen atoms found on Earth are the isotope that contains one proton and no neutrons: only one H atom in about 6,400 contains a deuterium nucleus. However, deuterium can be separated from the majority isotope, and it has many important uses in chemistry.

An isotope of hydrogen consisting of one proton and two neutrons, called **tritium**, is not found in the Earth's crust, but it can be produced in nuclear reactors. Unlike deuterium, tritium is radioactive. About half of the nuclei in a sample of tritium will decay in 12 years.

Nuclide Symbols

Each nuclide has a *mass number*, which is the *sum* of its number of protons and neutrons.

$$\text{\textbf{Mass Number} of a nucleus} = \textbf{Protons} + \textbf{Neutrons}$$

Example: All nuclei with six protons are named carbon. If a carbon nucleus has eight neutrons, the mass number of the isotope is 14.

A nuclide can be identified in two ways:

- By its number of protons and number of neutrons
- By its *nuclide symbol* (also termed its *isotope symbol*)

The nuclide symbol for an atom has two required parts: the *atom symbol* and the *mass number*, with the mass number written as a superscript in front of the atom symbol.

Examples: The three isotopes of hydrogen can be identified as

- One proton + no neutrons *or* as ^{1}H (named hydrogen-1 or a protium nucleus)
- One proton + one neutron *or* as ^{2}H (termed hydrogen-2 or a deuterium nucleus)
- One proton + two neutrons *or* as ^{3}H (called hydrogen-3 or a tritium nucleus)

Uranium has two isotopes that are important commercially and historically.

- ^{238}U, the most common naturally occurring isotope, contains 92 protons and 146 neutrons.
- ^{235}U, the isotope that is *split* in atomic bombs and nuclear power plants, contains 92 protons and 143 neutrons.

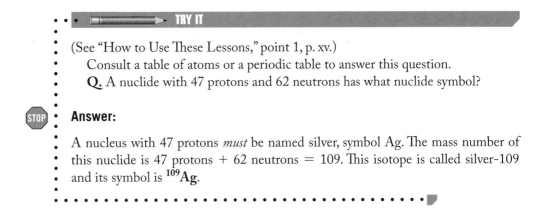

(See "How to Use These Lessons," point 1, p. xv.)

Consult a table of atoms or a periodic table to answer this question.

Q. A nuclide with 47 protons and 62 neutrons has what nuclide symbol?

STOP

Answer:

A nucleus with 47 protons *must* be named silver, symbol Ag. The mass number of this nuclide is 47 protons + 62 neutrons = 109. This isotope is called silver-109 and its symbol is ^{109}Ag.

Nuclide symbols may also be written with the nuclear charge below the mass number. This is called *A–Z notation*, illustrated for a tritium nucleus below. *A* is the symbol for mass number and *Z* represents nuclear charge.

$$^{3}_{1}\text{H}$$

Any nuclear particle that includes protons is by definition an atom, and since the atom symbol also identifies the number of protons in the nucleus, *Z* values are not required to identify an atom. However, for many subatomic particles, the nuclear charge is not the same as the number of protons, and showing the nuclear charge is helpful in identifying these particles. In addition, for problems in which we must balance nuclear reactions, knowing the nuclear charge is necessary, and showing *Z* is helpful.

PRACTICE

First learn the rules above, then complete these problems.

1. The charge on the nucleus of an atom is determined by its number of _____.

2. The mass number of a nucleus is determined by its number of _____.

3. Isotopes have the same number of _____ but different numbers of

    _____.

4. Of the three subatomic particles, the two with the highest mass are

    _____ and _____.

5. Write the nuclide (isotope) symbol for a single proton using *A–Z* notation.

6. Consulting a periodic table, fill in the blanks below.

Protons	Neutrons	Atomic Number	Mass Number	Nuclide Symbol
2	2			
	118	79		
82			206	
				^{242}Pu

Lesson 22.2

Radioactive Decay

Stable Nuclei

Protons (abbreviated p^+) have a positive electrical charge. If there is more than one proton in a nucleus, the like charges of the protons repel and the nucleus will have a tendency to fly apart.

Neutrons (abbreviated n^0) have zero electrical charge and they do not repel other types of particles or each other. Neutrons act in a complex way as the "glue" of the nucleus: if the *right* number of neutrons is mixed with the protons, the nucleus is stable.

For small nuclei, the neutron to proton ratio that results in a stable nucleus is about one to one. As the number of protons in nuclei increases, the number of neutrons needed to form a stable nucleus increases slightly faster: the n^0/p^+ ratio gradually increases.

Examples:

- All stable fluorine nuclei have 9 protons and 10 neutrons.
- Chlorine has two stable nuclei: both have 17 protons, while one has 18 and the other 20 neutrons.
- All lead atoms have 82 protons. The four stable isotopes of lead have 122, 124, 125, or 126 neutrons.

However, forming a stable nucleus is more complex than just adding more neutrons as "glue" or being in the right range of ratios. Certain combinations of protons and neutrons are stable, but others are not.

Radioactive Decay

A nucleus that is **radioactive** is between stable and unstable: it will have a tendency to *gradually* expel particles until a stable neutron and proton combination is achieved. The process of expelling particles from the nucleus is termed **radioactive decay**. Depending on the composition of the nucleus, the average time of radioactive decay may be seconds or billions of years.

Radioactive decay can be a powerful tool in the study of chemical reactions.

Example: For most atoms with fewer than 84 protons, stable isotopes exist, but radioactive isotopes also exist. The radioactive nuclei can either be found in nature or synthesized in nuclear reactors. A radioactive atom will undergo *decay* that we can detect, but in chemical reactions the radioactive and nonradioactive forms of the atom have essentially the same behavior.

By substituting a radioactive nucleus for a stable nucleus, we can "tag" the atoms in substances. Because we can detect the location of radioactive nuclei as they decay, we can track how they behave during chemical reactions, including reactions in biological systems. The use of *radioactive dyes* in medical imaging is one example of the importance of nuclear chemistry.

There are several types of radioactive decay but the two types that explain the decay of most radioactive nuclei are **alpha (α) decay** and **beta (β) decay**.

Alpha Decay

When a particle composed of two protons and two neutrons is ejected from a nucleus, it is termed an **alpha particle**. Because an alpha particle has the same structure as a helium-4 nucleus, it can be given the same isotopic symbol.

$$\alpha \text{ particle} = {}_2^4\text{He}$$

The process of ejecting an alpha particle from a nucleus is termed *alpha decay*. Alpha decay lowers the atomic number (and nuclear charge) of a nucleus by two and its mass number by four.

Example: The isotope U-238 undergoes radioactive decay by emitting an alpha particle. This nuclear reaction is written as

$$^{238}_{92}U \xrightarrow{\alpha} {}^{234}_{90}Th + {}^{4}_{2}He$$

Balancing Nuclear Reactions

Nuclear reactions balance by different rules than chemical reactions, but nuclear reactions balance easily. The rule is

> In nuclear reactions, both mass numbers and nuclear charge must be conserved.

This means that in a balanced nuclear reaction equation, on both sides of the arrow,

- The sum of the mass numbers (A values on top) must be the same
- The sum of the nuclear charges (Z values on the bottom) must be the same

The result is that nuclear reactions can be balanced by simple addition and subtraction.

Example: In the alpha decay of U-238, the mass numbers on both sides of the reaction equation total 238 and the nuclear charges total 92.

$$^{238}_{92}U \xrightarrow{\alpha} {}^{234}_{90}Th + {}^{4}_{2}He$$

 TRY IT

Apply the nuclear balancing rule to these problems.

Q. Write the symbol for the nucleus remaining after the alpha decay of plutonium-242.

$$^{242}_{94}Pu \xrightarrow{\alpha} \underline{\hspace{1.5cm}} + {}^{4}_{2}He$$

Answer:

The nuclear charges on the bottom must total 94 on both sides. The nuclear charge on the missing particle must be 92 which means the atom is uranium. The mass numbers on top add up to 242 on both sides, so the mass number of the missing nucleus must be 238. The balanced nuclear reaction is

$$^{242}_{94}Pu \xrightarrow{\alpha} {}^{238}_{92}U + {}^{4}_{2}He$$

Try another.

Q. Which isotope is produced by the alpha decay of radium-226?

Answer:

A key to balancing nuclear reactions is to write the known isotopes in *A–Z* notation. Begin there for radium-226.

Radium by definition has a nucleus with 88 protons, so this reaction begins

$$^{226}_{88}\text{Ra} \xrightarrow{\alpha}$$ Fill in the missing symbols.

In alpha decay, one product is always an alpha particle. Add its symbol on the right.

$$^{226}_{88}\text{Ra} \xrightarrow{\alpha} {}^{4}_{2}\text{He} +$$

Use the balancing rule to write the isotopic formula for the remaining particle: the nucleus left behind after the alpha particle is expelled.

$$^{226}_{88}\text{Ra} \xrightarrow{\alpha} {}^{4}_{2}\text{He} + {}^{222}_{86}\text{Rn}$$

After the decay, the nucleus has 86 protons, so it must be radon (Rn). For the mass numbers to balance, the Rn nucleus must have a mass number of 222.

PRACTICE   **A**

Answer these in your notebook.

1. Write the balanced equation for the alpha decay of radon-219.

2. How many protons and how many neutrons are in ^{219}Rn?

3. How many protons and how many neutrons are in the nucleus left behind after the alpha decay of radon-219? How many protons and neutrons are lost in the decay?

4. Lead-206 can be formed by the alpha decay of which radioactive isotope?

Beta Decay

Beta decay is a different type of radioactive decay. In beta decay, a neutron decays into a proton and an electron, and the electron is expelled from the nucleus at high speed. An electron formed in this manner is termed a **beta particle**.

- Because an electron has no protons and neutrons, its mass number is zero.
- Because an electron has a negative charge, when it is formed in the nucleus its "nuclear charge" is −1.

In nuclear reaction equations, a beta particle can be represented in these ways:

$$\beta \text{ particle} = {}^{0}_{-1}\beta \quad \text{or} \quad {}^{0}_{-1}\text{e}$$

Atoms must have a positive nuclear charge, but on a *sub*atomic particle (one without protons), the nuclear charge may be $+1, 0$, or $−1$.

In beta decay, the number of neutrons in a nucleus *decreases* by one, but the number of protons increases by one, so the mass number of the isotope stays the same.

Example: The equation for the beta decay of the radioactive isotope lead-210 can be written as

$$^{210}_{82}\text{Pb} \xrightarrow{\beta} {}^{0}_{-1}\text{e} + {}^{210}_{83}\text{Bi}$$

Before and after the reaction, the mass numbers total 210 and the nuclear charges total 82 on each side, so this is a balanced nuclear equation.

Note that the nucleus remaining after beta decay has lost a neutron but gained a proton, so its atomic *number* increases.

Using the rules for nuclear balancing, we can predict the structure and symbol for the products of beta decay.

TRY IT

Q. For the beta decay of carbon-14, write the isotope symbols for the two nuclear particles formed in the reaction.

STOP

Answer:
Begin by converting to *A–Z* notation: $^{14}_{6}\text{C} \xrightarrow{\beta}$

One product must be a beta particle. For its symbol, use a β or an e.

$^{14}_{6}\text{C} \xrightarrow{\beta} \,^{0}_{-1}\text{e} +$ Complete the balancing.

STOP

$$^{14}_{6}\text{C} \xrightarrow{\beta} \,^{0}_{-1}\text{e} + \,^{14}_{7}\text{N}$$

The isotope formed by the beta decay of carbon-14 is nitrogen-14.

PRACTICE   B

1. From memory, write the isotopic symbols for an alpha particle and a beta particle.

2. Write balanced equations for these decay reactions.

   a. $^{40}\text{K} \xrightarrow{\beta}$ b. $^{239}\text{Pu} \xrightarrow{\alpha}$

3. The isotope ^{131}I is used for the treatment of hyperthyroidism: a condition in which the thyroid gland produces too much thyroid hormone. In the body, iodine is absorbed by the thyroid gland. If iodine-131 is administered to a patient, its beta decay kills cells in the thyroid, resulting in a reduced level of thyroid hormone without surgery. Write the symbol for the nucleus produced by the beta decay of iodine-131.

4. Lead-206 can be produced by the beta decay of which nucleus?

5. In a part of what is termed a **radioactive decay series**, nucleus A with 88 protons and 140 neutrons can beta decay to form nucleus B. B can emit a high-speed electron to form nucleus C, which can α decay to form nucleus D. Write isotopic symbols for A, B, C, and D.

6. Radioactive atom A with 82 protons and 132 neutrons emits a beta particle to become atom B. Atom B emits an alpha particle to become atom C, which can emit a high speed electron from the nucleus to form atom D. Write the isotopic symbols for the nuclei in A, B, C, and D.

Lesson 22.3

Fission and Fusion

Fission

Nuclear fission is a reaction in which a nucleus divides into two smaller nuclei, both of which contain more than two protons. A fission reaction is often accompanied by the release of free neutrons, and those neutrons can collide with other nearby fissionable nuclei and cause them to split.

If this "splitting of atoms" begins in a sample of fissionable nuclei that is large enough to have **critical mass**, the result can be a **chain reaction** that releases large amounts of energy. If the chain reaction is not controlled, the result is an **atomic bomb**.

In a **nuclear power plant** (a type of **nuclear reactor**), a chain reaction is controlled by adding materials that absorb some of the free neutrons. The large amount of energy produced by a chain reaction is then released gradually, and the resulting heat can be harnessed to drive turbines that produce electricity.

One example of nuclear fission is the splitting of uranium-235. ^{235}U can fission in many ways, but a typical reaction is

$$^{235}_{92}U + ^{1}_{0}n \rightarrow ^{236}_{92}U \rightarrow ^{140}_{54}Xe + ^{94}_{38}Sr + 2\,^{1}_{0}n + 2 \times 10^{10} \text{ kJ/mol}$$

In this reaction, a U-235 nucleus is struck by a free neutron. The neutron is at first absorbed, but this unstable nucleus then splits into two smaller nuclei plus two free neutrons. Those two neutrons can collide with other fissionable nuclei to create a chain reaction.

Fission reactions produce amounts of energy per mole that are millions of times larger than that produced by chemical reactions such as the burning of fossil fuels, and processes that can produce energy at a controlled rate are valuable to society. However, a disadvantage of using fission for electricity generation is that the products include highly radioactive isotopes. Exposure to the radiation released by radioactive decay can cause cancer, and some of the waste products of fission remain significantly radioactive for thousands of years. A major issue in nuclear power generation is how to store waste products so that they cannot escape into the environment.

Isotopic Separation

Both U-235 and U-238 can be split, but in practice only U-235 is an effective fuel for chain reactions. For use in nuclear power plants or weapons, naturally occurring uranium must be **enriched**, meaning that the percentage of nuclei that are U-235 must be increased. In mined uranium ore, 99.3% of nuclei are U-238 and only 0.7% are U-235. To generate nuclear power, uranium must be enriched to at least 3% U-235. For nuclear weapons, uranium is enriched to at least 20% U-235, and over 50 kilograms of this "weapons-grade" uranium must be collected.

Since all isotopes, including U-235 and U-238, have the same tendency to react chemically, chemical reactions cannot effectively separate isotopes. However, substance particles with the lighter isotopes will move faster on average at a given temperature. For example, molecules of UF_6 gas that contain U-235 atoms diffuse (move away from a starting point) slightly faster than those containing U-238, and gaseous diffusion is one method that is used to separate isotopes. Because the heavier isotopes have a higher mass, in a given substance the molecules with the heavier isotope tend to be distributed toward the outside when spun in a circle at high speed, so that a centrifuge can separate uranium isotopes.

However, both gaseous diffusion and centrifugation of uranium-containing substances are slow and expensive processes. This makes it difficult (thankfully) to obtain the amount of enriched uranium needed to build a nuclear weapon.

PRACTICE   A

1. From memory, write the isotopic symbol for a free neutron.

2. Fill in the one missing isotopic symbol in this nuclear fission reaction.

$$^{235}_{92}\text{U} + ^{1}_{0}\text{n} \rightarrow ^{87}_{35}\text{Br} + \underline{\hspace{2cm}} + 6\,^{1}_{0}\text{n}$$

3. Fill in the missing isotopic symbol that is the first reactant in this fission equation.

$$\underline{\hspace{2cm}} + ^{1}_{0}\text{n} \rightarrow ^{91}_{38}\text{Sr} + ^{146}_{56}\text{Ba} + 3\,^{1}_{0}\text{n}$$

Fusion

Nuclear fusion is a reaction that combines two nuclei to make a larger one. The fusion of nuclei lighter than iron generally produces large amounts of energy. An example of a fusion reaction is

$$^{2}_{1}\text{H} + ^{3}_{1}\text{H} \rightarrow ^{4}_{2}\text{He} + ^{1}_{0}\text{n} + 2 \times 10^{9}\text{ kJ/mol}$$

In this reaction, two hydrogen nuclei are fused, and one product is a heavier helium nucleus. Fusion is the reaction that produces energy in stars, including our sun, and in hydrogen bombs.

The primary reaction that causes stars to "burn" (release energy) is the conversion of hydrogen to helium. At the extremely high temperatures and pressures found in stars, lighter nuclei can fuse to form heavier nuclei, and those nuclei can undergo successive fusion reactions. After long periods of making heavier nuclei, some stars become unstable and explode. Over time, the atoms scattered into space from exploded stars can accumulate due to gravitational attraction, forming new stars and planets. The atoms that coalesced to form our own planet billions of years ago are nuclei, or the decay products of nuclei, that were originally formed by fusion in a star.

When fusion combines hydrogen isotopes, the products are generally stable nuclei rather than long-lived radioactive isotopes. If the fusion of hydrogen could be slowed and contained in a nuclear reactor, the result could be energy generated without greenhouse gas production or radioactive waste. To produce the energy needed for our society, such fusion reactors could replace the burning of fossil fuels and fission-based nuclear power plants, both of which form products that can harm our environment. However, all of the nuclear reactors currently in use are based on nuclear fission. No practical way has yet been discovered to engineer a gradual release of the energy of nuclear fusion.

SUMMARY

So far in this chapter, we have learned the following four rules:

1. α particle $= ^{4}_{2}\text{He}$; β particle $= ^{0}_{-1}\beta$ *or* $^{0}_{-1}\text{e}$; neutron $= ^{1}_{0}\text{n}$

2. To balance nuclear reactions, write nuclide symbols in *A–Z* notation: mass number on top, nuclear charge on the bottom.

(continued)

3. In nuclear reactions, the mass numbers and nuclear charges must be conserved: each must *add* to give the same number on both sides of the reaction equation.

4. Fission splits a nucleus, fusion combines nuclei.

As needed, design flashcards that assist in moving these rules into long-term memory.

PRACTICE B

1. Briefly describe the difference between fission and fusion.

2. If a single nucleus is formed as the product of this reaction, write its isotope symbol.

$$_1^2H + _1^2H \rightarrow$$

3. In stars that are *red giants*, helium-4 can fuse with beryllium-8 to form a single nucleus. Write the equation for this reaction.

4. Assuming that *one* isotope symbol is missing from these equations, fill in the missing nuclide symbol, then write the name for this *type* of reaction.

Type:

a. $^{235}U + _0^1n \rightarrow \, ^{92}Kr + \qquad + 3\,_0^1n$ _____

b. $^{238}U \rightarrow \, ^4He +$ _____

c. $^1H + \, ^3H \rightarrow$ _____

d. $^{234}Th \rightarrow \, _{-1}^0e +$ _____

Lesson 22.4 **Radioactive Half-Life**

Fractions and Percentages: Review

To calculate a radioactive half-life, we will need to convert between fractions and percentages. Review these rules, then complete Practice A.

1. Fraction = $\dfrac{\text{quantity A}}{\text{quantity B}}$ and often equals $\dfrac{\text{part}}{\text{total}} = \dfrac{\text{smaller number}}{\text{larger number}}$

2. Fraction = decimal equivalent = numerator divided by denominator
   Example: $1/4 \equiv 0.25$
   In value and in meaning: fraction = decimal equivalent

3. A percentage is a fraction or its decimal equivalent multiplied by 100%.

Percent = $\dfrac{\text{part}}{\text{total}} \times 100\%$ = fraction $\times 100\%$ = (decimal equivalent) $\times 100\%$

4. To convert between a decimal equivalent and its corresponding percentage,
   - The decimal moves two places
   - The number representing the percentage is always 100 times larger than the decimal equivalent
     Examples: $0.25 = 25\%$; $0.54\% = 0.0054$

5. If a percent is WANTED, solve for the *fraction first*, then convert the resulting decimal equivalent to its percentage.

PRACTICE A

For additional review, see Lesson 10.5.

1. 1/8 has what decimal equivalent and what percentage?

2. 16% has what decimal equivalent?

3. 0.24% has what decimal equivalent?

4. 7.4/10,000 has what decimal equivalent and percentage?

5. After decay, if 6 parts remain out of an original 30,

   a. What fraction remains?   b. What percentage remains?

Half-Life

The **half-life** of a reactant (symbol $t_{1/2}$) is the time required for half of the reactant to be used up in a reaction. A radioactive nucleus has a characteristic half-life: in any sample, the time in which half of the nuclei decay is *constant*.

The rate of chemical reactions varies with changes in temperature, but in radioactive decay, a nuclear reaction, the time of a half-life does not vary significantly even over temperature ranges of 1000 K.

Some radioactive nuclides have a half-life of a few seconds; others have a half-life of billions of years. It is not possible to predict when any one nucleus will decay, but in any sample of more than a few hundred of a given nucleus, we can calculate how long it will take for any *percentage* of the nuclei to decay.

Half-Life Calculations for Simple Multiples

In calculations involving half-life, the two variables will generally be the *time* over which a given nuclide in a sample decays and the *percentage* of those nuclei that remain. If the time period for the decay is equal to either the half-life or a simple multiple of the half-life, calculations can be answered by mental arithmetic.

- If a sample of a given nucleus has decayed for a time equal to *one* half-life, half of the original nuclei have decayed and half remain.

  If the nuclei are in a sample that has a constant volume (which should be assumed unless other conditions are stated), half of the original *concentration* of the nucleus remains after one half-life.

- After *two* half-lives (double the time of the half-life), the number of nuclei remaining is half *of* the half that remained after the first half-life: half of $1/2 = 1/4$ (25%) of the original nuclei remain and 75% have decayed.

- At triple the half-life, 1/2 of 1/4 = 1/8 (12.5%) of the original nuclei remain.

In decay calculations, note that you must distinguish between quantities *decayed* and *remaining*.

- 100% − percentage decayed = percentage remaining
  Example: If 25% remains, 75% has decayed.
- 1.000 − fraction decayed = fraction remaining
  Example: If the fraction decayed is 0.35, the fraction remaining is 0.65.

TRY IT

Apply the rules above to the following problem.

Q. Fluorine-18, a radioactive isotope used in nuclear medicine, has a 1.8 hr half-life. How long will it take for 87.5% of the ^{18}F nuclei in a sample to decay?

Answer:

If 87.5% has decayed, 12.5% remains. How many half-lives are required? How much time would this be?

After two half-lives, 25% remains, so after three half-lives, 25% × 1/2 = 12.5% remains. 3 half-lives × 1.8 hr/half-life = **5.4 hr**

In radioactive decay calculations for these "simple multiple" cases, given the headings in the table below, you will need to be able to fill in the rest of the table from memory. This should not be difficult: note that in the two middle columns, each number is simply half of the one above.

For Radioactive Nuclei, at Time =	Fraction Remaining	Percentage Remaining	Percentage Decayed
0	1	100%	0%
One half-life	1/2	50%	50%
Two half-lives	1/4	25%	75%
Three half-lives	1/8	12.5%	87.5%

For half-life calculations that are not easy multiples, we will use the logic of this chart to make estimates to check answers.

PRACTICE   B

Write the table above until, given the top row, you can fill in four rows below from memory. Then complete these problems.

1. If 90.% of a sample has decayed, what fraction remains?

2. If the fraction of a sample that has decayed is 0.40, what percent remains?

3. The nucleus of plutonium-239 undergoes radioactive decay with a half-life of 24,400 years. In a sample with a constant volume containing ^{239}Pu,

a. After how many years will 25% of the original ^{239}Pu nuclei remain?

b. After how many half-lives will the [^{239}Pu] be 1/16th of its original concentration?

c. What percentage of the ^{239}Pu has decayed after exactly four half-lives?

Lesson 22.5   Natural Logarithms

Review of Base 10 Logs

In chemistry, **natural logarithms** are used to solve a variety of problems, including calculations involving the half-life of radioactive isotopes. The rules for natural logs parallel those for base 10 logs, but because our number system is based on 10, it is easier to learn the logic of the base 10 logs first.

   If you have not completed Lesson 20.6 on base 10 logarithms, do so now. If you have done that lesson, review the following summary, then complete the practice set below.

SUMMARY

Log Rules to Commit to Memory

1. A *logarithm* is simply an exponent: The power to which a base number is raised.

2. A logarithm answers the question: If a number is written as a base to a power, what is the power?

3. (log) buttons on a calculator find the *power* of a number written as 10 to a power.

4. The equation defining a **log** is **log 10^x ≡ x**. The *log* of 100 is 2.

5. Knowing the log of a number *x*, find the number by "taking the antilog" or "finding the inverse log." Perform the calculator operation **10$^{\log x}$**.
   Recite and repeat to remember: "10 to the log x equals x."

6. On a calculator, to convert a log value to a number,
   • Input the log value, then press (INV)(LOG) or (2nd)(LOG), or
   • Input the log, then press (10^x) *or* input 10, (x^y), input the log, (=)

7. To check log calculations: When a number is written in scientific notation, its power of 10 must agree with its base 10 logarithm within ± 1.

PRACTICE   A

Practice with the calculator you will use on tests. Complete every other problem now, and the rest in a later practice session. If you have problems with *any* of these, review Lesson 20.6 on base 10 logarithms.

1. Complete these without a calculator. Answer using integers.

   a. Log(10^{14}) =   b. Log(10) =   c. Log(1) =

(continued)

2. Answer the following with a calculator. For this question, round all answers to three *s.f.*

   a. $10^{3.2} =$ (in scientific notation):   (± 1?) _____

   b. $10^{-12.3} =$ (± 1?) _____

   c. $10^{-0.2} =$ (number): _____ (scientific notation):   (± 1?) _____

   d. $\text{Log}(2.0 \times 10^{14}) =$ (± 1?) _____

   e. $\text{Log}(2.0 \times 10^{-14}) =$ f. $\text{Log}(5.0) =$

   g. $\text{Log}(0.0050) =$ h. $\text{Log X} = 4.7, \text{X} =$

   i. $\text{Log A} = -8.2, \text{A} =$ j. $\text{Log D} = -0.50, \text{D} =$

   k. $\text{Log X} = 6.6, \text{X} =$ l. $10^{-9.5} =$

   m. $\text{Log}(3.0 \times 10^{-5}) =$

The Symbol *e*

In mathematical and scientific equations, the lowercase *e* is an abbreviation for a number: **2.7182818....** For calculations, the value $e = 2.718$ must be memorized.

The number *e* has many interesting mathematical properties. In science, *e* is found in many equations that predict natural phenomena. In these equations, *e* is the base for values expressed in the form e^x, and *e* is termed the **natural exponential**.

To solve calculations involving radioactive half-life, we will use both the natural exponential *e* and the natural log function **ln**. Let's consider calculations with *e* first.

Calculating with the Natural Exponential

TRY IT

Q. We know that e^1 equals what number?

Answer:

2.718.... Using 1 and the (e^x) button, write the key sequence that produces that answer for e^1 on your calculator.

- A standard TI-type calculator might use 1 (e^x)
- On an RPN scientific calculator, try 1 (enter) (e^x)

Q. Use your key sequence and write the answers to these.

   a. $e^2 =$ b. $e^{2.5} =$

   c. $e^{-1} =$ d. $e^{-2.5} =$

(Because the statistical basis for significant figures does not apply to logarithmic calculations, in these lessons we will use this general rule: During e and ln calculations, round fixed-decimal numbers and significands in answers that are not integers to three significant figures.)

Answers:

a. $e^2 = $ **7.39**
b. $e^{2.5} = $ **12.2**

Recall that to enter a negative number usually requires the $\boxed{+/-}$ key.

c. $e^{-1}\,(= 1/e = 1/2.718) = $ **0.368**   d. $e^{-2.5}\,(= 1/e^{2.5}) = $ **0.0821**

(Note: Some calculators use an **E** at the right side of the answer screen to show the power of **10** for numbers in scientific notation. This **E** is *not* the same as the symbol *e* for the natural exponential.)

Calculating Natural Logs

The **ln** function (the **natural log**) answers this question: if a number is written as e to a power, what is the power?

Just as by definition, **log** $10^x \equiv x$, the natural log definition is **ln $e^x \equiv x$** .

 TRY IT

Q. Use the natural log definition to do these with*out* a calculator.

a. $\ln e^0 = $
b. $\ln e^1 = $
c. $\ln e^{-4} = $

Answers:

a. $\ln e^0 = $ **0**
b. $\ln e^1 = $ **1**
c. $\ln e^{-4} = $ **−4**

Q. By definition, $\ln e\ = $ ____.

Answer:

$\ln e = \ln e^1 = $ **1**

Q. Try this one in your head: **ln(2.718)** should equal about ____.

Answer:

$\ln(2.718) \approx \ln e \approx \ln e^1 \approx $ **1**

Q. Now use your calculator for the same calculation: **ln(2.718)** $= $ ____.

Answer:

Is the calculator answer close to the mental arithmetic answer?

Write down the key sequence that works to solve **ln(2.718)** above. The same steps should take the natural log of any positive number.

TRY IT

Q. Using your calculator: $\ln(314) =$

Answer:

$\ln(314) = \mathbf{5.75}$

To *check* the calculation of an ln value, *after* writing it down, use the $\boxed{e^x}$ key or $\boxed{\text{INV}}$ $\boxed{\ln}$ keys and see whether you *return* to the number.

TRY IT

Q. Apply that rule as a check on these:

     a. $\ln(0.00500) =$ After writing the answer, use $\boxed{e^x}$. Check? _____

     b. $\ln(6.02 \times 10^{23}) =$ Check?

     c. $\ln(19.3 \times 10^{-15}) =$ Check?

Answers:

     a. $\ln(0.00500) = \mathbf{-5.30}$ b. $\ln(6.02 \times 10^{23}) = \mathbf{54.8}$

     c. $\ln(19.3 \times 10^{-15}) = \mathbf{-31.6}$

Note in *part c* that a calculator does *not* require the input of *scientific* notation. However, if you use the $\boxed{e^x}$ key to check your answer, it will likely return the original number *converted* to scientific notation.

PRACTICE   B

Round your answers to three significant figures.

1. $e^{2.0} =$ 2. $e^{-4.7} =$

3. $e^{-11} =$ 4. $\ln(42) =$

5. $\ln(0.0200) =$ 6. $\ln(9 \times 10^5) =$

7. $\ln(5.00 \times 10^{-4}) =$ 8. $\ln(10^{-4}) =$

Converting ln Values to Numbers

A base **10** definition is $\boxed{10^{\log x} \equiv x}$.
    A base *e* definition is $\boxed{e^{\ln x} \equiv x}$.
    Both of those definitions must be memorized. Noting their similarities will help.
    Using the base *e* definition above, for some calculations involving ln and *e* you will not need a calculator.

TRY IT

Q. $e^{\ln(11)} =$

Answer:

$e^{\ln(11)} = \mathbf{11}$

The equation $e^{\ln x} = x$ also means that if you know the ln value, to find the corresponding fixed-decimal *number*, make the ln value a power of e.

TRY IT

Q. If $\ln(X) = 1$, the fixed-decimal number X (solve in your head) is _____.

Answer:

If $\ln(X) = 1, X = e^{\ln X} = e^1 = \mathbf{2.718}\ldots$

Knowing that answer, do the same ln-to-number conversion on your calculator by taking the antilog.

TRY IT

Q. If $\ln(X) = 1$, the number X obtained using the calculator is _____.

Answer:

Input 1, then press (INV or 2nd) (ln) *or* press (e^x).

Write down or circle the key sequence that converted ln $= 1$ to the number 2.718....

Use your key sequence to convert the following ln values to numbers. Add a power to e, then fill in the blanks.

TRY IT

Q. If $\ln(X) = 6, X = e$ ____ $=$ (number): _____ $=$ (scientific notation): _____

Answer:

If $\ln(X) = 6, X = e^6 =$ (number): **403** $=$ (scientific notation): $\mathbf{4.03 \times 10^2}$

Q. Apply the same steps to these.

a. If $\ln(X) = -4.5; X = e$ ____ $=$ (number): _____
$=$ (scientific notation): _____

b. $\ln(X) = 57.2; X = e$ ____ $=$ (scientific notation): _____

c. $\ln[A] = 0.0300; [A] = e$ ____ $=$ (number and unit): _____

Answers:

   a. If $\ln(X) = -4.5, X = e^{-4.5} = 0.0111 = \mathbf{1.11 \times 10^{-2}}$

   b. $\ln(X) = 57.2, X = e^{57.2} = \mathbf{6.94 \times 10^{24}}$

   c. If $\ln[A] = 0.0300, [A] = e^{0.0300} = \mathbf{1.03\ M}$

Unlike base 10, in base e calculations there is no obvious correlation between the scientific notation exponent and the base e logarithm that helps in checking your answer. However, you can check by taking the ln of the number answer and see whether it returns to the original ln value.

Units and Logarithms

Note that in part c above, the unit expected for a concentration has been added. From a strict mathematical perspective, logarithms cannot be taken of values with units, and logarithm values do not have units. All precisely stated scientific relationships obey these rules.

However, some of the equations we write in chemistry are "shortcuts" that simplify more complex relationships in order to speed problem solving. When using such equations, special rules may be needed to assign units to answers. To make shortcut equations work in half-life calculations, our rules will include the following.

1. When taking the logarithm (using any base) of a value with units, write the result as a value without units.

2. If a WANTED unit is to be calculated based on a logarithm value, first (if needed) make all of the units supplied in the problem *consistent*, then attach the appropriate consistent unit to the answer.

The quantity involved most often in log calculations will be concentration in moles per liter. The rule will be this: If a $[x]$ is WANTED, attach moles/liter (M) to the answer.

TRY IT

Q. Apply those rules to the following problems. Write the answer as a number *or* in scientific notation. When in doubt, check answers as you go.

   a. $\ln[Z] = -12.5, [Z] =$

   b. $\ln[R] = -0.17, [R] =$

   c. $[D] = e^{-1.39}, [D] =$

   d. $\ln(0.250\ M) =$

   e. $\ln[A] = -2.63, [A] =$

Answers:

   a. $[Z] = e^{-12.5} = \mathbf{3.73 \times 10^{-6}\ M}$ (Attach the *unit* of concentration: mol/L.)

b. $[R] = e^{-0.17} = \mathbf{0.844\,M}$ c. $[D] = \mathbf{0.249\,M}$

d. $\ln(0.250\,M) = \mathbf{-1.39}$ (drop the unit)   e. $[A] = e^{-2.63} = \mathbf{0.0721\,M}$

If you get lost on a natural log calculation, a good strategy is to do a similar and simple base 10 mental computation, and then apply the same logic to the natural log case. Simple base 10 calculations can be solved in your head, and the formulas and steps for base 10 and base e calculations are parallel.

Converting between Base 10 and Natural Logs

A general rule for logarithms of any base is $\log_b(x) = \ln(x)/\ln(b)$ where **b** is the base. For base 10 logs, this equation becomes

$$\text{Log}_{10}(x) = \ln(x)/\ln(10) = \log_{10}(x) = \ln(x)/2.303$$

This relationship is generally memorized as $\mathbf{\ln(x) = 2.303\,\log(x)}$ or as "The *natural* log of a number is always 2.303 times higher than the base 10 log."

SUMMARY

Add these to the list of log rules at the beginning of this lesson, and commit these 14 rules to memory. Design and use flashcards as needed.

8. The symbol e is an abbreviation for a number with special properties: $e = \mathbf{2.718\ldots}$

9. The **ln** (natural log) function answers this question: If a number is written as e to a power, what is the power?

10. Knowing the ln, to find the number, take the *antilog*. On a calculator,
    - Input the ln value, then press (INV) (ln), or
    - Input the ln value, then press (e^x). An ln is simply an exponent of e.

11. When you encounter **log** or **ln** in calculations, it helps to write

$$\log 10^x = x \qquad \text{and} \qquad 10^{\log x} = x$$
$$\ln e^x = x \qquad \text{and} \qquad e^{\ln x} = x$$

Note the patterns. Note the logic: A log is an exponent.

12. $\mathbf{\ln(x) = 2.303\,\log(x)}$

13. When taking the logarithm (using any base) of a value with units, write that result as a value without units.

14. If a WANTED unit is to be calculated based on a logarithm value, first (if needed) make all of the supplied units *consistent*, then *attach* the appropriate consistent unit to the answer.

Design flashcards as needed to help in learning these rules.

PRACTICE C

Try the odd-numbered problems first. Complete the even-numbered problems for additional practice or pretest review. Round answers to three *s.f.*

1. $e^{5.2} =$

2. $e^{-1.7} =$

3. $e^{-20.75} =$

4. $\ln(1066) =$

5. $\ln(0.0050) =$

6. $\ln(3 \times 10^8) =$

7. $\ln(14.92 \times 10^{-6}) =$

8. $\ln e^{6.2} =$

9. $e^{\ln(42)} =$

10. If $\ln X = -6.8$, $X = e$ = (number in scientific notation):

11. If $\ln D = 7.4822$, $D =$

12. If $\ln X = -12.5$, $X =$

13. If $\log [A] = -9$, $[A] =$

14. If $\log x = 13.7$, $x =$

15. $\text{Log A} = -13.7$, $A =$

16. $10^{-11.7} =$

17. $\ln[B] = -13.7$, $[B] =$

18. $e^{-11.7} =$

19. $\ln(0.050 \text{ M}) =$

20. $e^{-0.693} =$

Solve the problems below in your notebook.

21. If $\log(x) = 5.0$, $\ln(x) =$

22. If $\ln(x) = 34.5$, $\log(x) =$

23. If $\ln(x) = -(0.075 \text{ day}^{-1})(4.0 \text{ days})$; $x = ?$

24. Given this equation: $\ln[A] = (-0.0173 \text{ s}^{-1})(t)$

    a. If $t = 20.0$ seconds, $[A] = ?$ b. If $[A] = 0.500 \text{ M}$, $t = ?$

25. Given this equation: $\ln[A] = (-0.0241 \text{ yr}^{-1})(t)$

    a. If $[A] = 0.0025 \text{ M}$, $t = ?$ b. If $t = 28.8$ years, $[A] = ?$

Lesson 22.6   Radioactive Half-Life Calculations

Rate Constants for Radioactive Decay

In half-life calculations that do not involve simple multiples, we can solve using **rate equations**. Each radioactive nucleus has a **rate constant** (**k**) for decay that is characteristic: a value that is constant. Different radioactive nuclei have different values for *k*.

The equation that predicts the decay rate for radioactive nuclei can be written as

$$\ln\left[\frac{[A]_t}{[A]_0}\right] = -kt$$

which can be abbreviated as **ln(fraction remaining) = −kt**

In the first equation, $[A]_t/[A]_0$ is the *fraction* of nuclei *remaining* at time = *t*, measured from *t* = 0.

Example: After one half-life, half (50%) of the original radioactive nuclei in a sample have decayed and **0.50** is the *fraction* remaining.

 TRY IT

Q. After two radioactive half-lives, the fraction remaining will be?

Answer:

After two half-lives, the percentage remaining is 25%(1/4), and the fraction remaining is **0.25**

Because the decaying nuclei are being used up over time, the value of [A] after time = *t* will be less than it was at time = **0**, so the value of the fraction must be less than one. This means that in decay calculations, *fractions* must have values between 0 and 1.00 (such as 0.25).

Radioactive half-life calculations often involve decimal equivalents or percentages, and in those cases the form of the equation above that includes (*fraction remaining*) is the more convenient to use.

 TRY IT

Apply the equation that includes *fraction remaining* to this problem.

Q. For the decay of a radioactive isotope with a short half-life, the value for its rate constant is + 0.00500 s^{-1}. What fraction of this isotope remains after 45 seconds?

Answer:

WANTED: Fraction remaining

Equation: ln(fraction remaining) = −*kt*

To solve with an equation, make a data table listing each term in the equation. To solve this equation for the fraction remaining, you will need to solve for **ln**(fraction remaining) first.

DATA: ln(fraction remaining) = ?

  k = 0.00500 s^{-1}

  t = 45 s

SOLVE: ln(fraction remaining) = −kt = −(0.00500 s^{-1}) (45 s)
      = **−0.225**

From ln(fraction) = −0.225, how can you calculate the fraction?

$$\text{Fraction remaining} = e^{\ln(\text{fraction})} = e^{-0.225} = \mathbf{0.80}$$

As one check on a decay calculation, recall that a *fraction remaining* must have a decimal equivalent value between 0 and +1, which 0.80 does.

Solving with Percentages

In the equation

$$\ln(\text{fraction remaining}) = -kt$$

the term *fraction remaining* is an abbreviation for the fraction $[A]_t / [A]_0$.

The units used to calculate this fraction can be units of concentration *or* any consistent units that are proportional to concentration, including the mass or number of particles in a sample that has a fixed volume.

However, to use the equation with the term (fraction remaining), you must calculate using fractions and not percentages. This means

- If a percentage is WANTED, you will need to solve for the *fraction* first.
- If a percentage is *given*, you must convert to its decimal equivalent fraction to substitute into the equation.

Be careful as well to distinguish between quantities *decayed* and *remaining*. Example: If 45% remains, 55% has decayed.

PRACTICE   A

Commit to memory the equation above that includes (*fraction remaining*), then complete each of these.

1. For a radioactive sample, −1.386 is the value for ln(fraction remaining).

   a. What fraction of the sample remains?   b. What percentage has decayed?

2. For the decay of a radioactive nucleus, if the rate constant of the reaction is $k = 0.04606 \text{ hr}^{-1}$, what percentage remains after 50.0 hr?

3. The plutonium-238 used in nuclear power supplies has a half-life of 87.7 yr: at that time, 50.0% of the original nuclei remain. Calculate the rate constant for the decay of this isotope.

Half-Life Calculations for Non-Simple-Multiples

For a radioactive nucleus, after a time equal to one half-life (symbol $t_{1/2}$), half of a sample has decayed and *half* remains. Substituting into the equation

$$\ln(\text{fraction remaining}) = -kt$$

at a time equal to one half-life, we can write

$$\ln(1/2) = -kt_{1/2}$$

TRY IT

Q. Solve the last equation above in symbols for half-life.

Answer:

One way of several to write the equation is $t_{1/2} = -\ln(1/2)\,/\,k$

This equation is one way to *define* radioactive half-life.

In these equations, the rate constant (k) and half-life ($t_{1/2}$) are variables: their numeric values will differ for different radioactive nuclei. The term $\ln(1/2)$ is a constant: it can be converted to a numeric value.

 TRY IT

Q. Use your calculator to convert $\ln(1/2)$ to a fixed decimal number.

$$\ln(1/2) =$$

Answer:

$\ln(1/2) = \ln(0.500) = \mathbf{-0.693}$

Q. Substitute this numeric value into the equation above that defines half-life and simplify.

Answer:

$$t_{1/2} = \frac{-\,(-0.693)}{k}$$

which simplifies to

$$t_{1/2} = \frac{\mathbf{0.693}}{k}$$

This last equation above is a form often listed in textbooks as a definition of radioactive half-life. From this form, it is clear that if you know the half-life, you can find the rate constant k, and if you know the rate constant you can find the half-life.

To solve decay calculations, we need equations that relate the fraction remaining, half-life, time, and rate constant. Several combinations of the equations above can be used, but the best equations are those that are easy to remember. In these lessons, we will follow this rule.

The Radioactive Decay Prompt

If a *radioactive decay* calculation includes a *half-life* and a *fraction* or *percentage* of a sample, and the answer cannot be calculated using simple multiples, write in the DATA and use

$$\ln(\text{fraction remaining}) = -kt \qquad \text{and} \qquad \ln(1/2) = -kt_{1/2}$$

Note that the second equation is simply a special case of the first: when the fraction remaining is 1/2, the time is equal to the half-life.

When using the prompt, we solve for the common variable k.

- First, solve for k using the equation that can be solved with the data provided.
- Then, use the k value to solve for the variable WANTED.

 TRY IT

Apply the two equations of the radioactive decay prompt to this problem.

Q. Iodine-131, the radioactive isotope used to treat thyroid disorders, has a half-life of 8.1 days. What percentage of an initial $[^{131}I]$ remains after 48 hours?

STOP   **Answer:**

WANTED: Percent $[\text{I-131}]_{48\,\text{hr}}$ = % remaining

DATA: $t_{1/2}$ = 8.1 **days** = radioactive half-life

48 hr = 2.0 **days** = t (For equations, must convert DATA to *consistent* units.)

Strategy: Write the equations that relate the symbols in the problem. For radioactive decay calculations that include half-life and fraction or *percentage*, if no simple multiples are evident, write

$$\ln(\text{fraction remaining}) = -kt \quad \text{and} \quad \ln(1/2) = -kt_{1/2}$$

$$\text{Percentage remaining} = \text{fraction remaining} \times 100\%$$

If needed, adjust your work and solve from here.

 The variable that links both of the prompt equations is k.

Since we know the half-life, we can solve the second equation for k.

Knowing k and t, the first prompt equation will find $\ln(\text{fraction remaining})$.

Knowing $\ln(\text{fraction remaining})$, the fraction remaining can be found with

$$\text{Fraction} = e^{\ln(\text{fraction})} \quad \text{and} \quad \text{percentage} = \text{fraction} \times 100\%$$

Apply those steps.

$$k = \frac{-\ln(1/2)}{t_{1/2}} = \frac{-(-0.693)}{t_{1/2}} = \frac{+0.693}{8.1 \text{ days}} = \mathbf{0.0856 \text{ day}^{-1}}$$

$$\ln(\text{fraction remaining}) = -kt = -(0.0856 \text{ day}^{-1})(2.0 \text{ days}) = \mathbf{-0.171}$$

$$\text{Fraction} = e^{\ln(\text{fraction})} = e^{-0.171} = 0.843 = \mathbf{84\%} \text{ of I-131 } \textit{remains} \text{ after 2.0 days}$$

PRACTICE   B

Add the equations of the decay prompt to your flashcards and practice the cards for this chapter, then try the problems below. Save one problem for your next practice session.

1. The rate constant for the decay of the tritium isotope of hydrogen is 0.0562 yr^{-1}. Calculate the half-life of tritium.

2. Strontium-90 is a radioactive nuclide found in **fallout**: dust particles in the cloud produced by the atmospheric testing of nuclear weapons. In chemical and biological

systems, strontium behaves much like calcium. If dairy cattle consume crops exposed to dust or rain containing fallout, dairy products containing calcium will also contain ^{90}Sr. Similar to calcium, ^{90}Sr will be deposited in the bones of dairy product consumers, including children. In part for this reason, most (but not all) nations conducting nuclear tests signed a 1963 treaty that banned atmospheric testing. Strontium-90 undergoes beta decay with a half-life of 28.8 yr. What percentage of an original [^{90}Sr] in bones will remain after 40.0 yr?

    a. Estimate the answer.   b. Calculate the answer.

    c. Write the equation for the beta decay of strontium-90.

3. The element polonium was first isolated by Dr. Marie Sklodowska Curie and named for her native Poland. Radioactive ^{210}Po is found in significant concentrations in tobacco. If 20.0% of ^{210}Po remains in a sample after 321 days of alpha decay,

    a. Estimate the half-life of ^{210}Po.

    b. Calculate a precise half-life of ^{210}Po. Compare it to your *part a* estimate.

4. In a sample of radon-222, 10.0% remains after 12.6 days of alpha decay.

    a. What is the composition of the radon-222 nucleus?

    b. Write the decay reaction.

    c. Estimate the half-life for ^{222}Rn.

    d. Calculate a precise half-life of ^{222}Rn. Compare to your *part c* estimate.

5. If the half-life of carbon-14 is 5,730 yr, what fraction of the original carbon-14 in a sample has decayed after 1,650 yr? Estimate, then calculate.

SUMMARY

1. α particle $= {}^{4}_{2}He$; β particle $= {}^{0}_{-1}\beta$ or ${}^{0}_{-1}e$, neutron $= {}^{1}_{0}n$

2. To balance nuclear reactions,

    a. Use A–Z notation: mass number (A) on top, nuclear charge (Z) on the bottom.

    b. Both mass numbers and nuclear charge must be conserved; each must *add* to give the same number before and after the reaction.

3. Fission splits a nucleus; fusion combines nuclei.

4. In the decay of radioactive isotopes, the half-life is constant.
    • After one half-life, 1/2 of the original number of nuclei remain.
    • At double the time of the half-life, 1/4 of the original nuclei remain.
    • At triple the half-life, 1/8 of the original nuclei remain.

(continued)

5. **The radioactive decay prompt:** If a *radioactive decay* calculation includes *half-life* and a *fraction* or *percentage* of a sample and the answer cannot be calculated using simple multiples, write in the DATA: $\ln(\text{fraction remaining}) = -kt$ and $\ln(1/2) = -kt_{1/2}$ and use the math of natural logs to solve.

6. In radioactive half-life calculations,
   - If you WANT a percentage, first find its decimal equivalent fraction.
   - Before using the equation with (*fraction* remaining), you must convert percentages to fractions.
   - The fractions must have decimal equivalent values between 0 and 1.00 (such as 0.25).
   - Percentage = fraction × 100%
   - Percentage remaining = 100% − percentage decayed
   - Fraction remaining = 1.000 − fraction decayed

ANSWERS

Lesson 22.1

Practice 1. Protons   2. Protons + neutrons

3. Same number of **protons**, different number of **neutrons**.   4. Protons and neutrons.

5. ^1_1H (A particle with one proton is always given the symbol H.)

6.

Protons	Neutrons	Atomic Number	Mass Number	Nuclide Symbol
2	2	2	4	^4He
79	118	79	197	^{197}Au
82	124	82	206	^{206}Pb
94	148	94	242	^{242}Pu

In writing the nuclide or isotope symbols for atoms, the nuclear charge below the mass number is optional.

Lesson 22.2

Practice A 1. $^{219}_{86}\text{Rn} \xrightarrow{\alpha} \,^4_2\text{He} + \,^{215}_{84}\text{Po}$ 2. **86 protons** and **133 neutrons**.

3. **84 protons** and **131 neutrons**: **2 protons and 2 neutrons** are always lost in alpha decay.

4. $\boxed{^{210}_{84}\text{Po}} \xrightarrow{\alpha} \,^4_2\text{He} + \,^{206}_{82}\text{Pb}$

Practice B 1. α particle $= \,^4_2\text{He}$ and β particle $= \,^0_{-1}\beta$ or $\,^0_{-1}\text{e}$

2a. $^{40}_{19}\text{K} \xrightarrow{\beta} \,^0_{-1}\text{e} + \,^{40}_{20}\text{Ca}$ 2b. $^{239}_{94}\text{Pu} \xrightarrow{\alpha} \,^4_2\text{He} + \,^{235}_{92}\text{U}$

3. $^{131}_{53}\text{I} \xrightarrow{\beta} \,^0_{-1}\text{e} + \boxed{^{131}_{54}\text{Xe}}$ 4. $\boxed{^{206}_{81}\text{Tl}} \xrightarrow{\beta} \,^0_{-1}\text{e} + \,^{206}_{82}\text{Pb}$

5. $\boxed{^{228}_{88}\text{Ra}} \xrightarrow{\beta} \,^0_{-1}\text{e} + \boxed{^{228}_{89}\text{Ac}} \xrightarrow{\beta} \,^0_{-1}\text{e} + \boxed{^{228}_{90}\text{Th}} \xrightarrow{\alpha} \,^4_2\text{He} + \boxed{^{224}_{88}\text{Ra}}$

　　　　A　　　　　　　　　B　　　　　　　　　C　　　　　　　　　D

6. $\boxed{^{214}_{82}\text{Pb}} \xrightarrow{\beta} {^{0}_{-1}\text{e}} + \boxed{^{214}_{83}\text{Bi}} \xrightarrow{\alpha} {^{4}_{2}\text{He}} + \boxed{^{210}_{81}\text{Tl}} \xrightarrow{\beta} {^{0}_{-1}\text{e}} + \boxed{^{210}_{82}\text{Pb}}$

      A   **B**   **C**   **D**

Lesson 22.3

Practice A   1. ${^{1}_{0}\text{n}}$ 2. ${^{235}_{92}\text{U}} + {^{1}_{0}\text{n}} \rightarrow {^{87}_{35}\text{Br}} + \boxed{^{143}_{57}\text{La}} + 6\,{^{1}_{0}\text{n}}$ 3. $\boxed{^{239}_{94}\text{Pu}} + {^{1}_{0}\text{n}} \rightarrow {^{91}_{38}\text{Sr}} + {^{146}_{56}\text{Ba}} + 3\,{^{1}_{0}\text{n}}$

Practice B   1. Fission splits a nucleus, fusion combines nuclei.

       2. ${^{1}_{1}\text{H}} + {^{1}_{2}\text{H}} \rightarrow \boxed{^{4}_{2}\text{He}}$ 3. ${^{4}_{2}\text{He}} + {^{8}_{4}\text{Be}} \rightarrow \boxed{^{12}_{6}\text{C}}$

       4a. ${^{235}_{92}\text{U}} + {^{1}_{0}\text{n}} \rightarrow {^{92}_{36}\text{Kr}} + \boxed{^{141}_{56}\text{Ba}} + 3\,{^{1}_{0}\text{n}},$ **fission**   4b. ${^{238}_{92}\text{U}} \rightarrow {^{4}_{2}\text{He}} + \boxed{^{234}_{90}\text{Th}},$ **alpha decay**

       4c. ${^{1}_{1}\text{H}} + {^{3}_{1}\text{H}} \rightarrow \boxed{^{4}_{2}\text{He}},$ **fusion**   4d. ${^{234}_{90}\text{Th}} \rightarrow {^{0}_{-1}\text{e}} + \boxed{^{234}_{91}\text{Pa}},$ **beta decay**

Lesson 22.4

Practice A   1. Decimal equivalent of 1/8 = **0.125**   Percentage = decimal equivalent $\times$ 100% = **12.5%**

       2. Decimal equivalent = percent/100% = 16%/100% = **0.16**

       3. Decimal equivalent = percent/100% = 0.24%/100% = **0.0024 = 2.4 $\times$ 10^{-3}**

       4. To divide by 10,000, move the decimal four times to the left.

          Decimal equivalent = 7.4/10,000 = **0.00074 = 7.4 $\times$ 10^{-4}**

          Percentage = decimal equivalent $\times$ 100% = 0.00074 $\times$ 100% = **0.074% = 7.4 $\times$ 10^{-2}%**

       5. Fraction = $\dfrac{\text{part}}{\text{total}} = \dfrac{6 \text{ remain}}{30} = \dfrac{1}{5}$ remains = **0.20 remains**

          Percentage = fraction $\times$ 100% = decimal equivalent $\times$ 100% = 0.20 $\times$ 100% = **20% remains**

Practice B   1. If 90.% has decayed, 10.% remains, and fraction remaining = 10%/100% = **0.10**

       2. If fraction decayed = 0.40, percentage decayed = 0.40 $\times$ 100% = 40.%, and percentage remaining = 100% $-$ 40.% = **60.%**

       3a. First-order half-life is constant. Half remains after one half-life, half of that half (25%) remains after two half-lives. Two half-lives = 2 $\times$ 24,400 yr = **48,800 yr.**

       3b. Half remains after one half-life, 1/4th after two, 1/8th after three, 1/16th after **four** half-lives.

       3c. 1/16th remains after **four** half-lives. 1/16 = 0.0625 = 6.25% remains, so **93.75% has decayed**.

Lesson 22.5

Practice A   1a. $\text{Log}(10^{14}) = $ **14**   1b. $\text{Log}(10) = \text{Log}(10^{1}) = $ **1**   1c. $\text{Log}(1) = \text{Log}(10^{0}) = $ **0**

       2a. $10^{3.2} = $ **1.58 $\times$ 10^3**   2b. $10^{-12.3} = $ **5.01 $\times$ 10^{-13}**   2c. $10^{-0.2} = $ **0.631 = 6.31 $\times$ 10^{-1}**

       2d. $\text{Log}(2.0 \times 10^{14}) = $ **14.3**   2e. $\text{Log}(2.0 \times 10^{-14}) = $ **-13.7**   2f. $\text{Log } 5.0 = $ **0.699**

       2g. $\text{Log}(0.0050) = $ **-2.30**   2h. $\text{Log } x = 4.7, x = $ **5.01 $\times$ 10^4**

       2i. $\text{Log } A = -8.2, A = $ **6.31 $\times$ 10^{-9}**   2j. $\text{Log } D = -0.50, D = $ **0.316**

       2k. $\text{Log } X = 6.6, X = $ **3.98 $\times$ 10^6**   2l. $10^{-9.5} = $ **3.16 $\times$ 10^{-10}**

       2m. $\text{Log}(3.0 \times 10^{-5}) = $ **-4.52**

Practice B   1. $e^{2.0} = $ **7.39**   2. $e^{-4.7} = $ **9.10 $\times$ 10^{-3}**   3. $e^{-11} = $ **1.67 $\times$ 10^{-5}**

       4. $\ln(42) = $ **3.74**   5. $\ln(0.020) = $ **-3.91**   6. $\ln(9 \times 10^5) = $ **13.7**

       7. $\ln(5.0 \times 10^{-4}) = $ **-7.60**   8. $\ln(10^{-4}) = \ln(1 \times 10^{-4}) = $ **-9.21**

Practice C   1. $e^{5.2} = $ **181**   2. $e^{-1.7} = $ **0.183**   3. $e^{-20.75} = $ **9.74 $\times$ 10^{-10}**

       4. $\ln(1066) = $ **6.97**   5. $\ln(0.0050) = $ **-5.30**   6. $\ln(3 \times 10^8) = $ **19.5**

       7. $\ln(14.92 \times 10^{-6}) = $ **-11.1**   8. $\ln e^{6.2} = $ **6.20**   9. $e^{\ln(42)} = $ **42.0**

       10. If $\ln X = -6.8; X = e^{-6.8} = $ **1.11 $\times$ 10^{-3}**

11. If $\ln D = 7.4822, D = \mathbf{1{,}780}$ 12. If $\ln X = -12.5, X = \mathbf{3.73 \times 10^{-6}}$

13. If $\log[A] = -9, [A] = \mathbf{1.00 \times 10^{-9}\,M}$ 14. If $\log x = 13.7, x = \mathbf{5.01 \times 10^{13}}$

15. $\text{Log } A = -13.7, A = \mathbf{2.00 \times 10^{-14}}$ 16. $10^{-11.7} = \mathbf{2.00 \times 10^{-12}}$

17. $\ln[B] = -13.7, [B] = \mathbf{1.12 \times 10^{-6}\,M}$ 18. $e^{-11.7} = \mathbf{8.29 \times 10^{-6}}$

19. $\ln(0.050\ M) = \mathbf{-3.00}$ (drop the unit)   20. $e^{-0.693} = \mathbf{0.500}$

21. If $\log(x) = 5.00, \ln(x) = ?$ $\ln(x) = 2.303 \log(x)$; $\ln(x) = (2.303)(5.00) = \mathbf{11.5}$

22. If $\ln(x) = 34.5, \log(x) = ?$ $\ln(x) = 2.303 \log(x)$; $\log(x) = 34.5/2.303 = \mathbf{15.0}$

23. If $\ln(x) = -(0.075\ \text{day}^{-1})(4.0\ \text{days}); x = ?$

$\ln(x) = -0.300(\text{day}^{-1})(\text{days}) = -0.300\ \text{day}^{0} = -0.300\ (1) = -0.300$

$x = e^{\ln(x)} = e^{-0.300} = \mathbf{0.741}$

24. Given: $\ln[A] = (-0.0173\ \text{s}^{-1})\,(t)$

   a. Strategy: to find [A], first solve for ln[A].

     $\mathbf{\ln[A] = (-0.0173\ \text{s}^{-1})\,(20.0\ \text{s}) = -0.346}$

     WANTED is [A] at 20.0 s. Known is ln[A] at 20.0 s $= -0.346$. Solve for [A].

     $[A] = e^{\ln[A]} = e^{-0.346} = \boxed{\mathbf{0.708\ M = [A]}}$ If a [x] is wanted, attach M as the unit.

   b. Strategy: Solve for t in symbols first.

     $t = \dfrac{\ln[A]}{-0.0173\ \text{s}^{-1}} = \dfrac{\ln[0.500\ M]}{-0.0173\ \text{s}^{-1}} = \dfrac{-0.693}{-0.0173\ \text{s}^{-1}} = \boxed{\mathbf{40.1\ s = t}}$

     Answer units: $1/\text{s}^{-1} = (\text{s}^{-1})^{-1} = \text{s}$

25. Given: $\ln[A] = (-0.0241/\text{yr})\,(t)$

   a. Strategy: Solve for t in symbols first.

     $t = \dfrac{\ln[A]}{-0.0241/\text{yr}} = \dfrac{\ln[0.0025]}{-0.0241/\text{yr}} = \dfrac{-5.99}{-0.0241/\text{yr}} = \boxed{\mathbf{249\ yr = t}}$

   b. Strategy: To find [A], first solve for ln[A].

     $\ln[A] = (-0.0241/\text{yr})\,(28.8\ \text{yr}) = \mathbf{-0.694}$

     WANTED is [A]. Known is ln[A] $= -0.694$. Solve for [A].

     $[A] = e^{\ln[A]} = e^{-0.694} = \boxed{\mathbf{0.500\ M = [A]}}$ If a [X] is wanted, attach M to answer.

Lesson 22.6

Practice A   1a.   WANTED: Fraction of sample remaining

       DATA: $\ln(\text{fraction remaining}) = \mathbf{-1.386}$

              Knowing a value for ln(fraction remaining), to find fraction remaining, use

$$\text{Fraction remaining} = e^{\ln(\text{fraction remaining})}$$

       SOLVE: Fraction remaining $= e^{\ln(\text{fraction remaining})} = e^{-1.386} = \mathbf{0.250}$

   1b. If the fraction remaining is 0.250, 25.0% remains, and the percentage decayed is **75.0%**.

   2.   WANTED: % of sample remaining. To find %, find *fraction* first.

       DATA: $0.04606\ \text{hr}^{-1} = k$

               $50.0\ \text{hr} = t$

     Strategy: The equation that relates the three terms is $\ln(\text{fraction remaining}) = -kt$

            Knowing k and t, ln(fraction) and then fraction can be found.

SOLVE: $\ln(\text{fraction remaining}) = -kt = -(0.04606 \text{ hr}^{-1})(50.0 \text{ hr}) = -2.303$

Fraction remaining $= e^{\ln(\text{fraction remaining})} = e^{-2.303} = \mathbf{0.100}$

Percentage remaining $=$ fraction $\times 100\% = \boxed{\mathbf{10.0\%}}$

3. WANTED: $k =$ the rate constant

Equation: $\ln(\text{fraction remaining}) = -kt$ (This is the equation we know that uses these variables.)

DATA: Fraction remaining $= 50.0\% = \mathbf{0.500}$ (The percentage must be converted to its fraction.)

$t = 87.7 \text{ yr}$

Solve the equation for the WANTED variable.

$$k = \frac{-\ln(\text{fraction remaining})}{t} = \frac{-\ln(0.500)}{87.7 \text{ yr}} = \frac{+0.693}{87.7 \text{ yr}} = \mathbf{0.00790 \text{ yr}^{-1}}$$

Practice B 1. WANTED: $t_{1/2}$ for tritium

DATA: $0.0562 \text{ yr}^{-1} = k$

Strategy: In radioactive decay calculations that include half-life and fraction or percentage, write

$\ln(\text{fraction remaining}) = -kt$ and $\ln(1/2) = -kt_{1/2}$

In this problem, only the second equation is needed to relate the symbols in the WANTED and DATA.

SOLVE: $t_{1/2} = \dfrac{\ln(1/2)}{-k} = \dfrac{-0.693}{-0.0562 \text{ yr}^{-1}} = \mathbf{12.3 \text{ yr}}$

Unit math: $1/\text{yr}^{-1} = (\text{yr}^{-1})^{-1} = \text{yr}$

2a. WANTED: *Estimate* of % [Sr-90] remaining after 40 yr. If one half-life is about 30 yr and 50% remains, and two half-lives is about 60 yr and 25% remains, then at 40 yr, **about . . . 40%** remains?

2b. WANTED: % [Sr-90] remaining at $t = 40.0$ yr

Strategy: In radioactive decay calculations that include half-life and fraction or percentage, write

$\ln(\text{fraction remaining}) = -kt$ and $\ln(1/2) = -kt_{1/2}$

DATA: $28.8 \text{ yr} = t_{1/2}$

$40.0 \text{ yr} = t$

Percentage $=$ fraction $\times 100\%$

From half-life, k can be found. From k and t, $\ln(\text{fraction})$ and then fraction can be found.

SOLVE: Starting from the equation that includes half-life: $\ln(1/2) = -kt_{1/2}$

$$k = \frac{-\ln(1/2)}{t_{1/2}} = \frac{-(-0.693)}{t_{1/2}} = \frac{0.693}{28.8 \text{ yr}} = \mathbf{0.02406 \text{ yr}^{-1}}$$

$\ln(\text{fraction remaining}) = -kt = -(0.02406 \text{ yr}^{-1})(40.0 \text{ yr}) = \mathbf{-0.9624}$

Fraction $= e^{\ln(\text{fraction})} = e^{-0.9624} = 0.382 = \boxed{\mathbf{38.2\%} \text{ Sr-90 remains after 40 yr}}$

Compare this to your estimate.

2c. $^{90}_{38}\text{Sr} \xrightarrow{\beta} {}^{0}_{-1}\text{e} + {}^{90}_{39}\text{Y}$

3a. Estimate: We know that 20% remains after 321 days. We also know that 50% remains after one half-life, and 25% after two half-lives.
20% is close to 25%, and when 25% remains, it would be about 300 days, and 25% is two half-lives, so one halflife is **about 150 days**.

3b. WANTED: $t_{1/2}$

DATA: 20.0% Po-210 remains

Fraction remaining = 20.0%/100% = 0.200

$t = 321$ days

See radioactive decay, half-life and fraction or percentage? Write

$$\ln(\text{fraction remaining}) = -kt \quad \text{and} \quad \ln(1/2) = -kt_{1/2}$$

Percentage = fraction × 100%

Knowing the fraction and t, k can be found from the first prompt equation. Half-life can then be found from the second equation. Solve for k in symbols first:

$$k = \frac{-\ln(\text{fraction})}{t} = \frac{-\ln(0.200)}{321 \text{ days}} = \frac{-(-1.61)}{321 \text{ days}} = \mathbf{5.01 \times 10^{-3} \, days^{-1}}$$

$$t_{1/2} = \frac{-\ln(1/2)}{k} = \frac{+0.693}{5.01 \times 10^{-3} \text{ days}^{-1}} = \boxed{\textbf{138 days} = \text{half-life of Po-210}}$$

Is this answer close to the estimate in *part a?*

4a. A radon-222 nucleus has **86 protons** and **136 neutrons**.

4b. $^{222}_{86}\text{Rn} \rightarrow \, ^{4}_{2}\text{He} + \, ^{218}_{84}\text{Po}$

4c. Estimate: 10% remains after 12.6 days. 12.5% remains after three half-lives, which is close to 10%.

If three half-lives is about 12 days, then one half-life would be **about 4 days**.

4d. WANTED: $t_{1/2}$

DATA: 90.0% Rn-222 has decayed, so 10.0% remains

$t = 12.6$ days

$\ln(\text{fraction remaining}) = -kt$ and $\ln(1/2) = -kt_{1/2}$

Fraction remaining = 10.0%/100% = **0.100**

SOLVE: Knowing the fraction and t, k can be found from the first equation. Half-life can then be found from the second equation.

$$k = \frac{-\ln(\text{fraction})}{t} = \frac{-\ln(0.100)}{12.6 \text{ days}} = \frac{-(-2.30)}{12.6 \text{ days}} = \mathbf{0.183 \, days^{-1}}$$

$$t_{1/2} = \frac{-\ln(1/2)}{k} = \frac{+0.693}{0.183 \text{ days}^{-1}} = \boxed{\textbf{3.79 days} = \textbf{half-life of Rn-222}}$$

Is this close to your estimate in *part a?*

5. Estimate: 0.50 is the fraction decayed after about 6,000 yr, so **about 0.15** is the fraction decayed in about one-third of that time?

Calculate:

WANTED: Fraction [C-14] decayed = 1.000 − fraction remaining

DATA: $t_{1/2} = 5,730$ yr

$t = 1,650$ yr

$\ln(\text{fraction remaining}) = -kt \quad \text{and} \quad \ln(1/2) = -kt_{1/2}$

From half-life, k can be found. From k and t, ln(fraction) and then fraction can be found.

SOLVE: $k = \dfrac{-\ln(1/2)}{t_{1/2}} = \dfrac{-(-0.693)}{t_{1/2}} = \dfrac{+0.693}{5,730 \text{ yr}} = \mathbf{1.21 \times 10^{-4} \, yr^{-1}}$

$\ln(\text{fraction remaining}) = -kt = -(1.21 \times 10^{-4} \text{ yr}^{-1})(1,650 \text{ yr}) = \mathbf{-0.1996}$

Fraction remaining $= e^{\ln(\text{fraction remaining})} = e^{-0.1996} = \mathbf{0.819}$

If 0.819 = fraction remaining, 1.000 − 0.819 = $\boxed{\mathbf{0.181} = \text{fraction C-14 } \textit{decayed}}$

Compare to your estimate.

Table of Atomic Masses

The atomic masses in the following table use fewer significant figures than most similar tables in college textbooks. By keeping the numbers simple, it is hoped that you will use mental arithmetic to do easy numeric cancellations and simplifications before you use a calculator for arithmetic.

Many calculations in these lessons have been set up so that you should not need a calculator at all to solve, if you look for *easy cancellations* first.

After any use of a calculator, use mental arithmetic and simple cancellations to *estimate* the answer, in order to catch errors in calculator use.

Atomic Masses

Element	Symbol	Atomic Number[1]	Atomic Mass[2]
Actinium	Ac	89	(227)
Aluminum	Al	13	27.0
Americium	Am	95	(243)
Antimony	Sb	51	121.8
Argon	Ar	18	39.9
Arsenic	As	33	74.9
Astatine	At	85	(210)
Barium	Ba	56	137.3
Berkelium	Bk	97	(247)
Beryllium	Be	4	9.01
Bismuth	Bi	83	209.0
Bohrium	Bh	107	(264)
Boron	B	5	10.8
Bromine	Br	35	79.9
Cadmium	Cd	48	112.4
Calcium	Ca	20	40.1
Californium	Cf	98	(251)
Carbon	C	6	12.0
Cerium	Ce	58	140.1
Cesium	Cs	55	132.9
Chlorine	Cl	17	35.5
Chromium	Cr	24	52.0
Cobalt	Co	27	58.9
Copernicium	Cn	112	(285)

(continued)

Element	Symbol	Atomic Number[1]	Atomic Mass[2]
Copper	Cu	29	63.5
Curium	Cm	96	(247)
Darmstadtium	Ds	110	(271)
Dubnium	Db	105	(262)
Dysprosium	Dy	66	162.5
Einsteinium	Es	99	(252)
Erbium	Er	68	167.3
Europium	Eu	63	152.0
Fermium	Fm	100	(257)
Fluorine	F	9	19.0
Francium	Fr	87	(223)
Gadolinium	Gd	64	157.3
Gallium	Ga	31	69.7
Germanium	Ge	32	72.6
Gold	Au	79	197.0
Hafnium	Hf	72	178.5
Hassium	Hs	108	(277)
Helium	He	2	4.00
Holmium	Ho	67	164.9
Hydrogen	H	1	1.008
Indium	In	49	114.8
Iodine	I	53	126.9
Iridium	Ir	77	192.2
Iron	Fe	26	55.8
Krypton	Kr	36	83.8
Lanthanum	La	57	138.9
Lawrencium	Lr	103	(262)
Lead	Pb	82	207.2
Lithium	Li	3	6.94
Lutetium	Lu	71	175.0
Magnesium	Mg	12	24.3
Manganese	Mn	25	54.9
Meitnerium	Mt	109	(268)
Mendelevium	Md	101	(258)
Mercury	Hg	80	200.6
Molybdenum	Mo	42	95.9
Neodymium	Nd	60	144.2
Neon	Ne	10	20.2
Neptunium	Np	93	(237)
Nickel	Ni	28	58.7
Niobium	Nb	41	92.9

Element	Symbol	Atomic Number[1]	Atomic Mass[2]
Nitrogen	N	7	14.0
Nobelium	No	102	(259)
Osmium	Os	76	190.2
Oxygen	O	8	16.0
Palladium	Pd	46	106.4
Phosphorus	P	15	31.0
Platinum	Pt	78	195.1
Plutonium	Pu	94	(244)
Polonium	Po	84	(209)
Potassium	K	19	39.1
Praseodymium	Pr	59	140.9
Promethium	Pm	61	(145)
Protactinium	Pa	91	(231)
Radium	Ra	88	(226)
Radon	Rn	86	(222)
Rhenium	Re	75	186.2
Rhodium	Rh	45	102.9
Roentgenium	Rg	111	(272)
Rubidium	Rb	37	85.5
Ruthenium	Ru	44	101.1
Rutherfordium	Rf	104	(261)
Samarium	Sm	62	150.4
Scandium	Sc	21	45.0
Seaborgium	Sg	106	(266)
Selenium	Se	34	79.0
Silicon	Si	14	28.1
Silver	Ag	47	107.9
Sodium	Na	11	23.0
Strontium	Sr	38	87.6
Sulfur	S	16	32.1
Tantalum	Ta	73	180.9
Technetium	Tc	43	(98)
Tellurium	Te	52	127.6
Terbium	Tb	65	158.9
Thallium	Tl	81	204.4
Thorium	Th	90	232.0
Thulium	Tm	69	168.9
Tin	Sn	50	118.7
Titanium	Ti	22	47.9
Tungsten	W	74	183.8
Uranium	U	92	238.0

(continued)

Element	Symbol	Atomic Number[1]	Atomic Mass[2]
Vanadium	V	23	50.9
Xenon	Xe	54	131.3
Ytterbium	Yb	70	173.0
Yttrium	Y	39	88.9
Zinc	Zn	30	65.4
Zirconium	Zr	40	91.2

[1]The atomic number is the number of **protons** in the nucleus of the atom.

[2]For radioactive atoms, ( ) is the mass number of the most stable isotope.

The Periodic Table of the Elements

1 1A	2 2A	3 3B	4 4B	5 5B	6 6B	7 7B	8 8B	9 8B	10 8B	11 1B	12 2B	13 3A	14 4A	15 5A	16 6A	17 7A	18 8A
1 1.008 H Hydrogen																	2 4.00 He Helium
3 6.94 Li Lithium	4 9.01 Be Beryllium											5 10.8 B Boron	6 12.07 C Carbon	7 14.0 N Nitrogen	8 16.0 O Oxygen	9 19.0 F Fluorine	10 20.2 Ne Neon
11 23.0 Na Sodium	12 24.3 Mg Magnesium											13 27.0 Al Aluminum	14 28.1 Si Silicon	15 31.0 P Phosphorus	16 32.1 S Sulfur	17 35.5 Cl Chlorine	18 39.9 Ar Argon
19 39.1 K Potassium	20 40.1 Ca Calcium	21 45.0 Sc Scandium	22 47.9 Ti Titanium	23 50.9 V Vanadium	24 52.0 Cr Chromium	25 54.9 Mn Manganese	26 55.8 Fe Iron	27 58.9 Co Cobalt	28 58.7 Ni Nickel	29 63.5 Cu Copper	30 65.4 Zn Zinc	31 69.7 Ga Gallium	32 72.6 Ge Germanium	33 74.9 As Arsenic	34 79.0 Se Selenium	35 79.9 Br Bromine	36 83.8 Kr Krypton
37 85.5 Rb Rubidium	38 87.6 Sr Strontium	39 88.9 Y Yttrium	40 91.2 Zr Zirconium	41 92.9 Nb Niobium	42 95.9 Mo Molybdenum	43 (98) Tc Technitium	44 101.1 Ru Ruthenium	45 102.9 Rh Rhodium	46 106.4 Pd Palladium	47 107.9 Ag Silver	48 112.4 Cd Cadmium	49 114.8 In Indium	50 118.7 Sn Tin	51 121.8 Sb Antimony	52 127.6 Te Tellurium	53 126.9 I Iodine	54 131.3 Xe Xenon
55 132.9 Cs Cesium	56 137.3 Ba Barium	57 138.9 La Lanthanum	72 178.5 Hf Hafnium	73 180.9 Ta Tantalum	74 183.8 W Tungsten	75 186.2 Re Rhenium	76 190.2 Os Osmium	77 192.2 Ir Iridium	78 195.1 Pt Platinum	79 197.0 Au Gold	80 200.6 Hg Mercury	81 204.4 Tl Thallium	82 207.2 Pb Lead	83 209.0 Bi Bismuth	84 (209) Po Polonium	85 (210) At Astatine	86 (222) Rn Radon
87 (223) Fr Francium	88 (226) Ra Radium	89 (227) Ac Actinium	104 (261) Rf Rutherfordium	105 (262) Db Dubnium	106 (266) Sg Seaborgium	107 (264) Bh Bohrium	108 (277) Hs Hassium	109 (268) Mt Meitnerium	110 (271) Ds Darmstadtium	111 (272) Rg Roentgenium	112 (285) Cn Copernicium						

58 140.1 Ce Cerium	59 140.9 Pr Praseodymium	60 144.2 Nd Neodymium	61 (145) Pm Promethium	62 150.4 Sm Samarium	63 152.0 Eu Europium	64 157.3 Gd Gadolinium	65 158.9 Tb Terbium	66 162.5 Dy Dysprosium	67 164.9 Ho Holmium	68 167.3 Er Erbium	69 168.9 Tm Thulium	70 173.0 Yb Ytterbium	71 175.0 Lu Lutetium
90 232.0 Th Thorium	91 (231) Pa Protactinium	92 238.0 U Uranium	93 (237) Np Neptunium	94 (244) Pu Plutonium	95 (243) Am Americium	96 (247) Cm Curium	97 (247) Bk Berkelium	98 (251) Cf Californium	99 (252) Es Einsteinium	100 (257) Fm Fermium	101 (258) Md Mendelevium	102 (259) No Nobelium	103 (262) Lr Lawrencium

Index